Digital Control and Estimation
A Unified Approach

PRENTICE HALL INFORMATION AND SYSTEM SCIENCES SERIES

Thomas Kailath, Editor

ANDERSON & MOORE	*Optimal Control*
ANDERSON & MOORE	*Optimal Filtering*
ÅSTRÖM & WITTENMARK	*Computer-Controlled Systems: Theory and Design*
GARDNER	*Statistical Spectral Analysis: A Nonprobabilistic Theory*
GOODWIN & SIN	*Adaptive Filtering, Prediction, and Control*
GRAY & DAVISSON	*Random Processes: A Mathematical Approach for Engineers*
HAYKIN	*Adaptive Filter Theory*
JAIN	*Fundamentals of Digital Image Processing*
JOHNSON	*Lectures on Adaptive Parameter Estimation*
KAILATH	*Linear Systems*
KUMAR & VARAIYA	*Stochastic Systems*
KUNG	*VLSI Array Processors*
KUNG, WHITEHOUSE, & KAILATH, EDS.	*VLSI and Modern Signal Processing*
KWAKERNAAK & SIVAN	*Signals and Systems*
LANDAU	*System Identification and Control Design Using PIM+ Software*
LJUNG	*System Identification*
MACOVSKI	*Medical Imaging Systems*
MIDDLETON & GOODWIN	*Digital Control and Estimation*
NARENDRA & ANNASWAMY	*Stable Adaptive Systems*
SASTRY & BODSON	*Adaptive Control: Stability, Convergence, and Robustness*
SOLIMAN & SRINATH	*Continuous and Discrete Signals and Systems*
SPILKER	*Digital Communications by Satellite*
WILLIAMS	*Designing Digital Filters*

Digital Control and Estimation
A Unified Approach

Richard H. Middleton
Graham C. Goodwin

Centre for Industrial Control Science
Department of Electrical Engineering
   and Computer Science
University of Newcastle
New South Wales, Australia

Prentice-Hall International, Inc.

ISBN 0-13-211798-3

This edition may be sold only in those countries to which it is consigned by Prentice-Hall International. It is not to be re-exported and it is not for sale in the U.S.A., Mexico, or Canada.

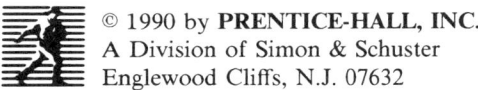 © 1990 by **PRENTICE-HALL, INC.**
A Division of Simon & Schuster
Englewood Cliffs, N.J. 07632

All rights reserved. No part of this book may be reproduced, in any form or by any means, without permission in writing from the publisher.

Printed in the United States of America

10 9 8 7 6 5 4 3 2 1

ISBN 0-13-211798-3

Prentice-Hall International (UK) Limited, *London*
Prentice-Hall of Australia Pty. Limited, *Sydney*
Prentice-Hall Canada Inc., *Toronto*
Prentice-Hall Hispanoamericana, S.A., *Mexico*
Prentice-Hall of India Private Limited, *New Delhi*
Prentice-Hall of Japan, Inc., *Tokyo*
Simon & Schuster Asia Pte. Ltd., *Singapore*
Editora Prentice-Hall do Brasil, Ltda., *Rio de Janeiro*
Prentice-Hall, Inc., *Englewood Cliffs, New Jersey*

Contents

Preface xv

1 Introduction 1
References 5

2 Continuous Time Models 7
2.1 Introduction 7
2.2 State Space Modeling 8
2.3 Linear State Space Models 12
    2.3.1 Compensating Nonlinearities 12
    2.3.2 Linear Approximation of Nonlinear Systems 17
2.4 Input-Output Models for Dynamic Systems 19
2.5 Models for Disturbances 20
2.6 Summary 22
2.7 References 23
2.8 Problems 23

3 Discrete Time Models 27
3.1 Introduction 27
3.2 Sampling 28
    3.2.1 Notation 28

    3.2.2 Analog-to-Digital Conversion 28
    3.2.3 Aliasing 28
 3.3 Signal Reconstruction 30
    3.3.1 Polynomial Interpolation 31
    3.3.2 Shannon Reconstruction Theorem 31
    3.3.3 Digital-to-Analog Conversion 32
 3.4 Discrete Models for Continuous Process 32
    3.4.1 General Models 32
    3.4.2 Linear State Space Models 33
   *3.4.3 Evaluation of the Matrix Exponential 36
 3.5 Discrete Time Operator Models 41
    3.5.1 The Shift Operator 42
    3.5.2 The Delta Operator 43
    3.5.3 Evaluation of Discrete Models in Delta Form 46
 3.6 Inherently Discrete Systems 47
 3.7 Summary 49
 3.8 References 50
 3.9 Problems 51

4 Transform Techniques 54

 4.1 Introduction 54
 4.2 Laplace Transforms 55
 4.3 Discrete Time Transforms 59
    4.3.1 Z-Transforms 59
    4.3.2 Delta Transforms 63
 4.4 Unified Transform Theory 65
 4.5 Inverse Transformations 76
 4.6 Summary 80
 4.7 References 81
 4.8 Problems 82

5 Transfer Functions 85

 5.1 Introduction 85
 5.2 Transfer Function 86
 5.3 Connection between Discrete and Continuous Transfer Functions 87
 5.4 Poles and Zeros 90
 5.5 Poles and Zeros of Sampled Data Systems 94
 5.6 Summary 104
 5.7 References 106
 5.8 Problems 106

6 Frequency Response 109

- 6.1 Introduction 109
- 6.2 Evaluation of the System Frequency Response 109
- 6.3 Interconnecting Continuous and Discrete Frequency Responses 113
  - 6.3.1 Approximation by Numerical Integration 113
  - 6.3.2 Frequency Domain Approximation 114
  - 6.3.3 Bilinear Transformation 115
  - 6.3.4 Step Invariance 117
- 6.4 Modeling of Time Delay Systems 118
  - 6.4.1 Transfer Functions of Delay 119
  - *6.4.2 Sampling of Time Delay Systems 120
  - 6.4.3 Rational Approximations to Delays 121
- 6.5 Summary 124
- 6.6 References 125
- 6.7 Problems 126

7 Classical Control System Analysis 128

- 7.1 Introduction 128
- 7.2 Stability for Linear Time Invariant Systems 129
- 7.3 Polynomial Stability Tests 132
- 7.4 Root Locus 138
- 7.5 Frequency Domain Stability Criteria 143
- 7.6 Advanced Frequency Domain Stability Analysis 147
  - 7.6.1 Robustness of Feedback Systems 147
  - 7.6.2 Systems with Nonlinearities 148
- 7.7 Tracking Performance 150
  - 7.7.1 Specifications 150
  - 7.7.2 System Type 152
  - 7.7.3 Internal Model Principle 153
- 7.8 Transient Performance 154
  - 7.8.1 Time Domain Specifications 154
  - 7.8.2 Frequency Domain Specifications 154
- 7.9 Compensation 159
  - 7.9.1 Phase Advance 159
  - 7.9.2 Phase Lag 163
  - 7.9.3 Lead-Lag 164
- 7.10 Multivariable Systems 166
  - 7.10.1 Multivariable Frequency Domain Stability Criterion 166
  - 7.10.2 Precompensation 170
- 7.11 Summary 172
- 7.12 References 173
- 7.13 Problems 174

8 Time Domain Analysis of Systems   179

- 8.1   Introduction   179
- 8.2   State Space Models   180
- 8.3   Reachability, Controllability, Stabilizability   180
- 8.4   Canonical Forms for Reachable Systems   186
  - 8.4.1   Controllability Form   187
  - 8.4.2   Controller Forms   188
  - 8.4.3   Right Matrix Fraction Descriptions   189
- 8.5   Controllability Gramian   191
- 8.6   Per Unit Values   195
- 8.7   Reconstructibility, Observability, and Detectability   196
- 8.8   Canonical Forms for Observable Systems   198
  - 8.8.1   Observability Form   198
  - 8.8.2   Observer Form   198
  - 8.8.3   Left Matrix Fraction Descriptions   199
- 8.9   The Observability Gramian   199
- 8.10 Canonical Structure, Minimal Models, Input-Output Equivalence   201
- 8.11 Hankel Singular Values and Balanced Realizations   202
- 8.12 Stability for Nonlinear Systems   204
  - 8.12.1 Basic Definitions   204
  - 8.12.2 Lyapunov Stability   208
- 8.13 Summary   212
- 8.14 References   213
- 8.15 Problems   214

9 Pole Assignment   219

- 9.1   Introduction   219
- 9.2   Pole Assignment by State Variable Feedback   219
- 9.3   State Observers   220
- 9.4   Dynamic Output Feedback   221
- 9.5   Pole Assignment for Systems with Disturbances   225
  - 9.5.1   The Diophantine Equation   227
  - 9.5.2   Solving the Diophantine Equation   229
  - 9.5.3   Numerical Issues for the Discrete Case   231
  - 9.5.4   Model Reference Control   232
- 9.6   Fractional Representation   234
  - 9.6.1   Fractional Models   234
  - 9.6.2   Representation of All Stabilizing Controllers   236
  - *9.6.3 Extension to Systems Defined on General Rings   242
  - 9.6.4   Multiinput, Multioutput Systems   244
  - *9.6.5 Multivariable Model Reference Control   251

9.7 Summary 257
9.8 References 258
9.9 Problems 259

10 Optimal State Estimation 263

10.1 Introduction 263
10.2 Nonrecursive Optimal Estimation 264
10.3 Sequential Processing 268
10.4 State Estimation for Linear Time Invariant Systems 273
10.5 Delta Form of the Optimal State Estimator 274
10.6 Unification of Continuous and Discrete Results 276
10.7 Properties of the Optimal Filter 283
10.8 Spectral Factorization 291
10.9 Characterization of All State Estimators 293
10.10 Alternative Optimization Criteria 295
    10.10.1 Mean Square Criterion Revisited 296
    10.10.2 Mini-Max Criteria 297
10.11 Robustness Considerations in Observer Design 299
10.12 Bandwidth Considerations in Observer Design 300
10.13 Summary 302
10.14 References 304
10.15 Problems 305

11 Optimal Control 309

11.1 Introduction 309
11.2 Finite Horizon LQ Optimal Control 310
11.3 Infinite Horizon LQ Optimal Control 313
11.4 Frequency Domain Interpretations of LQ Optimal Control 320
11.5 Internal Model Principle in LQ Optimal Control 325
11.6 Linear Quadratic Gaussian (LQG) Stochastic Optimal Control 327
11.7 Other Optimization Criteria 329
11.8 Summary 346
11.9 References 346
11.10 Problems 347

12 Parameter Estimation 351

12.1 Introduction 351
12.2 Linear Regressions for Dynamical Systems 352

            12.2.1   Discrete Time Shift Operator Linear Models   353
            12.2.2   Unified Linear Models   353
            12.2.3   Nonlinear Models   355
            12.2.4   Models Which Are Nonlinear in the Parameters   356
    12.3   Least Squares Estimation   356
    12.4   Recursive Least Squares   358
    12.5   Properties of the Recursive Least Squares Algorithm   360
            12.5.1   Prediction Error Properties   361
            12.5.2   Parameter Convergence   362
    12.6   Incorporation of Prior Knowledge   366
    12.7   Dealing with Noise in Parameter Estimation   367
            12.7.1   Deterministic Disturbances   368
            12.7.2   White Noise   369
            12.7.3   Colored Noise with Known Spectra   369
            12.7.4   Colored Noise with Unknown Spectra   370
    12.8   Dealing with Undermodeling in Parameter Estimation   376
            12.8.1   Algorithm Modifications   376
            12.8.2   Properties of Estimator with Deadzone   379
            12.8.3   Quantification of Undermodeling on Estimation   381
    12.9   Dealing with Parameter Time Variations   387
            12.9.1   Kalman Filter Formulation   388
            12.9.2   Exponential Data Weighting   388
            12.9.3   Gradient Algorithm   390
            12.9.4   Resetting Algorithm   390
            12.9.5   Exponential Forgetting and Resetting Algorithm   390
            12.9.6   A General Class of Algorithms   391
            12.9.7   Response Time Considerations   392
    12.10   Adaptive Control   394
            12.10.1 Certainty Equivalence   394
            12.10.2 An Illustrative Algorithm   395
            12.10.3 Electromechanical Servo Example   397
            12.10.4 Heat Exchanger Example   404
    12.11   Summary   406
    12.12   References   407
    12.13   Problems   409

13   Design Considerations   416

    13.1   Introduction   416
    13.2   The Case for High Gain Feedback   417
    13.3   Factors Limiting Feedback Bandwidth   421
    13.4   Frequency Domain Sensitivity Functions   423
    13.5   Feedback Design Guidelines   433
    13.6   Feedforward   441
    13.7   On-Line Estimation   450
    13.8   Summary   451

13.9 References 452
13.10 Problems 452

14 Implementation Issues in Digital Control 456

14.1 Introduction 456
14.2 Selection of Sampling Rate 456
    14.2.1 Effects of Slow Sampling 457
    14.2.2 Effects of Rapid Sampling 460
    14.2.3 Recommended Sampling Rate 461
14.3 Choice of Analog Interface Hardware 461
    14.3.1 Choice of Antialiasing Filter 461
    14.3.2 Choice of A/D Converters 464
    14.3.3 Choice of D/A Converters 464
14.4 Control Algorithm Packaging 467
    14.4.1 Integral Windup 467
    14.4.2 Bumpless Transfer 471
14.5 Numerical Issues 472
    14.5.1 Implementation in Delta Form 472
    14.5.2 Cascade and Parallel Realizations 474
    14.5.3 Numerical Advantages of Delta Implementations 476
14.6 Summary 480
14.7 References 480
14.8 Problems 481

15 An Industrial Case Study 483

15.1 Introduction 483
15.2 Roll Eccentricity Control for Strip Rolling Mills 483
    15.2.1 Introduction 483
    15.2.2 Rolling Mill Model 484
    15.2.3 Gaugemeter Control 486
    15.2.4 Eccentricity Control Scheme 487
    15.2.5 Design Considerations 488
    15.2.6 Experimental Results 496
15.3 Summary 497
15.4 References 498

APPENDIX A Hardware Aspects of Digital-to-Analog Interfacing 499

A.1 D/A Conversion 499
A.2 A/D Conversion 499
A.3 References 501

APPENDIX B Numerical Issues in Control 502

- B.1 Coefficient Representation 502
- B.2 Frequency Response Sensitivity 506
- B.3 Round-Off Noise 508

APPENDIX C Numerical Issues in Optimal Filtering 511

- C.1 The Discrete Riccati Difference Equation (DRDE) 511
- C.2 The Algebraic Riccati Equation 512
- C.3 References 515

APPENDIX D CAD Software for Control Systems Design 516

APPENDIX E Real-Time Software 518

- E.1 Terminology 518
- E.2 Real-Time Programming 520
- E.3 Proper Control Laws 521

APPENDIX F Convergence Analysis of a Typical Adaptive Control Algorithm 525

- F.1 The Algorithm 525
- F.2 Analysis of the Algorithm 527

Index 533

Preface

Many books already exist on the topics of digital control and estimation. The prospective reader of this book might well then ask, "why another book in this area?"

One problem with the existing literature is that it emphasizes the differences between discrete and continuous theory. This dichotomy is largely historical in nature and may not be the best approach from a pedagogical viewpoint. For example, shift operators and Z-transforms, which form the basis of most discrete time analyses, are inappropriate when used with fast sampling and have no continuous time counterpart. Our philosophy, as presented in this book, is that the continuous and discrete cases can, and should, be understood under a common framework. We show that this is facilitated if the shift operator is augmented with alternative forms including one which we call the delta operator. Using the latter operator, it becomes evident that all discrete time theory converges smoothly to the appropriate continuous results as the sampling rate increases. An additional, and somewhat unexpected, bonus arising from the use of the alternative operators is that numerical properties can be substantially improved relative to the more traditional shift operator.

Thus, this book presents continuous and discrete control and estimation theory in a unified fashion, highlighting the interrelationships between the two cases. Our firm belief is that this unified view of discrete and continuous theory is much richer and more informative than when either of the two are studied in isolation.

Another thrust of the book is to unify practical considerations with theoretical analysis. This is achieved by discussing implementation issues in detail and by presenting an industrial case study.

The book has a dual audience. Part of the book would be suitable for a first undergraduate course in digital control. The remainder would form the basis of one or more graduate courses in advanced control and estimation.

The prerequisite for the "undergraduate" portion of the text is an elementary mathematical background in Linear Algebra, Differential Equations, Calculus and Complex Numbers. The more advanced material depends upon additional background normally available to graduate students.

The authors would like to acknowledge those who assisted with the preparation of this book. First, the book would not have been possible without the support and understanding of our wives Ruth (Middleton) and Rosslyn (Goodwin). The book has also been used in both undergraduate and graduate courses at the University of Newcastle, Australia, and we would like to thank the students in these courses for their advice as the book took shape. A detailed solutions manual was prepared by Changyun Wen and Youyi Wang. Very helpful feedback was also obtained from Robert Bitmead, Robin Evans, Arie Feuer, Art Harvey, Peter Hippe, Konrad Hitz, Michel Kinnaert, Bengt Lennartson, Mario Salgado and Bjorn Wittenmark. The authors are also grateful for very helpful suggestions made by several anonymous reviewers.

The manuscript was expertly typed by Ildiko de Souza and the diagrams were prepared by Wanda Lis and Vilma Lucky.

Rick Middleton
Graham Goodwin

Digital Control and Estimation

A Unified Approach

1

Introduction

This book presents a fresh approach to the topics of digital control and estimation. A central theme is to unify continuous and discrete theory. Our belief is that it is instructive to view continuous time results as a suitable limiting case of the corresponding discrete results. This represents a divergence from previous presentations of discrete theory, which tended to highlight the differences between the two cases.

Typically, the topics of digital control and estimation are described in a different framework than that used for continuous time systems. As an example of the divergence between the approaches, consider the problem of stability. In the usual continuous time theory, we say that a linear system is stable if its poles lie in the left half-plane. However, in the traditional discrete time theory, we say that a linear system is stable if its poles lie inside the unit circle. It is difficult to see the connection between these results. We argue in this book that the apparent schism that has arisen between analog and digital theory can be traced to the widespread use of shift operators and Z-transforms in the discrete case. In this book, we introduce an additional operator, which we call the delta operator. This operator offers the same flexibility as does the shift operator in the description of discrete systems yet has several advantages over the latter, including:

- It highlights the similarities, rather than the differences, between discrete and continuous systems, thus allowing continuous insights to be applied to the discrete case.
- It introduces the sampling period as an exhibit parameter, thus allowing the effect of different choices for this parameter to be readily assessed.
- It allows a unified systems theory to be developed without needing to run a separate line of development for continuous and discrete.

- It allows most continuous time results to be obtained as a simple special case of the discrete results (by setting the sampling period to zero).
- It offers substantial numerical advantages in most cases of practical interest.

A brief outline of the content of the book is as follows. We begin in Chapter 2 by giving a brief overview of how simple models of physical systems can be derived from the laws of nature. This modeling typically involves the use of state space ideas. The important question of linearization of nonlinear models is also discussed in some detail, including such techniques as input–output transformations, feedback compensation, and linear approximation.

Chapter 3 describes the sampling process, including choice of sampling rate, and aliasing. We also show how models for sampled data systems can be derived from an underlying continuous time representation. The notion of the shift operator is introduced, and this is then extended to more general representations, including the delta operator. We also briefly discuss numerical issues leading to the conclusion that the delta operator has significant advantages over the shift form.

One of the most useful tools to have evolved for the analysis of linear estimation and control systems is that of transform techniques. The traditional techniques of Laplace and Z-transforms are discussed in Chapter 4. This is followed by the development of a unified transform theory, which allows both continuous and discrete systems to be treated simultaneously.

This leads us, in Chapter 5, to the topic of transfer function analysis. Particular emphasis is given to the notions of poles and zeros and their relationships for discrete and continuous systems. A distinctive feature of the chapter is the insightful discussion of the zeros that arise when a continuous time system is sampled.

Chapter 6 builds on the notion of transfer functions and introduces the important concept of system frequency response. We examine the interrelationship of continuous and discrete frequency responses. We also study systems having pure time delays.

Chapter 7 explores the analysis of control system performance using classical (that is, frequency domain) techniques. The discussion includes elementary notions of stability, the motivation for the use of feedback, as well as performance specifications in both the time and frequency domain for control systems.

In Chapter 8, we extend the methods of systems analysis to time domain techniques based on state space representations and matrix fraction descriptions. The notions of controllability, observability, minimal realizations, per unit values, canonical forms, and balanced realizations of systems are discussed.

This leads naturally to the discussion of state observers and state variable feedback in Chapter 9. We examine the interrelationship with frequency domain control systems analysis. We also introduce the ideas of fractional representations and their use in control systems analysis.

Chapter 10 discusses optimal state estimation. The Kalman filter is derived and its properties analyzed. Also, the class of all stable unbiased state estimators is described and used to motivate various robustness issues.

Chapter 11 describes the linear quadratic optimal control problem. Both time and frequency domain properties of the resultant control system are discussed. Also, other optimization criteria including H^∞ are described in some detail.

Chapter 12 shows how the parameters in models of systems can be estimated. Emphasis is given to the properties of the resultant estimation algorithms and methods for quantifying the estimation errors arising from nonideal factors, including noise and undermodeling. Adaptive control is also described.

Chapter 13 rounds out the discussion of analysis methods (frequency domain, pole assignment, optimal control, and so on) by discussing the implications of these methods in design. Various design constraints are studied, including minimum and maximum desirable bandwidth, desirable closed-loop pole locations, and frequency domain sensitivity.

Chapter 14 covers implementation issues such as the choice of sampling rate, antialiasing filter, finite word length effects, and numerical issues. Other practical issues including anti-integral windup and bumpless transfer are also discussed.

Chapter 15 describes a complete industrial case study. This problem shows how the various themes introduced in the book are brought together in a particular design example. The case study also illustrates the kind of extra considerations that are necessary to achieve success in a practical design problem.

The appendixes cover material that lies outside the main theme of the book but that is, nonetheless, useful background for someone who wants a deep understanding of digital control and estimation. The topics covered in the appendixes include:

Appendix A   Discussion of A/D and D/A conversion equipment.
Appendix B   Finite word length effects in digital control.
Appendix C   Finite word length effects in estimation.
Appendix D   A discussion of computer-aided design tools and their use.
Appendix E   A brief introduction to real-time software and operating systems.
Appendix F   Convergence analysis of a typical adaptive control algorithm.

The material in the book evolves from simple ideas through to more complex notions. However, some material in the book is not essential to the subsequent development and may be omitted on a first reading. We have marked this material by * in the text.

It is suggested that an undergraduate course on digital control (assuming the students have had one prior continuous time classical control course) can be constructed by covering the following material:

CHAPTER 2

Section 2.2 (include one or two examples of state space modeling)
Section 2.3 Linear state space models (concentrate on Section 2.3.2)
Section 2.4
Section 2.5

CHAPTER 3

Section 3.2
Section 3.3 (omit proof of Theorem 3.3.1)
Section 3.4 (omit section 3.4.3)
Section 3.5 (emphasize Section 3.5.2 and Example 3.5.3; omit Section 3.5.3)

CHAPTER 4

Sections 4.4 and 4.5 (emphasize that Laplace transforms essentially carry over to discrete time with corrections of the order of the sampling period)

CHAPTER 5

Section 5.2 (review)
Section 5.3
Section 5.4 (review)
Section 5.5 (emphasize Lemmas 5.5.1 and 5.5.2; expound Example 5.5.5)

CHAPTER 6

Section 6.2 (emphasize connection between continuous and discrete)

CHAPTER 7

Section 7.2 (emphasize Table 7.2.1)
Section 7.3 (revise)
Section 7.4 (revise)
Section 7.5 (revise)
Section 7.7 (revise)
Section 7.8 (revise)
Section 7.9 (revise)

CHAPTER 8

(If concepts of controllability and observability have been covered elsewhere, then most of this chapter will be review.)

Section 8.3
Section 8.4 (omit the multiinput case)
Section 8.5 (emphasize the connection between continuous and discrete in Lemma 8.5.1)
Sections 8.7, 8.8, 8.9 (emphasize duality and omit multioutput cases)
Section 8.10

CHAPTER 9

Section 9.2
Section 9.3
Section 9.4 (emphasize Theorem 9.4.2; sketch the proof)
Section 9.5 (omit Section 9.5.2)
Section 9.6 (emphasize Section 9.6.2; omit Sections 9.6.3, 9.6.4, and 9.6.5)

CHAPTER 11

Section 11.2 (briefly only)
Section 11.3 (emphasize equations (11.3.1) and (11.3.2))
Section 11.4 (emphasize Corollary 11.4.1)

CHAPTER 13

Section 13.2 (revise)
Section 13.3 (revise)
Section 13.4 (cover briefly; emphasize Lemma 13.4.1 but omit the proof; emphasize Corollary 13.4.1)
Section 13.5 (cover in detail)
Section 13.6 (cover in detail)

If extra time is available, it would also be useful to include some material from Chapter 14.

A postgraduate course on Advanced Control and Estimation can be constructed by reviewing Chapters 1 to 9 (with emphasis on the material omitted from the undergraduate course), plus Chapters 10 to 15. An alternative postgraduate course on Estimation and Adaptive Control can be based on Chapters 10 and 12.

We have taught the above course patterns at the University of Newcastle and have found them to be well appreciated by students.

We have found the unified approach presented in this book, including the delta operator representation, to be of major assistance in both theory and practice. It is our hope that, through this book, we are able to share our considerable enthusiasm for this approach with our readers.

REFERENCES

Additional information on the history of control can be found in:

DOEBELIN, E. O. (1985) *Control Systems: Principles and Design.* Wiley, New York.
MAYR, O. (1970) *The Origins of Feedback Control.* MIT Press, Cambridge, Mass.

Further background on estimation theory may be found in:

KAILATH, T. (1974) "A view of three decades of filtering theory," *IEEE Trans. Inf. Theory*, IT-20, 2, p. 146.

The advantages of digital control over analog control are discussed more fully in:

CASSELL, D. A. (1983) *Microcomputers and Modern Control Engineering*. Prentice-Hall, Englewood Cliffs, NJ.

For a brief introduction to delta operators, see:

MIDDLETON, R. H., and G. C. GOODWIN (1986) "Improved finite word length characteristics in digital control using delta operators," *IEEE Trans. Automatic Control*, AC-31, N. 11, pp. 1015–1021.

TSCHAUNER, J. (1963) *Introduction à la théorie des systèmes échantillonnés*. Dunod, Paris.

2

Continuous Time Models

2.1 INTRODUCTION

The main topic of this book will be the study of control and estimation theory and how this theory relates to physical processes. Before commencing a detailed study of control and estimation, it is therefore helpful to have an elementary appreciation of the dynamics of physical systems. In particular, it is helpful to review various mathematical descriptions for dynamical systems and to see how these descriptions can be arrived at from the laws of nature. This will be the topic of the current chapter.

In describing physical systems, it is useful to distinguish four types of variables: the inputs, the disturbances, the outputs, and the internal variables. We will use the term *input* to describe those variables that influence the operation of a system and which can be directly manipulated. The term *disturbance* will be used to describe other variables that influence the operation of a system but that cannot be directly manipulated. The *outputs* are those variables that are directly measured and whose behavior is a result of the action of the inputs and disturbances. Finally, the term *internal variable* will be used for any variable that is relevant to the behavior of a system. A special form of internal variable is the system state, which allows one to summarize the effect of all past inputs and disturbances on the system. The concept of system state forms the basis of a powerful modeling and analysis technique for both linear and nonlinear systems. We will study these models since they are central to much of modern control theory and have a close link with the fundamental physics of the process under study.

Many models derived from physical laws will exhibit nonlinear behavior. However, an extremely important subclass of state space models are those that exhibit linear behavior. We show how some nonlinear systems can be exactly converted to linear models. We also show how other nonlinear systems can be approximated by a linear model about an equilibrium position.

State space models generally have the form of a coupled set of first-order ordinary differential equations. An alternative to this model format is to use a single high-order differential equation relating the inputs directly to the outputs. We will call this description an input–output model. We will briefly describe this class of models and their relationship with state space descriptions.

Finally, we show how certain simple types of disturbances can also be described in state space and input–output form.

2.2 STATE SPACE MODELING

The state of a system can be loosely defined as a (nonunique) set of variables such that knowledge of these variables at some time, together with the future inputs and disturbances, is sufficient to allow determination of the system's future behavior. Thus the system state completely summarizes the influence of all past inputs and disturbances on the system.

In many practical cases it turns out that the state can be adequately represented by a finite-dimensional vector. We will designate these systems as *finite-dimensional systems*. For this class of systems, we will denote the dimension by the integer n and the system state by $x(t)$, which is an n vector. We will also denote the input to the system by $u(t)$, the disturbance by $d(t)$, and the output by $y(t)$. With this notation, we can describe the relationship between these various variables by a *state space* model as follows:

$$\frac{d}{dt}x(t) = f(x(t), u(t), d(t), t), \qquad x(t_0) = x_0 \qquad (2.2.1)$$

and

$$y(t) = g(x(t), u(t), d(t), t) \qquad (2.2.2)$$

The modeling problem, in state space form, is then one of:

(a) Determining an appropriate vector of state variables, x.
(b) Finding the functions f and g that describe the dynamic behavior of the system.

In many cases, *natural* state variables can be suggested. In electrical circuits, these would usually be inductor currents or capacitor voltages. In dynamic mechanisms, the natural states may be linear or angular positions and velocities. In fluid flow problems, the states may be fluid pressures, temperatures, and so on. In chemical systems, the natural state variables are often temperatures or chemical concentrations.

To find the functions f and g, we can employ physical laws such as Newton's law of motion, heat or mass balance equations, rate of reaction equations, and Kirchhoff's law for electrical circuits, plus the electrical properties of inductors and capacitors.

We will present next several examples taken from different fields that illustrate the derivation of state space models using the laws of nature.

Example 2.2.1. Electrical Circuit

Consider the electric circuit shown in Figure 2.2.1. A natural choice for the state variables in this case is the capacitor voltages. Application of Kirchhoff's laws gives

$$i_{C1} = (v_i - v_{C1}) \cdot \frac{1}{R1} + (v_{C2} - v_{C1}) \cdot \frac{1}{R2} \tag{2.2.3}$$

$$i_{C2} = (v_{C1} - v_{C2}) \cdot \frac{1}{R2} \tag{2.2.4}$$

Using $i = C(dv/dt)$, we then have

$$\dot{v}_{C1} = \frac{dv_{C1}}{dt} = -v_{C1}\left(\frac{1}{R1 \cdot C1} + \frac{1}{R2 \cdot C1}\right) + v_{C2}\frac{1}{C1 \cdot R2} + v_i\frac{1}{C1 \cdot R1} \tag{2.2.5}$$

$$\dot{v}_{C2} = \frac{dv_{C2}}{dt} = v_{C1}\left(\frac{1}{R2 \cdot C2}\right) - v_{C2}\left(\frac{1}{R2 \cdot C2}\right) \tag{2.2.6}$$

Thus if $x = [v_{C1} \; v_{C2}]^T$ and $u = v_i$, we can write

$$\dot{x} = \begin{bmatrix} -\frac{1}{R1 \cdot C1} - \frac{1}{R2 \cdot C1} & \frac{1}{C1 \cdot R2} \\ \frac{1}{R2 \cdot C2} & -\frac{1}{R2 \cdot C2} \end{bmatrix} x + \begin{bmatrix} \frac{1}{C1 \cdot R1} \\ 0 \end{bmatrix} u \tag{2.2.7}$$

$$y = [0 \; 1]x \tag{2.2.8}$$

which is in state space form [(2.2.1), (2.2.2)].

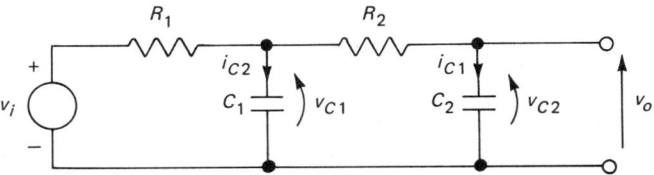

Figure 2.2.1 Example electrical system.

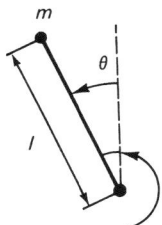

Figure 2.2.2 Example mechanical system.

Example 2.2.2. Torque-driven Inverted Pendulum

Consider the torque-driven inverted pendulum shown in Figure 2.2.2. We may select as state variables the angle θ of the rod about a vertical axis and ω, the angular velocity corresponding to θ. The input in this case is the torque applied to the mass and rod. Using Newton's second law of motion (in angular form), we have that

$$\left(\tfrac{1}{2}ml^2\right)\dot{\omega} = \tau + (mlg)\sin\theta \tag{2.2.9}$$

Letting $x_1 = \theta$ and $x_2 = \omega$, we then have

$$\dot{x}_1 = x_2 \tag{2.2.10}$$

$$\dot{x}_2 = \frac{2g}{l}\sin(x_1) + \frac{\tau}{\left(\tfrac{1}{2}ml^2\right)} \tag{2.2.11}$$

and

$$y = \theta = \begin{bmatrix} 1 & 0 \end{bmatrix} x \tag{2.2.12}$$

which is in state space form.

Example 2.2.3. Decalcification Plant

Consider the decalcification plant shown in Figure 2.2.3. The idea of the system is to reduce the concentration of calcium hydroxide present in the water by forming a calcium carbonate precipitate. Although not shown in Figure 2.2.3, we assume that some means for removing the precipitate is included in the system. In modeling this system, we will make the following simplifying assumptions: that the tank is completely stirred and that all the carbon dioxide injected dissolves in the water. The basic reaction in the system is

$$Ca(OH)_2 + CO_2 \Leftrightarrow CaCO_3 + H_2O \tag{2.2.13}$$

We further assume that the reaction proceeds only in the forward direction, that all calcium carbonate formed precipitates, and that the amount of water produced in the reaction (2.2.13) is insignificant. We will use the following notation

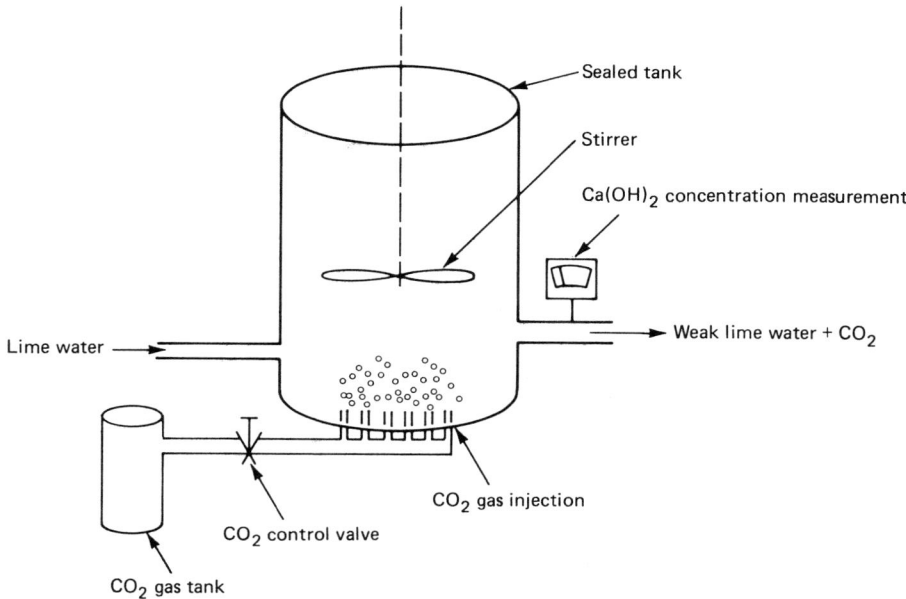

Figure 2.2.3 Example chemical system.

to describe the system:

x_1, x_2	State variables representing the concentrations of calcium hydroxide and carbon dioxide (respectively) in moles/liter in the tank.
k	Rate of inflow, in moles/second, of calcium hydroxide.
V	Total tank volume in liters.
r	Rate of production of calcium carbonate in moles/second.
c	Constant (in liters^2/(mole · second)) relating r to x_1 and x_2.
u	Input rate of flow of CO_2 in moles/second.
y	Calcium hydroxide concentration at the outlet (moles/liter).

We hypothesize that the rate of reaction is proportional to the product of the concentrations; that is,

$$r = c x_1 x_2 \tag{2.2.14}$$

Note that, in practice, the constant of proportionality c is strongly dependent on the temperature of the reactants. The rate of change of concentrations can be expressed as

$$\dot{x}_1 = \frac{k}{V} - \frac{r}{V} \tag{2.2.15}$$

and
$$\dot{x}_2 = \frac{u}{V} - \frac{r}{V} \tag{2.2.16}$$

Equations (2.2.14) to (2.2.16) can be written in state space form as

$$\dot{x}_1 = \frac{k}{V} - \frac{c}{V} x_1 x_2 \tag{2.2.17}$$

$$\dot{x}_2 = \frac{u}{V} - \frac{c}{V} x_1 x_2 \tag{2.2.18}$$

and
$$y = \begin{bmatrix} 1 & 0 \end{bmatrix} x \tag{2.2.19}$$

2.3 LINEAR STATE SPACE MODELS

In the previous sections we have given some examples of the development of state space models for dynamic systems. Some of these models have the special property that the functions f and g are linear in the variables; that is, $f(x, u, t) = Ax + Bu$ and $g(x, u, t) = Cx + Du$. We will call these models *linear state space models*. These systems have many useful properties and thus we will give them special emphasis in this book. For example, one key property satisfied by linear state space models is superposition. This principle can be stated as follows. Let $y_1(t)$ denote the response with $x(0) = 0$, $u(t) = u_1(t)$, $y_2(t)$ the response with $x(0) = 0$ and $u(t) = u_2(t)$, and $y_0(t)$ the response with $x(0) = x_0$ and $u(t) = 0$. Then the response to $x(0) = \alpha_0 x_0$, $u(t) = \alpha_1 u_1(t) + \alpha_2 u_2(t)$ is $y(t) = \alpha_0 y_0(t) + \alpha_1 y_1(t) + \alpha_2 y_2(t)$ for any scalars $\alpha_0, \alpha_1, \alpha_2$. This principle and other properties make linear systems particularly amenable to analysis. In subsequent chapters we will explore the rich theory available for this special class of systems.

Unfortunately, however, not all systems are linear, see, for example, Examples 2.2.2 and 2.2.3. These systems exhibit more complex behavior and are more difficult to analyze. For this reason, it is often desirable to transform a given nonlinear system into an associated linear system whenever this is feasible. We discuss next several methods for linking particular nonlinear systems to associated linear models.

2.3.1 Compensating Nonlinearities

Compensation techniques rely on the presence of special structures in the nonlinearities exhibited in (2.2.1) and (2.2.2) and are thus applicable only for certain classes of systems. Several illustrative examples are discussed next.

(a) **Input–output nonlinearities**

Consider the case where the system model (2.2.1) and (2.2.2) has the following structure:

$$\frac{d}{dt} x(t) = Ax(t) + B f_u(u) \tag{2.3.1}$$

$$y(t) = g_y(Cx(t)) \tag{2.3.2}$$

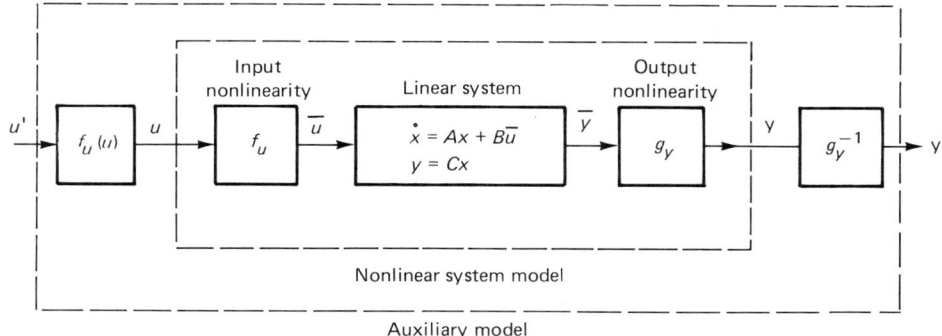

Figure 2.3.1 Diagram showing the concept of input–output nonlinearity compensation.

If f_u and g_y are invertible functions over some operating range, then we define auxiliary variables u' and y' by the equations

$$u = f_u^{-1}(u') \tag{2.3.3}$$

and

$$y' = g_y^{-1}(y) \tag{2.3.4}$$

Then, clearly, over the range of values for which f_u and g_y can be inverted, u' and y' can be related by the following linear model:

$$\frac{d}{dt}x(t) = Ax(t) + Bu'(t) \tag{2.3.5}$$

$$y'(t) = Cx(t) \tag{3.3.6}$$

This technique of nonlinearity compensation is depicted in Figure 2.3.1.
The following simple example illustrates how this technique may be used.

Example 2.3.1. Electric Water Heater

Let us suppose that the heat capacity of the water in the tank in Figure 2.3.2 is C and that, if ambient (outside) temperature is T_0, heat is lost to the outside at a rate equal to $k(T - T_0)$. As our state variable, let $x = T - T_0$ be the water temperature above ambient. Then we can show that an appropriate state space model is

$$\dot{x} = -\frac{k}{C}x + \frac{1}{CR}u^2 \tag{2.3.7}$$

and

$$y = x \tag{2.3.8}$$

Clearly, this model is linear except for the term u^2 in (2.3.7), and so, by defining an

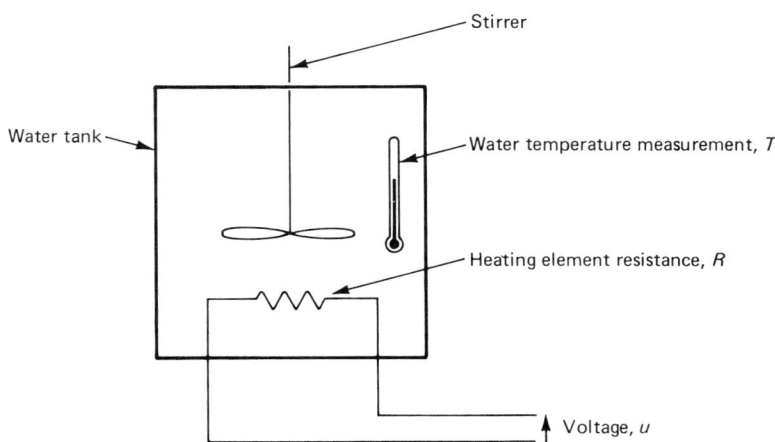

Figure 2.3.2 Electric water heater diagram for Example 2.3.1.

auxiliary input variable $p = u^2$, we can obtain the following linear model:

$$\dot{x} = -\frac{k}{C}x + \frac{1}{CR}p \quad (2.3.9)$$

$$y = x \quad (2.3.10)$$

Note that this auxiliary model is only valid over the operating range $p > 0$. ▽▽▽

(b) Output feedback nonlinearities

We now consider another compensation technique for obtaining a linear model when the nonlinear model has the following structure:

$$\frac{d}{dt}x(t) = Ax(t) + B[f_y(y) + u(t)] \quad (2.3.11)$$

$$y(t) = Cx(t) \quad (2.3.12)$$

The system is linear, save for the term f_y. We can then define the auxiliary input

$$u' = u + f_y(y) \quad (2.3.13)$$

which leads to the following linear auxiliary state space form, illustrated in Figure 2.3.3:

$$\frac{d}{dt}x(t) = Ax(t) + Bu'(t) \quad (2.3.14)$$

$$y(t) = Cx(t) \quad (2.3.15)$$

This technique is illustrated by the following example:

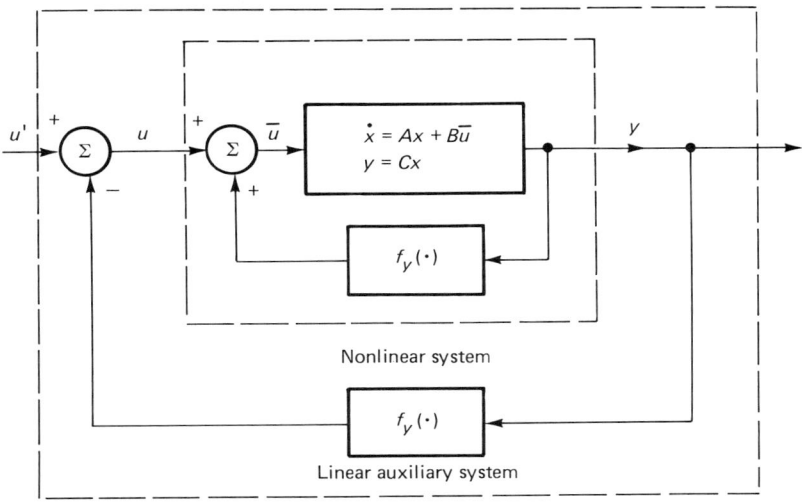

Figure 2.3.3 Block diagram illustrating linearization of a feedback nonlinearity.

Example 2.3.2. Inverted Pendulum (continued)

Consider again the torque-driven inverted pendulum of Example 2.2.2. Define an auxiliary input torque, τ', by the equation

$$\tau' = \tau + mlg \sin \theta \tag{2.3.16}$$

Then, from Equations (2.2.10)–(2.3.12), we can obtain the following linear auxiliary model:

$$\dot{x}_1 = x_2$$
$$\dot{x}_2 = \frac{\tau'}{\frac{1}{2}ml^2} \tag{2.3.17}$$

$$y = \theta = [1, \quad 0]x \tag{2.3.18}$$

Example 2.3.2 is, in fact, a one degree of freedom robot. The same linearization method can be applied to multiple degrees of freedom, in which case the method is known as computed torque control.

(c) Linearization by change of states

Consider a general state space model of the form

$$\frac{d}{dt}x = f(x, u) \tag{2.3.19}$$

Sometimes it is possible to find a nonlinear function of the states, say $y = \gamma(x)$ such that y and the first $n - 1$ derivatives of y with respect to time are independent of

u. In this case we can define a new state vector as
$$\bar{x}^T \triangleq (y, \dot{y}, \ldots, y^{n-1}) \quad (2.3.20)$$
Using this new state vector, the corresponding state equations become
$$\begin{aligned} \frac{d}{dt}\bar{x}_1 &= \bar{x}_2 \\ &\vdots \\ \frac{d}{dt}\bar{x}_{n-1} &= \bar{x}_n \\ \frac{d}{dt}\bar{x}_n &= y^n \triangleq \theta(\bar{x}, u) \end{aligned} \quad (2.3.21)$$

Finally, if the function $\theta(\bar{x}, u)$ is invertible with respect to u, then we can solve $\theta(\bar{x}, u) = u'$ for u in terms of u' and $\bar{x}$, leading to the following linear system:
$$\begin{aligned} \frac{d}{dt}\bar{x}_1 &= \bar{x}_2 \\ &\vdots \\ \frac{d}{dt}\bar{x}_{n-1} &= \bar{x}_n \\ \frac{d}{dt}\bar{x}_n &= u' \end{aligned} \quad (2.3.22)$$

This method is illustrated by the following example.

Example 2.3.3

Consider the following nonlinear system:
$$\dot{x}_1 = x_1 + x_2^3 + u \quad (2.3.23)$$
$$\dot{x}_2 = x_1 + u \quad (2.3.24)$$
We define $\bar{x}_1 = y$ as
$$\bar{x}_1 \triangleq y \triangleq x_1 - x_2 \quad (2.3.25)$$
Then
$$\bar{x}_2 \triangleq \dot{y} = \dot{x}_1 - \dot{x}_2 = x_2^3 \quad \text{(independent of } u\text{)} \quad (2.3.26)$$
Also,
$$\begin{aligned} \ddot{y} &= 3x_2^2 \dot{x}_2 \\ &= 3x_2^2(x_1 + u) \end{aligned} \quad (2.3.27)$$
Hence, defining $\bar{x}_1 \triangleq y$, $\bar{x}_2 \triangleq \dot{y}$, we have
$$\dot{\bar{x}}_1 = \bar{x}_2 \quad (2.3.28)$$
$$\dot{\bar{x}}_2 = 3\bar{x}_2^{2/3}(\bar{x}_1 + \bar{x}_2^{1/3}) + 3\bar{x}_2^{2/3} u \quad (2.3.29)$$

Finally, putting
$$u = \tfrac{1}{3}u'\bar{x}_2^{-2/3} - (\bar{x}_1 + \bar{x}_2^{1/3}) \tag{2.3.30}$$
we obtain the following linear model:
$$\dot{\bar{x}}_1 = \bar{x}_2 \tag{2.3.31}$$
$$\dot{\bar{x}}_2 = u' \tag{2.3.32}$$

▽▽▽

The techniques just described can be combined in various obvious ways. However, there still remain some difficult nonlinear systems that either cannot be exactly linearized or are difficult to exactly linearize (for example, the simple decalcification plant of Example 2.2.3 has a model that cannot be linearized using either technique a or b and that can be linearized by method c only if the output is differentiated and the process constants c, k, and V are known). In such cases the use of linear approximations to nonlinear systems may be a useful way of linking the nonlinear behavior to linear behavior.

2.3.2 Linear Approximation of Nonlinear Systems

Consider the case where the nonlinear system (2.2.1) and (2.2.2) is operating with small deviations about a fixed operating point (which is assumed to be an equilibrium point for the system). Let
$$x(t) = X + \Delta x(t) \tag{2.3.33}$$
$$u(t) = U + \Delta u(t) \tag{2.3.34}$$
and
$$y(t) = Y + \Delta y(t) \tag{2.3.35}$$
where X, U, and Y are the constants defining the operating point.

The assumption of equilibria ensures that
$$f(X, U) = 0 \tag{2.3.36}$$
$$Y = g(X, U) \tag{2.3.37}$$

Then, using a Taylor's series approximation, we have the following approximate model for the system (where we assume that f and g do not depend explicitly on time).
$$\frac{d}{dt}\Delta x(t) = \frac{d}{dt}x(t) = f(X + \Delta x(t), U + \Delta u(t))$$
$$\approx \left[\frac{\partial f}{\partial x}\right]\bigg|_{\substack{x=X\\u=U}} \Delta x(t) + \left[\frac{\partial f}{\partial u}\right]\bigg|_{\substack{x=X\\u=U}} \Delta u(t) \tag{2.3.38}$$
and
$$\Delta y(t) = g(X + \Delta x(t), U + \Delta u(t)) - Y$$
$$\approx \left[\frac{\partial g}{\partial x}\right]\bigg|_{\substack{x=X\\u=U}} \Delta x(t) + \left[\frac{\partial g}{\partial u}\right]\bigg|_{\substack{x=X\\u=U}} \Delta u(t) \tag{2.3.39}$$

Note that this technique of linearization requires the computation of various Jacobian matrices that arise when we differentiate a vector function with respect to a vector (for example, $\partial f/\partial x$ is the $n \times n$ matrix whose i, jth element is $\partial f_i/\partial x_j$). Note also that, by letting $A = [\partial f/\partial x]\big|_{\substack{x=X \\ u=U}}$ (and so on for B, C, and D), we can rewrite (2.3.38) and (2.3.39) in the standard linear state space form:

$$\frac{d}{dt}(\Delta x) \approx A(\Delta x) + B(\Delta u) \qquad (2.3.40)$$

and

$$(\Delta y) \approx C(\Delta x) + D(\Delta u) \qquad (2.3.41)$$

This technique is illustrated in the following example.

Example 2.3.4. Decalcification Plant

Consider the simple decalcification plant of Example 2.2.3 with a nominal operating point, $u = U = k$ and $y = X_1 = Y$. Then, using Equations (2.2.17)–(2.2.19) and (2.3.36), we can show that the nominal operating condition for x_2 is

$$X_2 = \frac{k}{cX_1} \qquad (2.3.42)$$

Then, rewriting (2.2.17), we have

$$(\Delta \dot{x}_1) = \dot{x}_1 = \frac{k}{V} - \frac{c}{V} x_1 x_2$$

$$\approx -\frac{c}{V} X_2 \Delta x_1 - \frac{c}{V} X_1 \Delta x_2 \qquad (2.3.43)$$

Similarly,

$$(\Delta \dot{x}_2) \approx -\frac{c}{V} X_2 \Delta x_1 - \frac{c}{V} X_1 \Delta x_2 + \frac{1}{V} \Delta u \qquad (2.3.44)$$

and thus an approximate linear model for the system is

$$\frac{d}{dt} \Delta x = \begin{bmatrix} -\frac{k}{V}\frac{1}{X_1} & -\frac{c}{V} X_1 \\ -\frac{k}{V}\frac{1}{X_1} & -\frac{c}{V} X_1 \end{bmatrix} \Delta x + \begin{bmatrix} 0 \\ \frac{1}{V} \end{bmatrix} \Delta u \qquad (2.3.45)$$

$$\Delta y = \begin{bmatrix} 1 & 0 \end{bmatrix} \Delta x \qquad (2.3.46)$$

In the next section, we consider an alternative to the state space approach based on input–output models.

2.4 INPUT–OUTPUT MODELS FOR DYNAMIC SYSTEMS

In an input–output model, we write the differential equations that describe a system as a higher-order differential equation relating input to output, rather than as a first-order vector differential equation involving system states. This type of description is sometimes more natural and bears a close connection to the transfer function for linear systems to be discussed later in Chapter 5.

We introduce the symbol ρ to represent the differential operator, $\rho \equiv d/dt$. In the general nonlinear case, an input–output model has the form

$$h(\rho^n y, \ldots, y; \rho^m u, \ldots, u, t) = 0 \tag{2.4.1}$$

In the case of linear time invariant systems, (2.4.1) simplifies to

$$A_n \rho^n y + A_{n-1} \rho^{n-1} y + \cdots + A_0 y = B_m \rho^m u + B_{m-1} \rho^{m-1} u + \cdots + B_0 u \tag{2.4.2}$$

That is,

$$A(\rho) y = B(\rho) u \tag{2.4.3}$$

where $A(\cdot)$ and $B(\cdot)$ are matrix polynomials in the operator ρ.

Note that linear input–output models and linear state space models give an equivalent description of the input–output properties of the system; that is, given any state space model, there exists an equivalent operator model relating y and u, and vice versa. For example, given a single-input, single-output model of the form (2.4.2) (where $a_n = 1$ and $m < n$ for simplicity), the following state space model can be readily shown to be equivalent to (2.4.2):

$$\frac{d}{dt} x = \begin{bmatrix} -a_{n-1} & 1 & 0 & \cdots & & 0 \\ -a_{n-2} & 0 & 1 & 0 & \cdots & 0 \\ \vdots & \vdots & & & & \vdots \\ \vdots & \vdots & & & & 0 \\ \vdots & \vdots & & & & 1 \\ -a_0 & 0 & \cdots & & \cdots & 0 \end{bmatrix} x + \begin{bmatrix} b_{n-1} \\ b_{n-2} \\ \vdots \\ \vdots \\ \vdots \\ b_0 \end{bmatrix} u \tag{2.4.4}$$

$$y = \begin{bmatrix} 1 & 0 & 0 & \cdots & \cdots & 0 \end{bmatrix} x \tag{2.4.5}$$

where we have used lowercase notation since A_i, B_i are scalars. To demonstrate this equivalence, we compute $\rho y, \rho^2 y, \ldots, \rho^n y$ from (2.4.4) and (2.4.5), and then show that

$$\left(\rho^n + a_{n-1} \rho^{n-1} + \cdots + a_0\right) y = \left(b_{n-1} \rho^{n-1} + \cdots + b_0\right) u \tag{2.4.6}$$

Further details may be found in Section 8.8.3 of Chapter 8.

The converse result, that any linear state space system can be written in input–output form, is also true and is left as an exercise for the reader (Problem 8).

The following examples illustrate how input–output models may be directly determined from physical principles.

Example 2.4.1

Consider again the torque-driven inverted pendulum of Example 2.2.2. From Newton's second law of motion, we have directly that

$$\left(\tfrac{1}{2}ml^2\right)\ddot{\theta} - mlg\sin\theta = \tau \qquad (2.4.7)$$

or

$$\rho^2\{\theta\} - \frac{2g}{l}\sin\theta = \frac{2}{ml^2}\tau \qquad (2.4.8)$$

which is in a nonlinear input–output form. ▽▽▽

Example 2.4.2

As a further example, consider the problem of modeling an armature voltage-controlled dc motor. Assuming a constant field current, and neglecting saturation, armature reaction, and the armature time constant, we have the following equations for the dynamic behavior of the dc motor:

$$V = RI_a + E_a \qquad (2.4.9)$$

$$E_a = k\dot{\theta} \qquad (2.4.10)$$

and

$$\tau_{\text{elec}} = J\ddot{\theta} = kI_a \qquad (2.4.11)$$

where V is the terminal voltage (input), I_a is the armature current, R is the armature resistance, θ is the motor shaft position, E_a is the back emf generated in the motor, k is the motor torque constant, J is the shaft inertia, and τ_{elec} is the torque produced by the motor. Equations (2.4.9)–(2.4.11) can be rearranged as

$$\left\{\rho^2 + \frac{k^2}{RJ}\rho\right\}\theta = \frac{k}{RJ}V \qquad (2.4.12)$$

▽▽▽

2.5 MODELS FOR DISTURBANCES

The previous descriptions of system models have neglected the effects of the environment within which the system operates. To obtain a more complete description, it is often helpful to adjoin an additional model that describes the influence of the environment.

A number of disturbance models are possible depending on the nature of the environment. A key ingredient in disturbance modeling is the extent to which the disturbance can be predicted based on past values. We will briefly discuss several classes of disturbances. These models can be combined with the system model to give a composite model for the system and its environment.

An important class of disturbances are those called *deterministic*. A deterministic disturbance is one that can be perfectly predicted into the future. Examples of such disturbances are dc offsets, drift at constant rate, and periodic disturbances including seasonal components. This kind of disturbance can usually be modeled by a homogeneous state space or input–output model. Examples of these models are

(a) DC offset: $d = $ constant; that is, the model is

$$\rho d = 0, \qquad d(0) = \text{constant} \tag{2.5.1}$$

(b) Drift at constant rate: $d = \alpha t + \beta$; that is, the model is

$$\rho^2 d = 0, \qquad d(0) = \beta, \qquad \rho d(0) = \alpha \tag{2.5.2}$$

(c) Sinusoidal disturbances: $d = G\cos(\omega_0 t + \phi)$; that is,

$$(\rho^2 + \omega_0^2)d = 0, \qquad d(0) = G\cos(\phi), \qquad \rho d(0) = \omega_0 G \sin(\phi) \tag{2.5.3}$$

These models have the general form

$$S(\rho)d = 0 \tag{2.5.4}$$

where

$$S(\rho) = \rho^n + s_{n-1}\rho^{n-1} + \cdots + s_0$$

An equivalent state space description can also be obtained by the techniques of the previous section. Thus, the model (2.5.4) can also be written as

$$\rho x = \begin{bmatrix} -s_{n-1} & 1 & & \\ \vdots & & \ddots & \\ & & & 1 \\ -s_0 & 0 & \cdots & 0 \end{bmatrix} x \tag{2.5.5}$$

$$d = \begin{bmatrix} 1 & 0 & \cdots & 0 \end{bmatrix} x \tag{2.5.6}$$

These models for disturbances can be combined with a system model to obtain a composite description. This is illustrated for a simple case in Example 2.5.1.

Example 2.5.1. Water Heater

Consider again the water heater of Example 2.3.1. In the previous derivation we assumed that the ambient temperature T_0 was constant. A better model could be to assume that the ambient temperature fluctuates daily and is thus describable as a sinusoidal disturbance. Letting x_1 denote the temperature above mean ambient temperature, then the following model is appropriate

$$\rho x_1 = -\frac{k}{c}x_1 + \frac{1}{c}P + \frac{k}{c}d \tag{2.5.7}$$

$$y = x_1$$

where d is modeled as a sine wave and satisfies

$$\rho \begin{bmatrix} x_2 \\ x_3 \end{bmatrix} = \begin{bmatrix} 0 & 1 \\ -\omega_0^2 & 0 \end{bmatrix} \begin{bmatrix} x_2 \\ x_3 \end{bmatrix} \tag{2.5.8}$$

$$d = \begin{bmatrix} 1 & 0 \end{bmatrix} \begin{bmatrix} x_2 \\ x_3 \end{bmatrix} \tag{2.5.9}$$

Combining (2.5.7)–(2.5.9) leads to the following composite model in state space form:

$$\rho \begin{bmatrix} x_1 \\ x_2 \\ x_3 \end{bmatrix} = \begin{bmatrix} -\dfrac{k}{c} & \dfrac{k}{c} & 0 \\ 0 & 0 & 1 \\ 0 & -\omega_0^2 & 0 \end{bmatrix} \begin{bmatrix} x_1 \\ x_2 \\ x_3 \end{bmatrix} + \begin{bmatrix} \dfrac{1}{c} \\ 0 \\ 0 \end{bmatrix} P \tag{2.5.10}$$

$$y = x_1 \tag{2.5.11}$$

Equations (2.5.10) and (2.5.11) are also equivalent to the following input–output model:

$$(\rho^2 + \omega_0^2)\left(\rho + \dfrac{k}{c}\right) y = (\rho^2 + \omega_0^2) \dfrac{1}{c} P \tag{2.5.12}$$

Note that it is not possible to cancel the term $(\rho^2 + \omega_0^2)$ from both sides of (2.5.12) since, if this is done, the effect of the disturbance is lost. (More will be said about this in Chapter 8.)

In the preceding discussion, we have restricted attention to simple deterministic disturbances. This covers many practical cases of interest. However, more general disturbance models are possible. For example, in Chapter 10, we will consider random disturbances that have the property that they are not perfectly predictable from their past.

2.6 SUMMARY

The key points covered in this chapter were:

- The concept of system state, which summarizes the effect of all past inputs and disturbances on future system behavior.
- Nonlinear state space models of the form

$$\dfrac{d}{dt} x(t) = f(x(t), u(t), d(t), t), \qquad x(t_0) = x_0 \tag{2.2.1}$$

$$y(t) = g(x(t), u(t), d(t), t) \tag{2.2.2}$$

- A special class of state space models, linear models, described by

$$\dfrac{d}{dt} x = Ax + Bu$$

$$y = Cx + Du$$

- The idea that many nonlinear systems can be converted to exact linear systems by various compensation techniques (Section 2.3.1).
- The idea that nonlinear state space models can be approximated by linear systems about an equilibrium point (Section 2.3.2).
- A discussion of input–output models and their properties (Section 2.4).
- A discussion of the equivalence between linear state space models and linear input–output models.
- A description of deterministic disturbances, including steps, ramps, and sine waves, and a discussion of their corresponding models.

Thus, Chapter 2 covered various aspects of the modeling of physical processes by continuous time models. In Chapter 3, we will take the next step of introducing the notion of sampling and how this affects the modeling process.

2.7 REFERENCES

Further discussion of physical models may be found in:

DOEBELIN, E. O. (1985) *Control Systems: Principles and Design*. Wiley, New York.

Good introductions to linear state space models are contained in:

CHEN, C. T. (1984) *Linear System Theory and Design*. Holt, Rinehart and Winston, New York.

FORTMANN, T. E., and K. L. HITZ (1977) *An Introduction to Linear Control Systems*. Marcel Dekker, New York.

KAILATH, T. (1980) *Linear Systems*. Prentice-Hall, Englewood Cliffs, N.J.

Additional information on input–output based models is given in:

KAILATH, T. (1980) *Linear Systems*. Prentice-Hall, Englewood Cliffs, N.J.

WOLOVICH, W. A. (1974) *Linear Multivariable Systems*. Springer-Verlag, New York.

2.8 PROBLEMS

1. Extend the model of Example 2.2.1 to include nonzero source resistance and nonzero load conductance (requires some electrical background).
2. Extend the model of Example 2.2.2 to include friction in the joint and the mass of the link (requires some mechanical background).
3. Extend the model of Example 2.2.3 to include reaction limiting by product inhibition (requires some chemical engineering background).
4. Consider the level control problem in Figure 2.8.P4. Assuming that $f = c\sqrt{y}$ and that the cross-sectional area of the tank is A, derive a nonlinear model for the system.

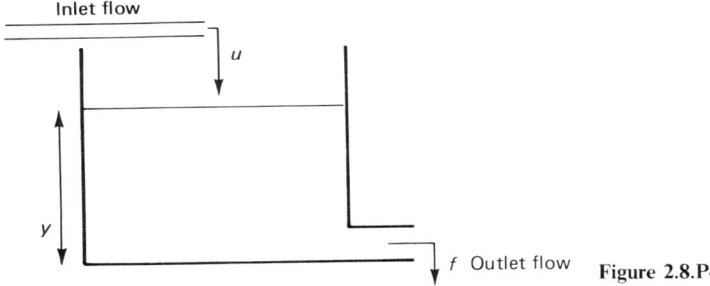

Figure 2.8.P4

5. Linearize the model in Problem 4 using both feedback linearization [Section 2.3.1(b)] and linear approximation [Section 2.3.2].
6. Extend Problems 4 and 5 (if possible) to the two coupled tanks shown in Figure 2.8.P6, where

$$f_1 = c_1\sqrt{y_1 - y_2}$$
$$f_2 = c_2\sqrt{y_2}$$

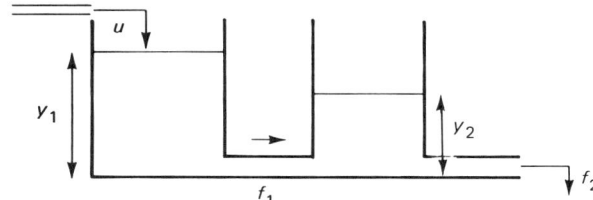

Figure 2.8.P6

7. (a) Show by successively eliminating states that (2.4.5) and (2.4.4) can be expressed in the form of (2.4.2).
   (b) Show that a model of the form (2.4.2) can be expressed in the form of (2.4.5) and (2.4.4) by appropriately defining x in terms of y, u and their derivatives.
8. Given a general single-input, single-output linear state space model of dimension n, show that there exists an equivalent nth-order input–output model. *Hint:* Find an expression for $\rho^i y$ in terms of $x, u, \ldots, \rho^{i-1} u$ using the state space model. Then use the Cayley–Hamilton theorem to eliminate x by adding the derivatives of y with appropriate weightings.
9. Convert the model in Example 2.4.2 to state space form as in Equation (2.4.4).
10. Can the nonlinear element in Figure 2.8.P10 be locally linearized? Give reasons for your answer.

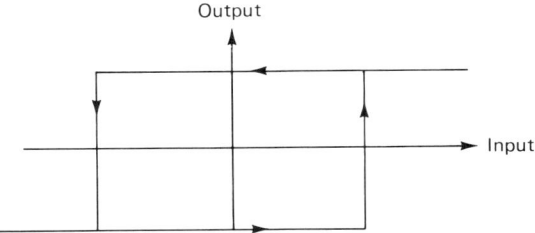

Figure 2.8.P10

11. Suggest a suitable deterministic disturbance model to add wind effects on the inverted pendulum model. Discuss some of the shortcomings of deterministic models in this instance.

12. Develop an input–output model for a disturbance of the form
$$d(t) = G_1 \sin(\omega_1 t + \phi_1) * G_2 \cos(\omega_2 t + \phi_2)$$

13. Suppose you are required to model the dynamic behavior of an automobile (driven in high gear) so that an automatic cruise control may be added. Suppose that the wind resistance of the vehicle (F_W in kN) is given by
$$F_W = -7 \times 10^{-5} S^2$$
where S is the car speed in meters/second. When in high gear, the engine force is (in kN)
$$F_E = 0.35\left(\frac{1 - 0.1S + 0.7\sqrt{T}}{1 + 0.1S}\right)$$
where T is the throttle (that is, accelerator) position in percent. ($T = 100$ is accelerator fully depressed, and $T = 0$ is fully released.) If the car has an effective mass M of 1 tonne (1000 kg), answer the following questions:
 (a) Find a nonlinear state space model for the system. (T is the input, and S is the measured output.) *Hint:* Newton's laws of motion give that $M\dot{S}$ = total force.
 (b) Find an approximate linear model for the system operating near a steady-state condition $S = 30$ m/s.
 (c) Show how a linear auxiliary system may be formed by defining a new input, V. Give an explicit expression for T in terms of V.
 (d) Describe qualitatively the effect on the model in part (a) of including the effect of gravity on the car, if the road is not flat.

14. Suppose for $x(0) = 0$, that is, zero initial conditions, the response of the linear time varying system
$$\dot{x}(t) = A(t)x(t) + B(t)u(t)$$
$$y(t) = C(t)x(t) + D(t)u(t)$$
for input $u(t) = u_1(t)$ is $x(t) = x_1(t)$ and $y(t) = y_1(t)$, and for $u(t) = u_2(t)$, the response is $x(t) = x_2(t)$ and $y(t) = y_2(t)$. Show that the system obeys superposition; that is, for any constants α_1, α_2, the response to $u(t) = \alpha_1 u_1(t) + \alpha_2 u_2(t)$ is $x(t) = \alpha_1 x_1(t) + \alpha_2 x_2(t)$, $y(t) = \alpha_1 y_1(t) + \alpha_2 y_2(t)$. *Hint:* Find $(d/dt)x(t)$ and $x(0)$.

15. Show that the following system is nonlinear (that is, does *not* obey superposition):
$$\dot{x} = -x + u^2$$
$$y = x$$
Hint: For $x(0) = 0$ and $u(t) = k$ (constant), $x(t) = k^2(1 - e^{-t})$; then consider $u_1(t) = +1$, $u_2(t) = -1$.

16. (a) Show that the following system is linear (from u to y) although it appears otherwise:
$$\dot{x} = -3x + (3x^{2/3})u$$
$$y = x^{1/3}$$
Hint: For $x(0) = 0$, show that $x(t) = \{\int_0^t e^{-(t-\tau)}u(\tau)\,d\tau\}^3$.
 (b) Show that the system has a nonlinear initial condition response. *Hint:* For $u(t) = 0$, $x(t) = e^{-3t}x(0)$. Thus find $y(t)$ for $u(t) = 0$.

17. Consider the mechanical system in Figure 2.8.P17. If the output y is $d_1 - d_2$, $v_1 = (d/dt)d_1$, $v_2 = (d/dt)d_2$, $a_1 = (d/dt)v_1$, and $a_2 = (d/dt)v_2$, which of the following sets of variables are valid states:
   (i) $x_1 = y$, $x_2 = v_1$
   (ii) $x_1 = d_1$, $x_2 = d_2$, $x_3 = v_1$, $x_4 = v_2$
   (iii) $x_1 = y$, $x_2 = v_2 - v_1$
   (iv) $x_1 = y^3$, $x_2 = \exp(v_2 - v_1)$

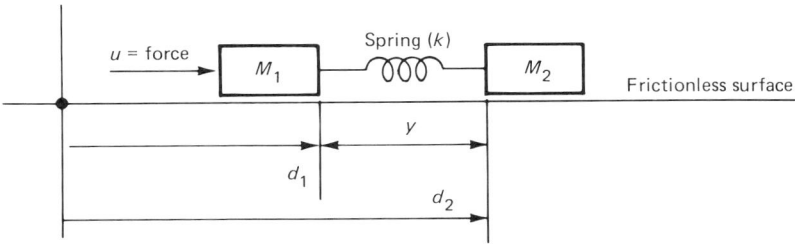

Figure 2.8.P17

18. Consider the electromagnetic system in Figure 2.8.P18. If the total reluctance of the magnetic circuit is $R = R_i + k_0 y$, and the windings are assumed to have zero resistance, find a state space model for the system. *Hint:*
   (i) Try using y, $\dot{y}$, and i as states.
   (ii) The induced emf is $u = N(d\phi/dt)$.
   (iii) The flux is $\phi = Ni/R$.
   (iv) The magnetic force is $\frac{1}{2}N^2 i^2 (d/dy)(1/R)$.

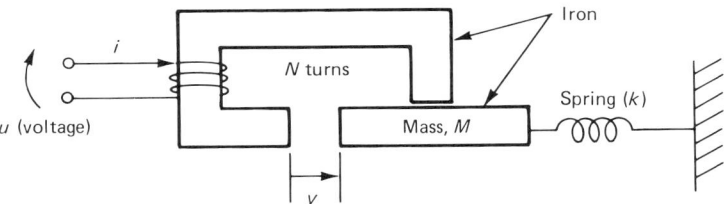

Figure 2.8.P18

19. Using states $x_1 = v_{c_1}$ and $x_2 = v_{c_2}$, find a linear state space model for the system in Figure 2.8.P19.

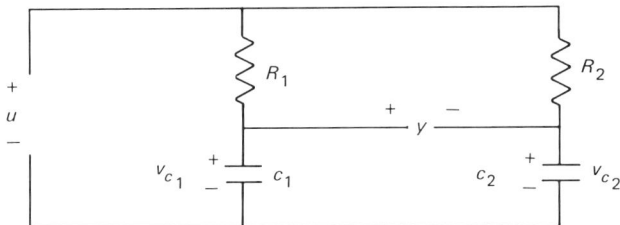

Figure 2.8.P19

26 Continuous Time Models   Chap. 2

3

Discrete Time Models

3.1 INTRODUCTION

We saw in Chapter 2 how continuous time physical systems can be modeled by sets of ordinary differential equations. Here we turn to the question of the interaction of a physical system with a computer. In particular, we wish to describe the model as seen by the computer. Thus, our objective is to relate sampled values of the output response to sampled values of the input. The interface between a computer and a continuous process is typically achieved with an analog to digital (A/D) converter and a digital to analog (D/A) converter, as shown in Figure 3.1.1.

The process of A/D conversion is basically one of sampling. However, we will see that it is also important, in general, that the signal be preconditioned prior to sampling. This preconditioning usually takes the form of a special low-pass filter (called an antialiasing filter) that removes high-frequency components from the

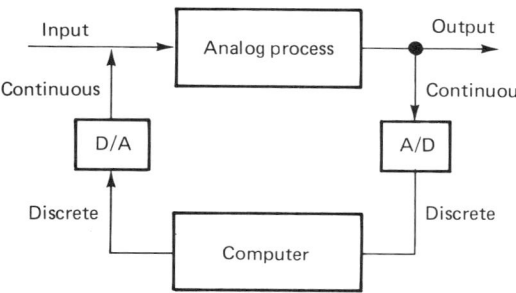

Figure 3.1.1 Computer interface to an analog process.

signal to be sampled. Unless this is done, these high-frequency components confuse the interpretation of the sampled signal. We will discuss this in detail in Section 3.2.

The process of D/A conversion is basically one of reconstructing an analog signal from discrete data. The most commonly used method of reconstruction is to interpolate between the given sample values. A simple method of reconstruction is to set the interpolated signal equal to the most recent digital value.

In the next section we examine the sampling process in more detail.

3.2 SAMPLING

We first introduce some notation that will be used in the remainder of the book.

3.2.1 Notation

In the sequel, we will need to refer to continuous time signals and to the sample values of these signals. We will use $y(\cdot)$ to denote a time function in continuous time; that is,

$$y(\cdot): \mathbb{R} \to \mathbb{R}^n \qquad (3.2.1)$$

where $\mathbb{R}$ denotes the reals and $\mathbb{R}^n$ denotes n-dimensional real space.

The corresponding sample values at interval Δ are simply

$$y(k\Delta), \quad \text{for } k = 0, 1, 2, \ldots \qquad (3.2.2)$$

When considering the discrete case in isolation, it will occasionally be convenient to drop the explicit dependence on Δ. We will then define

$$y_k = y(k\Delta), \quad \text{for } k = 0, 1, 2, \ldots \qquad (3.2.3)$$

3.2.2 Analog-to-Digital Conversion

The process of sampling, as outlined above, is generally realized in a device called an analog to digital converter (A/D converter). Such a device basically takes the sample values, usually spaced at regular time intervals, and converts the sampled analog voltage into a binary form suitable for use in the computer.

For completeness, we give some of the electronic hardware used to realize A/D converters in Appendix A.

3.2.3 Aliasing

When dealing with sampled signals, it is important to keep in mind that a loss of information is inherent in the sampling process. This can lead to results being misinterpreted unless certain precautions are taken.

One possible source of misinterpretation is that of frequency folding or *aliasing*. We illustrate the idea by considering a simple sinusoidal signal:

$$y(t) = G\cos(\omega_0 t) \qquad (3.2.4)$$

The sample values of this signal are
$$y_k = y(k\Delta) = G\cos(\omega_0 k\Delta), \quad k = 0, 1, \ldots \qquad (3.2.5)$$
Now consider a related set of signals defined as follows:
$$w(t) = G\cos[(\omega_0 + m\omega_s)t] \qquad (3.2.6)$$
$$v(t) = G\cos[(-\omega_0 + m\omega_s)t] \qquad (3.2.7)$$
where
$$\omega_s = \frac{2\pi}{\Delta} \quad \text{and} \quad m \text{ is any integer}$$
The sampled values of these signals are
$$w_k = G\cos[\omega_0 k\Delta + mk2\pi]$$
$$= G\cos[\omega_0 k\Delta] \qquad (3.2.8)$$
and
$$v_k = G\cos[-\omega_0 k\Delta + mk2\pi]$$
$$= G\cos[\omega_0 k\Delta] \qquad (3.2.9)$$
Note that we have
$$w_k = v_k = y_k \qquad (3.2.10)$$

We thus see that we obtain the same sample values for all signals whose frequencies (in radians per second) differ by, or add to, an integer multiple of $\omega_s = 2\pi/\Delta$. This phenomenon is known as *frequency folding* or *aliasing*. This is illustrated in Figure 3.2.1 where a 10- and 90-Hz signal are shown with a sampling period of 10 ms. Note that the sample values are the same for the two signals.

Because of aliasing problems, the frequency range of signals has to be limited to one band of length ω_s to avoid confusion. Usually this band is taken as the lowest frequency band, and this gives an allowable frequency range of $[-\omega_s/2, \omega_s/2]$ rad s^{-1}, where $\omega_s = 2\pi f_s$ and $f_s = 1/\Delta$ is the sampling frequency in hertz.

For the reader who is puzzled by the concept of negative frequencies, it may be helpful to think of a wave in the ocean. Negative frequencies in this context then correspond to a wave traveling in the opposite direction. Of course, if we fix the

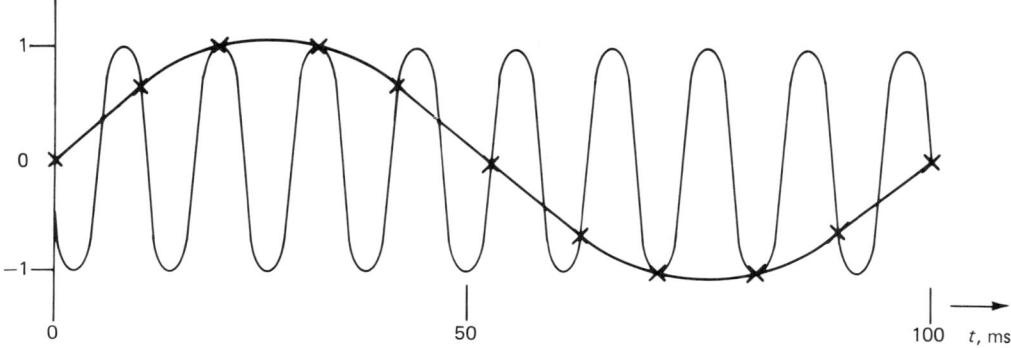

Figure 3.2.1 Example of aliasing.

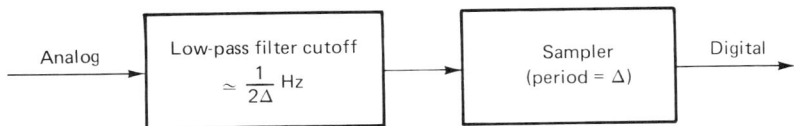

Figure 3.2.2 Practical A/D converter arrangement.

spatial point at which we observe the wave, then the time response will be identical for positive and negative frequencies (modulo a possible phase change).

Clearly, signals having frequency outside the range $[-\omega_s/2, \omega_s/2]$ are simply going to be misinterpreted as low-frequency signals. Thus, to avoid this misinterpretation, A/D converters are invariably preceded by low-pass filters (called antialiasing filters) that effectively eliminate all signals outside the range $[-\omega_s/2, \omega_s/2]$. These filters are especially important if high-frequency noise is present on the signal to be sampled. If this noise is not eliminated prior to sampling, then the components in the range $[\omega_s/2 + m\omega_s, \omega_s/2 + (m+1)\omega_s]$ will be folded on top of the range $[0, \omega_s/2]$ for $m = 0$ to ∞. This is clearly highly undesirable. Thus practical A/D converters are implemented as in Figure 3.2.2.

3.3 SIGNAL RECONSTRUCTION

The reverse process of going from sample values back to an analog signal is called *signal reconstruction*. Essentially, we have to fill in or interpolate the values of the signal between the given samples. We will discuss several ways of doing this.

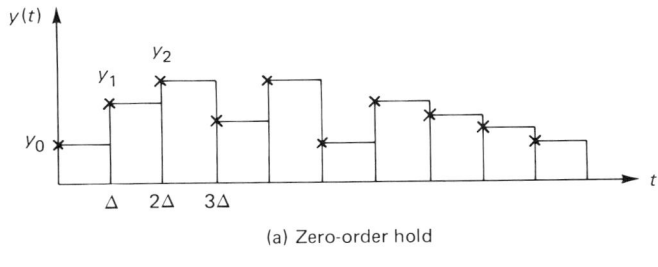

(a) Zero-order hold

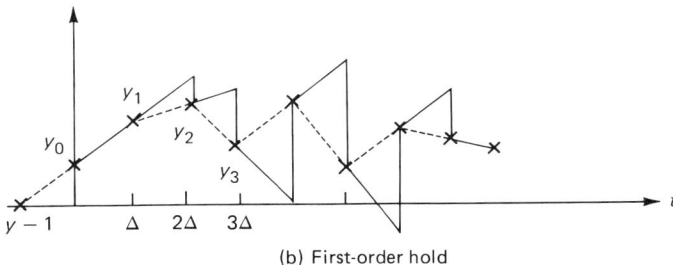

(b) First-order hold

Figure 3.3.1 Two types of polynomial interpolation.

3.3.1 Polynomial Interpolation

The basic idea of polynomial interpolation is to fit an nth-order polynomial through the last $n + 1$ samples and then to use the polynomial to interpolate the missing values. The process is called an nth-order hold. Figure 3.3.1 illustrates the idea for the case of zero-order and first-order holds.

3.3.2 Shannon Reconstruction Theorem

A natural question that arises in the context of polynomial reconstruction is what order polynomial is necessary to give perfect reconstruction. The answer to this question is straightforward in cases where the signal is described by a polynomial in time. However, we are often interested in a description of signals based on their constituent frequency content.

We have seen in Section 3.2 that any signal whose frequency content lies in the range $(-\omega_s/2, \omega_s/2)$ does not experience folding when sampled. A surprising fact is that for this class of signals it is actually possible to exactly reconstruct the continuous time signal from the samples. This is verified in the following result:

Theorem 3.3.1. Shannon's Reconstruction Theorem

Any signal $y(t)$ that consists of frequency components in the range $(-\omega_s/2, \omega_s/2)$ can be exactly reconstructed from the sample values $\{y(k\Delta)\}$, where Δ is the sampling period and ω_s is $2\pi/\Delta$. The appropriate reconstruction is

$$y(t) = \sum_{k=-\infty}^{\infty} y(k\Delta) \left[\frac{\sin[(\omega_s/2)(t - k\Delta)]}{(\omega_s/2)(t - k\Delta)} \right] \tag{3.3.1}$$

Proof: We will consider only a single sinusoidal component, since if the result holds in this case it is evident from the linearity of (3.3.1) that the result will also hold for a sum of such components. Furthermore, we will consider a complex exponential since, by taking real and imaginary parts, this covers both cosine and sine simultaneously. Thus consider

$$y(t) = e^{j\omega_0 t} \tag{3.3.2}$$

Also note that

$$\frac{\sin[(\omega_s/2)(t - k\Delta)]}{[(\omega_s/2)(t - k\Delta)]} = \frac{1}{\omega_s} \int_{-\omega_s/2}^{\omega_s/2} e^{j\omega(t - k\Delta)} \, d\omega \tag{3.3.3}$$

Hence, the right side of (3.3.1) becomes

$$\sum_{k=-\infty}^{\infty} y(k\Delta) \left[\frac{\sin[(\omega_s/2)(t - k\Delta)]}{(\omega_s/2)(t - k\Delta)} \right] = \frac{1}{\omega_s} \sum_{k=-\infty}^{\infty} e^{j\omega_0 k\Delta} \int_{-\omega_s/2}^{\omega_s/2} e^{j\omega(t - k\Delta)} \, d\omega$$

$$= \frac{1}{\omega_s} \int_{-\omega_s/2}^{\omega_s/2} e^{j\omega t} \sum_{k=-\infty}^{\infty} e^{jk\Delta(\omega_0 - \omega)} \, d\omega \tag{3.3.4}$$

Now, it is a standard result (see Problem 18) that the term $\sum_{k=-\infty}^{\infty} e^{jk\Delta(\omega_0 - \omega)}$ can be replaced by $(2\pi/\Delta)\sum_{n=-\infty}^{\infty} \delta(\omega_0 - \omega - n\omega_s)$, where δ is the Dirac delta function. Substituting this result into (3.3.4) gives

$$\sum_{k=-\infty}^{\infty} y(k\Delta) \left[\frac{\sin[(\omega_s/2)(t - k\Delta)]}{(\omega_s/2)(t - k\Delta)} \right]$$

$$= \frac{2\pi}{\Delta\omega_s} \int_{-\omega_s/2}^{\omega_s/2} e^{j\omega t} \sum_{k=-\infty}^{\infty} \delta(\omega_0 - \omega - n\omega_s) \, d\omega$$

$$= e^{j\omega_1 t} \tag{3.3.5}$$

where ω_1 is such that $\omega_1 \in (-\omega_s/2, \omega_s/2)$ and $\omega_0 - \omega_1 = n\omega_s$ for some integer n. Clearly, if $\omega_0 \in (-\omega_s/2, \omega_s/2)$, then $\omega_1 = \omega_0$ and the result follows. ▽▽▽

The preceding result shows that, in theory, it is possible to exactly reconstruct a continuous time signal from samples taken at interval Δ provided the frequency content of the signal is limited to the range $(-\omega_s/2, \omega_s/2)$, where $\omega_s = 2\pi/\Delta$. Actually, frequency components outside this range are simply folded back or aliased. This is evident from (3.3.5), which gives further insight into frequency folding and the information content of sampled data.

However, in practice, it is not sensible to use the preceding reconstruction technique. First, the formula (3.3.1) is complicated to evaluate and involves an infinite sum of both past and future samples. Second, the infinite sum is ill-conditioned since small numerical errors can be magnified into large errors in the reconstruction. Thus, we usually use some more simple approach, for example, those described in Section 3.3.1.

3.3.3 Digital to Analog Conversion

The term digital to analog converter (D/A converter) is used to describe practical circuits that implement signal reconstruction. Usually, only zero-order holds are used due to their simplicity. Appendix A gives some details of the hardware used to realize D/A conversions.

3.4 DISCRETE MODELS FOR CONTINUOUS PROCESS

In this section, we will show how models describing the sampled response of a system can be derived from the underlying continuous time model.

3.4.1 General Models

The basic idea used in relating the sampled output response $\{y(k\Delta), k = 0, 1, 2, \ldots\}$ to the sampled input $\{u(k\Delta), k = 0, 1, 2, \ldots\}$ is to solve the underlying continuous time equations over the sampling interval. For general nonlinear models, this cannot be achieved exactly. However, we can approximate the solution by various numerical techniques. For example, we can use Runge–Kutta procedures or predictor corrector formulas. The simplest such approximation is the Euler approximation, which approximates the solution to (2.2.1) as

$$x[(k+1)\Delta] = x[k\Delta] + \Delta f[x(k\Delta), u(k\Delta), d(k\Delta), k\Delta] \quad (3.4.1)$$

$$y(k\Delta) = g[x(k\Delta), u(k\Delta), d(k\Delta), k\Delta] \quad (3.4.2)$$

Similar results hold for other forms of numerical integration. All such formulas are approximations to the solution. However, in the special case of linear systems, we can obtain an exact description of the sampled response, as we now show.

3.4.2 Linear State Space Models

Consider a continuous time linear state space model of the form

$$\frac{d}{dt}x(t) = Ax(t) + Bu(t) \quad (3.4.3)$$

$$y(t) = Cx(t) \quad (3.4.4)$$

Let us assume that the system is in state $x(k\Delta)$ at the time $t = k\Delta$ and then compute the response for $t \geq k\Delta$. The exact solution of (3.4.3) is readily seen to be

$$x(t) = e^{A(t-k\Delta)}x(k\Delta) + \int_{k\Delta}^{t} e^{A(t-\tau)}Bu(\tau)\,d\tau \quad (3.4.5)$$

In equation (3.4.5), the notation e^{At} stands for the following matrix power series:

$$e^{At} \triangleq I + At + \frac{1}{2!}(At)^2 + \frac{1}{3!}(At)^3 + \cdots \quad (3.4.6)$$

This series is absolutely convergent for any finite At. Using (3.4.6), it is readily seen that

$$\frac{d}{dt}e^{At} = Ae^{At} = e^{At}A \quad (3.4.7)$$

Using (3.4.7), the validity of (3.4.5) is readily checked by substitution.

Now, say the input is generated by a zero-order hold; then

$$u(t) = u(k\Delta), \quad \text{for } k\Delta \leq t < (k+1)\Delta \quad (3.4.8)$$

Substituting (3.4.8) into (3.4.5) gives

$$x((k+1)\Delta) = e^{A\Delta}x(k\Delta) + \left[\int_{k\Delta}^{(k+1)\Delta} e^{A((k+1)\Delta-\tau)}B\,d\tau\right]u(k\Delta)$$

$$= e^{A\Delta}x(k\Delta) + \left[\int_{0}^{\Delta} e^{A(\Delta-\tau)}B\,d\tau\right]u(k\Delta) \quad (3.4.9)$$

Hence we immediately obtain the following model relating the sampled input to the sampled output:

$$x_{k+1} = Fx_k + Gu_k \quad (3.4.10)$$

$$y_k = Hx_k \quad (3.4.11)$$

where

$$x_k \triangleq x(k\Delta) \tag{3.4.12}$$
$$u_k \triangleq u(k\Delta) \tag{3.4.13}$$
$$y_k \triangleq y(k\Delta) \tag{3.4.14}$$
$$F \triangleq e^{A\Delta} \tag{3.4.15}$$
$$G = \int_0^\Delta e^{A(\Delta-\tau)} B\, d\tau \tag{3.4.16}$$
$$= A^{-1}(e^{A\Delta} - I)B, \quad \text{when } A \text{ is nonsingular}$$
$$H = C \tag{3.4.17}$$

In this derivation, note that, we have ignored the effect of the antialiasing filter in the output. This is justified if fast sampling is employed, since the filter bandwidth is then much wider than the system dynamics of interest. Alternatively, it is possible to include the dynamics of the filter in the derivation of the discrete model by including the filter in the continuous time state space model.

Example 3.4.1

As a simple example of the development of a discrete time model as in (3.4.10) and (3.4.11), consider the system shown in Figure 3.4.1. Choosing the state variable as the voltage on the capacitor, we obtain the following continuous state space model:

$$\frac{d}{dt}x(t) = -\frac{1}{RC}x(t) + \frac{1}{RC}u(t) \tag{3.4.18}$$

With sampling period Δ, the corresponding discrete model is readily seen to be as in (3.4.10) and (3.4.11), where F, G, and H are the following scalar quantities:

$$F = e^{-\Delta/RC} \tag{3.4.19}$$
$$G = 1 - e^{-\Delta/RC} \tag{3.4.20}$$
$$H = 1 \tag{3.4.21}$$

▽▽▽

The method just described can also be readily applied to deterministic disturbances. In fact, as we have seen in Section 2.5, these are usually modeled by linear continuous time models of the form (3.4.1) save that $u(t) \equiv 0$. We illustrate discrete time deterministic disturbances by the three examples discussed in Section 2.5.1.

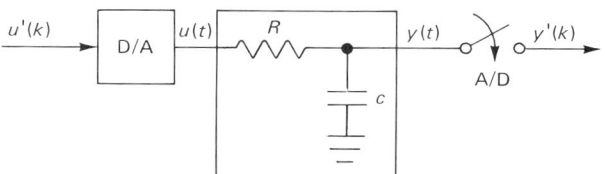

Figure 3.4.1 Simple discrete time system.

(a) DC Offset (d = constant)

The corresponding continuous time model is

$$\frac{d}{dt}x = 0, \quad x(0) = d \tag{3.4.22}$$

$$y = x \tag{3.4.23}$$

Solving (3.4.22) and (3.4.23) gives the following discrete time model:

$$x_{k+1} = x_k \tag{3.4.24}$$

$$y_k = x_k \tag{3.4.25}$$

(b) Drift at Constant Rate

$$d = \alpha t + \beta$$

The continuous time model is

$$\frac{d}{dt}x_1(t) = 0, \quad x_1(0) = \alpha \tag{3.4.26}$$

$$\frac{d}{dt}x_2(t) = x_1(t), \quad x_2(0) = \beta \tag{3.4.27}$$

$$y(t) = x_2(t) \tag{3.4.28}$$

Solving these equations gives

$$x_{1,k+1} = x_{1,k} \tag{3.4.29}$$

$$x_{2,k+1} = x_{2,k} + \Delta x_{1,k} \tag{3.4.30}$$

$$y_k = x_{2,k} \tag{3.4.31}$$

(c) Sinusoidal Disturbance

$$d = G\cos(\omega_0 t + \phi)$$

The continuous time model is

$$\frac{d}{dt}x_1 = x_2 \tag{3.4.32}$$

$$\frac{d}{dt}x_2 = -\omega_0^2 x_1 \tag{3.4.33}$$

$$y(t) = x_1(t) \tag{3.4.34}$$

Solving these equations gives

$$\begin{bmatrix} x_{1,k+1} \\ x_{2,k+1} \end{bmatrix} = \begin{bmatrix} \cos\omega_0\Delta & \frac{1}{\omega_0}\sin\omega_0\Delta \\ -\omega_0\sin\omega_0\Delta & \cos\omega_0\Delta \end{bmatrix} \begin{bmatrix} x_{1,k} \\ x_{2,k} \end{bmatrix} \tag{3.4.35}$$

$$y_k = x_{1,k} \tag{3.4.36}$$

*3.4.3 Evaluation of the Matrix Exponential

We have seen that, in the case of linear continuous time systems, the conversion to discrete time form depends on the evaluation of the matrix exponential defined in Equation (3.4.6). However, this formula is an infinite power series and thus some form of numerical or analytical procedure is necessary to evaluate matrix exponentials in practice. We outline next a selection of possible methods for doing this and discuss some of the related issues. In practice, a computer software package is used to evaluate the exponential of a matrix. However, it is important to be aware of the method(s) used within the package and its possible weaknesses.

Method 1. Truncated Power Series

An obvious method is to simply truncate the power series to a finite number of terms; that is,

$$e^{A\Delta} \simeq \sum_{k=0}^{N} \frac{(A\Delta)^k}{k!} \triangleq E_1(A\Delta) \tag{3.4.37}$$

This method works reasonably well provided $A\Delta$ is small compared to 1. The size of $A\Delta$ is usually measured by some *norm*. There are various ways of defining the norm of a matrix, but perhaps the simplest is the Frobenius norm, $\|\cdot\|_F$, defined as follows:

$$\|X\|_F \triangleq \left(\sum_{i,j} x_{ij}^2\right)^{1/2} \tag{3.4.38}$$

It can be readily shown (see Problem 19) that, provided $\|A\Delta\|_F < N + 1$, then the following conservative bound applies:

$$\|e^{A\Delta} - E_1(A\Delta)\|_F \leq \frac{\|(A\Delta)^{N+1}\|_F}{(N+1)!} \left\{ \frac{1}{1 - [\|A\Delta\|_F/(N+1)]} \right\} \tag{3.4.39}$$

This gives an upper bound on the error if the finite power series given in (3.4.37) is evaluated exactly. Note, however, that it is numerically difficult to evaluate this finite power series (3.4.37) when $\|A\Delta\|_F$ is large compared to 1. Thus, this method is generally useful only when $\|A\Delta\|_F$ is not large in comparison with 1.

Method 2. Eigenvalue Method

In this method, we first do an eigenvalue decomposition of the matrix A. We first consider the case of distinct eigenvalues. Then A can be written as

$$A = T\Lambda T^{-1} \tag{3.4.40}$$

where $\Lambda = \text{diag}(\lambda_1, \ldots, \lambda_n)$ and T is a matrix of eigenvectors. Then, using the property given in Problem 7(d), we show that

$$e^{A\Delta} = T(e^{\Lambda\Delta})T^{-1} = E_2(A\Delta) \tag{3.4.41}$$

It is easy to show that

$$e^{\Lambda\Delta} = \text{diag}(e^{\lambda_1\Delta},\ldots,e^{\lambda_n\Delta}) \qquad (3.4.42)$$

This method works well provided we have available a good eigenvalue decomposition algorithm and provided no two eigenvalues are close compared to the size of the maximum eigenvalue.

In the case of repeated (or almost coincident) eigenvalues, a Jordan canonical form can be used. In this case, (3.4.40) is replaced by

$$A = TJT^{-1} \qquad (3.4.43)$$

where T is a matrix of generalized eigenvectors and J has the structure

$$J = \begin{bmatrix} J_1 & 0 & \cdots & 0 \\ 0 & \ddots & & \vdots \\ \vdots & & \ddots & 0 \\ 0 & \cdots & 0 & J_m \end{bmatrix} \qquad (3.4.44)$$

where J_i is the Jordan block corresponding to the ith eigenvalues and has the structure

$$J_i = \begin{bmatrix} \lambda_i & 1 & 0 & \cdots & 0 \\ & \ddots & \ddots & & \vdots \\ 0 & & \ddots & \ddots & 0 \\ \vdots & & & \ddots & 1 \\ 0 & \cdots & & 0 & \lambda_i \end{bmatrix} \qquad (3.4.45)$$

Then (3.4.41) is replaced by

$$e^{A\Delta} = T(e^{J\Delta})T^{-1} \qquad (3.4.46)$$

where

$$e^{J\Delta} = \begin{bmatrix} e^{J_1\Delta} & 0 & \cdots & 0 \\ 0 & \ddots & & \vdots \\ \vdots & & \ddots & 0 \\ 0 & \cdots & 0 & e^{J_m\Delta} \end{bmatrix} \qquad (3.4.47)$$

and

$$e^{J_i\Delta} = e^{\lambda_i\Delta} \begin{bmatrix} 1 & \Delta & \frac{\Delta^2}{2!} & & \\ 0 & 1 & \ddots & \ddots & \\ \vdots & & \ddots & \ddots & \frac{\Delta^2}{2!} \\ & & & \ddots & \Delta \\ 0 & \cdots & & 0 & 1 \end{bmatrix} \qquad (3.4.48)$$

Method 3. Numerical Integration

It can be easily shown that the matrix exponential is the unique solution $\Psi(t)$ to the matrix differential equation:

$$\frac{d}{dt}\Psi(t) = A\Psi(t), \qquad \Psi(0) = I \tag{3.4.49}$$

Various types of numerical integration may then be used to solve (3.4.49), for example, multistep and Runge–Kutta. If, for example, we use a multistep Euler method, then the solution is equivalent to

$$e^{A\Delta} \simeq \left(I + A\frac{\Delta}{n}\right)^n = E_3(A\Delta) \tag{3.4.50}$$

Similarly, fourth-order Runge–Kutta with fixed step size is equivalent to

$$e^{A\Delta} \simeq \left(I + \left(\frac{A\Delta}{n}\right) + \frac{1}{2!}\left(\frac{A\Delta}{n}\right)^2 + \frac{1}{3!}\left(\frac{A\Delta}{n}\right)^3 + \frac{1}{4!}\left(\frac{A\Delta}{n}\right)^4\right)^n$$

Method 4. Cayley–Hamilton Approach

The Cayley–Hamilton theorem states that any square matrix satisfies its own characteristic equation. That is, there exists $\alpha_0 \ldots \alpha_{n-1}$ such that

$$A^n + \alpha_{n-1}A^{n-1} + \cdots + \alpha_0 I = 0 \tag{3.4.51}$$

Thus, we can show that, for any $m \geq n$, A^m can be written as a linear combination of $(I, A, \ldots, A^{n-1})$. It then follows that there exist scalar functions $\varphi_0(t), \ldots, \varphi_{n-1}(t)$ such that

$$e^{At} = \varphi_0(t)I + \varphi_1(t)A + \cdots + \varphi_{n-1}(t)A^{n-1} \tag{3.4.52}$$

Hence, the problem of finding e^{At} reduces to the problem of finding $\varphi_0(t) \ldots \varphi_{n-1}(t)$. Since $(d/dt)e^{At} = Ae^{At}$, then

$$\dot{\varphi}_0(t)I + \dot{\varphi}_1(t)A + \cdots + \dot{\varphi}_{n-1}(t)A^{n-1}$$
$$= \varphi_0(t)A + \varphi_1(t)A^2 + \cdots + \varphi_{n-1}(t)A^n \tag{3.4.53}$$

where $\dot{\varphi}_i(t) \triangleq (d/dt)\varphi_i(t)$. Then, using (3.4.51) and matching the coefficients of A^i, one possible solution is for $\varphi_i(t)$ to satisfy

$$\frac{d}{dt}\begin{bmatrix} \varphi_0(t) \\ \vdots \\ \vdots \\ \varphi_{n-1}(t) \end{bmatrix} = \begin{bmatrix} 0 & \cdots & \cdots & 0 & -\alpha_0 \\ 1 & \ddots & & & \vdots \\ 0 & \ddots & \ddots & & \vdots \\ \vdots & \ddots & \ddots & \ddots & \vdots \\ 0 & \cdots & 0 & 1 & -\alpha_{n-1} \end{bmatrix}\begin{bmatrix} \varphi_0(t) \\ \vdots \\ \vdots \\ \varphi_{n-1}(t) \end{bmatrix} \tag{3.4.54}$$

Since $e^{A0} = I$, we choose $\varphi_0(0) = 1$ and $\varphi_i(0) = 0$ for $i \neq 0$. We have now reduced the problem to that of solving a linear vector differential equation. This equation may be solved using numerical integration. Alternatively, due to the special structure of (3.4.54), various types of closed-form solutions may be found. For example, in the case where the eigenvalues of A, λ_i, are distinct, we can show that

$$\varphi_m(t) = \sum_{k=0}^{n-1} \frac{e^{\lambda_k t}\left(\lambda_k^{n-m-1} + \alpha_{n-1}\lambda_k^{n-m-2} + \cdots + \alpha_{m+1}\right)}{\prod_{l \neq k}(\lambda_k - \lambda_l)} \quad (3.4.55)$$

The preceding methods are representative of a large class of methods for evaluating the matrix exponential. For simple problems, virtually any method will work adequately. For problems where $\|A\Delta\|_F$ is large and/or $A\Delta$ has close eigenvalues, considerable caution is required. The references given at the end of the chapter contain useful discussion of the potential pitfalls. One idea that is sometimes helpful is to note that

$$e^{A\Delta} = \left(e^{A\Delta/n}\right)^n \quad (3.4.56)$$

Thus, we often select n as a power of 2, such that $(1/n)\|A\Delta\|_F$ is small.

The following simple example illustrates some of the features discussed.

Example 3.4.2

Consider the case where we wish to find e^A, where

$$A = \begin{bmatrix} 4 & 25.5 \\ -1 & -6.1 \end{bmatrix} \quad (3.4.57)$$

In this case, we can show that the correct value (to four decimal places) of the matrix exponential is

$$e^A = \begin{bmatrix} 2.1183 & 8.9271 \\ -0.3501 & -1.4175 \end{bmatrix} \quad (3.4.58)$$

We now consider how the four methods outlined previously may be used to calculate the matrix exponential.

Method 1

If we use $N = 10$ in (3.4.37) for this case, we obtain (using double-precision arithmetic) an error between e^A and $E_1(A)$ of approximately 1×10^{-5}. Note, however, that with $N = 10$ and using 12-bit floating-point arithmetic (that is, roughly three decimal places) we obtain

$$E_1(A) \approx \begin{bmatrix} 2.115 & 8.914 \\ -0.350 & -1.416 \end{bmatrix} \quad (3.4.59)$$

Thus we note that the truncated power series method gives an answer accurate to about 9 bits when calculated with 12-bit arithmetic, for this example.

Method 2

By analyzing the A matrix, we can show that the eigenvalues and eigenvectors of A are $\lambda_1 = -1$, $v_1 = [5.1, -1]^T$; $\lambda_2 = -1.1$, $v_2 = [5, -1]^T$. Thus

$$A = T\Lambda T^{-1}$$
$$= \begin{bmatrix} 5.1 & 5 \\ -1 & -1 \end{bmatrix} \begin{bmatrix} -1 & 0 \\ 0 & -1.1 \end{bmatrix} \begin{bmatrix} 10 & 50 \\ -10 & -51 \end{bmatrix} \qquad (3.4.60)$$

and

$$e^{At} = T \begin{bmatrix} e^{-t} & 0 \\ 0 & e^{-1.1t} \end{bmatrix} T^{-1} \qquad (3.4.61)$$

Substituting $t = 1$ in (3.4.61) reproduces the result given in (3.4.58). Note, however, that this eigenvalue–eigenvector analysis is numerically ill conditioned, due to the near singularity of T. For example, if we have a numerical error of 10^{-3} in one element of T, that is, we replace T by T', where

$$T' = \begin{bmatrix} 5.1 & 5 \\ -1.001 & -1 \end{bmatrix} \qquad (3.4.62)$$

then our approximation for e^A is

$$T' \begin{bmatrix} e^{-1} & 0 \\ 0 & e^{-1.1} \end{bmatrix} (T')^{-1} = \begin{bmatrix} 2.2123 & 9.3970 \\ -0.3689 & -1.5115 \end{bmatrix} \qquad (3.4.63)$$

This is accurate only to about 3 bits (that is, 10%).

Method 3

If, for example, we use a fourth-order Runge–Kutta method with eight steps, this is equivalent to finding

$$E_3(A) = \left[I + \frac{A}{8} + \frac{1}{2!}\left(\frac{A}{8}\right)^2 + \frac{1}{3!}\left(\frac{A}{8}\right)^3 + \frac{1}{4!}\left(\frac{A}{8}\right)^4 \right]^8$$
$$= \begin{bmatrix} 2.1183 & 8.9270 \\ -0.3501 & -1.4175 \end{bmatrix} \qquad (3.4.64)$$

which is accurate to about 1.0×10^{-4}. Note, however, that if 12-bit arithmetic is used in the computation, the result becomes

$$E_3(A) \approx \begin{bmatrix} 2.107 & 8.875 \\ -0.348 & -1.408 \end{bmatrix} \qquad (3.4.65)$$

which is accurate to only 5.4×10^{-2} (or about 7 bits). It should be noted, however, that the number of computations required in this method is substantially less than for any of the other methods considered here.

Method 4

In this case, we can show that the characteristic polynomial of A is

$$P_A(\lambda) = \lambda^2 + 2.1\lambda + 1.1$$
$$= (\lambda + 1.1)(\lambda + 1) \tag{3.4.66}$$

Then, from (3.4.55),

$$\varphi_0(t) = \frac{e^{-t}(-1 + 2.1)}{-1 - (-1.1)} + \frac{e^{-1.1t}(-1.1 + 2.1)}{-1.1 - (-1)}$$
$$= 11e^{-t} - 10e^{-1.1t} \tag{3.4.67}$$

and

$$\varphi_1(t) = \frac{e^{-t}}{-1 - (-1.1)} + \frac{e^{-1.2t}}{-1.1 - (-1)}$$
$$= 10(e^{-t} - e^{-1.1t}) \tag{3.4.68}$$

Thus

$$e^A = \varphi_0(1)I + \varphi_1(1)A$$
$$= 0.7180I + 0.3501A \tag{3.4.69}$$

Note that the above calculations would be numerically ill-conditioned due to the close eigenvalues. ▽▽▽

In the preceding discussion we have concentrated on the evaluation of $e^{A\Delta}$, whereas the complete solution of (3.4.9) also depends on evaluation of integrals of the type (3.4.16). A simple way of doing this is to note that, for A nonsingular,

$$\Gamma(\Delta) = \int_0^\Delta e^{At} dt = A^{-1}(e^{A\Delta} - I) \tag{3.4.70}$$

Alternatively, $\Gamma(\Delta)$ can be directly evaluated using extensions of methods 1 through 4. These extensions are based on the observation that $\Gamma(\Delta)$ satisfies

$$\Gamma(0) = 0 \tag{3.4.71}$$

and

$$\frac{d}{dt}\Gamma(t) = e^{At} = A\Gamma + I \tag{3.4.72}$$

(see Problem 20.)

3.5 DISCRETE TIME OPERATOR MODELS

When working with discrete time models, it is often convenient to use operator notation rather than to show explicitly the time dependencies as in (3.4.1). We will describe several possible operators next.

3.5.1 The Shift Operator

One way of describing discrete models is to use the forward shift operator q, defined by

$$qx_k \triangleq x_{k+1} \tag{3.5.1}$$

Using this operator, the linear discrete state space model (3.4.10) and (3.4.11) can be written compactly as

$$qx_k = Fx_k + Gu_k \tag{3.5.2}$$
$$y_k = Hx_k \tag{3.5.3}$$

▽▽▽

A simple example illustrating the use of the forward shift operator, q, is given next.

Example 3.5.1. Sinusoidal Disturbance

From the first line of (3.4.35), we have

$$(q - \cos \omega_0 \Delta) x_{1,k} = \left(\frac{1}{\omega_0} \sin \omega_0 \Delta\right) x_{2,k} \tag{3.5.4}$$

while the second line gives

$$(q - \cos \omega_0 \Delta) x_{2,k} + (\omega_0 \sin \omega_0 \Delta) x_{1,k} = 0 \tag{3.5.5}$$

Operating on (3.5.4) by $(q - \cos \omega_0 \Delta)$ gives

$$(q - \cos \omega_0 \Delta)^2 x_{1,k} - \left(\frac{1}{\omega_0} \sin \omega_0 \Delta\right)(q - \cos \omega_0 \Delta) x_{2,k} = 0 \tag{3.5.6}$$

Substituting (3.5.5) into (3.5.6) gives

$$(q - \cos \omega_0 \Delta)^2 x_{1,k} + (\sin \omega_0 \Delta)^2 x_{1,k} = 0 \tag{3.5.7}$$

Simplifying and using (3.4.36) gives

$$\left(q^2 - (2 \cos \omega_0 \Delta) q + 1\right) y_k = 0 \tag{3.5.8}$$

The reader should check (Problem 10) using elementary trigonometrical relationships that (3.5.8) is indeed valid when

$$y_k = G \sin(\omega_0 \Delta k + \phi) \tag{3.5.9}$$

▽▽▽

Equation (3.5.8) is a special case of a general operator model for purely deterministic sequences of the form

$$D(q) y = 0 \tag{3.5.10}$$

where, in this case, $D(q)$ has the special form

$$D(q) = q^2 - (2 \cos \omega_0 \Delta) q + 1 \tag{3.5.11}$$

3.5.2 The Delta Operator

The shift operator q, as defined in Section 3.5.1, is widely used to describe discrete time systems. However, a disadvantage of the operator q is that it is not at all like the continuous time operator d/dt. Heuristically, we might suspect that a better correspondence is obtained between continuous and discrete time if the shift operator is replaced by a difference operator that is more like a derivative. We will see later that this is indeed the case. With this in mind, we define the delta operator, as the following forward difference:

$$\delta \triangleq \frac{q-1}{\Delta} \tag{3.5.12}$$

The equivalent form of equation (3.5.1) then is

$$\delta x_k = \frac{x_{k+1} - x_k}{\Delta} = \frac{x(k\Delta + \Delta) - x(k\Delta)}{\Delta} \tag{3.5.13}$$

Note that the relationship between δ and q is a simple linear function, and thus δ offers the same flexibility in the modeling of discrete time systems as does q. The choice of which operator to use is therefore a function of the particular application. Generally, q leads to simpler expressions and emphasizes the sequential nature of sampled signals. On the other hand, since δ is a difference, it leads to models that are more like models in d/dt. Thus continuous time insights can sometimes be used in discrete design. Also, we will see presently that the numerical properties of δ models are generally superior to those of shift operator models.

We illustrate the use of the delta operator and its relationship to continuous time in the following example.

Example 3.5.2

Substituting (3.5.12) into (3.5.11) gives the following delta model for a sinusoidal disturbance:

$$L(\delta) d_k = 0 \tag{3.5.14}$$

where

$$L(\delta) = \left. \frac{D(q)}{\Delta^2} \right|_{q = \Delta\delta + 1}$$

$$= \frac{(\Delta\delta + 1)^2 - (2\cos\omega_0\Delta)(\Delta\delta + 1) + 1}{\Delta^2}$$

$$= \delta^2 + \frac{2(1 - \cos\omega_0\Delta)}{\Delta} \delta + \frac{2 - 2\cos\omega_0\Delta}{\Delta^2}$$

Note that for $\omega_0\Delta$ small (that is, fast sampling) $L(\delta)$ is approximately $\delta^2 + \omega_0^2$, which should be compared with the corresponding continuous time model $(d^2/dt^2) + \omega_0^2$.

The following example further illustrates the link between continuous and discrete systems achieved by use of the delta operator in the latter case. It also exemplifies the numerical advantages of the delta form over the shift form.

Example 3.5.3

Consider the design of a fourth-order Butterworth low-pass filter with a cutoff frequency of 1 rad s^{-1}. (Those readers who are unfamiliar with Butterworth filters can consult any book on signal processing, including those given in the references to this chapter.) This filter can be described by the following input–output model:

$$(\rho^4 + 2.6131\rho^3 + 3.4142\rho^2 + 2.6131\rho + 1)y = u \quad (3.5.15)$$

where $\rho \triangleq (d/dt)$.

If we convert this to state space form and then discretize assuming a zero-order hold input and a sampling period of 0.2 s, we obtain the following discrete input–output model in shift form:

$$(q^4 - 3.47882q^3 + 4.56798q^2 - 2.6809q + 0.59296)y_k$$
$$= 10^{-3}(0.06q^3 + 0.5936q^2 + 0.5347q + 0.0438)u_k \quad (3.5.16)$$

This model can be implemented in a standard state space form as

$$qx_k = \begin{bmatrix} 3.4788 & -4.5680 & 2.6809 & -0.5930 \\ 1 & 0 & 0 & 0 \\ 0 & 1 & 0 & 0 \\ 0 & 0 & 1 & 0 \end{bmatrix} x_k + \begin{bmatrix} 1 \\ 0 \\ 0 \\ 0 \end{bmatrix} u_k \quad (3.5.17)$$

$$y_k = 10^{-3}[0.06 \quad 0.5936 \quad 0.5347 \quad 0.0438]x_k \quad (3.5.18)$$

Note that the coefficients of this model bear no resemblance to those of the continuous time model (3.5.15).

The equivalent δ form of (3.5.16) is

$$(\delta^4 + 2.6059\delta^3 + 3.2883\delta^2 + 2.3269\delta + 0.77)y_k$$
$$= (0.0003\delta^3 + 0.0193\delta^2 + 0.2377\delta + 0.77)u_k \quad (3.5.19)$$

This may be implemented in a standard δ operator form as

$$q\bar{x}_k = \bar{x}_k + \Delta \left(\begin{bmatrix} -2.6059 & -3.2883 & -2.3269 & -0.77 \\ 1 & 0 & 0 & 0 \\ 0 & 1 & 0 & 0 \\ 0 & 0 & 1 & 0 \end{bmatrix} \bar{x}_k + \begin{bmatrix} 1 \\ 0 \\ 0 \\ 0 \end{bmatrix} u_k \right) \quad (3.5.20)$$

$$y_k = [0.0003 \quad 0.0193 \quad 0.2377 \quad 0.77]\bar{x}_k \quad (3.5.21)$$

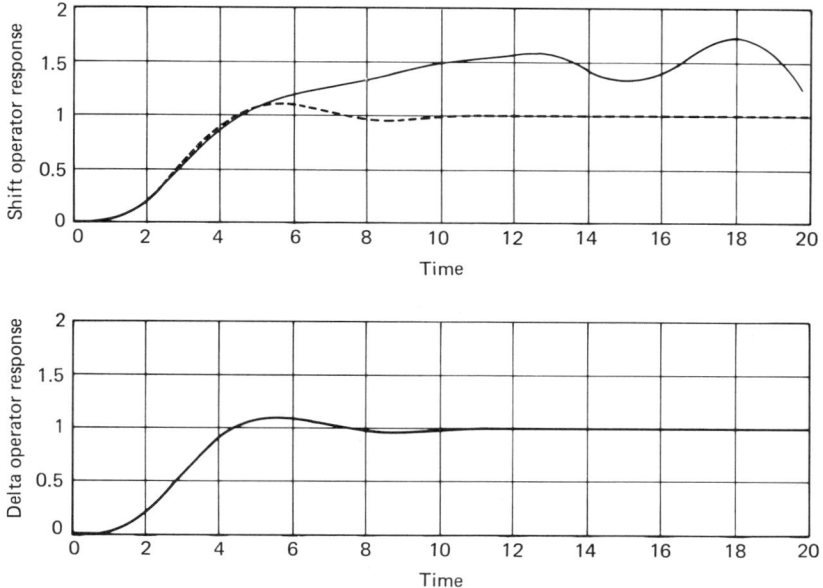

Figure 3.5.1 Twelve-bit floating-point step response for Example 3.5.3.

Note that the coefficients in the δ model, as in (3.5.19) or (3.5.20), (3.5.21), show a close resemblance to the coefficients in the continuous model (3.5.15).

To compare the numerical properties of the two digital filter implementations given in (3.5.17), (3.5.18), and (3.5.20), (3.5.21), these were simulated on a computer using floating-point arithmetic with a 12-bit mantissa. The results for a unit step with zero initial conditions are shown in Figure 3.5.1, where they are compared with the exact solution using infinite precision. From this figure it can be seen that the delta form has vastly superior numerical properties relative to the shift form. (Actualy, one of our postgraduate students spent a week trying to work out why his digital low-pass filter was not giving the expected result. He suspected a programming error, but in fact the problem was precisely the numerical difficulty alluded to here and was immediately resolved when he converted to delta form.) ▽▽▽

In the preceding example, it was pointed out that the shift form of the discrete time model bore little or no resemblance to the continuous time model. On the other hand, the coefficients in the delta form were very close to the corresponding coefficients in the continuous model. This observation is generally true and indicates that the delta form may be more insightful than the alternative shift form. For example, it is relatively easy to check if the delta model is approximately the correct answer for a given discretization of a continuous system, but this is extremely difficult, if not impossible, for shift models. This idea is illustrated in the following example.

Sec. 3.5   Discrete Time Operator Models

Example 3.5.4

Consider the following three continuous time systems expressed in terms of $D = d/dt$:

(C1) $(D^2 + D + 1)y = u$

(C2) $(D^2 + 10D - 1)y = u$

(C3) $(D^2 - D + 10)y = u$

Each of these systems is discretized with sampling interval $\Delta = 0.02$ s. The resulting discrete time models (not necessarily in the same order) are:

SHIFT FORM

(S1) $(q^2 - 1.98q + 0.98)y = 10^{-4}(1.99q + 1.97)u$

(S2) $(q^2 - 2.02q + 1.02)y = 10^{-4}(2.01q + 2.03)u$

(S3) $(q^2 - 1.82q + 0.82)y = 10^{-4}(1.87q + 1.75)u$

DELTA FORM

(D1) $(\delta^2 - 0.808\delta + 10.1)y = (0.0101\delta + 1.01)u$

(D2) $(\delta^2 + 9.05\delta - 0.906)y = (0.0094\delta + 0.906)u$

(D3) $(\delta^2 + 1.01\delta + 0.99)y = (0.01\delta + 0.99)u$

We wonder how many readers were able to correctly guess that (S1), (S2), and (S3) corresponded to (C1), (C3), (C2), respectively. On the other hand, it is very easy to see that (D1), (D2), and (D3) correspond to (C3), (C2), (C1), respectively. Thus the advantages of the delta form are self-evident.

3.5.3 Evaluation of Discrete Models in Delta Form

Clearly, one way of evaluating a discrete delta form model is to first find a shift form model and then to convert to the delta model by using the replacement.

$$q = 1 + \Delta\delta \qquad (3.5.22)$$

In state space form, this leads to the result

$$\delta x = F'x + G'u \qquad (3.5.23)$$
$$y = H'x \qquad (3.5.24)$$

where

$$F' = \frac{F - I}{\Delta}, \qquad G' = \frac{G}{\Delta}, \qquad H' = H \qquad (3.5.25)$$

and where F, G, and H are the matrices in the shift model as given in (3.5.2) and (3.5.3).

While the form given in (3.5.25) is technically correct, this is not the best way to evaluate the delta model, since the numerical problems associated with evaluation of the shift form are carried over to the delta form. A better method is to directly evaluate the delta form from the continuous time state space equations as follows:

$$F' = \Omega A \qquad (3.5.26)$$

and

$$G' = \Omega B \qquad (3.5.27)$$

where

$$\Omega = \frac{1}{\Delta} \int_0^\Delta e^{A\tau} d\tau$$

$$= I + \frac{A\Delta}{2!} + \frac{A^2\Delta^2}{3!} + \cdots \qquad (3.5.28)$$

This result follows directly by considering a power series expansion of $F' = (e^{A\Delta} - I)/\Delta$. The evaluation of Ω can now be performed directly using the principles outlined in Section 3.4.3. An interesting observation from (3.5.26) and (3.5.27) is that they reveal the close connection between delta domain models and the underlying continuous time models since $\Omega \to I$ as $\Delta \to 0$.

In the remainder of the book we will use shift- and delta-based operator models. However, we will tend to emphasize the delta form because of the advantages illustrated. In Chapter 14, we will give a more detailed analysis of the numerical properties brought out in Example 3.5.3.

3.6 INHERENTLY DISCRETE SYSTEMS

In the preceding sections, we have shown how discrete time models arise from the sampling of continuous time processes. However, there are situations in which the system under study is inherently discrete in nature. This discrete nature may arise because the output is available only in sampled form or because the input is constrained to be in sampled form (for example, piecewise constant or a sequence of pulses). We briefly discuss several examples of these classes of systems next.

Example 3.6.1

A common example of an inherently discrete electrical system is a system controlled by a switching regulator, such as that shown in Figure 3.6.1, which works by opening and closing the switch, SW, in a controlled manner. One popular scheme for controlling the switching is to use a pulse width modulation scheme. In this scheme, the switch is opened periodically and is closed for a controlled percentage of each period. The waveform $v(t)$ for a pulse width modulation scheme is illustrated in Figure 3.6.2. The input, in this case, is the fraction α_i of the period Δ for which the switch is closed. The output would be the current i in the resistor.

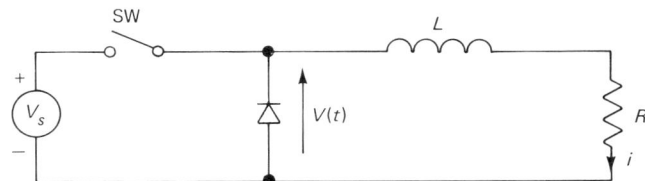

Figure 3.6.1 Example of switching regulator.

Clearly, the control variable α_i is a sequence, not a continuous variable, while the output current i is a continuous variable.

Using elementary circuit properties, we have

$$v(t) = L\frac{di}{dt} + Ri \qquad (3.6.1)$$

The solution to this differential equation may be shown to be

$$i(t) = i(t_0)e^{-(t-t_0)/\tau} + \frac{1}{L}\int_{t_0}^{t} e^{-(t-T)/\tau} v(T)\, dT \qquad (3.6.2)$$

where $\tau = L/R$ is the electrical time constant.

Thus, if we wish to relate sampled values of the current i, we note that

$$i(k\Delta + \Delta) = e^{-\Delta/\tau} i(k\Delta) + \frac{1}{L}\int_{k\Delta + \frac{\Delta}{2} - \frac{\alpha\Delta}{2}}^{k\Delta + \frac{\Delta}{2} + \frac{\alpha\Delta}{2}} e^{-((k+1)\Delta - T)/\tau} V_s\, dT$$

$$= e^{-\Delta/\tau} i(k\Delta) + \frac{V_s}{R} e^{-\Delta/2\tau}[e^{\alpha_k \Delta/2\tau} - e^{-\alpha_k \Delta/2\tau}] \qquad (3.6.3)$$

For Δ/τ small [which is usually the case so that the ripple between samples in $i(t)$ is small], we obtain the following approximate expression:

$$qi_k \approx e^{-\Delta/\tau} i_k + \left[\frac{\Delta V_s}{L} e^{-\Delta/2\tau}\right]\alpha_k \qquad (3.6.4)$$

or

$$\delta i_k \approx -\frac{1}{\tau}i_k + \frac{V_s}{L}\alpha_k \qquad (3.6.5)$$

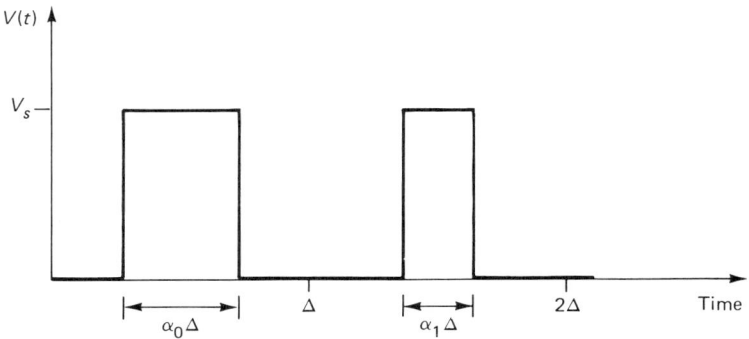

Figure 3.6.2 Switching waveforms for a pulse width modulation scheme.

The approximate model (3.6.5) or its continuous time counterpart

$$(1 + D\tau)i \approx \frac{V_s}{R}\alpha \tag{3.6.6}$$

is often used as the basis for controller designs in practical pulse width modulation circuits. ▽▽▽

Example 3.6.1 typifies a class of systems where the input is constrained to be sequential in nature. Another example of such a system is the control of fuel injection and/or ignition timing in an internal combustion engine.

Another class of inherently sampled systems is those in which samples only of the output are available. These types of systems are common in chemical processes where the cost of continuous measurement is impractical. Often, however, samples of a product may be taken and analyzed off-line to allow determination of the output. For this class of systems, it is important to remember that the underlying process is continuous in nature; however, practical constraints dictate a sampling strategy. In particular, the discussion in Section 3.2.3 regarding aliasing should be borne in mind.

3.7 SUMMARY

The main points covered in this chapter were:

- The idea that continuous time systems can be interconnected to a computer via D/A and A/D converters.
- The concept of frequency folding or aliasing in which frequency components outside the range $(-\omega_s/2, \omega_s/2)$ are folded back into this range when a signal is sampled.
- The use of low-pass filters prior to sampling to avoid frequency folding.
- The notion of polynomial interpolation and the special cases of zero-order and first-order holds.
- Shannon's reconstruction theorem, which states that any signal that consists of frequency components in the range $(-\omega_s/2, \omega_s/2)$ can be exactly reconstructed from the sampled values.
- A description of how approximate discrete time models for nonlinear continuous time processes can be obtained by numerical integration procedures.
- An in-depth treatment of discrete time models for continuous time linear systems.
- A discussion of methods for evaluating the matrix exponential that occurs in the discretization of linear continuous time state space models.
- The use of operators to describe discrete time models in a compact notation.
- The introduction of the shift operator q, defined by

$$qx_k \triangleq x_{k+1}$$

- The introduction of the delta operator δ, defined by

$$\delta x_k \triangleq \frac{x_{k+1} - x_k}{\Delta}$$

- A brief discussion of some of the advantages of the delta operator, including connections between continuous and discrete time and numerical issues.
- A discussion of systems that are inherently discrete time.

Thus, in general terms, Chapter 3 studied various aspects of sampling and the related topic of the modeling of discrete time systems. In Chapter 4, we introduce a powerful tool for studying dynamical systems in both continuous and discrete time.

3.8 REFERENCES

Sampling and aliasing are covered in:

OPPENHEIM, A. V., and R. W. SCHAFER (1975) *Digital Signal Processing.* Prentice-Hall, Englewood Cliffs, N.J.

OPPENHEIM, A. V., and A. S. WILLSKY (1983) *Signals and Systems.* Prentice-Hall, Englewood Cliffs, N.J.

More information on intersample and ripple problems is contained in:

DESOUZA, C. E. and G. C. GOODWIN (1984) "Intersample variances in discrete minimum variance control." *IEEE Trans. Auto. Control*, August, pp. 759–761.

LENNARTSON, B. (1986) "An investigation of the intersample variance for linear stochastic control," 25th CDC., Athens, pp. 1770–1775.

For elementary treatments of discrete time state space models and shift operators refer to:

ÅSTRÖM, K. J., and B. WITTENMARK (1984) *Computer Controlled Systems, Theory and Design.* Prentice-Hall, Englewood Cliffs, N.J.

FRANKLIN, G. F., and J. D. POWELL (1980) *Digital Control of Dynamic Systems.* Addison-Wesley, Reading, Mass.

KUO, B. C. (1980) *Digital Control Systems.* Holt, Rinehart and Winston, New York.

For further properties of matrix exponentials, see:

FORTMANN, T. E., and K. L. HITZ (1977) *An Introduction to Linear Control Systems.* Marcel Dekker, New York.

MOLER, C., and C. V. LOAN (1978) "Nineteen dubious ways to compute the exponential of a matrix," *SIAM Review*, vol. 20, no. 4, pp. 801–836.

Further background on deterministic and stochastic disturbances is contained in:

ÅSTRÖM, K. J. (1970) *Introduction to Stochastic Control Theory.* Academic Press, New York.

GOODWIN, G. C., and K. S. SIN (1984) *Adaptive Filtering, Prediction and Control.* Prentice-Hall, Englewood Cliffs, N.J.

For further reading on delta operators, see:

AGARWAL, R. C., and C. S. BURRUS (1975) "New recursive digital filter structures having very low sensitivity and roundoff noise." *IEEE Trans. Circuits Syst.*, vol. CAS-22, no. 12 (application of δ operators in digital filters).

HILDEBRAND, F. B. (1956) *Introduction to Numerical Analysis.* McGraw-Hill, New York (see sections on first divided difference operator).

HORI, N., P. N. NIKIFORUK, and K. KANAI (1988) "On a discrete-time system expressed in the Euler operator," *Proc. American Control Conf.*, Atlanta, GA, June, pp. 873–878 (discusses some of the advantages of the delta operator, here referred to as the Euler operator).

KARWOSKI, R. J. (1979) "Introduction to the Z transforms and its derivation." TRW psi Products, El Segundo, CA, tutorial paper (here the δ operator is used to motivate Z-transforms).

MIDDLETON, R. H., and G. C. GOODWIN (1986) "Improved finite word length characteristics in digital control using delta operators." *IEEE* AC-31, no. 11, pp. 1015–1021 (properties of δ operator in digital control).

MORI, T. and others (1987) "A class of discrete time models for a continuous time system." *Proc. American Control Conf.*, Minneapolis, MN, June, pp. 953–957 (linking continuous and discrete models).

ORLANDI, G., and G. MARTINELLI (1984) "Low sensitivity recursive digital filters obtained via the delay replacement." *IEEE Trans. Circuits Syst.*, vol. CAS-31, no. 7 (numerical aspects of δ operator in digital filters).

PETERKA, V. (1986) "Control of uncertain processes: applied theory and algorithms." *Kybernetika*, vol. 22, pp. 1–102 (this paper gives an extensive discussion of δ operators, including design software for control-related problems).

SALGADO, M. E., R. H. MIDDLETON, and G. C. GOODWIN (1987) "Connection between continuous and discrete Riccati equations with applications to Kalman filtering." *Proc. IEE*, Part D (interpretations of δ modeling in Riccati equations and optimal filtering).

TSCHAUNER, J. (1963) *Introduction à la théorie des systémes échantillon.* Dunod, Paris (note especially the link to continuous time systems).

3.9 PROBLEMS

1. Estimate the difference in maximum A/D conversion time between two converters both using D/A feedback, that is, staircase (linear search) and successive approximation (binary search) (see Appendix A).

2. Use superposition to explain the operation of the circuit in Figure A.1.1.

3. If the resistor values in a b-bit resistor ladder, as in Figure A.1.1, are independent and have an accuracy of $\pm x\%$, evaluate the maximum conversion error.

4. Repeat Problem 3 where the absolute accuracy of the resistors is again $\pm x\%$; however, the relative accuracy is $\pm y\%$. (Relative accuracy is a specification of the maximum error in the ratio of any two resistor values in the ladder.)

5. Modify the development in Section 3.2.3 when a phase shift is added to the signal in (3.2.5) to give

$$y(t) = G\cos(\omega_0 t + \phi)$$

6. If a system has multiple outputs that require A/D conversion, suggest a hardware arrangement that will eliminate timing skew in the sampling process.
7. Prove the following properties of the matrix exponential defined in (3.4.6)
   (a) $e^{A0} = I$
   (b) $e^{At}e^{As} = e^{A(t+s)}$
   (c) $(e^A)^{-1} = e^{-A}$
   (d) $e^{TAT^{-1}} = Te^A T^{-1}$
8. Verify that (3.4.5) is the solution to (3.4.3). (Note that linearity implies the existence and uniqueness of the solution.)
9. Show how G in (3.4.16) can be evaluated when A is singular.
10. Use trigonometrical relationships to directly verify that (3.5.8) is true for the signal given in (3.5.9).
11. Consider an arbitrary discrete time periodic disturbance of period p samples. Show that this can be modeled as in (3.5.10) and give an explicit expression for $D(q)$.
12. Consider the model (2.4.12) that describes a dc servo system. Convert the model to state space from where the states are θ and $\dot{\theta}$. Convert the model to discrete state form assuming a sampling period of Δ and a zero-order hold input. Express the discrete state space model in both shift and delta form.
13. Rederive the model of the form (3.4.9) assuming a first-order hold rather than zero-order hold on the input. *Hint:* As a first step, derive a model of the form

$$x'(k+1) = Fx'(k) + G_0 u(k) + G_1 u(k-1)$$

Introduce additional states to deal with the term $u(k-1)$.

14. What is the delta operator model for a constant drift rate disturbance? Compare this with the corresponding continuous time model.
15. Derive an expression for $x'(k)$ in terms of $x'(0)$ and $u(0), \ldots, u(k-1)$ for equation (3.4.10).
16. (a) Consider a continuous time system as in (3.4.3) and (3.4.4), with zero-order hold input and sampling period Δ. Show that the sampled inputs, outputs, and states obey the following delta operator model:

$$\delta x = \left(\frac{e^{A\Delta} - I}{\Delta}\right) x + \left[\frac{1}{\Delta} \int_0^\Delta e^{A\tau} d\tau\right] Bu$$

$$y = Cx$$

   (b) Discuss what happens to the model in part (a) as $\Delta \to 0$.
17. Consider a dc motor with fixed field current and input $u \equiv$ armature voltage. Let $y \equiv$ rotor angle, J be the total shaft inertia, L be the armature inductance, and k be the back emf constant (or torque constant) for the given field current.
   (a) Find a state space model for this system.
   (b) If $L/R \ll JR/4k^2$, show that the eigenvalues of the A matrix in the model are at (approximately) $0, -R/L, -k^2/JR$.
   (c) If $k = 25$ mV s/rad, $J = 40$ kg (cm)^2, $R = 1.3$ Ω, and $L = 10$ mH, find the diagonalized system (that is, find the eigenvalues and eigenvectors).
   (d) Thus find a discrete time state space model for the system if sampled with zero-order hold input and sampling frequency 100 Hz.

*18. Show that

$$\sum_{k=-\infty}^{\infty} e^{jk\Delta(\omega_0-\omega)} = \frac{2\pi}{\Delta} \sum_{n=-\infty}^{\infty} \delta(\omega_0 - \omega - n\omega_s)$$

where $\delta(\cdot)$ is the Dirac delta function. *Hint:* Consider

$$\lim_{N\to\infty} \sum_{k=-N}^{N} e^{jk\alpha} \quad (\alpha = \Delta(\omega_0 - \omega))$$

Show that

$$\sum_{k=-N}^{N} e^{jk\alpha} = \frac{\sin\left(\frac{2N+1}{2}\alpha\right)}{\sin(\alpha/2)} \triangleq h(N,\alpha)$$

Note that the height of $h(N,\alpha)$ for $\alpha = 2n\pi$ is $(2N+1)$. Also, show that

$$I_N \triangleq \int_{2n\pi - \frac{\pi}{N+1}}^{2n\pi + \frac{\pi}{N+1}} h(N,\alpha)\, d\alpha = 2\pi$$

19. Prove the error bound (3.4.39) for the truncated power series method for calculating the matrix exponential. *Hint:*

$$\|e^{A\Delta} - E_1(A\Delta)\|_F = \left\| \sum_{k=N+1}^{\infty} \frac{(A\Delta)^k}{k!} \right\|_F$$

$$\leq \sum_{k=N+1}^{\infty} \frac{\|(A\Delta)^{N+1}\|_F}{k!} \|A\Delta\|_F^{k-N-1}$$

20. With $\Gamma(\Delta) \triangleq \int_0^\Delta e^{At}\, dt$, show that:
    (a) $\Gamma(0) = 0$
    (b) $(d/dt)\Gamma(t) = e^{At}$
    (c) $e^{At} = A\Gamma(t) + I$
    (d) $\Gamma(t) = tI + \frac{t^2}{2!}A + \frac{t^3}{3!}A^2 + \cdots$
    (e) If $A = T^{-1}\Lambda T$, and $X(\Delta) \triangleq \int_0^\Delta e^{\Lambda t}\, dt$, then $\Gamma(\Delta) = T^{-1}X(\Delta)T$.

21. Find the equivalent to equations (3.4.1) and (3.4.2) if a second-order Runge–Kutta method is used for numerical integration (rather than the Euler approximation) for the continuous time nonlinear differential equation (2.2.1).

4

Transform Techniques

4.1 INTRODUCTION

In the previous two chapters we have shown that the dynamic equations governing many systems of interest can be expressed as differential or difference equations. In control theory, we are interested in the behavior of dynamic systems, and so the properties of the solution of differential or difference equations will emerge as an important topic.

In this chapter we will examine the use of transform techniques to study the behavior of linear systems. The underlying idea of this approach is to transform the time domain differential or difference equations into frequency domain linear algebraic equations. The solution of these algebraic equations may then be computed and the time domain solution found. This transformation into the frequency domain yields many important concepts, such as poles, zeros, and transfer functions, as we will see later.

A key point to be made in this chapter is that the one transform method, Laplace transforms, essentially applies to both the continuous and discrete time cases, provided the latter case is properly formulated using delta operators. To be precise, the discrete case requires corrections to the Laplace transform of the order of the sampling period. Full details of the necessary adjustments are given in the chapter. However, these corrections can be ignored in gaining an initial intuitive appreciation of the discrete case. Moreover, they become insignificant when fast sampling is employed.

4.2 LAPLACE TRANSFORMS

Consider a function, $f(t)$, that is zero for $t < 0$. Then with $f(t)$ we associate the function $F(s)$ [which we call the Laplace transform of $f(\cdot)$; that is, $F(s) = L\{f(t)\}$] defined by

$$F(s) \triangleq \int_{0^-}^{\infty} f(t) e^{-st} \, dt \qquad (4.2.1)$$

where 0^- denotes the limit of zero approached from the negative direction (similarly for 0^+).

It can be shown that, if there exist constants m and σ such that $|f(t)| \le m e^{\sigma t}$, then the integral in (4.2.1) converges absolutely for all complex numbers s that satisfy

$$\text{Re}\{s\} > \sigma \qquad (4.2.2)$$

The region in the complex plane described by (4.2.2) is termed the region of convergence for the Laplace transform. It turns out that the Laplace transform, if it exists, gives a unique representation of any continuous function. In other words, suppose $f_1(t)$ and $f_2(t)$ are continuous functions that have the same Laplace transform [that is, they satisfy (4.2.3)]

$$\int_{0^-}^{\infty} f_1(t) e^{-st} = F_1(s) = F_2(s) = \int_{0^-}^{\infty} f_2(t) e^{-st} \qquad (4.2.3)$$

for all s in some region of convergence, then

$$f_1(t) = f_2(t), \qquad \text{for all } t \ge 0 \qquad (4.2.4)$$

The following example shows how we can find the Laplace transform of simple functions.

Example 4.2.1

(a) The Laplace transform of $f(t) = t$ is

$$F(s) = \int_{0^-}^{\infty} t e^{-st} \, dt$$

$$= \left[-\frac{t e^{-st}}{s} - \frac{e^{-st}}{s^2} \right]_{0^-}^{\infty}$$

$$= \frac{1}{s^2}, \qquad \text{for Re}\{s\} > 0 \qquad (4.2.5)$$

(b) The Laplace transform of $g(t) = e^{at}$ is

$$G(s) = \int_{0^-}^{\infty} e^{at} e^{-st} \, dt$$

$$= \left[\frac{e^{(a-s)t}}{a-s} \right]_{0^-}^{\infty}$$

$$= \frac{1}{s-a}, \quad \text{for Re}\{s-a\} > 0 \qquad (4.2.6)$$

In this example, we have derived the Laplace transform of two simple functions. Further Laplace transforms are given in Table 4.2.1.

Some important properties of Laplace transforms are given in the following theorem.

Theorem 4.2.1. Laplace Transform Properties

(i) Linearity

For any scalar α_1, α_2,

$$\mathbf{L}\{\alpha_1 f(t) + \alpha_2 g(t)\} = \alpha_1 \mathbf{L}\{f(t)\} + \alpha_2 \mathbf{L}\{g(t)\}$$

(ii) Differentiation

$$\mathbf{L}\left\{\frac{d}{dt}[f(t)]\right\} = s\mathbf{L}\{f(t)\} - f(0^-)$$

TABLE 4.2.1 TRANSFORMS OF COMMON TIME DOMAIN FUNCTIONS

Time domain function	Description	Laplace transform
$i(t)$	Impulse	1
$1(t) = \begin{cases} 1, & t \geq 0 \\ 0, & t < 0 \end{cases}$	Unit step	$\frac{1}{s}$
t	Ramp	$\frac{1}{s^2}$
t^2	Parabola	$\frac{2}{s^3}$
e^{At}	Exponential	$(sI - A)^{-1}$
te^{At}		$(sI - A)^{-2}$
$t^2 e^{At}$		$2(sI - A)^{-3}$
$\sin \omega t$	Sine wave	$\frac{\omega}{s^2 + \omega^2}$
$\cos \omega t$	Cosine wave	$\frac{s}{s^2 + \omega^2}$

(iii) **Integration**

$$L\left\{\int_0^t f(\tau)\,d\tau\right\} = \frac{1}{s}L\{f(t)\}$$

(iv) **Frequency Differentiation**

$$L\{tf(t)\} = -\frac{d}{ds}[L\{f(t)\}]$$

(v) **Frequency Integration**

$$L\left\{\frac{f(t)}{t}\right\} = \int_s^\infty F(u)\,du$$

where

$$F(u) = [L\{f(t)\}]|_{s=u}$$

(vi) **Time Shifting**

$$L\{1(t-T)f(t-T)\} = e^{-sT}L\{f(t)\}$$

where $1(\cdot)$ denotes the unit step function,

$$1(t) = \begin{cases} 1, & t \geq 0 \\ 0, & t < 0 \end{cases}$$

and

$$T > 0 \in \mathbb{R}$$

(vii) **Frequency Shifting**

$$L\{e^{at}f(t)\} = F(s-a)$$

where

$$F(s) = L\{f(t)\}$$

(viii) **Convolution**

Let $H(s) = G(s)F(s)$, where $F(s) = L\{f(t)\}$ and $G(s) = L\{g(t)\}$; then

$$H(s) = L\left\{\int_0^t f(t-\tau)g(\tau)\,d\tau\right\}$$

assuming that $f(t) = g(t) = 0$ for $t < 0$.

(ix) **Inverse Transformation**

For $t > 0$, the inverse transform $f(t)$ of $F(s)$ can be found by

$$f(t) = \frac{1}{2\pi j}\int_{s=\sigma-j\infty}^{s=\sigma+j\infty} F(s)e^{st}\,ds$$

where σ is chosen sufficiently large, that is, such that the contour of integration is in the region of convergence of $F(s)$.

(x) **Initial Value Theorem**

$$\lim_{t \to 0^+} f(t) = \lim_{s \to \infty} \{sF(s)\}$$

provided the limit exists.

(xi) **Final Value**

If $\lim_{t \to \infty} f(t)$ exists, then

$$\lim_{t \to \infty} f(t) = \lim_{s \to 0} \{sF(s)\}$$

(xii) **Complex Convolution**

$$\mathcal{L}\{f_1(t)f_2(t)\} = \frac{1}{2\pi j} \int_{u=\sigma-j\infty}^{u=\sigma+j\infty} F_1(u) F_2(s-u)\, du$$

where σ is in the region of convergence of both $f_1(t)$ and $f_2(t)$.

Proof: We defer the proof of the theorem until later when we will present it as a special case of a more general result (Theorem 4.4.1). ▽▽▽

Example 4.4.2 illustrates how Laplace transforms can be used to solve linear differential equations.

Example 4.2.2

Consider the following system:

$$\dot{x} = -x + u, \qquad x(0) = 1 \qquad (4.2.7)$$

$$u(t) = t, \qquad \text{for } t \geq 0 \qquad (4.2.8)$$

Taking the Laplace transform of (4.2.7) gives

$$sX(s) - x(0) = -X(s) + \frac{1}{s^2}, \qquad x(0) = 1 \qquad (4.2.9)$$

Hence

$$X(s) = \frac{s^2 + 1}{s^2(s+1)} \qquad (4.2.10)$$

We will defer until Example 4.5.1 the final step of obtaining the corresponding time domain function. ▽▽▽

Example 4.2.2 can be extended to general state space models as follows: Consider the linear continuous time state space model

$$\dot{x} = Ax + Bu \qquad (4.2.11)$$

$$y = Cx \qquad (4.2.12)$$

Let $X(s)$, $U(s)$, and $Y(s)$ denote the Laplace transforms of $x(t)$, $u(t)$, and $y(t)$, respectively. Then, using Theorem 4.2.1, we can rewrite (4.2.11) and (4.2.12) in the transform domain as

$$sX(s) - x(0^-) = AX(s) + BU(s) \tag{4.2.13}$$

and

$$Y(s) = CX(s) \tag{4.2.14}$$

Thus we have that

$$X(s) = (sI - A)^{-1}(BU(s) + x(0^-)) \tag{4.2.15}$$

and

$$Y(s) = C(sI - A)^{-1}BU(s) + C(sI - A)^{-1}x(0^-) \tag{4.2.16}$$

We see that the linear differential time domain equation (4.2.11) has been reduced to a linear algebraic equation (4.2.15). The term $(C(sI - A)^{-1}B)$, often called the *transfer function*, completely describes the input–output behavior if there are zero initial conditions. More will be said about this function in Chapter 5.

Also, comparing (4.2.11) with (4.2.13) we see that, apart from the effect of initial conditions, we simply replace d/dt by s in going from the time domain description (4.2.11) to the Laplace domain description of (4.2.13).

4.3 DISCRETE TIME TRANSFORMS

4.3.1. Z-Transforms

The traditional discrete time equivalence of the Laplace transform of a function is the Z-transform of a sequence. Given a sequence, $\{f_t\}$, we define the one-sided Z-transform $F(z)$ as

$$F(z) = \mathbf{Z}\{f_t\} = \sum_{t=0}^{\infty} f_t z^{-t} \tag{4.3.1}$$

In an analogous manner to the Laplace transform, if $|f_t| < me^{\sigma t}$ for all t, then the Z-transform converges absolutely for all complex z that belong to the region

$$|z| > e^{\sigma} \tag{4.3.2}$$

Also, if f_{1_t} and f_{2_t} have the same Z-transform for all z in some region of convergence, then it follows that f_{1_t} and f_{2_t} are equal for all nonnegative t.

Example 4.3.1 illustrates the evaluation of the Z-transform for some simple sequences, which are the discrete equivalents to those covered in Example 4.2.1.

Example 4.3.1

(a) Find the Z-transform of the sequence $\{f_k = k\}$:

$$F(z) = \sum_0^\infty kz^{-k}$$
$$= z^{-1} + z^{-2} + z^{-3} + \cdots$$
$$\quad + z^{-2} + z^{-3} + \cdots$$
$$\quad + z^{-3} + \cdots$$
$$= \sum_1^\infty z^{-k}(1 + z^{-1} + z^{-2} + \cdots)$$
$$= \sum_1^\infty \frac{z^{-k}}{1 - z^{-1}}, \quad \text{for } |z| > 1$$
$$= \frac{z}{(z-1)^2}, \quad \text{for } |z| > 1 \tag{4.3.3}$$

(b) Find the Z transform of the sequence $\{f_k = \sigma^k\}$:

$$G(z) = \sum_0^\infty \sigma^k z^{-k}$$
$$= \frac{1}{1 - \sigma z^{-1}}, \quad \text{for } |z| > |\sigma|$$
$$= \frac{z}{z - \sigma} \tag{4.3.4}$$

Table 4.3.1 gives the Z-transforms of several common sequences.
Several important properties of the Z-transform are given in the following theorem.

Theorem 4.3.1. Z-Transform Properties

(i) Linearity

For any scalar α_1, α_2,
$$\mathbf{Z}\{\alpha_1 f_k + \alpha_2 g_k\} = \alpha_1 \mathbf{Z}\{f_k\} + \alpha_2 \mathbf{Z}\{g_k\}$$

(ii) Shift
$$\mathbf{Z}\{qf_k\} = z[\mathbf{Z}\{f_k\} - f_0]$$
where q is the shift operator $\{qf_k\} \triangleq \{f_{k+1}\}$

(iii) Summation
$$\mathbf{Z}\left\{\sum_{l=0}^k f_l\right\} = \frac{z}{z-1}\mathbf{Z}\{f_k\}$$

TABLE 4.3.1 Z-TRANSFORMS OF COMMON SEQUENCES

Time domain function	Description	Z-transform
$p_k \triangleq \begin{cases} 1, & k=0 \\ 0, & k \neq 0 \end{cases}$	Pulse	1
$1_k \triangleq 1, k \geq 0$	Unit step	$\dfrac{z}{z-1}$
k	Ramp	$\dfrac{z}{(z-1)^2}$
k^2	Parabola	$\dfrac{z(z+1)}{(z-1)^3}$
A^k	Exponential	$z(zI - A)^{-1}$
kA^k		$zA(zI - A)^{-2}$
$k^2 A^k$		$zA(zI + A)(zI - A)^{-3}$
$\sin(\omega k)$	Sine wave	$\dfrac{z \sin \omega}{z^2 - 2(\cos \omega)z + 1}$
$\cos(\omega k)$	Cosine wave	$\dfrac{z(z - \cos \omega)}{z^2 - 2(\cos \omega)z + 1}$

(iv) Frequency Differentiation

$$Z\{kf_k\} = -z\frac{d}{dz}Z\{f_k\}$$

(v) Frequency Integration

$$Z\left\{\frac{1}{k}f_k\right\} = \int_z^\infty \frac{F(w)}{w}\,dw$$

where

$$F(w) \triangleq Z\{f_k\}|_{z=w}$$

(vi) Time Shifting

$$Z\{1(k-N)f_{k-N}\} = z^{-N}Z\{f_k\}$$

(vii) Frequency Shifting

$$Z\{\sigma^k f_k\} = F(\sigma^{-1}z)$$

where

$$F(z) = Z\{f_k\}$$

(viii) Convolution

Let $H(z) = G(z)F(z)$, where $F(z) = Z\{f_k\}$ and $G(z) = Z\{g_k\}$. Then

$$H(z) = Z\left\{\sum_{l=0}^{k} f_l g_{k-l}\right\}$$

(ix) Inverse Transformation

For $k > 0$, the inverse transform f_k of $F(z)$ can be found by

$$f_k = \frac{1}{2\pi j} \oint \frac{F(z) z^k \, dz}{z}$$

where the contour of integration circles the origin once in the anticlockwise direction and lies within the region of convergence of $F(z)$.

(x) Initial Value Theorem

$$f_0 = \lim_{z \to \infty} \{F(z)\}$$

(xi) Final Value

If $\lim_{k \to \infty} f_k$ exists, then

$$\lim_{k \to \infty} f_k = \lim_{z \to 1} \{(z - 1) F(z)\}$$

(xii) Complex Convolution

$$Z\{f_k g_k\} = \frac{1}{2\pi j} \oint F(u) G\left(\frac{z}{u}\right) \frac{du}{u}$$

where $F(z) = Z\{f_k\}$, $G(z) = Z\{g_k\}$, and the contour of integration encircles the origin once anticlockwise and lies in the region of convergence of both $F(z)$ and $G(z)$.

Proof: We will defer the proof of this result since it is a special case of Theorem 4.4.1. ∇∇∇

Example 4.3.2 illustrates how Z-transforms can be used to solve linear difference equations.

Example 4.3.2

Consider the following discrete time system:

$$x_{k+1} = 0.5 x_k + u_k, \qquad x_0 = 1 \qquad (4.3.5)$$

$$u_k = k, \qquad \text{for } k \geq 0 \qquad (4.3.6)$$

Taking the Z-transform of (4.3.5) gives

$$z X(z) - z x_0 = 0.5 X(z) + \frac{z}{(z - 1)^2}$$

Hence

$$X(z) = \frac{z(z^2 - 2z + 2)}{(z - 0.5)(z - 1)^2} \qquad (4.3.7)$$

Again we defer the final step of getting the corresponding time domain function until Example 4.5.2. ▽▽▽

Example 4.3.2 can be generalized as follows: Consider the discrete time equation

$$qx_k = Ax_k + Bu_k \quad (4.3.8)$$

$$y_k = Cx_k \quad (4.3.9)$$

Then letting $X(z)$, $U(z)$, and $Y(z)$ be the Z-transforms of x_k, u_k, and y_k, respectively, (4.3.8) and (4.3.9) may be transformed to

$$zX - zx_0 = AX + BU \quad (4.3.10)$$

$$Y = CX \quad (4.3.11)$$

These equations can be rearranged in the form

$$X = (zI - A)^{-1}(BU + zx_0) \quad (4.3.12)$$

and

$$Y = C(zI - A)^{-1}(BU + zx_0) \quad (4.3.13)$$

Thus we see that the system output may be formed, in the transform domain, as a linear algebraic equation in terms of the input and initial state.

If we compare equation (4.3.10) with equation (4.3.8), then we see that, apart from the effect of initial conditions, we essentially replace q by z in going from the difference equation form to the Z-domain form of the equation. Because of this fact, the symbols q and z are often used interchangeably. However, we will prefer to keep q for the shift operator and z for the corresponding transform variable.

4.3.2 Delta Transforms

In Section 4.3.1 we have seen that there is a close connection between the forward shift operator q and the Z-transform variable z. In Chapter 3, we defined an alternative discrete time operator, the delta operator defined by

$$\delta = \frac{q-1}{\Delta} \quad (4.3.14)$$

In view of the connection between q and z, by analogy we define a new transform variable γ associated with δ as $\gamma = (z-1)/\Delta$. We can then define the corresponding transform using the Z-transform as

$$F'_\Delta(\gamma) = F(z)|_{z=\Delta\gamma+1} \quad (4.3.15)$$

Using Equation (4.3.1), we see that $F'_\Delta(\gamma)$ can be defined directly in terms of the sequence $\{f_k\}$ as

$$F'_\Delta(\gamma) = \sum_{k=0}^{\infty} f_k(1 + \Delta\gamma)^{-k} \quad (4.3.16)$$

For reasons that will become clear later, we introduce a further scaling in (4.3.16) by a factor Δ. This leads us, finally, to the definition of the delta transform as

$$\mathcal{D}(f_k) = F_\Delta(\gamma) \triangleq \Delta F'_\Delta(\gamma) = \Delta \sum_{k=0}^{\infty} f_k (1 + \Delta\gamma)^{-k} \qquad (4.3.17)$$

A table of transforms similar to Table 4.3.1 can readily be obtained for delta transforms by simply noting that

$$F_\Delta(\gamma) = \Delta F(z)|_{z=\Delta\gamma+1} \qquad (4.3.18)$$

For example, the delta transform of 1_k is easily seen to be

$$\frac{\Delta(\Delta\gamma + 1)}{\Delta\gamma} = \frac{1 + \Delta\gamma}{\gamma}$$

A complete set of delta transforms will be presented in the next section.

It is also a trivial matter to restate the properties given in Theorem 4.3.1 for the alternative delta transform. For example, property (ii) can be restated as follows.

Corollary 4.3.1

$$\mathcal{D}(\delta f_k) = \gamma \mathcal{D}(f_k) - f_0(1 + \Delta\gamma) \qquad (4.3.19)$$

Proof:

$$\mathcal{D}(\delta f_k) = \mathcal{D}\left(\frac{qf_k - f_k}{\Delta}\right)$$

$$= \frac{1}{\Delta}[\mathcal{D}(qf_k) - \mathcal{D}(f_k)]$$

$$= [Z(qf_k) - Z(f_k)]|_{z=\Delta\gamma+1}$$

$$= \Delta\gamma Z(f_k) - (\Delta\gamma + 1)f_0$$

$$= \gamma \mathcal{D}(f_k) - f_0(1 + \Delta\gamma) \qquad (4.3.20)$$

▽▽▽

The preceding result can be used directly to transform discrete time equations expressed in δ form to the delta domain. Specifically, let us consider

$$\delta x_k = Fx_k + Gu_k, \qquad x_0 \text{ given} \qquad (4.3.21)$$

$$y_k = Cx_k \qquad (4.3.22)$$

Letting $X(\gamma), U(\gamma), Y(\gamma)$ be the delta-transforms of x_k, u_k, and y_k, respectively, then (4.3.21), (4.3.22) can be transformed to

$$\gamma X - x_0(1 + \Delta\gamma) = FX + GU \qquad (4.3.23)$$

$$Y = CX \qquad (4.3.24)$$

Comparing (4.3.23) with (4.3.21) we see that, apart from the effect of initial conditions, we replace δ by γ in going from the delta form of the difference equation to the delta transform domain. Thus γ and δ are linked in the same way that z and q are.

In the next section, we will see that there is a further property of delta transforms that makes them very useful in systems theory: the Laplace transform can be recovered from the delta transform by simply setting the sampling period to zero. This fact allows us to present a unified transform theory that covers both discrete and continuous cases simultaneously.

4.4 UNIFIED TRANSFORM THEORY

To facilitate presentation of a unified transform theory, we will first introduce the following notation:

(i) $y(t)$ will denote a function of time when we are referring to continuous time or a sequence when referring to discrete time. In other words, $y(\cdot)$ is a mapping from real time, $t \in \Omega \subset \mathbb{R}$ to $\mathbb{R}$, where

$$\Omega = \mathbb{R} \text{ (continuous time)}$$
$$\Omega = \left\{ t : \frac{t}{\Delta} \in \mathbb{Z} \right\} \text{(discrete time)} \tag{4.4.1}$$

$\mathbb{R}$ denotes the real numbers and $\mathbb{Z}$ denotes the integers. Thus the time argument of $y(t)$ must be an integer number of sample periods (in discrete time) or a real number (continuous time).

(ii) ρ will be called the *generalized derivative* and will denote d/dt in continuous time or δ in discrete time.

(iii) $\mathbf{S}_{t_1}^{t_2} f(\tau) \, d\tau$ denotes, in continuous time, the Riemann integral:

$$\int_{\tau=t_1}^{\tau=t_2} f(\tau) \, d\tau$$

and in discrete time the lower Riemann sum,

$$\Delta \sum_{k=\frac{t_1}{\Delta}}^{k=\frac{t_2}{\Delta}-1} f(k\Delta)$$

where t_1 and $t_2 \in \Omega$. We will also use

$$\mathbf{S}_{t_1}^{\infty} f(\tau) \, d\tau$$

to denote

$$\lim_{T \to \infty} \mathbf{S}_{t_1}^{T} f(\tau) \, d\tau$$

Note that our use of the symbol $\mathbf{S}$ to denote both integration and summation completes an interesting historical circle, since the origin of the integration symbol $\int$, is a medieval long *s*, the initial letter of summa, or sum, and Σ is the Greek capital sigma.

(iv) $E(A, t)$ (where $A \in \mathbb{C}^{n \times n}$ and $t \in \Omega$) will be called the generalized matrix exponential and will denote:

$$e^{At}, \quad \text{in continuous time}$$

or

$$(I + A\Delta)^{t/\Delta}, \quad \text{in discrete time}$$

where $\mathbb{C}^{n \times n}$ denotes the set of $n \times n$ matrices with complex entries.

(v) $1(t)$ will denote the unit step function; that is, for both continuous and discrete time,

$$1(t) = \begin{cases} 1 & t \geq 0 \\ 0, & t < 0 \end{cases}$$

(vi) $i(t)$ will denote the unit pulse function; that is, $i(t)$ is the dirac delta function in continuous time and in discrete time, we have

$$i(t) = \begin{cases} \dfrac{1}{\Delta}, & \text{for } t = 0 \\ 0, & \text{otherwise} \end{cases}$$

In discrete time, it will also be convenient to use delayed versions of $1(t)$, and thus we also define

$$1^+(t) = 1(t - \Delta)$$

Using the preceding notation, we can unify the description of transform theory. We first define the *generalized transform* $F(\gamma)$ of a function, $f(t)$, as follows:

$$F(\gamma) = T\{f(t)\} \triangleq \int_{0^-}^{\infty} f(\tau) E(\gamma, -\tau) \, d\tau \tag{4.4.2}$$

where $\gamma \in C \subset \mathbb{C}$ [where C is the region of convergence; that is, the generalized integral in (4.4.2) converges absolutely for all $\gamma \in C$]. This definition can be extended to include matrix valued functions in a trivial fashion.

Note that the above generalized transform corresponds to:

(a) The Laplace transform in continuous time.
(b) The delta transform in discrete time, defined in (4.3.17).

In manipulating the operators S and ρ and the function $E(\gamma, t)$, and so on, the following facts are often helpful.

Lemma 4.4.1

(i) **Differentiation of Product**

(a) $\rho[x(t)y(t)] = [\rho x(t)] y(t) + x(t)[\rho y(t)]$
$\qquad\qquad\qquad\quad + \Delta [\rho x(t)][\rho y(t)]$

(b) $\rho[x(t)y(t)] = [\rho x(t)]y(t) + x(t+\Delta)[\rho y(t)]$

(c) $\rho[x(t)y(t)] = [\rho x(t)]y(t+\Delta) + x(t)[\rho y(t)]$

(ii) **Differentiation of Inverse**

$$\rho[A^{-1}(t)] = -A^{-1}(t+\Delta)[\rho A(t)]A^{-1}(t)$$
$$= -A^{-1}(t)[\rho A(t)]A^{-1}(t+\Delta)$$

(iii) **Derivative of Integral**

$$\rho\left[\underset{a}{\overset{t}{S}} f(\tau)\,d\tau\right] = f(t)$$

$$\rho\left[\underset{t}{\overset{b}{S}} f(\tau)\,d\tau\right] = -f(t)$$

(iv) **Integral of Derivative**

$$\underset{a}{\overset{b}{S}} [\rho f(t)]\,dt = f(b) - f(a)$$

(v) **Semigroup Property of Exponentials**

$$E(A,t)E(A,\tau) = E(A, t+\tau)$$

(vi) **Properties of Exponentials**

$$E(a,t)E(b,-t) = E\left(\frac{a-b}{1+b\Delta}, t\right)$$

$$E(a,t)E(b,t) = E(a+b+ab\Delta, t)$$

$$E(a,-t) = E\left(-\frac{a}{1+a\Delta}, t\right)$$

$$E(a,t) = E\left(-\frac{a}{1+a\Delta}, -t\right)$$

(vii) **Derivative of Exponential**

$$\rho[E(A,t)] = AE(A,t) = E(A,t)A$$
$$\rho[E(A,-t)] = -AE(A,-t-\Delta) = -E(A,-t-\Delta)A$$

(viii) **Integration by Parts**

$$\underset{a}{\overset{b}{S}}[\rho f(t)]g(t)\,dt = f(b)g(b) - f(a)g(a) - \underset{a}{\overset{b}{S}} f(t+\Delta)\rho\{g(t)\}\,dt$$

Proof: We leave the proof of these results as an exercise for the reader to gain familiarity with the unified notion. ▽▽▽

The following simple example illustrates how the preceding results can be used to obtain the generalized transforms of simple functions.

Sec. 4.4 Unified Transform Theory

Example 4.4.1

(a) Find the generalized transform of the function $f(t) = t$.

$$F(\gamma) = \int_{0^-}^{\infty} tE(\gamma, -t)\, dt$$

$$= \int_0^{\infty} t\rho\left[-\frac{1}{\gamma}E(\gamma, -t+\Delta)\right] dt, \quad \text{using Lemma 4.4.1(vii)}$$

$$= \left[-\frac{t}{\gamma}E(\gamma, -t+\Delta)\right]_{t=0}^{t=\infty}$$

$$+ \int_0^{\infty} \rho\{t\}\frac{1}{\gamma}E(\gamma, -t)\, dt, \quad \text{using Lemma 4.4.1(viii)}$$

$$= 0 + \int_0^{\infty} -\frac{1}{\gamma^2}\rho\{E(\gamma, -t+\Delta)\}\, dt, \quad \text{using Lemma 4.4.1(vii)}$$

$$= \left[-\frac{E(\gamma, -t+\Delta)}{\gamma^2}\right]_{t=0}^{t=\infty}$$

$$= \frac{1+\Delta\gamma}{\gamma^2} \tag{4.4.3}$$

TABLE 4.4.1 GENERALIZED TRANSFORMS OF COMMON TIME DOMAIN FUNCTIONS

Time domain function	Description	Generalized transform
$i(t)$	Impulse	1
$1(t)$	Unit step	$\dfrac{1+\Delta\gamma}{\gamma}$
t	Ramp	$\dfrac{1+\Delta\gamma}{\gamma^2}$
t^2	Parabola	$\dfrac{(1+\Delta\gamma)(2+\Delta\gamma)}{\gamma^3}$
$E(A,t)$	Exponential	$(1+\Delta\gamma)(\gamma I - A)^{-1}$
$tE(A,t)$		$(1+\Delta\gamma)(I + A\Delta)(\gamma I - A)^{-2}$
$t^2 E(A,t)$		$(1+\Delta\gamma)(2+\Delta\gamma)(I+\Delta A)(I+A\Delta)(\gamma I - A)^{-3}$
$\sin \omega t$	Sine wave	$\dfrac{(1+\Delta\gamma)\omega\,\text{sinc}(\omega\Delta)}{\gamma^2 + \Delta\phi(\omega,\Delta)\gamma + \phi(\omega,\Delta)}$
		where $\text{sinc}(\omega\Delta) \triangleq \dfrac{1}{\omega\Delta}\sin(\omega\Delta)$
		and $\phi(\omega,\Delta) \triangleq \dfrac{2(1-\cos\omega\Delta)}{\Delta^2}$
$\cos \omega t$	Cosine wave	$\dfrac{(1+\Delta\gamma)\left(\gamma + \frac{\Delta}{2}\phi(\omega,\Delta)\right)}{\gamma^2 + \Delta\phi(\omega,\Delta)\gamma + \phi(\omega,\Delta)}$

(b) Find the generalized transform of $g(t) = E(a, t)$.

$$G(\gamma) = S\int_0^\infty E(a, t) E(\gamma, -t) \, dt$$

$$= S\int_0^\infty E\left(\frac{a - \gamma}{1 + \Delta\gamma}, t\right) dt, \quad \text{using Lemma 4.4.1(vi)}$$

$$= S\int_0^\infty \frac{1 + \Delta\gamma}{a - \gamma} \rho\left\{E\left(\frac{a - \gamma}{1 + \Delta\gamma}, t\right)\right\} dt, \quad \text{using Lemma 4.4.1(vii)}$$

$$= \left(\frac{1 + \Delta\gamma}{a - \gamma}\right)\left[E\left(\frac{a - \gamma}{1 + \Delta\gamma}, t\right)\right]_{t=0}^{t=\infty}$$

$$= \frac{1 + \Delta\gamma}{\gamma - a} \quad (4.4.4)$$

▽▽▽

Additional transform pairs are given in Table 4.4.1. By comparing Table 4.4.1 for the generalized transforms with Table 4.2.1 for the Laplace transforms, we see that the Laplace transform for the given functions can be simply obtained by taking the limit $\Delta \to 0$ in the generalized transform. The following results show that this conclusion holds for a very wide class of time functions.

Lemma 4.4.2

Suppose $f(\cdot)$ ($\mathbb{R} \to \mathbb{R}$) satisfies

(i) $\exists\ c_1, c_2 \in \mathbb{R}$ s.t. $|f(t)| \leq c_1 e^{c_2 t}\ \forall\ t$; and
(ii) For any $\gamma > c_2$, $f(t) e^{-\gamma t}$ is Riemann integrable.

Then, for any $\gamma > c_2$,

$$\lim_{\Delta \to 0} \{F_\Delta(\gamma)\} = F_s(s)\big|_{s=\gamma}$$

$$= \int_0^\infty f(t) e^{-\gamma t} \, dt \quad (4.4.5)$$

where the integral in (4.4.5) is a Riemann integral.

Proof: Conditions (i) and (ii) guarantee that the Riemann integral in (4.4.5) exists. We will now show that the difference between the left side and right side in (4.4.5) is zero.

$$\left|\lim_{\Delta \to 0} \{F_\Delta(\gamma)\} - F_s(\gamma)\right| = \left|\lim_{\Delta \to 0} \Delta \sum_{k=0}^\infty f(k\Delta)\left((1 + \gamma\Delta)^{-k} - e^{-\gamma k\Delta}\right)\right| \quad (4.4.6)$$

Let $\beta \triangleq 1/\Delta \ln(1 + \gamma\Delta)$. Then from (4.4.6) we have

$$\left| \lim_{\Delta \to 0} \{ F_\Delta(\gamma) \} - F_s(\gamma) \right|$$

$$\leq \lim_{\Delta \to 0} \Delta \sum_{k=0}^{\infty} |f(k\Delta)|(1 + \gamma\Delta)^{-k}|1 - e^{-(\beta - \gamma)k\Delta}|$$

$$\leq \lim_{\Delta \to 0} \Delta \sum_{k=0}^{\infty} \left(|f(k\Delta)|(1 + \gamma\Delta)^{-k}|\beta - \gamma|k\Delta \right) \left| \frac{1 - e^{(\beta - \gamma)k\Delta}}{(\beta - \gamma)k\Delta} \right|$$

$$\leq \lim_{\Delta \to 0} \Delta \sum_{k=0}^{\infty} \left(|f(k\Delta)|(1 + \gamma\Delta)^{-k}|\beta - \gamma|k\Delta \right)$$

$$\leq \left\{ \lim_{\Delta \to 0} |\beta - \gamma| \right\} \left\{ \lim_{\Delta \to 0} \Delta \sum_{k=0}^{\infty} k\Delta |f(k\Delta)|(1 + \gamma\Delta)^{-k} \right\} \quad (4.4.7)$$

$\lim_{\Delta \to 0}(\beta - \gamma)$ is zero in view of the definition of β. We have thus reduced the problem to that of establishing the convergence of a limiting sum. From (i) we have

$$\lim_{\Delta \to 0} \left\{ \Delta \sum_{k=0}^{\infty} k\Delta |f(k\Delta)|(1 + \gamma\Delta)^{-k} \right\} \leq \lim_{\Delta \to 0} \left\{ c_1 \Delta \sum_{k=0}^{\infty} k\Delta e^{c_2 k\Delta}(1 + \gamma\Delta)^{-k} \right\}$$

$$(4.4.8)$$

Since $\gamma > c_2$, let $c_3 = (\gamma + c_2)/2$; then $c_2 < c_3 < \gamma$, and for Δ sufficiently small, $e^{c_3 \Delta} \leq 1 + \gamma\Delta$. Using this fact in (4.4.8) gives

$$\lim_{\Delta \to 0} \left\{ \Delta \sum_{k=0}^{\infty} k\Delta |f(k\Delta)|(1 + \gamma\Delta)^{-k} \right\} \leq \lim_{\Delta \to 0} \left\{ \Delta c_1 \sum_{k=0}^{\infty} k\Delta e^{c_2 k\Delta - c_3 k\Delta} \right\}$$

$$= c_1 \int_0^\infty t e^{(c_2 - c_3)t} \, dt \quad (4.4.9)$$

Since $c_3 > c_2$, the integral in (4.4.9) is finite, and the desired result follows. ▽▽▽

Condition (i) in Lemma 4.4.2 is a requirement that γ be in the region of convergence of the Laplace transform. The second condition in Lemma 4.4.2 is essentially a regularity condition on $f(\cdot)$. If this condition is not satisfied, then the discrete sequence $\{f(k\Delta)\}$ will not converge to the continuous time function $f(t)$ even with arbitrarily fast sampling.

The following example illustrates the preceding result for a function of time not given in Table 4.4.1.

Example 4.4.1. (Continued)

Part (c)

Find the delta transform of $f(t) = e^{-t} \cos 2t$, $t \geq 0$, for $\Delta = 1, 0.1,$ and 0. Note that we can write $f(t)$ as

$$f(t) = \tfrac{1}{2} [E(\alpha, t) + E(\alpha^*, t)]$$

where
$$\alpha \triangleq \frac{e^{-(1+2j)\Delta} - 1}{\Delta}$$

so
$$F(\gamma) = \frac{1}{2}(1 + \Delta\gamma)\left[\frac{1}{(\gamma - \alpha)} + \frac{1}{(\gamma - \alpha^*)}\right]$$
$$= \frac{(1 + \Delta\gamma)\left(\gamma - \frac{e^{-\Delta}\cos 2\Delta - 1}{\Delta}\right)}{\gamma^2 - 2\left(\frac{e^{-\Delta}\cos 2\Delta - 1}{\Delta}\right)\gamma + \frac{1 - 2e^{-\Delta}\cos 2\Delta + e^{-2\Delta}}{\Delta^2}}$$

Thus, for $\Delta = 1$, we have
$$F(\gamma) = \frac{(\gamma + 1)(\gamma - 1.1531)}{(\gamma^2 - 2.3062\gamma + 1.4415)}$$

or, for $\Delta = 0.1$,
$$F(\gamma) = \frac{(0.1\gamma + 1)(\gamma + 1.1320)}{\gamma^2 + 2.2640\gamma + 4.5129}$$

or, for $\Delta \to 0$,
$$\frac{e^{-\Delta}\cos 2\Delta - 1}{\Delta} \to -1 \quad \text{and} \quad \left(\frac{1 - 2e^{-\Delta}\cos 2\Delta + e^{-2\Delta}}{\Delta^2}\right) \to 5$$

and
$$F(\gamma) = \frac{\gamma + 1}{\gamma^2 + 2\gamma + 5}$$

It is readily verified that this final result is indeed the corresponding Laplace transform. ▽▽

Using the definition given in (4.4.2), we can now establish the following theorem concerning the generalized transform. (Note that this theorem includes Theorems 4.2.1 and 4.3.1 as special cases.)

Theorem 4.4.1. Transform Properties

(i) Linearity

For any scalar α_1, α_2,
$$\boldsymbol{T}\{\alpha_1 f(t) + \alpha_2 g(t)\} = \alpha_1 \boldsymbol{T}\{f(t)\} + \alpha_2 \boldsymbol{T}\{g(t)\}$$

(ii) Differentiation
$$\boldsymbol{T}\{\rho[f(t)]\} = \gamma \boldsymbol{T}\{f(t)\} - f(0^-)(1 + \Delta\gamma)$$

(iii) **Integration**
$$T\left\{\int_0^t f(\tau)\,d\tau\right\} = \frac{1}{\gamma}T\{f(t)\}$$

(iv) **Frequency Differentiation**
$$T\{tf(t)\} = -(\Delta\gamma + 1)\frac{d}{d\gamma}[T\{f(t)\}]$$

(v) **Frequency Integration**
$$T\left\{\frac{f(t)}{t}\right\} = \int_\gamma^\infty \frac{F(u)}{1+\Delta u}\,du$$

where
$$F(u) = [T\{f(t)\}]|_{\gamma=u}$$

(vi) **Time Shifting**
$$T\{1(t-T)f(t-T)\} = E(\gamma, -T)T\{f(t)\}$$

where $1(\cdot)$ denotes the unit step function,
$$1(t) = \begin{cases} 1, & t \geq 0 \\ 0, & t < 0 \end{cases}$$

and
$$T > 0 \in \Omega$$

(vii) **Frequency Shifting**
$$T\{E(a,t)f(t)\} = F\left(\frac{\gamma - a}{1 + \Delta a}\right)$$

where
$$F(\gamma) = T\{f(t)\}$$

(viii) **Convolution**

Let $H(\gamma) = G(\gamma)F(\gamma)$, where $F(\gamma) = T\{f(t)\}$ and $G(\gamma) = T\{g(t)\}$; then
$$H(\gamma) = T\left\{\int_0^{t+\Delta} f(t-\tau)g(\tau)\,d\tau\right\}$$

assuming $f(t) = g(t) = 0$ for $t < 0$.

(ix) **Inverse Transformation**

For $t > 0$, the inverse transform $f(t)$ of $F(\gamma)$ can be found by
$$f(t) = \frac{1}{2\pi j}\oint F(\gamma)E(\gamma, t-\Delta)\,d\gamma$$

where the contour of integration C is inside the region of convergence of $F(\gamma)$ and encircles all singularities of $F(\gamma)$ once in the anticlockwise sense.

(x) Initial Value Theorem

$$\lim_{t \to 0^+} f(t) = \lim_{\gamma \to \infty} \left\{ \frac{\gamma F(\gamma)}{(1 + \gamma \Delta)} \right\}$$

provided the limit exists.

(xi) Final Value

If $\lim_{t \to \infty} f(t)$ exists, then

$$\lim_{t \to \infty} f(t) = \lim_{\gamma \to \infty} \{\gamma F(\gamma)\}$$

(xii) Complex Convolution

Suppose that the transform of $f_1(t)$ is $F_1(\gamma)$ with a region of convergence, C_1, and $f_2(t)$ has transform $F_2(\gamma)$ with region of convergence C_2; then for any γ outside the contour C,

$$T\{f_1(t)f_2(t)\} = \frac{1}{2\pi j} \oint_C \frac{F_1(u) F_2\left(\frac{\gamma - u}{1 + \Delta u}\right)}{(1 + \Delta u)} du$$

where $C \in C_1 \cap C_2$ and encircles the stability boundary once in the anticlockwise direction.

Proof: (Note that this proof includes the proofs of Theorems 4.2.1 and 4.3.1. Various parts of the proofs are left to the reader as exercises. See Problem 5.)

(ii) From the definition of the generalized exponential, we can show that

$$\rho E(\gamma, -t) = \frac{-\gamma}{1 + \gamma \Delta} E(\gamma, -t) \quad (4.4.10)$$

and, from the definition of ρ,

$$\rho\{f(t)g(t)\} = (1 + \rho\Delta)\{f(t)\}\rho\{g(t)\} + \rho\{f(t)\}g(t),$$

$$\text{using Lemma 4.4.1(i)} \quad (4.4.11)$$

Then

$$T\{\rho f(t)\} = \underset{0^-}{\overset{\infty}{S}} E(\gamma, -t)[\rho f(t)]\, dt$$

$$= \underset{0^-}{\overset{\infty}{S}} (1 + \gamma\Delta)[(1 + \rho\Delta) E(\gamma, -t)][\rho f(t)]\, dt$$

$$= \underset{0^-}{\overset{\infty}{S}} (1 + \gamma\Delta)[\rho\{E(\gamma, -t)f(t)\} - \rho\{E(\gamma, -t)\}f(t)]\, dt$$

$$= (1 + \gamma\Delta)[E(\gamma, -t)f(t)]_{0^-}^{\infty} + \underset{0^-}{\overset{\infty}{S}} \gamma E(\gamma, -t) f(t)\, dt,$$

$$\text{using Lemma 4.4.1(iv)}$$

$$= \gamma T\{f(t)\} - (1 + \gamma\Delta) f(0^-) \quad (4.4.12)$$

as required.

(iv) From the definition of the generalized exponential, we can also show that

$$\frac{d}{d\gamma}\{E(\gamma,-t)\} = \frac{-t}{1+\Delta\gamma}E(\gamma,-t) \quad (4.4.13)$$

Thus

$$(1+\Delta\gamma)\frac{d}{d\gamma}[T\{f(t)\}] = (1+\Delta\gamma)\frac{d}{d\gamma}\left\{\int_{0^-}^{\infty} E(\gamma,-t)f(t)\,dt\right\}$$

$$= -\int_{0^-}^{\infty} E(\gamma,-t)tf(t)\,dt$$

$$= -T\{tf(t)\} \quad (4.4.14)$$

(vi)

$$T\{1(t-a)f(t-a)\} = \int_{a^-}^{\infty} E(\gamma,-t)f(t-a)\,dt$$

$$= E(\gamma,-a)\int_{a^-}^{\infty} E(\gamma,-(t-a))f(t-a)\,dt$$

$$= E(\gamma,-a)T\{f(t)\} \quad (4.4.15)$$

where we have used the semigroup property of exponentials [Lemma 4.4.1(v)].

(viii)

$$H(\gamma) = \left(\int_{-\infty}^{\infty} E(\gamma,-\tau)g(\tau)\,d\tau\right)\left(\int_{-\infty}^{\infty} E(\gamma,-x)f(x)\,dx\right)$$

$$= \left(\int_{0^-}^{\infty} E(\gamma,-\tau)g(\tau)\,d\tau\right)\left(\int_{0^-}^{\infty} E(\gamma,-x)f(x)\,dx\right) \quad (4.4.16)$$

where we have used the properties of an exponential and the fact that $f(t)$ and $g(t)$ are zero for $t < 0$. Replacing $(t - \tau)$ in (4.4.15) by x, we then have

$$H(\gamma) = \left(\int_{-\infty}^{\infty} E(\gamma,-\tau)g(\tau)\,d\tau\right)\left(\int_{-\infty}^{\infty} E(\gamma,-x)f(x)\,dx\right)$$

$$= \left(\int_{0^-}^{\infty} E(\gamma,-\tau)g(\tau)\,d\tau\right)\left(\int_{0^-}^{\infty} E(\gamma,-x)f(x)\,dx\right) \quad (4.4.17)$$

where we have again used the fact that f and g are zero for all negative times. The desired result follows directly from (4.4.17).

(x) Using the transform of a derivative rule, (ii), we can show that

$$\int_{0^-}^{\infty} \frac{E(\gamma,-t)}{1+\gamma\Delta}(\rho f(t))\,dt = \frac{\gamma F(\gamma)}{1+\Delta\gamma} - f(0^-) \quad (4.4.18)$$

Now

$$\lim_{\gamma\to\infty}\int_{0^-}^{\infty} \frac{E(\gamma,-t)}{1+\gamma\Delta}(\rho f(t))\,dt$$

$$= \lim_{\gamma\to\infty}\left\{\int_{0^+}^{\infty} \frac{E(\gamma,-t)}{1+\gamma\Delta}(\rho f(t))\,dt + \int_{0^-}^{0^+} \frac{E(\gamma,-t)}{1+\gamma\Delta}(\rho f(t))\,dt\right\}$$

$$= f(0^+) - f(0^-) \quad (4.4.19)$$

From (4.4.19) and (4.4.18), the desired result follows.

(xii) $\quad T\{f_1(t)f_2(t)\} = \int_0^\infty E(\gamma, -t)\left(\frac{1}{2\pi j}\oint F_1(u)E(u, t-\Delta)\,du\right)f_2(t)\,dt$

where we have used part (ix).

$$T\{f_1(t)f_2(t)\} = \frac{1}{2\pi j}\oint \frac{F_1(u)}{1+\Delta u}\left(\int_0^\infty E(\gamma, -t)E(u, t)f_2(t)\,dt\right)du$$

$$= \frac{1}{2\pi j}\oint \frac{F_1(u)}{1+\Delta u} F_2\left(\frac{\gamma - u}{1+\Delta u}\right)du$$

where we have used part (vii). ▽▽▽

The key point about Theorem 4.4.1 is that it treats continuous and discrete simultaneously. In particular, all Laplace transform results are simply obtained by setting $\Delta = 0$ and all Z-transform results can be obtained by setting $\gamma = (z-1)/\Delta$ and scaling by Δ appropriately.

Example 4.4.2

Consider the following system:

$$\rho x(t) = -x(t) + u(t), \qquad x(0) = 1 \qquad (4.4.20)$$

$$u(t) = t, \qquad \text{for } t \geq 0 \qquad (4.4.21)$$

Taking transforms of both sides, we obtain

$$\gamma X(\gamma) - (1+\Delta\gamma)x(0) = -X(\gamma) + \frac{1+\Delta\gamma}{\gamma^2} \qquad (4.4.22)$$

Thus

$$X(\gamma) = \frac{(1+\Delta\gamma)(1+\gamma^2)}{(\gamma+1)\gamma^2} \qquad (4.4.23)$$

The corresponding time domain function will be evaluated in Example 4.5.3. ▽▽▽

The extension of Example 4.4.2 to general state space models is as follows. Consider the general state space equation

$$\rho x = Ax + Bu, \qquad x_0 \text{ given} \qquad (4.4.24)$$

Taking transforms gives

$$X(\gamma) = (\gamma I - A)^{-1}((1+\Delta\gamma)x_0 + BU(\gamma)) \qquad (4.4.25)$$

$$Y(\gamma) = CX(\gamma) \qquad (4.4.26)$$

Thus, as in Section 4.3.2, we see that, apart from initial conditions, we simply replace ρ by γ in going from the time domain to the generalized transform domain.

4.5 INVERSE TRANSFORMATIONS

Thus far we have considered various properties of transforms and have indicated how the transform of a sequence or function may be obtained. An important issue, which we have yet to consider, is how the inverse transform may be calculated. Theorem 4.4.1(ix) may be used to calculate the inverse transform; however, the evaluation of the complex integral is quite difficult in general.

In cases where the transform can be expressed as a ratio of polynomials in γ, it is usually much easier to use a partial fractions expansion. Suppose $F(\gamma)$ can be expressed as

$$F(\gamma) = \frac{N(\gamma)}{D(\gamma)} = \frac{n_n \gamma^n + n_{n-1} \gamma^{n-1} + \cdots + n_0}{\gamma^n + d_{n-1} \gamma^{n-1} + \cdots + d_0} \qquad (4.5.1)$$

where we have assumed that the degree of the numerator is not greater than the degree of the denominator. This condition on the degrees will almost always be satisfied in practice, since, otherwise, $f(t)$ contains derivatives of impulses if $f(t)$ is a continuous function, or nonzero values for $t < 0$ if $f(t)$ is a sequence.

We may rewrite $F(\gamma)$ as

$$F(\gamma) = n_n + \frac{N'(\gamma)}{D(\gamma)} \qquad (4.5.2)$$

where

$$N'(\gamma) = N(\gamma) - n_n D(\gamma)$$

We now factorize $D(\gamma)$ into its n complex roots; that is, let

$$D(\gamma) = (\gamma - p_1)(\gamma - p_2) \ldots (\gamma - p_n) \qquad (4.5.3)$$

The method from this point is simplest in the case where the zeros of $D(\gamma)$, that is, p_i, are distinct. For the case of distinct roots, we may write

$$F(\gamma) = n_n + \sum_{k=1}^{n} \frac{\alpha_k}{\gamma - p_k} \qquad (4.5.4)$$

where the residues α_k may be calculated as follows:

$$\alpha_k = \left\{ \frac{N'(\gamma)}{D(\gamma)} \cdot (\gamma - p_k) \right\} \bigg|_{\gamma = p_k} \qquad (4.5.5)$$

The problem of inverting the transform has now been reduced to the inversion of the simple functions given in equation (4.5.4). This is straightforward in view of the uniqueness and linearity of transforms. Thus, using line 1 and line 5 of Table 4.4.1 together with the time shift property of transforms [Theorem 4.4.1(vi)], we immediately see that

$$f(t) = n_n i(t) + \sum_{k=1}^{n} \alpha_k 1^+(t) E(p_k, t - \Delta) \qquad (4.5.6)$$

At this stage the inverse transformation is complete; however, some of the constants α_k and p_k may be complex. Assuming that the coefficients of $D(\gamma)$ are real [which will always be the case if $f(t)$ is real valued], then for any p_k that is complex there exists some $k^* \in \{1, 2, \ldots, n\}$ such that $p_k^* = p_{k^*}$. From (4.5.5) it also follows that $\alpha_k^* = \alpha_{k^*}$.

In the case of these complex conjugate pairs, it is usual to combine the terms due to k and k^* in (4.5.6) in the following way:

$$1^+(t)\alpha_k E(p_k, t - \Delta) + 1^+(t)\alpha_{k^*} E(p_{k^*}, t - \Delta)$$
$$= 1^+(t)\alpha_k E(p_k, t - \Delta) + 1^+(t)\alpha_k^* E(p_k^*, t - \Delta)$$
$$= 1^+(t) 2\,\text{Re}\{\alpha_k E(p_k, t - \Delta)\}$$
$$= 1^+(t) 2\,\text{Re}\left\{\frac{\alpha_k}{1 + \Delta p_k}(E(p_k, t))\right\} \quad (4.5.7)$$

We then let $\alpha_k/(1 + \Delta p_k) = a_k + jb_k$ and $E(p_k, t) = e^{\sigma_k t} e^{j\omega_k t}$, where $\sigma_k + j\omega_k = (1/\Delta)\ln(1 + \Delta p_k)$. Then from (4.5.7) we have that

$$1^+(t)\alpha_k E(p_k, t - \Delta) + 1^+(t)\alpha_{k^*} E(p_{k^*}, t - \Delta)$$
$$= 1^+(t) 2 e^{\sigma_k t}(a_k \cos \omega_k t - b_k \sin \omega_k t) \quad (4.5.8)$$

In the case where $D(\gamma)$ has repeated zeros, the same principles apply, although the algebra is more complicated. We give details for completeness; however, there is no loss of continuity by proceeding directly to the example to follow. Say we have

$$D(\gamma) = (\gamma - p_1)^{\nu_1}(\gamma - p_2)^{\nu_2}\ldots(\gamma - p_m)^{\nu_m} \quad (4.5.9)$$

where $\sum_{i=1}^{m}\nu_i = n$. We then write

$$F(\gamma) = n_n + \frac{N'(\gamma)}{D(\gamma)}$$
$$= n_n + \frac{\alpha_1(\gamma)}{(\gamma - p_1)^{\nu_1}} + \frac{\alpha_2(\gamma)}{(\gamma - p_2)^{\nu_2}} + \cdots + \frac{\alpha_m(\gamma)}{(\gamma - p_m)^{\nu_m}} \quad (4.5.10)$$

where $\alpha_k(\gamma)$ is a polynomial in γ of degree $(\nu_k - 1)$. Note that a zero of $D(\gamma)$, p_k, of multiplicity ν_k will, in general, give rise to terms of the form $t^i E(p_k, t - i\Delta)$, where i ranges from $0, 1, \ldots, (\nu_{k-1})$.

We then decompose each of the terms in (4.5.10) as follows:

$$\frac{\alpha_k(\gamma)}{(\gamma - p_k)^{\nu_k}} = \frac{\beta_{1k}}{(\gamma - p_k)} + \frac{\beta_{2k}}{(\gamma - p_k)^2} + \frac{\beta_{3k}}{(\gamma - p_k)^3} + \cdots + \frac{\beta_{\nu_k k}}{(\gamma - p_k)^{\nu_k}} \quad (4.5.11)$$

From (4.5.11) it can be seen that, provided $p_k \neq -(1/\Delta)$ [the case $p_k = -(1/\Delta)$ is left as an exercise; see Problem (7)],

$$\frac{\alpha_k(\gamma)}{(\gamma - p_k)^{\nu_k}} = T\left\{1^+(t)\sum_{i=1}^{\nu_k}\beta_{ik}\prod_{j=1}^{i-1}\frac{t - j\Delta}{j}E(p_k, t - i\Delta)\right\} \quad (4.5.12)$$

In the case where some p_k are complex, we again group them into complex conjugate pairs to obtain an expression involving real quantities only.

A final point in evaluating inverse transforms is that Table 4.4.1 is usually most conveniently used when going from time to transform domain. Table 4.5.1 is

TABLE 4.5.1 INVERSE TRANSFORMS OF COMMON GENERALIZED TRANSFORMS

Generalized transform	Description	Time domain function
1	Impulse	$i(t)$
$\dfrac{1}{\gamma}$	Delayed step	$1^+(t) \triangleq 1(t - \Delta)$
$\dfrac{1}{\gamma^2}$	Delayed ramp	$1^+(t)(t - \Delta)$
$\dfrac{1}{\gamma^3}$	Delayed parabola	$\frac{1}{2} 1^+(t)(t - \Delta)(t - 2\Delta)$
$(\gamma I - A)^{-1}$	Delayed exponential	$1^+(t) E(A, t - \Delta)$
$(\gamma I - A)^{-2}$		$1^+(t)(t - \Delta) E(A, t - 2\Delta)$
$(\gamma I - A)^{-3}$		$\frac{1}{2} 1^+(t)(t - \Delta)(t - 2\Delta) E(A, t - 3\Delta)$
$\dfrac{1}{\gamma^2 + \Delta \phi \gamma + \phi}$, where $0 < \Delta^2 \phi < 4$	Delayed sine wave	$\dfrac{1}{\omega \operatorname{sinc}(\omega \Delta)} 1^+(t) \sin(\omega(t - \Delta))$, $\omega = \dfrac{1}{\Delta} \cos^{-1}\left(1 - \dfrac{\Delta^2 \phi}{2}\right)$
$\dfrac{\gamma}{\gamma^2 + \Delta \phi \gamma + \phi}$, where $0 < \Delta^2 \phi < 4$	Delayed cosine wave	$1^+(t)\left[\cos(\omega(t - \Delta)) - \dfrac{\Delta \phi}{2\omega \operatorname{sinc} \omega \Delta} \sin(\omega(t - \Delta))\right]$, $\omega = \dfrac{1}{\Delta} \cos^{-1}\left(1 - \dfrac{\Delta^2 \phi}{2}\right)$

readily obtained from Table 4.4.1 using the properties given in Theorem 4.4.1, but is sometimes more convenient when going from the transform domain back to the time domain.

We will next illustrate the preceding techniques by some simple examples. We first complete Examples 4.2.2, 4.3.2, and 4.4.2.

Example 4.5.1. Example 4.2.2 continued

In Example 4.2.2 we found that [Equation (4.2.10)]

$$X(s) = \frac{s^2 + 1}{s^2(s + 1)} \tag{4.5.13}$$

Using a partial fraction expression,

$$X(s) = \frac{2}{s + 1} + \frac{1}{s^2} - \frac{1}{s} \tag{4.5.14}$$

Finally, using Table 4.2.1, we have

$$x(t) = 2e^{-t} + t - 1 \tag{4.5.15}$$

Example 4.5.2. Example 4.3.2 continued

In Example 4.3.2 we found that [Equation (4.3.7)]

$$X(z) = \frac{z(z^2 - 2z + 2)}{(z - 0.5)(z - 1)^2} \qquad (4.5.16)$$

Looking at the results given in Table 4.3.1, we see that most of the transform pairs have z in the numerator. Hence, in evaluating the partial fraction expansion, it is most convenient to work with $X(z)/z$. Using a partial fraction expansion in (4.5.16) gives

$$\frac{X(z)}{z} = \frac{5}{z - 0.5} - \frac{4}{z - 1} + \frac{2}{(z - 1)^2} \qquad (4.5.17)$$

Finally, using Table 4.3.1, we have

$$x_k = 5(0.5)^k - 4(1_k) + 2k, \qquad k \geq 0 \qquad (4.5.18)$$

▽▽▽

Example 4.5.3. Example 4.4.2 Continued

In Example 4.4.2 we found that [Equation (4.4.22)]

$$X(\gamma) = \frac{(1 + \Delta\gamma)(1 + \gamma^2)}{(\gamma + 1)\gamma^2} \qquad (4.5.19)$$

Retaining the numerator factor of $(1 + \Delta\gamma)$ and then doing a partial fraction expansion gives

$$X(\gamma) = \frac{2(1 + \Delta\gamma)}{(\gamma + 1)} + \frac{1 + \Delta\gamma}{\gamma^2} - \frac{1 + \Delta\gamma}{\gamma} \qquad (4.5.20)$$

Finally, using Table 4.4.1,

$$x(t) = 2E(-1, t) + t - 1 \qquad (4.5.21)$$

▽▽▽

Example 4.5.4 illustrates the use of generalized transforms and their inverses to solve a difference equation.

Example 4.5.4

Consider the sequence defined by

$$x_{t+2} = \tfrac{1}{2}(x_{t+1} + x_t), \qquad x_0 = 0, \quad x_1 = 1, \quad \Delta = 1 \qquad (4.5.22)$$

We can rewrite Equation (4.5.22) in operator notation as
$$(1 + 2\rho + \rho^2)x_t = \tfrac{1}{2}((1+\rho)x_t + x_t), \qquad \rho = \delta$$
or
$$(\rho^2 + \tfrac{3}{2}\rho)x_t = 0 \tag{4.5.23}$$

Let $y_t = \rho x_t$ and $Y(\gamma)$ denote the delta transform of y_t. Then, from (4.5.23), $Y(\gamma)$ must satisfy
$$\gamma Y(\gamma) - (1+\gamma)y_0 + \tfrac{3}{2}Y(\gamma) = 0 \tag{4.5.24}$$
where
$$y_0 = (\rho x_t)|_{t=0} = 1$$
So
$$Y(\gamma) = \frac{\gamma + 1}{\gamma + (3/2)} \tag{4.5.25}$$

We rewrite (4.5.25) as
$$X(\gamma) = \frac{1}{\gamma}Y(\gamma) + \frac{1 + \Delta\gamma}{\gamma}x_0$$
$$= \frac{2/3}{\gamma} + \frac{1/3}{(\gamma + 1.5)} \tag{4.5.26}$$

Then, from Table 4.5.1, it is clear that
$$x_t = 1^+(t)\tfrac{2}{3} + \tfrac{1}{3}1^+(t)E(-1.5, t-1)$$
$$= \frac{1^+(t)}{3}\left(2 + (-0.5)^{t-1}\right)$$
$$= \frac{1^+(t)}{3}\left(2 - 2(-0.5)^t\right) \tag{4.5.27}$$

Thus the first few values of x_t are
$$x_t = 0, 1, 0.5, 0.75, 0.625, 0.6875, 0.65635, \ldots \tag{4.5.28}$$

4.6 SUMMARY

This chapter covered the following points:

- Definition of Laplace transform of continuous time functions as
$$F(s) \triangleq \int_{0^-}^{\infty} f(t)e^{-st}\,dt \tag{4.2.1}$$
- A table of Laplace transforms (Table 4.2.1).
- Properties of Laplace transforms in Theorem 4.2.1.

- Definition of Z-transforms of sequences

$$F(z) = \sum_{t=0}^{\infty} f_t z^{-t} \tag{4.3.1}$$

- A table of Z-transforms (Table 4.3.1).
- Properties of Z-transforms (Theorem 4.3.1).
- A definition of delta transforms:

$$F_\Delta(\gamma) = \Delta \sum_{k=0}^{\infty} f_k (1 + \Delta\gamma)^{-k} \tag{4.3.17}$$

$$= \Delta F(z)|_{z=\Delta\gamma+1} \tag{4.3.18}$$

- A unified treatment of transform theory using delta transforms and based on the following unified notation:

$$\rho = \begin{cases} \dfrac{d}{dt}, & \text{in continuous time} \\ \delta, & \text{in discrete time} \end{cases}$$

$$S = \begin{cases} \int, & \text{in continuous time} \\ \Delta\Sigma, & \text{in discrete time} \end{cases}$$

$$E(A, t) = \begin{cases} e^{At}, & \text{in continuous time} \\ (I + A\Delta)^{t/\Delta}, & \text{in discrete time} \end{cases}$$

- Manipulations using the unified notation were discussed in Lemma 4.4.1.
- A table of generalized transforms, as given in Table 4.4.1.
- A proof that Laplace transforms are a limiting case of delta transforms (Lemma 4.4.2).
- Properties of generalized transforms (Theorem 4.4.1).
- Inverse transforms using partial fraction expansions.

Thus, in brief, this chapter covered both continuous and discrete transform theory. A key point was that both continuous and discrete can be treated simultaneously by use of generalized transforms.

4.7 REFERENCES

Laplace transforms are covered in virtually every book on classical control. See, for example:

DI STEFANO, J. J., A. R. STUBBERUD, and I. J. WILLIAMS (1976) *Feedback and Control Systems*, Schaum's Outline Series. McGraw-Hill, New York.

Z-transforms are covered in books on sampled data control, including:

ÅSTRÖM, K. J. and B. WITTENMARK (1984) *Computer Controlled Systems, Theory and Design*. Prentice-Hall, Englewood Cliffs, N.J.

FRANKLIN, G. F. and J. D. POWELL (1980) *Digital Control of Dynamic Systems*. Addison-Wesley, Reading, Mass.

KUO, B. C. (1980) *Digital Control Systems*. Holt, Rinehart and Winston, New York.

As far as we are aware, no previous book deals with unified transform theory. However, our view is that this is the most insightful way to describe the sampled data case.

4.8 PROBLEMS

1. Find the Laplace transforms of the following time functions:
   (a) $f_1(t) = \begin{cases} 1, & t > 0 \\ 0, & t \leq 0 \end{cases}$
   (b) $f_2(t) = \sin(\omega t)$
   (c) $f_3(t) = \begin{cases} 1 & \text{if } [t] \text{ is even} \\ -1, & \text{if } [t] \text{ is odd} \end{cases}$

2. Find the Z-transform of the following sequences:
   (a) $f_{1_t} = \begin{cases} 1, & t > 0 \\ 0, & t \leq 0 \end{cases}$
   (b) $f_{2_t} = \sin(\omega t)$
   (c) $f_{3_t} = (1, -1, 1, -1, 1, -1, \ldots)$

3. Derive the following properties of $E(A, t)$:
   (a) $E(A, 0) = I$
   (b) $E(A, t)E(A, s) = E(A, t + s)$
   (c) Provided $(I + A\Delta)$ is nonsingular, $E(A, t)^{-1} = E(A, -t)$
   (d) $E(TAT^{-1}, t) = TE(A, t)T^{-1}$
   (e) $\lim_{\Delta \to 0} E(A, t) = e^{At}$ (pointwise, for any A, t)

4. Show the following properties of the generalized derivative and integral:
   (a) $\rho(xy) = (\rho x)y + x(\rho y) + \Delta(\rho x)(\rho y)$
   (b) $\rho(x^{-1}y) = [(1 + \rho\Delta)x]^{-1}[\rho y - (\rho x)x^{-1}y]$
       Hint: First show that $\rho(x^{-1}) = -[(1 + \rho\Delta)x]^{-1}\rho x[x]^{-1}$)
   (c) $\rho\left\{S\!\!\int_0^t f(\tau)\, d\tau\right\} = f(t)$
   (d) $S\!\!\int_{t_1}^{t_2} \rho\{f(t)\}\, dt = f(t_2) - f(t_1)$
   (e) $\rho E(A, t) = AE(A, t) = E(A, t)A$

5. Establish the parts of Theorem 4.4.1 for which proofs have not been explicitly given in the text.

6. Verify the generalized transforms given in Table 4.4.1.

7. Show that the inverse transform of $1/[(1 + \Delta\gamma)^n]$ is $i(n\Delta)$, where $i(\cdot)$ is the impulse function defined in Section 4.4.(v).

8. The Fibonacci numbers are known to satisfy the following recursion:
$$y_{k+2} = y_{k+1} + y_k, \quad \text{with } y_0 = 0, \quad y_1 = 1$$

   Find an explicit expression for y_k using transform techniques.

9. (a) Show that Equation (3.4.35) can be expressed in ρ operator form as

$$\rho x = \begin{bmatrix} \frac{\Delta}{2}\phi(\omega_0, \Delta) & \text{sinc}(\omega_0 \Delta) \\ -\omega_0^2 \text{sinc}(\omega_0 \Delta) & \frac{\Delta}{2}\phi(\omega_0, \Delta) \end{bmatrix} x$$

where ϕ and sinc are defined in Table 4.4.1.
(b) Examine the corresponding differential equation for $\Delta = 0$ and comment.
(c) Contrast the answer obtained in part (b) with that obtained by letting $\Delta = 0$ in (3.4.35). Comment on the result.

10. Use generalized transform techniques to solve the model given in Problem 9(a) for arbitrary $x(0)$.

11. Show that for $j > 0$

$$T^{-1}\{(\gamma I - A)^{-j}\} = 1^+(t)\left(\prod_{l=1}^{j-1} \frac{t - l\Delta}{l}\right) E(A, t - j\Delta)$$

Hint: Use induction on j and Theorem 4.4.1(iv).

12. Find the delta transform of the sequence (with $\Delta = 1$)

$$f_t = 1, 1, -1, -1, \ldots$$

by two methods:
(a) Evaluating the integral (4.4.2)
(b) Decomposing f_t into a sum of sinusoids with fundamental frequency $\pi/2$ rad/sec and using Table 4.4.1

13. Prove the following generalization of theorem 4.4.1(ii):

$$T\{\rho^n f(t)\} = \gamma^n T\{f(t)\}$$
$$- (1 + \Delta\gamma)\left\{\overset{(n-1)}{f}(0) + \gamma \overset{(n-2)}{f}(0) + \cdots + \gamma^{n-1} f(0)\right\}$$

where $\overset{(i)}{f}(0)$ denotes $\rho^i f(t)\big|_{t=0}$.

*14. Prove the following unified version of Parseval's theorem:

$$S\int_0^\infty f^2(t)\, dt = \frac{1}{2\pi} \int_{\omega=-\pi/\Delta}^{\omega=+\pi/\Delta} \left| F\left(\frac{e^{j\omega\Delta} - 1}{\Delta}\right) \right|^2 d\omega$$

provided the region of convergence of $F(\gamma)$ includes $|1 + \Delta\gamma| = 1$. *Hints:* (a) Use Theorem 4.4.1(xii) with the contour C as the circle, $|1 + \Delta u| = 1$, to find $T\{f^2(t)\}$. (b) Make the change of variable, $u = (e^{j\omega\Delta} - 1)/\Delta$ and use the final value theorem.

15. Find the inverse transform of the following complex functions, for $\Delta = 0.1$
(a) $\dfrac{1}{\gamma + 1}$
(b) $\dfrac{1}{(\gamma + 1)^2}$
(c) $\dfrac{1}{\gamma^2 + 0.1\gamma + 1}$
(d) $\dfrac{\gamma}{(\gamma + 10)^2}$

16. Consider a general linear state space system of the form
$$px = Ax + Bu, \quad x(0) = x_0$$
$$y = Cx + Du$$

Show that the solution to this equation is
$$x(t) = E(A, t)x_0 + \int_0^t E(A, t - \tau - \Delta) Bu(\tau)\, d\tau$$
$$y(t) = Cx(t) + Du(t)$$

using two methods:
(a) Direct verification (time domain)
(b) Transform methods (frequency domain)

17. Verify the properties given in Lemma 4.4.1 not treated previously in Problem 4.

18. If $F(\gamma) = \dfrac{\gamma^2 + \gamma - 2}{\gamma^3 + 2\gamma^2 + \gamma}$, find $f(t)$ for $\Delta = 0.5, 0.1,$ and 0.

5

Transfer Functions

5.1 INTRODUCTION

One key feature of the transform techniques described in Chapter 4 is that they convert linear differential or difference equations into algebraic equations. The quantity that links the system input to the system output in the transform domain is commonly known as the system *transfer function*. This will be the subject of the current chapter. Because of the simple algebraic connection between the inputs and outputs through the transfer function, deep insights can be obtained into system behavior by study of the transfer function.

An important concept relating to a system's transfer function is that of poles and zeros. These form an important basis for discussing the properties of systems.

We will address these issues in a unified framework treating both continuous and discrete cases simultaneously while showing their common features. This unified treatment is made possible by the use of the unified notation introduced in Chapter 4. In particular, we will use ρ to denote either d/dt (in continuous time) or δ (in discrete time). We will also use the unified transform variable γ as in Section 4.4. With this notation we need not make any distinction between the continuous and discrete cases. Indeed, almost all the results presented in the remainder of the book hold equally for continuous or discrete time. When this is the case, we will usually not explicitly mention whether sampling is used or not since both cases will be automatically covered (for example, $\Delta = 0$ gives the result for continuous time). In some rare cases, we may wish to quote a result that holds only for continuous time, in which case we will specifically mention this restriction and we will use the operator d/dt and the associated transform variable s. In other cases, we may quote, for completeness, results for discrete time shift operator models. These results do not apply to the limiting continuous time case, and thus we will indicate this explicitly and use the shift operator q and the associated transform variable z.

5.2 TRANSFER FUNCTION

To provide motivation for the concept of transfer function, consider the following general linear state space model:

$$\rho x = Ax + Bu, \quad x(0) = x_0 \tag{5.2.1}$$

and

$$y = Cx \tag{5.2.2}$$

Taking transforms, the solution of these equations can be readily seen to be

$$X(\gamma) = (\gamma I - A)^{-1}\{(1 + \Delta\gamma)x_0 + BU(\gamma)\} \tag{5.2.3}$$

$$Y(\gamma) = CX(\gamma) \tag{5.2.4}$$

where $X(\gamma)$, $Y(\gamma)$, and $U(\gamma)$ denote the transform of $x(t)$, $y(t)$, and $u(t)$ respectively.

From (5.2.3), (5.2.4) we have that for zero initial conditions

$$Y(\gamma) = G(\gamma)U(\gamma) \tag{5.2.5}$$

where

$$G(\gamma) = C(\gamma I - A)^{-1}B \tag{5.2.6}$$

The quantity $G(\gamma)$ is known as the system transfer function and completely defines the input–output behavior for zero initial conditions.

Note also that the transfer function is independent of the selection of state variables, x. In particular, if $x' = Tx$, where T is a nonsingular transformation matrix, then the state space model (5.2.1) can be rewritten as:

$$\rho x' = A'x' + B'u \tag{5.2.7}$$

$$y = C'x' \tag{5.2.8}$$

where

$$A' = TAT^{-1} \tag{5.2.9}$$

$$B' = TB \tag{5.2.10}$$

and

$$C' = CT^{-1} \tag{5.2.11}$$

It can then be seen that

$$G(\gamma) = C(\gamma I - A)^{-1}B = C'(\gamma I - A')^{-1}B' \tag{5.2.12}$$

and thus the transfer function is invariant with respect to linear transformations of the state variables.

The transfer function also has a close connection with input–output based models, as we now show. Suppose $y(t)$ and $u(t)$ are related by the input–output model

$$D(\rho)y(t) = N(\rho)u(t) \qquad (5.2.13)$$

where $D(\rho)$ and $N(\rho)$ are polynomial operators in ρ; then, using the result in Problem 13 in Chapter 4 the transfer function relation $Y(\gamma)$ to $U(\gamma)$ is easily seen to be

$$G(\gamma) = D(\gamma)^{-1}N(\gamma) \qquad (5.2.14)$$

In fact, any system that has a finite-dimensional state space model has a rational transfer function matrix, that is, one that can be expressed as a matrix whose entries are ratios of polynomials in γ.

An important observation is that the transfer function is simply the transform of the system response corresponding to the input $U(\gamma) = 1$ with zero initial conditions. This response is called the *impulse response* since, from Table 4.4.1, $U(\gamma) = 1$ implies that $u(t)$ is the unit impulse function; that is, in continuous time $u(t)$ is the Dirac delta function, and in the delta discrete domain $u(k\Delta) = 1/\Delta$ (for $k = 0$) and 0 (otherwise).

It then also follows from (5.2.5) and Theorem 4.4.1(viii) that the system output for zero initial conditions can be expressed as the following convolution:

$$y(t) = \underset{0}{\overset{t+\Delta}{S}} h(t-\tau)u(\tau)\,d\tau \qquad (5.2.15)$$

where $h(\tau)$ is the system impulse response.

5.3 CONNECTION BETWEEN DISCRETE AND CONTINUOUS TRANSFER FUNCTIONS

We have seen in section (3.4.2) that there is a simple connection between a continuous time state space model and the corresponding discrete time state space model obtained by sampling the output with zero-order hold input. We will now see how this idea extends to the case where transfer function models are used to describe the system.

Lemma 5.3.1

Consider a continuous time system having scalar transfer function $G(s)$. The corresponding discrete time transfer function obtained by sampling the output with a zero-order hold input is:

(a) **Shift Domain**

$$G'(z) = \frac{z-1}{z} \mathbf{Z}\left[\mathbf{L}^{-1}\left\{\frac{1}{s}G(s)\right\}\right] \qquad (5.3.1)$$

(b) δ Domain

$$G''(\gamma) = \frac{\gamma}{1 + \gamma\Delta} T\left[L^{-1}\left\{\frac{1}{s}G(s)\right\}\right] \tag{5.3.2}$$

Proof: We will only prove part (b), since part (a) follows similarly. As pointed out at the end of the previous section, the discrete transfer function is simply the transform of the system response to a discrete impulse of the form

$$\bar{u}(t) = \begin{cases} \frac{1}{\Delta}, & 0 \le t < \Delta \\ 0, & \text{otherwise} \end{cases} \tag{5.3.3}$$

since we have a zero-order hold input. Using the convolution theorem, Theorem 4.4.1(viii), we have that the response to the input (5.3.3) is given as in (5.2.15) by

$$\bar{y}(t) = \int_0^t h(t - \tau)\bar{u}(\tau)\, d\tau$$

$$= \frac{1}{\Delta} \int_{t-\Delta}^t h(\tau)\, d\tau \tag{5.3.4}$$

where $h(t)$ is the continuous time impulse response of the system; that is,

$$h(t) = L^{-1}\{G(s)\} \tag{5.3.5}$$

To find the discrete transfer function, we simply need to find the discrete transform of $\bar{y}(t)$:

$$G''(\gamma) = T\{\bar{y}(t)\}$$

$$= T\left\{\frac{1}{\Delta}\int_{t-\Delta}^t h(\tau)\, d\tau\right\}$$

$$= E(\gamma, -\Delta) T\left\{\frac{1}{\Delta}\int_t^{t+\Delta} h(\tau)\, d\tau\right\}, \quad \text{using Theorem 4.4.1(vi)}$$

$$= \frac{1}{1 + \gamma\Delta} T\left\{\delta\left[\int_0^t h(\tau)\, d\tau\right]\right\}$$

$$= \frac{\gamma}{1 + \gamma\Delta} T\left\{\int_0^t h(\tau)\, d\tau\right\}, \quad \text{using Theorem 4.4.1(ii)}$$

$$= \frac{\gamma}{1 + \gamma\Delta} T\left\{L^{-1}\left[\frac{G(s)}{s}\right]\right\}, \quad \text{using Theorem 4.2.1(iii)} \tag{5.3.6}$$

▽▽▽

The result in Lemma 5.3.1 simply says that we first find the step response of the system in the time domain and then take its discrete transform and finally multiply by $\gamma/(1 + \gamma\Delta)$. This has the effect of setting the discrete time step response equal to the sampled continuous time step response. For this reason, the mapping is usually given the name *step invariance*. This methodology is illustrated in Example 5.3.1.

Example 5.3.1

Consider the sampled data RC network of Example 3.4.1. The continuous time transfer function is

$$G(s) = \frac{1}{1 + s\tau} \tag{5.3.7}$$

where $\tau = RC$. (Note that this may be readily established without using a state space description.) The step response of the system is clearly

$$L^{-1}\left\{\frac{1}{s}G(s)\right\} = 1 - e^{-t/\tau} \tag{5.3.8}$$

With sampling period Δ seconds, we have

$$T\{1 - e^{-t/\tau}\} = \frac{1 + \Delta\gamma}{\gamma} - \frac{1 + \Delta\gamma}{\gamma + ((1 - e^{-\Delta/\tau})/\Delta)} \tag{5.3.9}$$

We then have that the discrete time transfer function obtained by the step invariance modeling in the δ domain is

$$G''(\gamma) = \frac{\gamma}{1 + \Delta\gamma}\left\{\frac{1 + \Delta\gamma}{\gamma} - \frac{1 + \Delta\gamma}{\gamma + [(1 - e^{-\Delta/\tau})/\Delta]}\right\}$$

$$= 1 - \frac{\gamma}{\gamma + [(1 - e^{-\Delta/\tau})/\Delta]}$$

$$= \frac{(1 - e^{-\Delta/\tau})/\Delta}{\gamma + ((1 - e^{-\Delta/\tau})/\Delta)} \tag{5.3.10}$$

In terms of the Z-transform, this transfer function is equivalent to

$$G'(z) = \frac{1 - e^{-\Delta/\tau}}{z - e^{-\Delta/\tau}} \tag{5.3.11}$$

Using either (5.3.10) or (5.3.11), it is clear that this answer is identical to that obtained in Example 3.4.1. ▽▽▽

The mapping described in Lemma 5.3.1 is tabulated in Table 5.3.1 for commonly occurring continuous time transfer functions. This table allows us to find rapidly the discrete time transfer function corresponding to typical continuous time transfer functions with zero order hold input. Further results of a similar nature can be readily derived on noting that the relationships (5.3.1) and (5.3.2) are linear. Thus, if

$$G(s) = \sum_{i=1}^{n} k_i G_i(s) \tag{5.3.12}$$

then

$$G''(\gamma) = \sum_{i=1}^{n} k_i G_i''(\gamma) \tag{5.3.13}$$

In particular, if we use a partial fraction for $G(s)$, that is, for the case of

TABLE 5.3.1 TABLE OF ZERO-ORDER HOLD DISCRETE TIME EQUIVALENTS

$G(s)$	$G''(\gamma)$
$\dfrac{1}{s}$	$\dfrac{1}{\gamma}$
$\dfrac{1}{s^2}$	$\dfrac{(\Delta/2)\gamma + 1}{\gamma^2}$
$\dfrac{1}{s^3}$	$\dfrac{\frac{1}{6}(\Delta\gamma)^2 + \Delta\gamma + 1}{\gamma^3}$
$\dfrac{1}{s+a}$	$\dfrac{\alpha}{a(\gamma+\alpha)}, \quad \alpha = \dfrac{1-e^{-a\Delta}}{\Delta}$
$\dfrac{1}{(s+a)^2}$	$\dfrac{(\alpha - ae^{-a\Delta})\gamma + \alpha^2}{a^2(\gamma+\alpha)^2}, \quad \alpha = \dfrac{1-e^{-a\Delta}}{\Delta}$
$\dfrac{\omega^2}{s^2+\omega^2}$	$\dfrac{[(\Delta/2)\gamma + 1]\phi}{\gamma^2 + \Delta\phi\gamma + \phi}, \quad \phi = \dfrac{2 - 2\cos\omega\Delta}{\Delta^2}$
$\dfrac{s}{s^2+\omega^2}$	$\dfrac{\text{sinc}(\omega\Delta)\gamma}{\gamma^2 + \Delta\phi\gamma + \phi}, \quad \text{sinc}(x) = \dfrac{\sin x}{x}$
	$\phi = \dfrac{2 - 2\cos\omega\Delta}{\Delta^2}$
$\dfrac{\omega}{(s+a)^2 + \omega^2}$	$\dfrac{\left[\dfrac{\omega(1-e^{-a\Delta}\cos\omega\Delta) - ae^{-a\Delta}\sin\omega\Delta}{\Delta}\right]\gamma + \omega\left[\dfrac{1 - 2e^{-a\Delta}\cos\omega\Delta + e^{-2a\Delta}}{\Delta^2}\right]}{(a^2+\omega^2)\left(\gamma^2 + \left[\dfrac{2 - 2e^{-a\Delta}\cos\omega\Delta}{\Delta}\right]\gamma + \left[\dfrac{1 - 2e^{-a\Delta}\cos\omega\Delta + e^{-2a\Delta}}{\Delta^2}\right]\right)}$
$\dfrac{s}{(s+a)^2 + \omega^2}$	$\dfrac{[e^{-a\Delta}(\sin\omega\Delta/\omega\Delta)]\gamma}{\gamma^2 + \left[\dfrac{2 - 2e^{-a\Delta}\cos\omega\Delta}{\Delta}\right]\gamma + \left[\dfrac{1 - 2e^{-a\Delta}\cos\omega\Delta + e^{-2a\Delta}}{\Delta^2}\right]}$

nonrepeated roots

$$G(s) = \sum_{i=1}^{n} \frac{r_i}{s - p_i} \tag{5.3.14}$$

then using (5.3.13) and entry 4 from Table 5.3.1 we have

$$G''(\gamma) = \sum_{i=1}^{n} \left\{ \frac{r_i}{p_i} \frac{[(e^{p_i\Delta} - 1)/\Delta]}{\gamma + [(1 - e^{p_i\Delta})/\Delta]} \right\} \tag{5.3.15}$$

5.4 POLES AND ZEROS

In the previous sections we have seen that linear time invariant finite-dimensional systems can be described by rational transfer function matrices. We begin with the case of a scalar input and a scalar output. In this case, the transfer function matrix also reduces to a scalar quantity. The zeros of the numerator and denominator of this quantity are called the zeros and poles of the transfer function, respectively.

More specifically we have:

Definition 5.4.1. Poles and Zeros (Scalar Case)

Given a scalar rational transfer function, $G(\gamma) = N(\gamma)/D(\gamma)$, where N is of degree m and D is of degree n, the n *poles* of the transfer function are the n roots of the equation $D(\gamma) = 0$, and the m *zeros* of the transfer function are the m roots of the equation $N(\gamma) = 0$. ▽▽▽

We next give some interpretations to poles and zeros. For simplicity, we will first treat the case where no cancellations occur between the poles and zeros.

The poles of a transfer function can be interpreted as follows. Suppose $G(\gamma)$ has a pole at $\gamma = p$; then, for appropriate initial conditions and with zero input, $Y(\gamma) = (1 + \Delta\gamma)/(\gamma - p)$. In other words, the poles of a system describe outputs that may occur when there is no input.

Example 5.4.1 illustrates this for a continuous time case.

Example 5.4.1

Consider the continuous time system described by

$$\dot{x} = \begin{bmatrix} -1 & 2 \\ 1 & 0 \end{bmatrix} x + \begin{bmatrix} 1 \\ 0 \end{bmatrix} u \qquad (5.4.1)$$

$$y = \begin{bmatrix} 1 & 1 \end{bmatrix} x \qquad (5.4.2)$$

It can be shown that this system has transfer function

$$G(s) = \frac{s+1}{s^2 + s - 2} \qquad (5.4.3)$$

From (5.4.3), we note that $G(s)$ has a zero at $s = -1$ and poles at $s = 1, -2$. Then, with $u(t) \equiv 0$, the transform of the output $Y(s)$ is given by

$$Y(s) = C(sI - A)^{-1} x_0 \qquad (5.4.4)$$

[see Equation (4.2.16)].

Then, with $x_0 = [0.5 \ 0.5]^T$, we can show that $Y(s) = 1/(s - 1)$; that is, $y(t) = e^t$. Also, with $x_0 = [2 \ -1]^T$, we can show that $Y(s) = 1/(s + 2)$; that is, $y(t) = e^{-2t}$. ▽▽▽

Conversely, if $G(\gamma)$ has a zero at $\gamma = \xi$, then for appropriate initial conditions the input with transform $U(\gamma) = (1 + \Delta\gamma)/(\gamma - \xi)$ gives no output (see Problem 3). In other words, the zeros of a system describe inputs that may give no output.

Example 5.4.1. (Revisited)

Consider the case where $u(t) = e^{-t}$; that is, $U(s) = 1/(s + 1)$ and $x_0 = [\frac{1}{2} \ -\frac{1}{2}]^T$. It then follows that

$$C(sI - A)^{-1} BU(s) = \frac{1}{s^2 + s - 2} \qquad (5.4.5)$$

and
$$C(sI - A)^{-1}x_0 = \frac{-1}{s^2 + s - 2} \tag{5.4.6}$$
and so
$$Y(s) = C(sI - A)^{-1}BU(s) + C(sI - A)^{-1}x_0$$
$$= 0 \tag{5.4.7}$$
That is,
$$y(t) = 0$$

▽▽▽

The previous statements regarding inputs, outputs, and their relationship to poles and zeros can also be understood in the following way. A pole at $\gamma = p$ implies that with zero input there can be an output of the form $Y(\gamma) = 1/(\gamma - p)$. This implies that the gain of the mode $1/(\gamma - p)$ is infinite. More precisely, if the input contains a term of the form $E(p, t)$, then the output will contain a term of the form $tE(p, t)$. Thus the gain associated with the term $E(p, t)$ tends to infinity as $t \to \infty$. Conversely, a zero at $\gamma = \xi$ implies that the gain of the system to a mode $U(\gamma) = 1/(\gamma - \xi)$ will be zero. In other words, if the input contains a term of the form $E(\xi, t)$, the output will not contain a term of that form.

So far we have considered the poles and zeros as real numbers, while not explicitly excluding complex poles or zeros. When $G(\gamma)$ has complex zeros or poles, these must occur in complex conjugate pairs; that is,

$$G(\xi) = 0 \Leftrightarrow G(\xi^*) = 0 \tag{5.4.8}$$
and
$$G^{-1}(p) = 0 \Leftrightarrow G^{-1}(p^*) = 0 \tag{5.4.9}$$
where * denotes complex conjugate.

In this case, we usually group the complex zeros or poles in conjugate pairs and note (from Chapter 4) that they correspond to oscillating modes with an exponential envelope. Example 5.4.2 illustrates the kind of system behavior obtained with complex poles and zeros.

Example 5.4.2

Consider a continuous time system described by

$$\dot{x} = \begin{bmatrix} -1 & -1 & -1 \\ 1 & 0 & 0 \\ 0 & 1 & 0 \end{bmatrix} x + \begin{bmatrix} 1 \\ 0 \\ 0 \end{bmatrix} u \tag{5.4.10}$$

$$y = \begin{bmatrix} 1 & 2 & 2 \end{bmatrix} x \tag{5.4.11}$$

The transfer function can be shown to be (see Problem 4)

$$G(s) = C(sI - A)^{-1}B = \frac{s^2 + 2s + 2}{s^3 + s^2 + s + 1} \tag{5.4.12}$$

the poles are at $s = -1, \pm j$ and the zeros are at $s = -1 \pm j$. Consider the

response of the system to an input $u(t) = e^{-t}\sin(t)$ with initial conditions
$x_0 = [-4/5, 3/5, -1/5]^T$.
Then $U(s) = 1/(s^2 + 2s + 2)$ and

$$C(sI - A)^{-1}x_0 = \frac{-1}{s^3 + s^2 + s + 1}$$

It thus follows that $Y(s)$ [and hence $y(t)$] is identically zero. Similarly, if $u(t) = e^{-t}\cos(t)$, $U(s) = (s + 1)/(s^2 + 2s + 2)$ and $x_0 = [2/5, -1/5, 2/5]^T$, then $Y(s) = 0$ (see Problem 5). Thus we see that the system gain to any mode of the form $u(t) = e^{-t}\cos(t + \phi)$ is zero.

In considering the poles at $s = \pm j$, let $u(t) = \sin(t)$. Then $U(s) = 1/(s^2 + 1)$ and, for zero initial conditions,

$$Y(s) = \frac{s^2 + 2s + 2}{(s+1)(s^2+1)^2} = \frac{1/4}{s+1} + \frac{(-1/4s^3 + 1/4s^2 + 1/4s + 7/4)}{(s^2+1)^2}$$

$$= \frac{1/4}{s+1} + \left(-\frac{3}{4}\right)\frac{s^2-1}{(s^2+1)^2} + \frac{2s}{s^2+1} - \frac{1}{4}\frac{s}{s^2+1} + \frac{1}{s^2+1}$$

(5.4.13)

It then follows that

$$y(t) = \tfrac{1}{4}e^{-t} - \tfrac{3}{4}t\cos(t) + \tfrac{1}{4}t\sin(t) - \tfrac{1}{4}\cos(t) + \sin(t) \quad (5.4.14)$$

Note that the output $y(t)$ contains terms of the form $t\sin(t)$ and $t\cos(t)$; that is, the gain to an input, $u(t) = \sin(t)$, goes to infinity.

Another property of the poles and zeros of a transfer function relates to the state space modeling of a system. For a general state space model, we have as in (5.2.6) that

$$G(\gamma) = C(\gamma I - A)^{-1}B$$
$$= \frac{C\,\mathrm{adj}(\gamma I - A)B}{\det(\gamma I - A)} \quad (5.4.15)$$

where $\det(\cdot)$ denotes determinant and $\mathrm{adj}(\cdot)$ denotes adjoint. In (5.4.15), $\det(\gamma I - A)$ is a polynomial in γ of degree n, where $A \in \mathbb{R}^{n \times n}$ and so the poles of G are the eigenvalues of A. $C\,\mathrm{adj}(\gamma I - A)B$ is also a polynomial of degree less than n and clearly gives the zeros of $G(\gamma)$. However, the relationship between C, A, and B and the zeros of $G(\gamma)$ can only be simply stated in special cases.

We next briefly extend Definition 5.4.1 to the multiinput, multioutput case. For simplicity, we treat only the square case where the number of inputs is equal to the number of outputs.

Definition 5.4.2. Poles and Zeros (Square Multivariable Case)

Given a square multivariable transfer function $G(\gamma) = D(\gamma)^{-1}N(\gamma)$, the *poles* of the transfer function are the roots of the equation $\det D(\gamma) = 0$, and the *zeros* of transfer function are the roots of the equation $\det N(\gamma) = 0$.

5.5 POLES AND ZEROS OF SAMPLED DATA SYSTEMS

In the examples in the previous section we used continuous time systems. We next turn to an in-depth study of the poles and zeros of sampled data systems.

The relationship between the poles of a continuous time system and the sampled data poles, when a zero-order hold input is used, is straightforward, as is shown in the following result.

Lemma 5.5.1. Poles of Sampled Data Systems

A continuous time system with poles at $s = p_i$, $i = 1, \ldots, n$, when sampled with period Δ gives a discrete time system with poles at $\gamma = (e^{p_i \Delta} - 1)/\Delta$ or, for Z-domain models,

$$z = e^{p_i \Delta}, \quad i = 1, \ldots, n$$

Proof: See Problem 6.   ▽▽

That this should be the case is intuitively reasonable, as we will now show. The poles of a system correspond to modes in a system that occur when there are no inputs. The case of no input to a system is identical regardless of whether the input is continuous or whether the input is obtained from a zero-order hold, as is the case for discrete systems. Thus, the sampled data modes with zero input must be simply the sampled version of the corresponding continuous time modes with zero input.

The situation regarding the zeros of a sampled data system is, however, much more complex. It is certainly *not* true, in general, that zeros at $s = \xi_i$, $i = 1, \ldots, m$, map to zeros at $\gamma = (e^{\xi_i \Delta} - 1)/\Delta$, as is illustrated by Example 5.5.1.

Example 5.5.1

Consider a continuous time system with transfer function

$$G(s) = \frac{K(s+1)}{s^2} \tag{5.5.1}$$

Using the methods of Section 5.3, the discrete transfer function in delta form is given by

$$G_d(\gamma) = \frac{K\{[1 + (\Delta/2)]\gamma + 1\}}{\gamma^2} \tag{5.5.2}$$

Thus the continuous time system has a zero at $s = -1$, while the discrete time zero is at

$$\gamma = \frac{-1}{1 + (\Delta/2)} \neq \frac{e^{-\Delta} - 1}{\Delta}$$

▽▽

The fact that the zeros do not map in the same way as the poles can be explained as follows. The continuous time zeros correspond to input modes that will excite no output (if the initial conditions are appropriate). Because the system input is from a zero-order hold in the discrete time case, it can never reproduce the continuous time zero modes (save for the very special case of a zero at $s = 0$; see Problem 12).

The preceding argument suggests why the zeros do not map in exactly the same manner as the poles. However, it also suggests that it is rare (no matter what the initial conditions are) that the continuous time output should be zero for any nonzero, zero-order hold discrete input. While this is also true in general, it does still allow for the continuous output to be zero *at the sampling instants*, thus giving discrete time zeros. This phenomenon is illustrated by the following reexamination of Example 5.5.1.

Example 5.5.1. (Revisited)

The continuous time system (5.4.1) can be represented in state space form as

$$\dot{x}(t) = \begin{bmatrix} 0 & 0 \\ 1 & 0 \end{bmatrix} x(t) + \begin{bmatrix} 1 \\ 0 \end{bmatrix} u(t) \tag{5.5.3}$$

$$y(t) = K\begin{bmatrix} 1 & 1 \end{bmatrix} x(t) \tag{5.5.4}$$

The response of this system for a zero-order hold input with discrete input $u_d(k\Delta)$, with $\Delta = 0.5$, is described by

$$\delta x(k\Delta) = \begin{bmatrix} 0 & 0 \\ 1 & 0 \end{bmatrix} x(k\Delta) + \begin{bmatrix} 1 \\ 0.25 \end{bmatrix} u_d(k\Delta) \tag{5.5.5}$$

$$y(k\Delta) = K\begin{bmatrix} 1 & 1 \end{bmatrix} x(k\Delta) \tag{5.5.6}$$

For $\Delta = 0.5$, the discrete time zero is at $\gamma = -0.8$. Consider then the discrete time input with transform

$$U(\gamma) = -\frac{1 + \Delta\gamma}{\gamma + 0.8} \tag{5.5.7}$$

That is, $u_d(t)$ has sample values

$$u_{d,k} = -\{1, 0.6, 0.36, 0.216, \dots\} \tag{5.5.8}$$

and

$$x(0)^T = \begin{bmatrix} 1.25 & -1.25 \end{bmatrix}$$

From (5.5.3) and noting that $u(t) = -1$ for $0 \le t < \Delta$, we have

$$x_1(t) = x_1(0) + \int_0^t u(\tau)\, d\tau = +1.25 - t \tag{5.5.9}$$

$$x_2(t) = x_2(0) + \int_0^t x_1(\tau)\, d\tau = -1.25 + 1.25t - 0.5t^2 \tag{5.5.10}$$

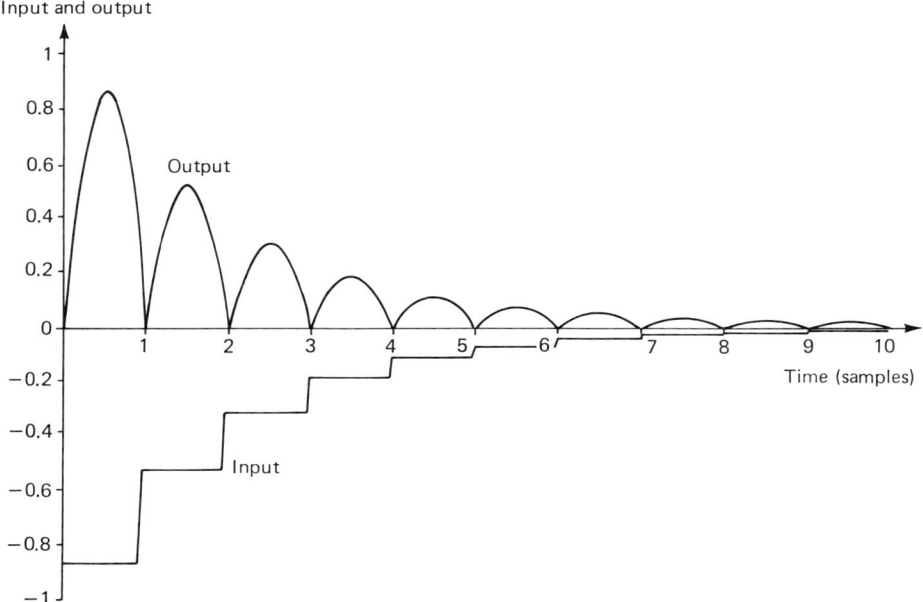

Figure 5.5.1 Sampled version of continuous zero.

and

$$y(t) = K(-0.5t^2 + 0.25t), \quad \text{for } 0 \le t \le \Delta = 0.5 \tag{5.5.11}$$

Thus $y(\Delta) = 0$, $x_1(\Delta) = 0.75$, and $x_2(\Delta) = -0.75$. Thus we note that $x(\Delta) = 0.6x(0)$ and $u_d(\Delta) = 0.6u_d(0)$. Thus the cycle repeats itself indefinitely and the sampled response $y(k\Delta)$ is always zero. Figure 5.5.1 shows the continuous time input and output under the preceding conditions with $K = 32$. ▽▽▽

A slightly more perplexing property of sampled data systems is that a continuous system having n poles and $m(< n)$ zeros, in general, gives a discrete time system with n poles and $(n - 1)$ zeros. A heuristic reason as to why this should be follows. Consider a scalar (continuous or discrete) rational transfer function $G(\gamma) = N(\gamma)/D(\gamma)$. Then it follows that the (generalized) derivative of the step response of $G(\gamma)$, evaluated at $t = 0^+$, is zero if and only if m [= degree of $N(\gamma)$] $\le n - 2$, where n = degree of $D(\gamma)$. (This result may be established using the initial value theorem and the transform of a derivative rule given in Theorem 4.4.1.) Thus, if a continuous time system has $m \le n - 2$, this implies that $\dot{s}(0) = s(0) = 0$, where $s(t)$ denotes the continuous step response. For a discrete time system, however, $m \le n - 2$ implies $\delta s_d(0) = s_d(0) = 0$, which in turn implies that $s_d(\Delta) = 0$, where $s_d(k)$ denotes the discrete step response. It is clear that if the sampled continuous time step response is equal to the continuous step response, that is, $s_d(k\Delta) = s(k\Delta)$, then $m \le n - 2$ (sampled data degrees) implies that the continuous time step response is exactly zero at $t = \Delta$. We submit that while this is possible it rarely occurs in practice (unless the continuous time system includes pure time delays). See also Problems 8 and 9.

The preceding discussion is illustrated in Example 5.5.2.

Example 5.5.2

Let

$$G(s) = \frac{-6(s-1)}{(s+1)(s+2)(s+3)} \tag{5.5.12}$$

Then the response of the system is

$$\frac{1}{s}G(s) = \frac{1}{s} - \frac{6}{s+1} + \frac{9}{s+2} - \frac{4}{s+3} \tag{5.5.13}$$

$$= L\{1 - 6e^{-t} + 9e^{-2t} - 4e^{-3t}\} \tag{5.5.14}$$

$$= L\{(1-e^{-t})^2(1-4e^{-t})\} \tag{5.5.15}$$

From (5.5.15), it is clear that the step response is zero only for $t = 0$ or $t = \ln(4)$. Thus, unless the sampling period is exactly $\ln(4)$ seconds, we expect $G_d(\gamma)$ to have two zeros and three poles. Otherwise, the discrete step response and the continuous step response would have to be zero at $t = \Delta$. In fact, we can show that

$$G_d(\gamma) = \frac{\gamma^2(6p_1 - 9p_2 + 4p_3) + \gamma(-3p_1p_2 + 10p_1p_3 - 5p_2p_3) + p_1p_2p_3}{(\gamma + p_1)(\gamma + p_2)(\gamma + p_3)} \tag{5.5.16}$$

where

$$p_1 = \frac{1-e^{-\Delta}}{\Delta}, \quad p_2 = \frac{1-e^{-2\Delta}}{\Delta}, \quad p_3 = \frac{1-e^{-3\Delta}}{\Delta}$$

Thus $G_d(\gamma)$ will always have two zeros except when

$$6p_1 - 9p_2 + 4p_3 = \frac{1}{\Delta}[6(1-e^{-\Delta}) - 9(1-e^{-2\Delta}) + 4(1-e^{-3\Delta})]$$

$$= \frac{1}{\Delta}[1 - 6e^{-\Delta} + 9e^{-2\Delta} - 4e^{-3\Delta}]$$

$$= 0 \tag{5.5.17}$$

Comparing (5.5.17) with (5.5.14), we confirm that for $\Delta \neq \ln(4)$ the sampled data system has two zeros. ▽▽▽

Thus we see that continuous time systems generally lead to sampled data models with relative degree 1 (relative degree = degree of denominator − degree of numerator), regardless of the relative degree of the continuous time system. We will refer to the extra zeros in discrete time systems as *sampling zeros*. (Note that the classification of zeros as sampling zeros or zeros due to continuous zeros is not always straightforward. Lemma 5.5.2 will clarify this connection.)

The nature of the sampling zeros is illustrated in Example 5.5.3.

Example 5.5.3

Consider a continuous time system with transfer function

$$G(s) = \frac{\frac{1}{2}}{(s+1)(s+\frac{1}{2})} = \frac{-1}{s+1} + \frac{1}{s+\frac{1}{2}} \tag{5.5.18}$$

One possible state space model for this system is

$$\dot{x} = \begin{bmatrix} -1 & 0 \\ 0 & -0.5 \end{bmatrix} x + \begin{bmatrix} 1 \\ 1 \end{bmatrix} u \tag{5.5.19}$$

and

$$y = \begin{bmatrix} -1 & 1 \end{bmatrix} x \tag{5.5.20}$$

With zero-order hold and sampling period of 1 sec, the discrete time model can be shown to be

$$\delta x(k\Delta) = \begin{bmatrix} (e^{-1}-1) & 0 \\ 0 & e^{-0.5}-1 \end{bmatrix} x(k\Delta) + \begin{bmatrix} (1-e^{-1}) \\ 2(1-e^{-1/2}) \end{bmatrix} u(k\Delta) \tag{5.5.21}$$

$$y(k\Delta) = \begin{bmatrix} -1 & 1 \end{bmatrix} x(k\Delta) \tag{5.5.22}$$

The discrete time transfer function is then

$$G_d(\gamma) = \begin{bmatrix} -1 & 1 \end{bmatrix} \begin{bmatrix} \dfrac{1}{\gamma + (1-e^{-1})} & 0 \\ 0 & \dfrac{1}{\gamma + (1-e^{-0.5})} \end{bmatrix} \begin{bmatrix} (1-e^{-1}) \\ 2(1-e^{-1/2}) \end{bmatrix}$$

$$= \frac{\gamma(1 - 2e^{-1/2} + e^{-1}) + (1 - e^{-1} - e^{-0.5} + e^{-1.5})}{\gamma^2 + (2 - e^{-1} - e^{-0.5})\gamma + (1 - e^{-1} - e^{-0.5} + e^{-1.5})} \tag{5.5.23}$$

(Of course, this result can be obtained directly from Table 5.3.1.) Thus this transfer function has a zero at

$$\gamma = -1.6065 \tag{5.5.24}$$

Figure 5.5.2 shows the continuous time response for a zero-order hold version of a discrete input corresponding to the mode described in (5.5.24), that is, with delta transform,

$$U(\gamma) = \frac{1 + \gamma\Delta}{\gamma + 1.6065} \tag{5.5.25}$$

with output scaled and appropriate initial conditions. ▽▽▽

As illustrated by the previous examples, sampling zeros do not give zero continuous time output. They give instead an output that is zero at the sampling instants. One way of viewing some sampling zeros is to consider them as oscillatory modes, at half the sampling frequency, with an exponential envelope. If the phase

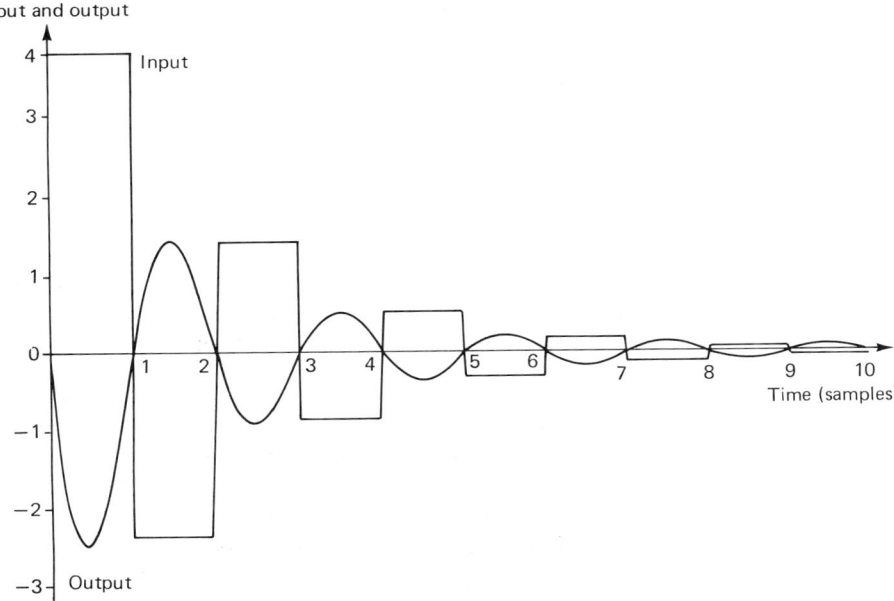

Figure 5.5.2 Sampled data zeros example.

shift to this mode is appropriate, then the output of this mode will be zero when sampled. Example 5.5.3 may be considered in this light, as we now show.

***Example 5.5.3.** (Revisited)

The system input as shown in Figure 5.5.2 can be decomposed using a Fourier series as

$$u(t) = (-0.6065)^k, \quad \text{where } t \in [k\Delta, (k+1)\Delta] \tag{5.5.26}$$

$$= e^{-0.5t} \begin{Bmatrix} e^{0.5(t-[t])}, & \text{if } [t] \text{ is even} \\ -e^{0.5(t-[t])}, & \text{if } [t] \text{ is odd} \end{Bmatrix} \tag{5.5.27}$$

$$= e^{-0.5t} \sum_{\substack{n \text{ odd} \\ n=1}}^{\infty} \left\{ \left(\frac{1 + \ln(0.5)}{0.5 + n^2 \pi^2} \right) (0.5 \cos(\pi t) + n\pi \sin(n\pi t)) \right\} \tag{5.5.28}$$

Considering the first term, $n = 1$, to dominate since the terms decrease as $1/n$, we have

$$u(t) \approx \text{Re} \left\{ \frac{ke^{-0.5t}}{0.5 + j\pi} e^{-j\pi t} \right\} \tag{5.5.29}$$

The steady-state continuous response of $G(s)$ to the modes described in (5.5.29) is

$$y_{ss}(t) \approx \text{Re}\left\{\frac{k}{0.5 + j\pi}G(-0.5 - j\pi)e^{-(0.5+j\pi)t}\right\} \quad (5.5.30)$$

$$= k\,\text{Re}\left\{\frac{e^{-(0.5+j\pi)t}}{(0.5 + j\pi)(0.5 - j\pi)(-j\pi)}\right\} \quad (5.5.31)$$

$$= k'e^{-0.5t}\sin \pi t \quad (5.5.32)$$

Thus we see that, when sampled, the steady-state response due to this mode is zero. [Note that, in this case, this happens to be exactly true; in general, however, it will only be approximately true due to the neglect of various terms in (5.5.28)]. ▽▽▽

In most systems, the sampling zeros will lie on the negative real axis, to the left of $-1/\Delta$ in the complex γ plane. One major exception to this occurs when the continuous time system contains complex poles at $s = \sigma \pm j\omega$, where folding or aliasing occurs. Aliasing was introduced in Section 3.2.3 in the context of sampled continuous time signals. We use the term aliasing here to denote situations where the mode corresponding to a pole at $s = \sigma \pm j\omega$ is folded, that is, when $\omega\Delta > \pi$. As noted in Section 3.2.3, aliasing is undesirable and in practice we would normally raise the sampling frequency so that aliasing does not occur. Example 5.5.4 illustrates how complex sampling zeros may arise when folding occurs.

*Example 5.5.4

Consider a continuous time system with transfer function

$$G(s) = \frac{1}{s(s^2 + 1)} \quad (5.5.33)$$

When sampled with period Δ seconds, this system gives the following discrete time transfer function:

$$G_d(\gamma) = \frac{(1 - \text{sinc}(\Delta))\gamma^2 + 2\Delta C(\Delta)\gamma + 2C(\Delta)}{\gamma(\gamma^2 + 2\Delta C(\Delta)\gamma + 2C(\Delta))} \quad (5.5.34)$$

where

$$\text{sinc}(\Delta) = \frac{\sin(\Delta)}{\Delta} \quad \text{and} \quad C(\Delta) = \frac{1 - \cos \Delta}{\Delta^2}$$

It follows that the discrete time (sampling) zeros will be complex if and only if

$$4\Delta^2 C^2(\Delta) < 8C(\Delta)(1 - \text{sinc}(\Delta)) \quad (5.5.35)$$

or, equivalently,

$$-\Delta(1 + \cos \Delta) + 2 \sin \Delta < 0 \quad (5.5.36)$$

It follows that the sampling zeros are complex if and only if [from (5.5.36)] $\Delta > \pi$. For $\Delta \leq \pi$, we can show that $G_d(\gamma)$ has two real zeros, $\gamma = -\xi_1(\Delta), -\xi_2(\Delta)$, where $\xi_1(\Delta)$ and $\xi_2(\Delta) > 1/\Delta$ for all $\Delta < \pi$. ▽▽▽

It turns out that the classification of the zeros of sampled data systems may be easily performed when rapid sampling is used. Lemma 5.5.2 shows how this may be done.

Lemma 5.5.2

Consider a continuous time scalar transfer function

$$G(s) = \frac{K\prod_{i=1}^{m}(s - \xi_i)}{\prod_{i=1}^{n}(s - p_i)}$$

where $m < n$; then the corresponding discrete time delta transfer function has $\leq (n - 1)$ zeros, ξ'_i, that satisfy:

(i)
$$\lim_{\Delta \to 0} \{\xi'_i\} = \lim_{\Delta \to 0} \frac{e^{\xi_i \Delta} - 1}{\Delta} = \xi_i, \quad i = 1 \ldots m \quad (5.5.37)$$

(ii)
$$\lim_{\Delta \to 0} \{(\gamma - \xi'_{m+1})(\gamma - \xi'_{m+2}) \ldots (\gamma - \xi'_{n-1})\} = \Delta^{-(r-1)} B_r(\Delta\gamma) \quad (5.5.38)$$

where $r = n - m$, and $B_r(x)$ denotes a polynomial of degree $(r - 1)$. The first few values of B_r are given next:

$$B_1(\Delta\gamma) = 1$$
$$B_2(\Delta\gamma) = \Delta\gamma + 2$$
$$B_3(\Delta\gamma) = (\Delta\gamma)^2 + 6(\Delta\gamma) + 6$$
$$B_4(\Delta\gamma) = (\Delta\gamma)^3 + 14(\Delta\gamma)^2 + 36(\Delta\gamma) + 24$$
$$B_5(\Delta\gamma) = (\Delta\gamma)^4 + 30(\Delta\gamma)^3 + 150(\Delta\gamma)^2 + 240(\Delta\gamma) + 120$$

and
(iii)
$$\lim_{\Delta \to 0} \{\xi'_i\} = -\infty, \quad \text{for } i = m + 1, \ldots, n - 1$$

Proof: A full proof of this result is given in Åström, Hagander, and Sternby (1984). We will motivate the result by establishing the convergence of the discrete (delta domain) transfer function to the continuous time transfer function.

Suppose the underlying continuous system is described in state space form as

$$\dot{x} = Ax + Bu$$
$$y = Cx$$

The continuous time transfer function is

$$G(s) = C(sI - A)^{-1}B$$

The corresponding delta operator discrete time model is

$$\delta x = Fx + Gu$$
$$y = Cx$$

where

$$F = \frac{e^{A\Delta} - I}{\Delta}$$

$$G = \frac{1}{\Delta}\int_0^{\Delta} e^{A\tau} B\, d\tau$$

Hence the discrete transfer function is

$$G'(\gamma) = C(\gamma I - F)^{-1} G$$

Then

$$\lim_{\Delta \to 0} G' = \lim_{\Delta \to 0} C\left(\gamma I - \frac{e^{A\Delta} - I}{\Delta}\right)^{-1} \frac{1}{\Delta}\int_0^{\Delta} e^{A\tau}\, d\tau B$$
$$= C(\gamma I - A)^{-1} B$$
$$= G_c(\gamma)$$

▽▽▽

The key point of the preceding theorem is that with fast sampling the continuous time zeros ξ_i are approximately mapped in the delta domain to $(e^{\xi_i \Delta} - 1)/\Delta$, and the extra zeros introduced by the sampling process are in the far left half-plane. This is apparent when we use a delta operator (or delta transform) based model, but is obscured when we use a shift operator (or Z-transform) based model. Example 5.5.5 illustrates these points.

Example 5.5.5

Consider a continuous time system with transfer function

$$G(s) = \frac{s^2 + s + 1}{s(s + 1)(s^2 + 4)} \tag{5.5.39}$$

The poles and zeros of the corresponding discrete time models for $\Delta = 0.0, 0.1, 0.2, \ldots, 1.0$ are shown in Figure 5.5.3 for a delta-based model and in Figure 5.5.4 for a shift-based model. Note that in the case of the shift operator model the sampling zero shows very little variation with sampling rate, while the remaining poles and zeros vary significantly. On the other hand, in the delta domain the poles and zeros move only slightly, except for the extra sampling zero, which moves into the far left half-plane.

▽▽▽

Lemma 5.5.2 shows a major advantage of delta operator models over shift operator models (as illustrated in Example 5.5.5). For delta models the significant poles and zeros are largely independent of sampling rate and converge to their continuous time counterparts as $\Delta \to 0$. Note that this is also true of the sampling zeros, which go to $-\infty$ as $\Delta \to 0$, which can be interpreted in similar fashion to the

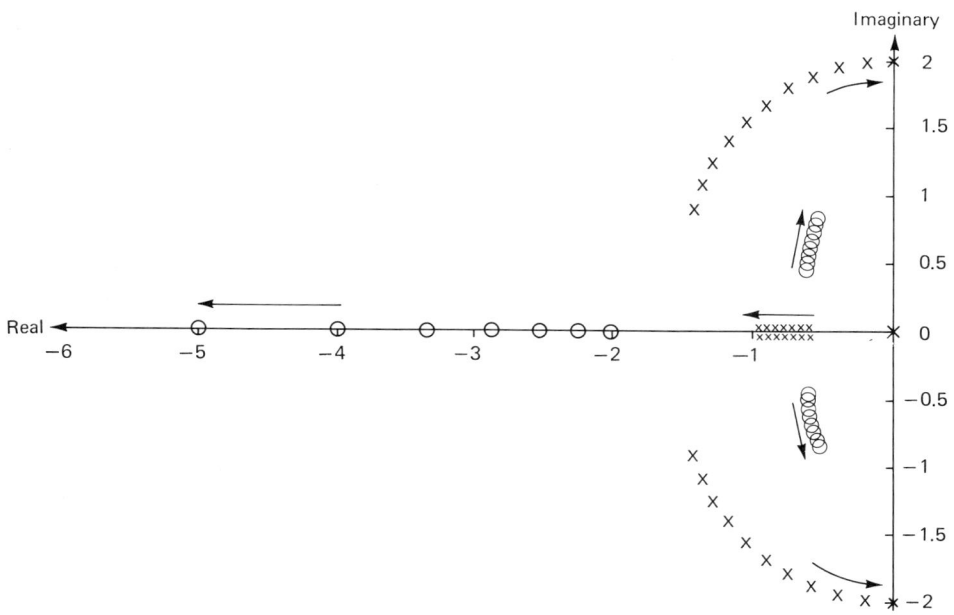

Figure 5.5.3 Delta domain sampled poles and zeros.

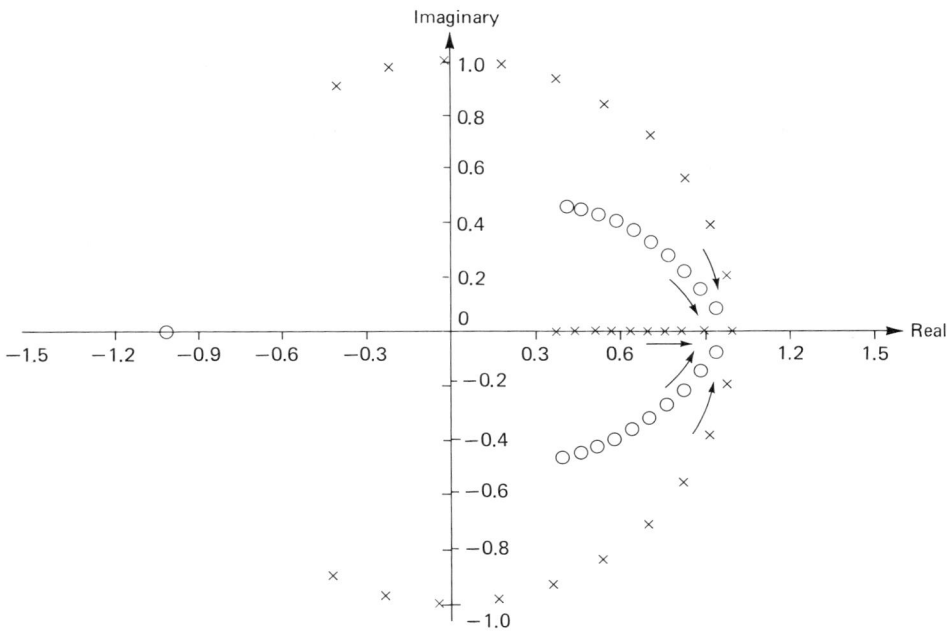

Figure 5.5.4 Z domain sampled poles and zeros.

Sec. 5.5 Poles and Zeros of Sampled Data Systems

continuous time relative degree, which defines zeros at ∞. On the other hand, the poles and zeros of shift operator models are heavily dependent on Δ. Moreover, as $\Delta \to 0$, the significant poles and zeros converge to the point $1 + j0$ independently of the underlying system, and the sampling zeros converge to fixed locations in the Z-plane. These latter Z domain results have, in the past, led to considerable misunderstandings, which are nicely resolved by the delta formulation.

5.6 SUMMARY

The key points made in this chapter were:

- The system transfer function $G(\gamma)$ links the transform of the input to the transform of the output:
$$Y(\gamma) = G(\gamma)U(\gamma) \qquad (5.2.5)$$
- There is a simple relationship between state space models and transfer functions:
$$G(\gamma) = C(\gamma I - A)^{-1}B \qquad (5.2.6)$$
- There is a simple relationship between input–output models and transfer functions:
$$G(\gamma) = D(\gamma)^{-1}N(\gamma) \qquad (5.2.14)$$
- The system transfer function is the transform of the system impulse responses.
- The system output can be computed with zero initial conditions using a convolution:
$$y(t) = \underset{0}{\overset{t+\Delta}{S}} h(t - \tau)u(\tau)\,d\tau \qquad (5.2.15)$$
- The discrete transfer functions corresponding to the continuous time transfer $G(s)$ can be readily computed as:

SHIFT DOMAIN:
$$G'(z) = \frac{z-1}{z} Z\left[L^{-1}\left\{ \frac{1}{s} G(s) \right\} \right] \qquad (5.3.1)$$

DELTA DOMAIN:
$$G''(\gamma) = \frac{\gamma}{1+\gamma\Delta} T\left[L^{-1}\left\{ \frac{1}{s} G(s) \right\} \right] \qquad (5.3.2)$$

- A table of common continuous and discrete transfer functions was given (Table 5.3.1).
- The relationship between continuous and discrete transfer functions is linear; that is, if
$$G(s) = \sum_{i=1}^{n} k_i G_i(s) \qquad (5.3.12)$$

then
$$G''(\gamma) = \sum_{i=1}^{n} k_i G_i''(\gamma) \tag{5.3.13}$$

- The poles and zeros of a transfer function were defined as the zeros of the denominator and numerator, respectively.
- The poles of a transfer function correspond to modes for which the system has infinite gain. They also correspond to modes in the output when there is no input.
- The zeros of a transfer function correspond to modes for which the system has zero gain. They also correspond to modes in the input that produce zero output.
- The poles of a transfer function correspond to the eigenvalues of the A matrix in the associated state space form.
- The poles $\{p_i'\}$ of a sampled data system with zero-order hold input are simply related to the poles of the associated continuous time system (Lemma 5.5.1):

$$p_i' = \frac{e^{p_i \Delta} - 1}{\Delta} \text{ (delta)} \quad \text{or} \quad p_i' = e^{p_i \Delta} \text{(shift)}$$

- The sampled data zeros can be split into two categories: (1) those that can be thought of as being mappings of continuous time zeros, and (2) those that arise from the sampling process (called the sampling zeros).
- The zeros γ_i that arise from the mapping of continuous time zeros ξ_i can be approximated (for small Δ) as follows:

DELTA:

$$\xi_i' \simeq \frac{e^{\xi_i \Delta} - 1}{\Delta} \xrightarrow[\Delta \to 0]{} \xi_i$$

SHIFT:

$$\xi_i' \simeq e^{\xi_i \Delta} \xrightarrow[\Delta \to 0]{} 1 \tag{5.5.37}$$

- The sampling zeros converge (as $\Delta \to 0$) to points in the far left plane for the delta case and to finite locations in the left half plane in the shift case.
- The delta domain transfer function converges (as $\Delta \to 0$) to the underlying continuous time transfer function (Lemma 5.5.2).
- The shift domain transfer function converges (as $\Delta \to 0$) to the following singular result (Lemma 5.5.2):

$$\frac{K(z-1)^m B_{n-m}(z-1)}{(z-1)^n}$$

Thus, in brief, this chapter introduced the concepts of transfer functions, poles, and zeros and studied the connection between the corresponding results for discrete and continuous time systems. The latter connection has been shown to be very simple for delta domain discrete models. Indeed, in modeling discrete time system, we have two alternatives to consider. On the one hand, if we use shift operators, then the system's poles and zeros will vary markedly with the sampling period, but the extra zeros arising from the sampling process will converge to fixed finite locations as $\Delta \to 0$. On the other hand, if we use delta operators, then the system's poles and zeros are approximately invariant with respect to the sampling period, but the extraneous zeros arising from the sampling process vary with Δ and converge to $-\infty$ as $\Delta \to 0$. We firmly believe that the latter option is preferable, since it keeps (nearly) constant the key system poles and zeros, while allowing the extraneous sampling zeros to move off to $-\infty$ as the sampling rate increases, thus giving full consistency with the continuous case.

5.7 REFERENCES

Additional material on transfer functions can be found in any elementary book on systems theory, for example:

CHEN, C. T. (1984) *Linear System Theory and Design*. Holt, Rinehart and Winston, New York.

DI STEFANO, J. J., A. R. STUBBERUD, and I. J. WILLIAMS (1976) *Feedback and Control Systems*, Schaum's Outline Series. McGraw-Hill, New York.

Also, the topic of transfer functions is often associated with elementary courses on circuit theory; see, for example:

NILSSON, J. W. (1983) *Electric Circuits*. Addison-Wesley, Reading, Mass.

Poles and zeros are treated in the preceding books, as well as in specialist texts, for example,

MADDOCK, R. J. (1982) *Poles and Zeros in Electrical and Control Engineering*. Holt, Rinehart and Winston, New York.

Further information on zeros of sampled systems is contained in:

ÅSTRÖM, K. J., P. HAGANDER, and J. STERNBY (1984) "Zeros of sampled systems." *Automatica*, vol. 20, pp. 21-39.

GAWTHROP, P. J. (1980) "Hybrid self tuning control." *Proc. IEE*, Part D, vol. 127, part 5.

GOODWIN, G. C., R. LOZANO-LEAL, D. Q. MAYNE, and R. H. MIDDLETON (1986) "Rapprochement between continuous and discrete model reference adaptive control." *Automatica*, vol. 22, no. 2, pp. 199-207.

5.8 PROBLEMS

1. Consider the system
$$D(\rho)y(t) = N(\rho)u(t)$$

Suppose $D(p) = 0$ for some p. Show that there exist initial conditions on $y(t)$ such that, for $u(t) = 0$ for all t, then
$$Y(\gamma) = \frac{1 + \Delta\gamma}{\gamma - p}$$

2. Show that (5.4.3) is the transfer function of the system (5.4.1), (5.4.2).
3. Show that if $G(\gamma)$ has a zero at $\gamma = \xi$ then, for appropriate initial conditions, the input with transform $U(\gamma) = (1 + \Delta\gamma)/(\gamma - \xi)$ gives no output.
4. Consider the system given in (5.4.10), (5.4.11). For this system show that

$$\text{adj}[sI - A] = \begin{bmatrix} s^2 & -(s+1) & -s \\ s & s(s+1) & -1 \\ 1 & s+1 & s^2+s+1 \end{bmatrix}$$

$$\det[sI - A] = s^3 + s^2 + s + 1$$

and

$$C(sI - A)^{-1} = C\frac{\text{adj}[sI - A]}{\det[sI - A]}$$

$$= \frac{[(s^2 + 2s + 2), (2s^2 + 3s + 1), (2s^2 + s)]}{s^3 + s^2 + s + 1}$$

Hence, or otherwise, verify (5.4.12).

5. Consider again the system of Problem 4; verify that if

$$U(s) = \frac{s+1}{s^2 + 2s + 2} \quad \text{and} \quad x_0 = [-2/5, -1/5, 2/5]^T$$

then $Y(s) = 0$.

6. Verify Lemma 5.5.1. *Hint:* Use the fact that the poles of a transfer function are the eigenvalues of the A matrix, together with results from Chapter 3.
7. Consider the system of Equation (5.5.1). Show that the corresponding discrete transfer function is (5.5.2).
8. Consider a system having transfer function $G(\gamma) = N(\gamma)/D(\gamma)$, where degree $D(\gamma)$ − degree $N(\gamma) = l$ (l is called the relative degree). Let $s(t)$ denote the step response of this system for zero initial conditions. Show that $\rho^i s(t)|_{t=0} = 0$ for $i = 0, \ldots, l-1$. *Hint:* Use the initial condition theorem of generalized transforms.
9. Suppose $y(t)$ satisfies $\delta^i y(t)|_{t=0} = 0$ for $i = 0, \ldots, l-1$; show that $y(i\Delta) = 0$ for $i = 0, \ldots, l-1$.
10. Verify Figure 5.5.3 by evaluating the discrete zeros corresponding to the continuous system given in (5.5.39).
11. Consider a true system, $H(s) = 1/[s^2(1 + 0.2s)]$, and an approximate system, $\hat{H}(s) = 1/s^2$. With 1-Hz sampling rate, find the discrete time poles and zeros for the true and approximate systems.
12. Show that if a continuous time system has a zero at $s = 0$ (that is, in state space form $CA^{-1}B = 0$) then the sampled data version also has a zero at $\gamma = 0$. *Hint:* A delta domain transfer function has a zero at $\gamma = 0$ if the dc gain is zero. The dc gain for the system

$$\delta x = Fx + Gu$$
$$y = Cx$$

is $CF^{-1}G$.

13. Let

$$\rho x = \begin{bmatrix} -1 & -1 \\ 1 & -1 \end{bmatrix} x + \begin{bmatrix} 1 \\ 1 \end{bmatrix} u$$

$$y = \begin{bmatrix} 1 & 0 \end{bmatrix} x$$

Find the transfer function, poles, and zeros for this system.

14. For the system in Problem 13, find an $x(0)$ such that, for $u(t) = 1$, $y(t) = 0$. *Hint:* Use transform techniques, and $U(\gamma) = (1 + \Delta\gamma)/\gamma$.

*15. Find the transfer function, poles, and zeros for the system

$$\rho x = \begin{bmatrix} -1 & -1 \\ 1 & -1 \end{bmatrix} x + \begin{bmatrix} 1 & -1 \\ 1 & 1 \end{bmatrix} u$$

$$y = \begin{bmatrix} 1 & 0 \\ 0 & 1 \end{bmatrix} x$$

16. For a continuous time system with transfer function

$$G(s) = \frac{s + 2}{s^2(s^2 + 1)}$$

find the discrete time transfer in both shift and delta domains for sampling periods $\Delta = 0.03$, $\Delta = 0.1$, and $\Delta = 0.3$.

17. Consider a system in the following state space form:

$$\rho x = \begin{bmatrix} 0 & 1 \\ -1 & 0 \end{bmatrix} x + \begin{bmatrix} 1 \\ 0 \end{bmatrix} u$$

$$y = \begin{bmatrix} 1 & -1 \end{bmatrix} x$$

Find the transfer function and the poles and zeros for this system.

18. Convert the model of Problem 17 into a discrete time model, using the step invariance technique, for $\Delta = 0.5, 0.1$, and 0.01. Find the poles and zeros at each of these sampling rates and comment.

19. Consider a linear continuous time state space model where the eigenvalues of the A matrix are distinct. Show that the step invariance method leads to the same discrete time model as is obtained by evaluating $e^{A\Delta}$. *Hint:* Because the eigenvalues of A are distinct, there exists a nonsingular similarity transformation such that the A matrix is diagonal.

6

Frequency Response

6.1 INTRODUCTION

In Chapter 5 we saw that the transfer function of a linear system plays a key role in the description of the input–output behavior. An important class of inputs is those that can be described by a single sinusoid. We will show in this chapter that the response due to such an input is very easily calculated from the system transfer function. We call the corresponding response the *system frequency response*. The system frequency response offers deep insights into the performance of linear systems. Some of these insights will be explored in this and subsequent chapters.

6.2 EVALUATION OF SYSTEM FREQUENCY RESPONSE

Consider a linear single-input, single-output system having transfer function $G(\gamma)$ and corresponding impulse response $g(t)$. We wish to compute the response of the system with zero initial conditions to the following input:

$$u(t) = A \cos(\omega t + \phi) \quad (6.2.1)$$

To simplify our derivation, we note that $u(t)$ can be computed as the real part of a complex exponential; that is,

$$u(t) = \text{Re}\{u'(t)\} \quad (6.2.2)$$

where

$$u'(t) = A \exp\{j(\omega t + \phi)\} \quad (6.2.3)$$

We will then find the response to $u'(t)$ and take the real part of the answer. Using result (5.2.15), we have

$$y(t) = \text{Re}\left[\int_0^{t+\Delta} g(\tau)u'(t-\tau)\,d\tau\right]$$

$$= \text{Re}\left[Ae^{j(\omega t + \phi)}\int_0^{t+\Delta} g(\tau)e^{-j\omega\tau}\,d\tau\right]$$

$$= \text{Re}\left[Ae^{j(\omega t + \phi)}\int_0^{t+\Delta} g(\tau)E(\beta, -\tau)\,d\tau\right] \tag{6.2.4}$$

where $\beta = (e^{j\omega\Delta} - 1)/\Delta$. This suggests that, for large t, the steady-state solution can be expressed as

$$y_{ss}(t) = \text{Re}[Ae^{j(\omega t + \phi)}G(\beta)] \tag{6.2.5}$$

since, by definition,

$$G(\beta) = \int_0^\infty g(\tau)E(\beta, -\tau)\,d\tau \tag{6.2.6}$$

Equation (6.2.5) can be verified on noting that $\lim_{t\to\infty}(y(t) - y_{ss}(t))$ is zero, provided the integral (6.2.6) exists.

If we express $G[(e^{j\omega\Delta} - 1)/\Delta]$ in terms of its magnitude and phase as

$$G\left(\frac{e^{j\omega\Delta} - 1}{\Delta}\right) = |\overline{G}|e^{j\theta} \tag{6.2.7}$$

then equation (6.2.5) becomes

$$y_{ss}(t) = \text{Re}\{|\overline{G}|e^{j\theta}Ae^{j\phi}e^{j\omega t}\}$$

$$= |\overline{G}|A\cos(\omega t + \phi + \theta) \tag{6.2.8}$$

In summary, we see that the steady-state frequency response is a sinusoid having the same frequency as the input but having its output increased by a factor $|\overline{G}|$ and its phase increased by a factor θ, where

$$|\overline{G}|e^{j\theta} \triangleq G(\beta) \tag{6.2.9}$$

Equation (6.2.8) has a simple physical interpretation: the transfer function is evaluated (as a complex number) on the contour $\gamma = (e^{j\omega\Delta} - 1)/\Delta$. We call this contour the *stability boundary* for reasons to be made clear in Chapter 7. This is shown in Figure 6.2.1 for the case of δ operators. For small $\omega\Delta$, note that $(e^{j\omega\Delta} - 1)/\Delta \simeq j\omega$, which is the well-known result for continuous time systems.

Other properties of the frequency response are as follows. First, the response at $-\omega$ is the conjugate of the response at ω; that is,

$$G\left(\frac{e^{-j\omega\Delta} - 1}{\Delta}\right) = G^*\left(\frac{e^{+j\omega\Delta} - 1}{\Delta}\right) \tag{6.2.10}$$

Second, in the discrete time case, the frequency response is periodic in ω with period

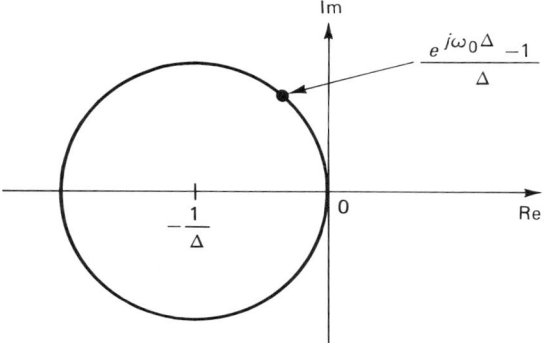

Figure 6.2.1 Stability boundary for discrete time system.

ω_s, where $\omega_s = 2\pi/\Delta$; that is,

$$G\left(\frac{e^{j\omega_1\Delta} - 1}{\Delta}\right) = G\left(\frac{e^{j\omega_2\Delta} - 1}{\Delta}\right), \quad \text{for } \omega_1 - \omega_2 = n\omega_s \tag{6.2.11}$$

This is the same result as that discussed in Section 3.2.3 concerning aliasing.

The frequency response of the plant is conveniently expressed in graphical form by plotting the magnitude and phase of

$$G'(\omega) = G\left(\frac{e^{j\omega\Delta} - 1}{\Delta}\right) \triangleq |\bar{G}|e^{j\theta} \tag{6.2.12}$$

as ω varies from 0 to ∞. This information is typically represented in one of two equivalent forms.

(a) Generalized Bode Diagram

Here the magnitude and phase are plotted against ω in two curves, as shown in Figure 6.2.2.

(b) Generalized Nyquist Diagram

In Figure 6.2.3 the magnitude and phase are plotted on a polar diagram with ω as the parameter.

The evaluation of the frequency response can be given a simple interpretation as follows: let us write the transfer function in factored form as

$$H(\gamma) = \frac{k\prod_{i=1}^{m}(\gamma - \xi_i)}{\prod_{k=1}^{n}(\gamma - p_k)} \tag{6.2.13}$$

where $\xi_1 \ldots \xi_m$ and $p_1 \ldots p_n$ are the zeros of the numerator and denominator

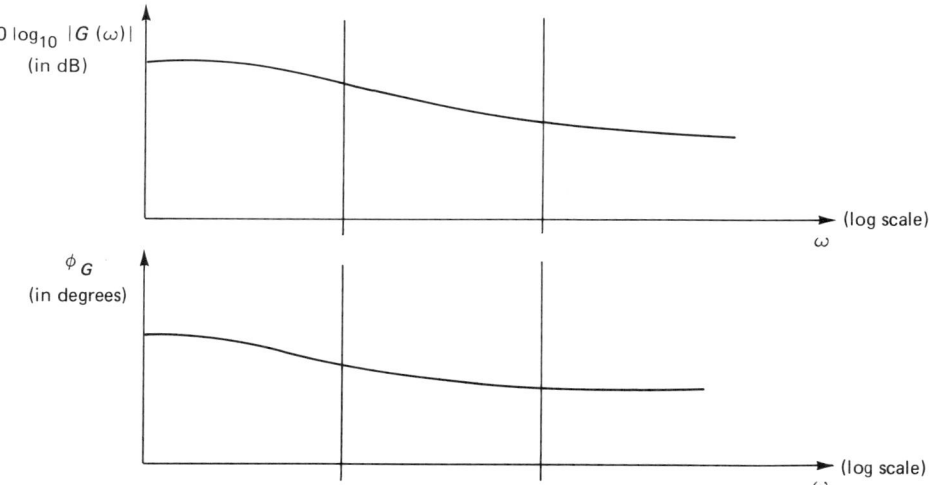

Figure 6.2.2 Generalized bode diagram.

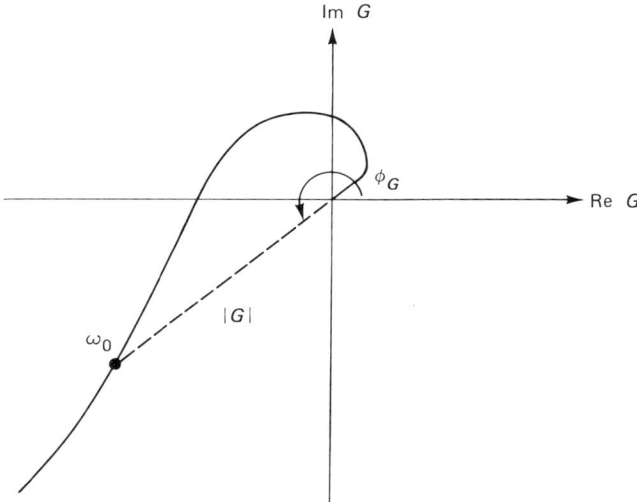

Figure 6.2.3 Generalized Nyquist diagram.

polynomials, respectively. The frequency response at ω is then simply the complex number:

$$H(\beta) = \frac{k\prod_{i=1}^{m}(\beta - \xi_i)}{\prod_{k=1}^{n}(\beta - p_k)}, \qquad \beta = \frac{e^{j\omega\Delta} - 1}{\Delta} \qquad (6.2.14)$$

In particular, the magnitude of the response is

$$|H(\beta)| = \frac{k\prod_{i=1}^{m}|\beta - \xi_i|}{\prod_{k=1}^{n}|\beta - p_k|} \qquad (6.2.15)$$

That is, it is equal to the product of k and the distances from β to the zeros divided by the product of the distances from β to the poles. A similar result holds for the phase (see Problem 2).

6.3 INTERCONNECTING CONTINUOUS AND DISCRETE FREQUENCY RESPONSES

We next turn to the problem of relating continuous and discrete frequency responses. This is of importance in many problems in system theory. For example, a common problem that arises in the design of sampled data systems is how to obtain a discrete time system whose frequency response approximates that of a given continuous time system. For example, we might design a continuous time bandpass filter and then aim to approximate this by a sampled data filter at some specified sampling rate. Note that equality of the continuous and discrete frequency responses is not achieved by the step invariance method discussed in Section 5.3 since the zero order hold on the input distorts the sine waves. We outline next the principal methods that can be used to obtain approximate equality of the continuous and discrete frequency responses.

6.3.1 Approximation by Numerical Integration

Suppose we have a continuous time system of the form

$$D(\rho)y(t) = N(\rho)u(t) \tag{6.3.1}$$

This can be decomposed into a set of parallel first-order systems by partial fraction expansion. Consider, then, one of the first-order subsystems:

$$\frac{d}{dt}x + ax = bu \tag{6.3.2}$$

The solution of (6.3.2) can be written in integral form as

$$x((k+1)\Delta) = x(k\Delta) + \int_{k\Delta}^{(k+1)\Delta}(-ax + bu)\,dt \tag{6.3.3}$$

We then consider the evaluation of the integral in (6.3.3) by various methods.

(a)   Rectangular Rule

If we approximate the integral on the right side of (6.3.3) using the rectangular rule, we obtain

$$x((k+1)\Delta) \simeq x(k\Delta) + \Delta\{-ax(k\Delta) + bu(k\Delta)\} \tag{6.3.4}$$

or, in delta form,

$$\delta x(k\Delta) = \frac{x((k+1)\Delta) - x(k\Delta)}{\Delta} \simeq -ax(k\Delta) + bu(k\Delta) \tag{6.3.5}$$

Comparing (6.3.2) with (6.3.5) suggests that the discrete time model can be obtained by replacing d/dt by δ. Thus, in terms of the corresponding transfer function, we replace s by γ. This leads to:

Rectangular Rule Approximation

$$H_R'(z) = H(s)|_{s=(z-1)/\Delta} \quad (Z \text{ domain}) \tag{6.3.6}$$

or, equivalently,

$$H_R''(\gamma) = H(s)|_{s=\gamma} \quad (\delta \text{ domain}) \tag{6.3.7}$$

(b) Trapezoidal Rule

The preceding description suggests that we might obtain a more accurate discrete approximation by using a higher-order integration procedure. For example, the trapezoidal rule gives

$$x((k+1)\Delta) \simeq x(k\Delta) + \frac{\Delta}{2}\{-ax(k\Delta) + bu(k\Delta) \\ - ax((k+1)\Delta) + bu((k+1)\Delta)\} \tag{6.3.8}$$

It follows that the corresponding discrete time transfer function in the Z domain is

$$H(z) = \frac{(\Delta/2)b(z+1)}{(z-1) + (a\Delta/2)(z+1)}$$

$$= \frac{b}{\frac{2}{\Delta}\frac{(z-1)}{(z+1)} + a} \tag{6.3.9}$$

This suggests the following approximation:

Trapezoidal Rule Approximation

This is also known as bilinear transformation (see Section 6.3.3).

$$H_T'(z) = H(s)|_{s=\frac{2}{\Delta}\frac{z-1}{z+1}} \quad (Z \text{ domain}) \tag{6.3.10}$$

or, equivalently,

$$H_T''(\gamma) = H(s)|_{s=\frac{\gamma}{1+(\Delta/2)\gamma}} \quad (\delta \text{ domain}) \tag{6.3.11}$$

6.3.2 Frequency Domain Approximation

We saw in Section 6.2 that the frequency response of a linear system is obtained in continuous time by replacing s by $j\omega$ and in discrete time by replacing z by $e^{j\omega\Delta}$ and γ by $(e^{j\omega\Delta} - 1)/\Delta$. Thus equality of the frequency responses can be obtained at every frequency in the range $-(\omega_s/2) \leq \omega \leq (\omega_s/2)$, by the following rule:

$$H_F'(z) = H(s)|_{s=(1/\Delta)\ln z} \quad (Z \text{ domain}) \tag{6.3.12}$$

or, equivalently,

$$H_F''(\gamma) = H(s)|_{s=(1/\Delta)\ln(\Delta\gamma+1)} \quad (\delta \text{ domain}) \tag{6.3.13}$$

Unfortunately, the transfer functions $H_F'(z)$ and $H_F''(\gamma)$ as obtained here are not rational functions. Thus we might envisage obtaining a rational approximation by approximating $s = (1/\Delta)\ln(z)$ (or $z = e^{s\Delta}$). The topic of approximating $e^{s\Delta}$ by a rational function will be explored in detail in Section 6.4.3. Here we consider two very simple approximations in the Z domain:

$$e^{s\Delta} \simeq 1 + s\Delta = z \tag{6.3.14}$$

and

$$e^{s\Delta} = \frac{e^{s\Delta/2}}{e^{-s\Delta/2}} \simeq \frac{1 + (s\Delta/2)}{1 - (s\Delta/2)} = z \tag{6.3.15}$$

Note that these two approximations are exact at $s = 0$ and are good approximations for $s = j\omega$, where $\omega \ll 1/\Delta$.

Using (6.3.14) and (6.3.15), we see that we have rederived the approximations obtained in Section 6.3.1; (6.3.14) gives the rectangular rule approximation and (6.3.15) gives the trapezoidal rule approximation. We thus see that the rectangular rule and trapezoidal rule both give good correspondence between the continuous and discrete frequency responses at low frequency (that is, $\Delta|s| \ll 1$).

6.3.3 Bilinear Transformation

A generalization of the trapezoidal rule approximation is the following bilinear transformation:

$$s = c\frac{z - 1}{z + 1} \tag{6.3.16}$$

where the constant c is used to match the frequency responses of the continuous and discrete filters at some point. The corresponding approximation is

$$H_B'(z) = H(s)\big|_{s = \frac{c(z-1)}{(z+1)}} \quad (Z \text{ domain}) \tag{6.3.17}$$

The frequency response of $H_B'(z)$ is obtained as

$$H_B'(e^{j\omega\Delta}) = H\left(\frac{c(e^{j\omega\Delta} - 1)}{(e^{j\omega\Delta} + 1)}\right) \tag{6.3.18}$$

Clearly, $H_B'(e^{j\omega\Delta})$ is equal to $H(j\omega)$ at $\omega = 0$ for all values of c.

Additional insight can also be obtained into the choice of c given in equation (6.3.10), that is, $c = 2/\Delta$, by noting that $\partial H_B'/\partial \omega|_{\omega=0} = \partial H/\partial \omega|_{\omega=0}$ when $c = 2/\Delta$. Thus the choice $c = 2/\Delta$ gives reasonable matching for low frequencies, but exact matching only at $\omega = 0$.

An alternative way of fixing the constant c is to match the frequency response at some particular frequency (in addition to $\omega = 0$). Thus we require

$$H_B'(z)\big|_{z = e^{j\omega\Delta}} = H(s)\big|_{s = j\omega} \tag{6.3.19}$$

or, using (6.3.17), we require

$$H(s)\big|_{s = \frac{c(e^{j\omega\Delta} - 1)}{(e^{j\omega\Delta} + 1)}} = H(s)\big|_{s = j\omega} \tag{6.3.20}$$

This is clearly achieved if

$$\frac{c(e^{j\omega\Delta} - 1)}{e^{j\omega\Delta} + 1} = j\omega \qquad (6.3.21)$$

that is, if

$$c = \omega \cot \frac{\omega\Delta}{2} \qquad (6.3.22)$$

We illustrate this by a simple example.

Example 6.3.1

Suppose we want to approximate a low-pass second-order Butterworth response having 3-dB cuttoff frequency of 1 rad sec^{-1} = $(1/2\pi)$ Hz, using a sampled data filter having a sampling rate of $(10/2\pi)$ Hz [that is, $\Delta = (2\pi/10)$]. The continuous time filter transfer function is

$$H(s) = \frac{1}{s^2 + 1.414s + 1} \qquad (6.3.23)$$

To achieve equality of the frequency responses at $\omega = 1$, we require that c be chosen as c^* where

$$c^* = 1 \cot\left(\frac{2\pi}{20}\right)$$

$$= 3.078 \qquad (6.3.24)$$

Hence the appropriate Z-domain transfer function is

$$H'_B(z) = H(s)\Big|_{s = \frac{c^*(z-1)}{(z+1)}}$$

$$= \frac{0.06746(z^2 + 2z + 1)}{z^2 - 1.143z + 0.4128} \qquad (6.3.25)$$

or, equivalently, in the δ domain

$$H''_B(\gamma) = \frac{0.06746\gamma^2 + 0.4294\gamma + 0.6835}{\gamma^2 + 1.364\gamma + 0.6835} \qquad (6.3.26)$$

By design, the frequency response of this transfer function is $0.7 \underline{/-90°}$ at $(1/2\pi)$ Hz. This can be contrasted with the result obtained using the rectangular approximation, which in the δ domain gives

$$H''_R(\gamma) = \frac{1}{\gamma^2 + 1.414\gamma + 1} \qquad (6.3.27)$$

which has a frequency response of $1.27 \underline{/-106°}$ at $(1/2\pi)$ Hz.

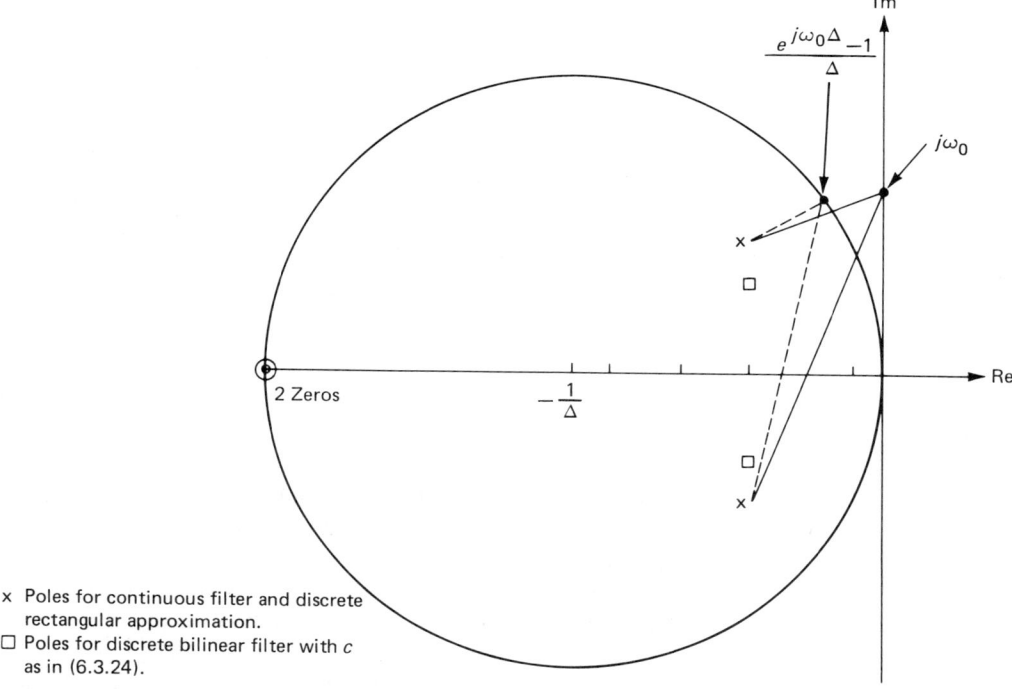

x Poles for continuous filter and discrete rectangular approximation.
□ Poles for discrete bilinear filter with c as in (6.3.24).

Figure 6.3.1 Frequency response calculation for low-pass filter.

These results can be given a simple graphical interpretation using the method described at the end of Section 6.2. The various results are shown in Figure 6.3.1. Note that the rectangular approximation gives a poor result in this case due to the use of a low sampling rate relative to the bandwidth of the filter and the fact that the system poles are well off the negative $j\omega$ axis.

6.3.4 Step Invariance

Another technique for obtaining a discrete approximation to a continuous frequency response is to use the step invariance method outlined in Section 5.3. Note that this method gives the exact discrete transfer function when the input is generated by a zero-order hold. However, for more general inputs, the method can be viewed as a way of obtaining an approximation to the required frequency response.

As shown in Section 5.3, the appropriate mappings are

$$H'_s(z) = \frac{z-1}{z} \mathbf{Z}\left\{ \mathbf{L}^{-1}\left\{ \frac{H(s)}{s} \right\} \right\} \quad (Z \text{ domain}) \quad (6.3.28)$$

$$H''_s(\gamma) = \frac{\gamma}{1+\gamma\Delta} \mathbf{T}\left\{ \mathbf{L}^{-1}\left[\frac{H(s)}{s} \right] \right\} \quad (\delta \text{ domain}) \quad (6.3.29)$$

Example 6.3.2

Consider again the low-pass filter design of Example 6.3.1. The step invariance form of the low-pass filter in the Z domain is

$$H'_S(z) = \frac{0.1453z + 0.1078}{z^2 - 1.158z + 0.4112} \qquad (6.3.30)$$

The poles of this transfer function are very close to those obtained by the bilinear transformation method. The main difference is that the two zeros at -1 produced by the bilinear method have been replaced by a single zero at -0.74. The frequency response of (6.3.30) at $(1/2\pi)$ Hz is $0.696 \underline{/-108°}$. (Compare with the results obtained in Example 6.3.1.) ∇∇∇

6.4 MODELING OF TIME DELAY SYSTEMS

In Section 6.3 we implicitly assumed that the underlying continuous time model gives rise to a rational transfer function in s. An important class of systems that do not have a rational transfer function are those containing pure time delays, for example, those that involve transportation of materials or where measurement of the output can only be obtained some time after the actual output has occurred. An example of this arises in the thickness control for a steel rolling mill (Figure 6.4.1).

The radius of the rolls in a mill is typically on the order of 0.5 m, whereas the thickness of the sheet metal leaving a mill is typical in the range from 0.05 mm to 10 mm. Thus it is not practical to infer the exit thickness of the sheet from the position of the rolls. An X-ray gauge is thus employed so that accurate thickness measurements may be made. Clearly, the gauge cannot be located directly at the point where the metal leaves the mill, and so it is located l meters, say, downstream of the rolls.

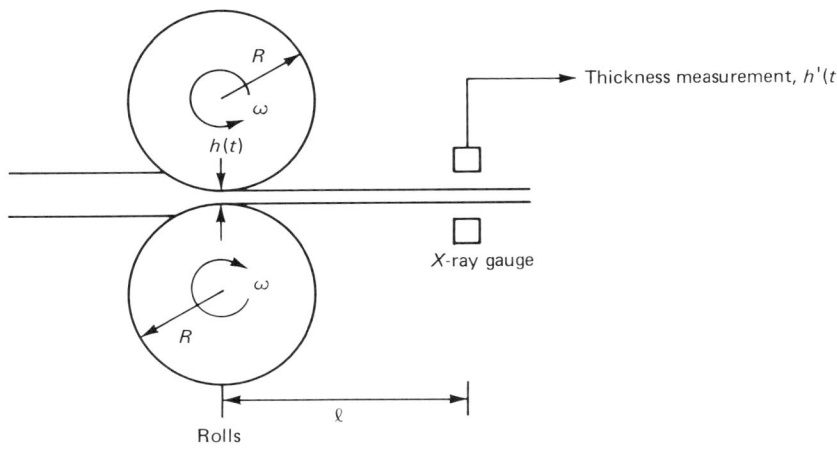

Figure 6.4.1 Time delay in steel rolling mill.

The thickness measurement, $h'(t)$, is thus a delayed version of the exit thickness, $h(t)$; that is, $h'(t) = h[t - (l/R\omega)]$. (For definitions of the variables, see Figure 6.4.1.)

6.4.1 Transfer Function of Delay

A time domain model for a linear system followed by a time delay, τ, in measurement of the output can thus be seen to be

$$\rho x(t) = Ax(t) + Bu(t) \tag{6.4.1}$$

$$y(t + \tau) = Cx(t) \tag{6.4.2}$$

Alternatively, in some systems the time delay may be associated with the input actuator, and the appropriate model is then

$$\rho x'(t) = Ax'(t) + Bu(t - \tau) \tag{6.4.3}$$

$$y(t) = Cx'(t) \tag{6.4.4}$$

(note that the states have a different interpretation in the equivalent models given above). With zero initial conditions, including $y(t) = 0$ for $t < 0$, and taking the transform of (6.4.1) and (6.4.2), we have

$$\gamma X(\gamma) = AX(\gamma) + BU(\gamma) \tag{6.4.5}$$

and

$$E(\gamma, \tau)Y(\gamma) = CX(\gamma) \tag{6.4.6}$$

The transfer function relating $Y(\gamma)$ to $U(\gamma)$ is thus

$$G(\gamma) = E(\gamma, -\tau)C[\gamma I - A]^{-1}B \tag{6.4.7}$$

The same result is obtained for the input delay model of (6.4.3), (6.4.4). Thus, we then think of a process of delaying a signal by τ seconds as being equivalent to multiplication of the transfer function by $E(\gamma, -\tau)$. We therefore immediately see that the frequency response of a pure delay of τ seconds is

$$E(\gamma, -\tau)\Big|_{\gamma = \frac{e^{j\omega\Delta}-1}{\Delta}} = e^{-j\omega\tau} \tag{6.4.8}$$

Thus the magnitude is unity and the phase increases linearly with frequency.

In a continuous time system, the transfer function of a pure delay is not rational. The treatment of time delays in continuous time systems thus requires a little additional caution, compared with the treatment of rational transfer functions.

In discrete time, it makes no sense to talk about delaying a sequence, $y(t)$, by τ, unless τ is an integral number of samples; that is, $\tau = N\Delta$ (for some $N \in \mathbb{Z}^+$). In this case, the transfer function of the delay reduces to

$$G''(\gamma) = E(\gamma, -\tau) = (1 + \gamma\Delta)^{-N} \quad \text{and equivalently} \quad G'(z) = z^{-N} \tag{6.4.9}$$

Thus we see that in *discrete time* a time delay does give rise to a rational transfer function. It should be noted that, unless the sampling rate is slow compared with the delay, the order of the transfer function may be large.

*6.4.2 Sampling of Time Delay Systems

We now address the issue of discrete time models for systems containing time delays that are not a multiple of the sampling period. Consider the case where we are sampling, with zero-order hold, a continuous time system including a time delay as in (6.4.3) and (6.4.4). Also, let

$$\tau = N\Delta + \eta \tag{6.4.10}$$

where $N \in \mathbb{Z}^+ \cup \{0\}$ and $0 \leq \eta < \Delta$. Then, proceeding as in Chapter 3, we have

$$x'((l+1)\Delta) = e^{A\Delta}x'(l\Delta) + \int_{l\Delta}^{(l+1)\Delta} e^{A((l+1)\Delta - t)} Bu(t - \tau)\, dt \tag{6.4.11}$$

$$= e^{A\Delta}x'(l\Delta) + \left\{\int_{\eta}^{\Delta} e^{A(\Delta - v)} B\, dv\right\} u((l - N)\Delta)$$

$$+ \left\{\int_{0}^{\eta} e^{A(\Delta - v)} B\, dv\right\} u((l - N - 1)\Delta) \tag{6.4.12}$$

From (6.4.12), it follows that

$$\delta x'(l\Delta) = \left(\frac{e^{A\Delta} - I}{\Delta}\right) x'(l\Delta) + \left\{\frac{1}{\Delta}\int_{\eta}^{\Delta} e^{A(\Delta - v)} B\, dv\right\} u((l - N)\Delta)$$

$$+ \left\{\frac{1}{\Delta}\int_{0}^{\eta} e^{A(\Delta - v)} B\, dv\right\} u((l - N - 1)\Delta) \tag{6.4.13}$$

This equation has the form

$$\delta x'(l\Delta) = Fx'(l\Delta) + G_1 u((l - N - 1)\Delta) + G_0 u((l - N)\Delta) \tag{6.4.14}$$

and

$$y(l\Delta) = Cx'(l\Delta) \tag{6.4.15}$$

Taking transforms of (6.4.14) and (6.4.15), we see that with zero initial conditions

$$\gamma X(\gamma) = FX(\gamma) + G_1(1 + \gamma\Delta)^{-(N+1)} U(\gamma) + G_0(1 + \gamma\Delta)^{-N} U(\gamma) \tag{6.4.16}$$

and

$$Y(\gamma) = CX(\gamma) \tag{6.4.17}$$

so

$$Y(\gamma) = C[\gamma I - F]^{-1}(G_1 + (1 + \gamma\Delta)G_0)(1 + \gamma\Delta)^{-(N+1)} U(\gamma) \tag{6.4.18}$$

Example 6.4.1 illustrates the manipulations given above.

Example 6.4.1

Consider a continuous time system with transfer function

$$G(s) = \frac{s + 1}{s^2} e^{-0.8s} \tag{6.4.19}$$

One possible state space model that has this transfer function is

$$\dot{x}(t) = \begin{bmatrix} 0 & 0 \\ 1 & 0 \end{bmatrix} x(t) + \begin{bmatrix} 1 \\ 0 \end{bmatrix} u(t - 0.8) \qquad (6.4.20)$$

$$y(t) = \begin{bmatrix} 1 & 1 \end{bmatrix} x(t) \qquad (6.4.21)$$

The aim of the exercise is to find the discrete time transfer function when the system is sampled with $\Delta = 0.5$ second. In this case, $e^{A\Delta}$ is particularly easy to evaluate since $A^2 = 0$; so

$$e^{A\Delta} = \begin{bmatrix} 1 & 0 \\ 0.5 & 1 \end{bmatrix} \qquad (6.4.22)$$

$$\frac{e^{A\Delta} - I}{\Delta} = \begin{bmatrix} 0 & 0 \\ 1 & 0 \end{bmatrix} \qquad (6.4.23)$$

$$\frac{1}{\Delta} \int_0^\eta e^{A(\Delta - v)} B \, dv = \frac{1}{0.5} \int_0^{0.3} \begin{bmatrix} 1 & 0 \\ (0.5 - v) & 1 \end{bmatrix} \begin{bmatrix} 1 \\ 0 \end{bmatrix} dv$$

$$= \begin{bmatrix} 0.6 \\ 0.21 \end{bmatrix} \qquad (6.4.24)$$

and

$$\frac{1}{\Delta} \int_\eta^\Delta e^{A(\Delta - v)} B \, dv = \begin{bmatrix} 0.4 \\ 0.04 \end{bmatrix} \qquad (6.4.25)$$

So

$$Y(\gamma) = \begin{bmatrix} 1 & 1 \end{bmatrix} \begin{bmatrix} \gamma & 0 \\ -1 & \gamma \end{bmatrix}^{-1} \left\{ \begin{bmatrix} 0.6 \\ 0.21 \end{bmatrix} + (1 + 0.5\gamma) \begin{bmatrix} 0.4 \\ 0.04 \end{bmatrix} \right\} (1 + 0.5\gamma)^{-2} U(\gamma) \qquad (6.4.26)$$

The discrete time transfer function is then

$$G_d(\gamma) = \begin{bmatrix} 1 & 1 \end{bmatrix} \begin{bmatrix} \gamma & 0 \\ 1 & \gamma \end{bmatrix} \left\{ \begin{bmatrix} 0.6 \\ 0.21 \end{bmatrix} + (1 + 0.5\gamma) \begin{bmatrix} 0.4 \\ 0.04 \end{bmatrix} \right\} \frac{1}{\gamma^2 (1 + 0.5\gamma)^2} \qquad (6.4.27)$$

$$= \begin{bmatrix} \gamma + 1 & \gamma \end{bmatrix} \begin{bmatrix} 1 + 0.2\gamma \\ 0.25 + 0.02\gamma \end{bmatrix} \frac{1}{\gamma^2 (1 + 0.5\gamma)^2} \qquad (6.4.28)$$

$$= \frac{0.88\gamma^2 + 5.8\gamma + 4}{\gamma^2 (\gamma + 2)^2} \qquad (6.4.29)$$

▽▽▽

6.4.3 Rational Approximations to Delays

In the previous sections we examined models of pure time delays. In continuous time, these are not rational functions. Here we show how these functions can be approximated by a rational transfer function. Approximations of these sorts and their corresponding limitations turn out to be important later when we consider control system design, since many control system design techniques are based around rational transfer functions.

One way to perform the approximation is described in the following steps. First note that we can write

$$e^{-sT} = \frac{e^{-(sT)/2}}{e^{(sT/2)}} \qquad (6.4.30)$$

We then use the fact that

$$\lim_{h \to 0} (1 + sTh)^{1/h} = e^{sT} \qquad (6.4.31)$$

So, for n some integer,

$$e^{+(sT)/2} \approx \left(1 + \frac{sT}{2n}\right)^n \qquad (6.4.32)$$

and

$$e^{-(sT)/2} \approx \left(1 - \frac{sT}{2n}\right)^n \qquad (6.4.33)$$

Thus one rational approximation to a time delay is

$$e^{-sT} \approx \frac{\left(1 - \dfrac{sT}{2n}\right)^n}{\left(1 + \dfrac{sT}{2n}\right)^n} \triangleq H_n(s) \qquad (6.4.34)$$

This approximation is accurate provided $|sT/2n| \ll 1$. In terms of frequency response, it thus follows that $H_n(s)$ gives a good approximation up to a limited frequency. The frequency response of the approximation, $H_n(s)$, can be seen to be

$$H_n(j\omega) = e^{-2jn \tan^{-1}(\omega T/2n)} \qquad (6.4.35)$$

Thus we note that this approximation gives the correct gain for all frequencies and gives approximately the correct phase for low frequencies; that is, $\omega \ll 2n/T$.

It turns out that a better approximation may be obtained by solving the following problem: Given $H(s)$ an analytic function in the complex variable s, find the coefficients $a_1, a_2, \ldots, b_n$ in

$$H_n(s) = \frac{b_0 + b_1 s + \cdots + b_n s^n}{1 + a_1 s + \cdots + a_n s^n} \qquad (6.4.36)$$

such that we match the first $2n$ derivatives of $H(s)$ at $s = 0$; that is, we solve the following equations for $b_0, \ldots, b_n$ and $a_1, \ldots, a_n$.

$$H_n(0) = H(0)$$
$$H_n'(0) = H'(0)$$
$$\vdots$$

and

$$H_n^{(2n)}(0) = H^{(2n)}(0)$$

where $H^{(i)}(\cdot)$ denotes the ith derivative.

This method for approximating functions is called Padé approximation. [Note that it can occasionally happen that although $H(s)$ is stable the corresponding approximation is not stable; see Chapter 7 for definitions of stability.] The corresponding approximations to $H(s) = e^{-sT}$ for $n = 1, 2, 3, 4$ are

$$H_1(s) = \frac{1 - \frac{sT}{2}}{1 + \frac{sT}{2}} \qquad (6.4.37)$$

$$H_2(s) = \frac{1 - \frac{sT}{2} + \frac{s^2T^2}{12}}{1 + \frac{sT}{2} + \frac{s^2T^2}{12}} \qquad (6.4.38)$$

$$H_3(s) = \frac{1 - \frac{sT}{2} + \frac{s^2T^2}{10} - \frac{s^3T^3}{120}}{1 + \frac{sT}{2} + \frac{s^2T^2}{10} + \frac{s^3T^3}{120}} \qquad (6.4.39)$$

$$H_4(s) = \frac{1 - \frac{sT}{2} + \frac{3}{28}s^2T^2 - \frac{s^3T^3}{84} + \frac{s^4T^4}{1680}}{1 + \frac{sT}{2} + \frac{3}{28}s^2T^2 + \frac{s^3T^3}{84} + \frac{s^4T^4}{1680}} \qquad (6.4.40)$$

Once again, note that these approximations give the correct gain at all frequencies and the correct phase over a limited bandwidth. Figure 6.4.2 shows the

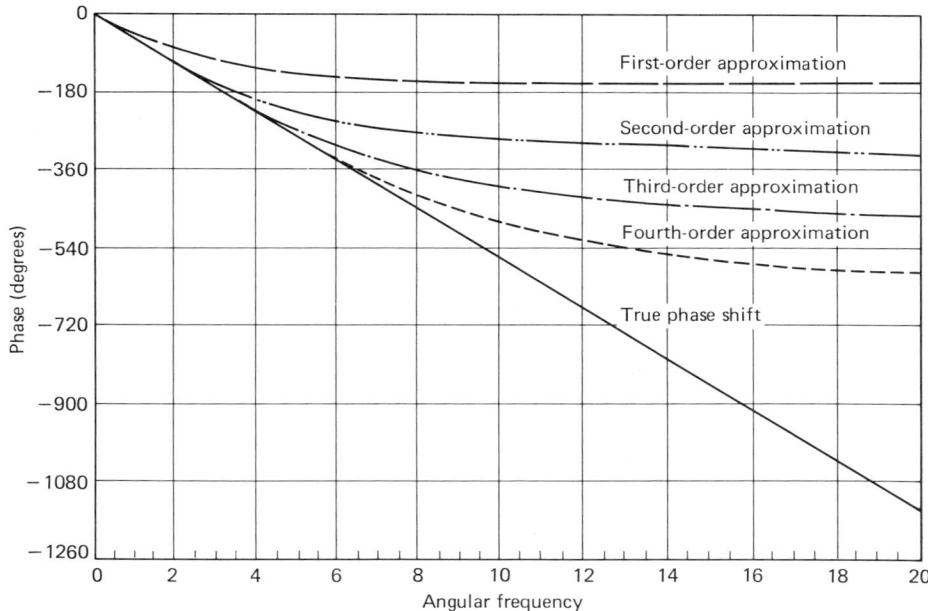

Figure 6.4.2 Phase versus frequency for 1 to 4 orders Padé approximations to a time delay.

phase response (in degrees) versus ωT for the preceding four transfer functions. Note that these Padé approximations may be generated using a finite number of terms from the following infinite continued fraction expansion (due to Euler) for the exponential:

$$e^x = 1 + \cfrac{1}{\left(\cfrac{1}{x} - 1\right) + \cfrac{1}{1 + \cfrac{1}{\left(\cfrac{3}{x} - 1\right) + \cfrac{1}{1 + \cfrac{1}{1 + \cfrac{1}{\left(\cfrac{5}{x} - 1\right) + \cfrac{1}{1 \cdot \cdot \cdot}}}}}} \tag{6.4.41}$$

For example,

$$e^{-sT} \approx H_{p,2}(s) = 1 + \cfrac{1}{\left(-\cfrac{1}{sT} - 1\right) + \cfrac{1}{1 + \cfrac{1}{1 + \cfrac{1}{\left(-\cfrac{3}{sT} - 1\right) + \cfrac{1}{1 + \cfrac{1}{1}}}}}} \tag{6.4.42}$$

In summary, this section has shown that, over a *limited range of frequencies*, $\omega T < 2n$, a rational approximation can give an accurate description of a time delay.

6.5 SUMMARY

The key points of this chapter were:

- The steady-state frequency response to the input $u(t) = A\cos(\omega t + \phi)$ of a system having transfer function $G(\gamma)$ is

$$y_{ss}(t) = |G(\beta)|A\cos(\omega t + \phi + \theta) \tag{6.2.8}$$

where

$$G(\beta) = |G(\beta)|e^{j\theta} \tag{6.2.7}$$

and

$$\beta = \frac{e^{j\omega\Delta} - 1}{\Delta}$$

- The frequency response $G(\beta)$ can be displayed on a Bode diagram or a Nyquist plot.
- A continuous time frequency response can be approximated by a discrete transfer function obtained by various transfer function mappings; for example

	Z domain	δ domain
Rectangular rule (6.3.6), (6.3.7)	$H'_R(z) = H(s)\big|_{s=\frac{z-1}{\Delta}}$	$H''_R(\gamma) = H(s)\big|_{s=\gamma}$
Trapezoidal rule (6.3.10), (6.3.11)	$H'_T(z) = H(s)\big|_{s=\frac{2}{\Delta}\left[\frac{z-1}{z+1}\right]}$	$H''_T(\gamma) = H(s)\big|_{s=\frac{\gamma}{1+\frac{\Delta}{2}\gamma}}$
Exact matching (6.3.12), (6.3.13)	$H'_F(z) = H(s)\big|_{s=\frac{1}{\Delta}\ln(z)}$	$H''_F(\gamma) = H(s)\big|_{s=\frac{1}{\Delta}\ln(\Delta\gamma+1)}$
Bilinear (6.3.17)	$H'_B(z) = H(s)\big|_{s=c\left[\frac{z-1}{z+1}\right]}$	$H''_B(\gamma) = H(s)\big|_{s=\frac{c\gamma}{\frac{2}{\Delta}+\gamma}}$
Step invariant (6.3.28), (6.3.29)	$H'_s(z) = \frac{z-1}{z}\mathbf{Z}\left\{\mathbf{L}^{-1}\left[\frac{H(s)}{s}\right]\right\}$	$H''_s(\gamma) = \frac{\gamma}{1+\gamma\Delta}\mathbf{T}\left\{\mathbf{L}^{-1}\left[\frac{H(s)}{s}\right]\right\}$

- The transfer function of a time delay of τ seconds is $E(\gamma, -\tau)$.
- In continuous time, the transfer function of a time delay is irrational.
- Various rational approximations of a pure time delay are possible and give accurate frequency response over a limited range of frequencies.

6.6 REFERENCES

Additional material on filter design and frequency response in continuous time is contained in:

CHEN, W. K. (1986) *Passive and Active Filters: Theory and Implementation*. Wiley, New York.

VANVALKENBURG, M. E. (1982) *Analog Filter Design*. Holt, Rinehart and Winston, New York.

Digital filter design is covered in:

FRANKLIN, G. F., and J. D. POWELL (1980) *Digital Control of Dynamic Systems*. Addison-Wesley, Reading, Mass.

ROBERTS, R. A., and C. T. MULLIS (1987) *Digital Signal Processing*. Addison-Wesley, Reading, Mass.

A deeper treatment of rational and irrational functions is contained in:

DAVIS, P. J. (1975) *Interpolation and Approximation*. Dover, New York.

Use of rational approximations in control is further discussed in:

GAWTHROP, P. J., and N. T. NIHTILA (1985) "Identification of time delays using a polynomial identification method." *Systems and Control Letters*, vol. 5, pp. 267–271.

SALGADO, M. E., C. E. DE SOUZA, and G. C. GOODWIN (1988) "Considerations on time delay modeling in control design." *IFAC Symposium on System Identification and System Parameter Estimation*. Beijing.

SHAMASH, Y. (1974) "Stable reduced-order models using Padé type approximations." *IEEE Trans. Automatic Control*, October, pp. 615–616.

DE SOUZA, C. E., G. C. GOODWIN, D. Q. MAYNE, and M. PALANISWAMI (1988) "An adaptive control algorithm for linear systems having unknown time delay." *Automatica*, vol. 24, no. 3, pp. 327–341.

6.7 PROBLEMS

1. Consider the following continuous time system:

$$G(s) = \frac{1}{s+1}$$

   (a) Find the corresponding sampled data transfer function in delta form.
   (b) Plot the corresponding Bode diagram for different values of Δ and compare with the continuous time result.
   (c) Repeat part (b) for a Nyquist diagram.

2. Show that the phase response of a linear system is given by the sum of the angles from a point on the stability boundary, β_0, to the zeros minus the sum of the corresponding angles to the poles.

3. Verify that (6.3.22) is the solution to (6.3.21).

4. Obtain the appropriate approximation, as in Section 6.3.1, by numerical integration if Simpson's rule is used.

5. Repeat Problem 4 for fourth-order Runge–Kutta.

6. Repeat Example 6.4.1 using step invariance. *Hint:* First show that the continuous time step response is

$$s(t) = \begin{cases} 0, & t < 0.8 \\ \frac{(t-0.8)^2}{2} + (t-0.8), & t \geq 0.8 \end{cases}$$

   which gives rise to a discrete time step response of

$$s_d(t) = 1 - (t-1)\left(\frac{(t-1)^2}{2} + 1.2(t-1) + 0.22\right), \quad \text{for } t \in \Omega$$

7. Consider a discrete time delay of 1 second; that is, the time delay has a transfer function $E(\gamma, -1)$, where the sampling period is $1/n$ seconds. Find a first-order approximation, $H_a(\gamma)$, to this time delay that has the following properties:

   (a) $H_a(\gamma)$ is all pass; that is,

$$\left| H_a\left(\frac{e^{j\omega\Delta} - 1}{\Delta}\right) \right| = 1, \quad \text{for all } \omega$$

   (b) $H_a(0) = E(0, -1)$, and

   (c) $\left.\dfrac{d}{d\gamma} H_a(\gamma)\right|_{\gamma=0} = \left.\dfrac{d}{d\gamma} E(\gamma, -1)\right|_{\gamma=0}$

Compare the frequency responses for various n. What happens to the approximation as the sampling rate increases ($n \to \infty$)?

8. (a) Design a second-order continuous time bandpass filter to cover the frequency range from 40 to 60 Hz.
   (b) Convert the filter to discrete form with a sampling rate of 300 Hz using all the techniques described in this chapter.
   (c) Compare the frequency responses of the preceding designs.
9. Compare the frequency responses of the discrete transfer functions found in Examples 6.3.1 and 6.3.2.
10. Compute the value for $e^{j(\pi/2)}$ and compare this and first- to fourth-order Padé approximations to $e^{j(\pi/2)}$.
11. Show that the Nyquist plot for an integrator, $1/\gamma$, is the line with real part $-\Delta/2$.
12. Show that the Nyquist plot for a transfer function of the form $(b_1\gamma + b_0)/(\gamma + a_0)$ is a circle.

7

Classical Control Systems Analysis

7.1 INTRODUCTION

In previous chapters, we introduced a number of basic tools that are useful for analyzing linear dynamic systems. In this chapter we add the important notion of feedback to these concepts. Feedback is frequently helpful in modifying the behavior of a system to achieve certain objectives. The key issues that arise in a feedback system are those of stability, tracking performance, and transient performance. We will treat each of these issues in this chapter.

The basic idea of a stable system is that the response should decay to zero when the input is removed and that a bounded input should produce a bounded output response. Clearly, such properties are highly desirable for any control system. In this chapter, we will study stability for linear systems from both frequency and time domain perspectives, with special emphasis on the effects of feedback.

We next examine the tracking performance of a linear feedback system. In particular, we look at the steady-state errors that result from polynomial or sinusoidal set points.

We also look at the problem of transient performance. The notions of rise time, overshoot, and settling time for step inputs in the set point are introduced. These time domain concepts are then linked to the associated frequency domain characteristics by various rules of thumb.

Finally, the preceding notions are extended to simple classes of multiinput, multioutput systems.

The concepts covered in this chapter are usually referred to as classical control systems analysis. This subject has received extensive treatment, especially in the continuous time case. A novel aspect of our treatment here is the unified presentation of continuous and discrete results. Also, we emphasize the close connection between the discrete and continuous cases, including the direct design of digital compensators using classical techniques.

7.2 STABILITY FOR LINEAR TIME INVARIANT SYSTEMS

For simplicity, we will consider an input–output model of the form

$$D(\rho)y(t) = N(\rho)u(t) \tag{7.2.1}$$

where we assume that $D(\rho)$ and $N(\rho)$ have no common factors. (The more general case will be studied in Chapter 8 using more advanced ideas.)

We now introduce a simple definition of stability.

Definition 7.2.1. Asymptotic Stability

The system (7.2.1) is said to be asymptotically stable if the output $y(t)$ for any initial conditions and zero input satisfies

$$\lim_{t \to \infty} y(t) = 0 \tag{7.2.2}$$

▽▽▽

This simple definition of stability can be related to the location of the poles of the system. This is done in Lemma 7.2.1, which gives a necessary and sufficient condition for asymptotic stability.

Lemma 7.2.1

The system (7.2.1) is asymptotically stable if and only if the roots, p_i, of $D(\gamma) = 0$ (that is, the poles of the system) satisfy

$$\frac{\Delta}{2}|p_i|^2 + \text{Re}\{p_i\} < 0 \tag{7.2.3}$$

where $\text{Re}\{\cdot\}$ denotes real part.

Proof: Taking transforms of (7.2.1) and using Theorem 4.4.1(ii), it follows that the zero input initial condition response is given by

$$D(\gamma)Y(\gamma) = (1 + \Delta\gamma)\{y_0^{(n-1)} + \gamma y_0^{(n-2)} + \cdots + \gamma^{n-1}y_0^{(0)}$$
$$+ d_{n-1}(y_0^{(n-2)} + \gamma y_0^{(n-3)} + \cdots + \gamma^{n-2}y_0^{(0)}) \cdots + d_1 y_0^{(0)}\} \tag{7.2.4}$$

where $D(\rho) = \rho^n + d_{n-1}\rho^{n-1} + \cdots + d_0$ and $y_0^{(i)}$ denote $\rho^i y(t)$ evaluated at $t = 0$.

For simplicity, we consider the case of nonrepeated poles. (The reader is asked to consider the more general case in Problem 1.) Performing a partial fraction expansion in (7.2.4) gives

$$Y(\gamma) = \sum_{i=1}^{n} \frac{\alpha_i(1 + \Delta\gamma)}{\gamma - p_i} \tag{7.2.5}$$

where any combination of α_i can be generated by appropriate choice of initial conditions. The corresponding output response is then

$$y(t) = \sum_{i=1}^{n} \alpha_i E(p_i, t) \tag{7.2.6}$$

From (7.2.6), $y(t)$ goes to zero if and only if $E(p_i, t)$ goes to zero for $i = 1, \ldots, n$. The result in (7.2.3) is immediate in continuous time. In discrete time, (7.2.3) is equivalent to $|1 + \Delta p_i| < 1$ (see Problem 12). The result then follows from the definition of $E(p_i, t)$ in discrete time. ▽▽▽

The discrete time case of Lemma 7.2.1 immediately implies the following result for shift operator discrete time models.

Corollary 7.2.1

In the discrete time shift operator domain, the system $D(q)y(k) = N(q)u(k)$ is asymptotically stable if and only if the poles, p_i, of the system satisfy

$$|p_i| < 1 \tag{7.2.7}$$

▽▽▽

Note that a major advantage of the delta form is that the corresponding discrete stability domain converges to the continuous time stability domain as $\Delta \to 0$! This is not true of the shift operator stability region.

The stability regions (the regions in the complex plane satisfying Lemma 7.2.1 and Corollary 7.2.1) for different operators are given in Table 7.2.1.

We next explore further implications of the preceding necessary and sufficient condition.

Lemma 7.2.2

Subject to (7.2.3), we have that for any initial condition a bounded input produces a bounded response.

Proof: First consider the case of zero initial conditions. Condition (7.2.3) ensures that the impulse response $h(t)$ satisfies

$$|h(t)| \leq K E(\lambda, t) \tag{7.2.8}$$

TABLE 7.2.1 STABILITY REGIONS FOR VARIOUS OPERATORS

Operator	Transform Variable	Stability Region	Stability Boundary	Diagram
$\dfrac{d}{dt}$	s (laplace transform)	$\text{Re}\{s\} < 0$	$\text{Re}\{s\} = 0$	
δ	γ (delta transform)	$\|1 + \Delta\gamma\| < 1$ or $\dfrac{\Delta}{2}\|\gamma\|^2 + \text{Re}\{\gamma\} < 0$	$\|1 + \Delta\gamma\| = 1$ or $\dfrac{\Delta}{2}\|\gamma\|^2 + \text{Re}\{\gamma\} = 0$	
q	z (Z-transform)	$\|z\| < 1$	$\|z\| = 1$	

where $-1/\Delta < \lambda < 0$ (see Problem 2). Now, using the convolution theorem [Theorem 4.4.1(viii)],

$$y(t) = \underset{0}{\overset{t+\Delta}{S}} h(\tau) u(t - \tau)\, d\tau \tag{7.2.9}$$

Hence

$$|y(t)| \leq \underset{0}{\overset{t+\Delta}{S}} |h(t)| A\, d\tau \tag{7.2.10}$$

where

$$A = \sup_{0 \leq t < \infty} |u(t)|$$

The result then follows on substituting (7.2.8) into (7.2.10).

The case of nonzero initial conditions can be similarly treated using linearity.

▽▽▽

7.3 POLYNOMIAL STABILITY TESTS

The stability tests given in Section 7.2 involve finding the zeros of a polynomial. For $n > 4$, a general analytical solution to this problem does not exist and numerical techniques must be employed to find the zeros. In this section we give various tests based solely on the coefficients of a polynomial that characterize the possible locations of the zeros of that polynomial. In this way, stability can be examined analytically without needing to explicitly evaluate the zeros. We will give stability tests useful for the three different operators we have used.

Lemma 7.3.1

Consider a polynomial, $p(r) = r^n + p_{n-1}r^{n-1} + \cdots + p_0$. Then necessary (though *not* sufficient) conditions for the zeros to be in the stability region are:

(i) Continuous

$p(s) = s^n + p_{n-1}s^{n-1} + \cdots + p_0$ has all positive coefficients.

(ii) Discrete (Delta Operator)

$p(\gamma)$ has all positive coefficients and $(-1)^n p\left[-\frac{2}{\Delta} - \gamma\right] \triangleq p_\gamma(\gamma)$ has all positive coefficients.

(iii) Discrete (Shift Operator)

$p_{z_1}(z) = p(z - 1)$ and $p_{z_2}(z) = (-1)^n p(-1 - z)$ should both have all positive coefficients.

Proof:

(i) Suppose the n zeros of $p(s)$, $\xi_1, \ldots, \xi_n$, satisfy

$$\text{Re}\{\xi_i\} < 0 \qquad (7.3.1)$$

Then, grouping the complex ξ_i into conjugate pairs, we can write

$$p(s) = [(s - \xi_1)(s - \xi_1^*)] \cdots [(s - \xi_l)(s - \xi_l^*)](s - \xi_{l+1}) \cdots (s - \xi_m) \qquad (7.3.2)$$

$$= [s^2 - 2\text{Re}\{\xi_1\}s + |\xi_1|^2] \cdots [s^2 - 2\text{Re}\{\xi_l\}s + |\xi_l|^2]$$
$$\times (s - \xi_{l+1}) \cdots (s - \xi_m) \qquad (7.3.3)$$

In view of (7.3.1), all factors in (7.3.2) have positive, real coefficients, and so $p(s)$ must have all positive coefficients.
Parts (ii) and (iii) follow similarly. ▽▽▽

The converse to Lemma 7.3.1 is not true, in general, as illustrated by the following continuous time counterexample.

Let

$$p(s) = s^3 + 2s^2 + 2s + 40 \tag{7.3.4}$$

Then, clearly, $p(s)$ has coefficients that are all positive; however,

$$p(s) = (s + 4)(s - 1 + j3)(s - 1 - j3) \tag{7.3.5}$$

and so $p(s)$ is not a stable polynomial.

The following partial converse to Lemma 7.3.1 is true, however.

Lemma 7.3.2

(i) Continuous

Suppose $p(s) = s^n + p_{n-1}s^{n-1} + \cdots + p_0$ has all positive coefficients; then, for any real $x \geq 0$, $p(x) \neq 0$.

(ii) Delta

Similarly, if $p_\gamma(\gamma)$ and $p(\gamma)$ [see Lemma 7.3.1(ii)] have all positive coefficients, then, for real $x \geq 0$ or $x \leq -(2/\Delta)$, $p(x) \neq 0$.

(iii) Z-Domain

If $p_{z_1}(z)$ and $p_{z_2}(z)$ [see Lemma 7.3.1(iii)] have all positive coefficients, then, for all real x, $|x| \geq 1$, $p(x) \neq 0$.

Proof:

(i) x real and positive together with all positive coefficients gives $p(x) > 0$, so clearly $p(x)$ cannot equal zero.
(ii) For $x \geq 0$, we simply use (i) above. For $x \leq -(2/\Delta)$ we consider $y = -(2/\Delta) - x$ (that is, $y \geq 0$). Then $(-1)^n p(x) = p_\gamma(y) > 0$.
(iii) Follows as in (ii). ▽▽▽

Lemma 7.3.2 gives a sufficient condition for there to be no unstable *real* poles in a system. However, one common form of instability in a control system is an expanding oscillatory mode, corresponding to complex closed loop poles that lie outside the stability region. Thus the test given in Lemma 7.3.2 is not adequate for this problem. We therefore need necessary and sufficient conditions for stability. The following theorem gives necessary and sufficient conditions for the continuous time case.

Theorem 7.3.1. Routh Criterion (Continuous Time)

Consider the polynomial $p(s)$:

$$p(s) = s^n + p_{n-1}s^{n-1} + \cdots + p_0$$

Form the following Routh array:

	Column 1	**Column 2**	$\cdots$
Row 1:	$1 = a_{11}$	$p_{n-2} = a_{12}$	$p_{n-4} = a_{13} \cdots$
Row 2:	$p_{n-1} = a_{21}$	$p_{n-3} = a_{22}$	$p_{n-5} = a_{23} \cdots$
Row 3:	a_{31}	a_{32}	$a_{33} \cdots$
$\vdots$	$\vdots$	$\vdots$	$\vdots$
Row (n):	$a_{n,1}$	$0 \cdots$	
Row ($n+1$):	$a_{n+1,1}$	$0 \cdots$	

where, for $i > 2$,

$$a_{ij} = a_{(i-2),(j+1)} - \frac{a_{(i-2),1} a_{(i-1),(j+1)}}{a_{(i-1),1}} \quad (7.3.6)$$

Then

(i) The zeros of $p(s)$ all have real part < 0 if and only if $a_{i1} > 0$ for $i = 1, 2, \ldots (n+1)$.

(ii) The number of right half-plane (that is, unstable) zeros of $p(s)$ is equal to the number of sign changes in the first column of the Routh array. (Special procedures may be used in cases where an element is zero; see Gantmacher, 1959).

Proof: The proof of this theorem is beyond the scope of this text; see Gantmacher (1959) for details. ▽▽▽

Example 7.3.1

Consider the polynomial given in Equation (7.3.4). The Routh array for this polynomial is

	Column 1	**Column 2**
Row 1	1	2
Row 2	2	40
Row 3	−18	0
Row 4	40	

Clearly, the polynomial is unstable with two roots in the right half-plane. ▽▽▽

The Routh array gives us a simple method of determining whether or not a continuous time system is stable. One method for determining the stability of a discrete time system is to use a conformal mapping that maps the appropriate discrete stability region into the left half-plane. We may then apply the Routh criterion to the resultant polynomial.

For the shift operator, an appropriate mapping is

$$w = \frac{z-1}{z+1} \quad \text{or} \quad z = \frac{1+w}{1-w} \tag{7.3.7}$$

We leave it to the reader to verify that (7.3.7) transforms the unit circle in the z plane into the left half, w, plane.

Thus, given $p(z)$ in the Z-domain, we introduce

$$\tilde{p}(w) = (1-w)^n p\left(\frac{1+w}{1-w}\right) \tag{7.3.8}$$

and perform the standard Routh test on $\tilde{p}(w)$.

For the delta operator, an appropriate mapping is

$$v = \frac{\gamma}{1 + (\Delta\gamma)/2} \quad \text{or} \quad \gamma = \frac{v}{1 - (\Delta v)/2} \tag{7.3.9}$$

Given $p(\gamma)$, we let

$$\bar{p}(v) = \left(1 - \frac{\Delta v}{2}\right)^n p\left(\frac{v}{(1 - \Delta v)/2}\right) \tag{7.3.10}$$

then $\bar{p}(v)$ is an nth-order polynomial. Furthermore, if $\bar{p}(v)$ has all zeros in the left half-plane, it follows that $p(\gamma)$ has all zeros in the circle centered on $-1/\Delta$, radius $1/\Delta$.

We then perform the standard Routh test on $\bar{p}(v)$. This test reduces to the appropriate test on the continuous time system as $\Delta \to 0$. This again shows that the delta domain results reduce to the continuous results by simply setting the sampling period to zero.

Example 7.3.2

Consider the following delta domain polynomial, where $\Delta = 0.1$:

$$p(\gamma) = \gamma^3 + 4\gamma^2 + 9\gamma + 10 \tag{7.3.11}$$

Using the transformation (7.3.10), we obtain

$$\bar{p}(v) = 0.821(v^3 + 3.866v^2 + 9.132v + 12.177) \tag{7.3.12}$$

The Routh array for this polynomial is

1	9.132
3.866	12.177
5.982	0
12.177	0

The polynomial is therefore stable. [In fact, it can be verified that the roots of (7.3.11) are at $-1 \pm 2j$ and -2.]

The preceding method gives an entirely satisfactory way of testing for discrete time stability. There do exist methods, however, similar to the Routh text that can be applied directly to discrete polynomials. The following theorem, due to Jury, shows how this may be performed in the case of shift domain polynomials.

Theorem 7.3.2. Jury's Criterion for Shift Domain Polynomials

Consider $p(z) = z^n + p_{n-1}z^{n-1} + \cdots + p_0$ and form the following table:

$$\begin{array}{llll} a_{11} = 1 & a_{12} = p_{n-1} & \cdots & a_{1n} = p_0 \\ b_{11} = a_{1n} & b_{12} = a_{1(n-1)} & \cdots & b_{1n} = a_{11} \\ a_{21} & a_{22} & & \\ b_{21} & b_{22} & & \\ \vdots & & & \\ a_{n1} & & & \end{array}$$

where

$$a_{ij} = a_{(i-1)j} - b_{(i-1)j}\alpha_{(i-1)} \tag{7.3.13}$$

$$\alpha_j = \frac{a_{jn}}{a_{j1}} \tag{7.3.14}$$

$$b_{ij} = a_{i(n-j-i+2)} \tag{7.3.15}$$

Then $p(z)$ has all zeros within the unit circle if and only if $a_{i1} > 0$ for $i = 1, 2, \ldots, (n-1)$.

Proof: The proof of this theorem is also beyond the scope of the current text; see Jury and Blanchard (1961) for details. ▽▽▽

Example 7.3.3

Consider the following Z-domain polynomial:

$$p(z) = (z - 1.5)(z - 0.5)z = z^3 - 2z^2 + 0.75z \tag{7.3.16}$$

The Jury table for this polynomial is

$$\begin{array}{llll} 1.0 & -2.0 & 0.75 & 0 \\ 0 & 0.75 & -2.0 & 1.0 \\ 1.0 & -2.0 & 0.75 & \\ 0.75 & -2.0 & 1.0 & \\ 0.438 & -0.5 & & \\ -0.5 & 0.438 & & \\ -0.134 & & & \\ -0.134 & & & \end{array}$$

Thus the polynomial has one unstable zero. ▽▽▽

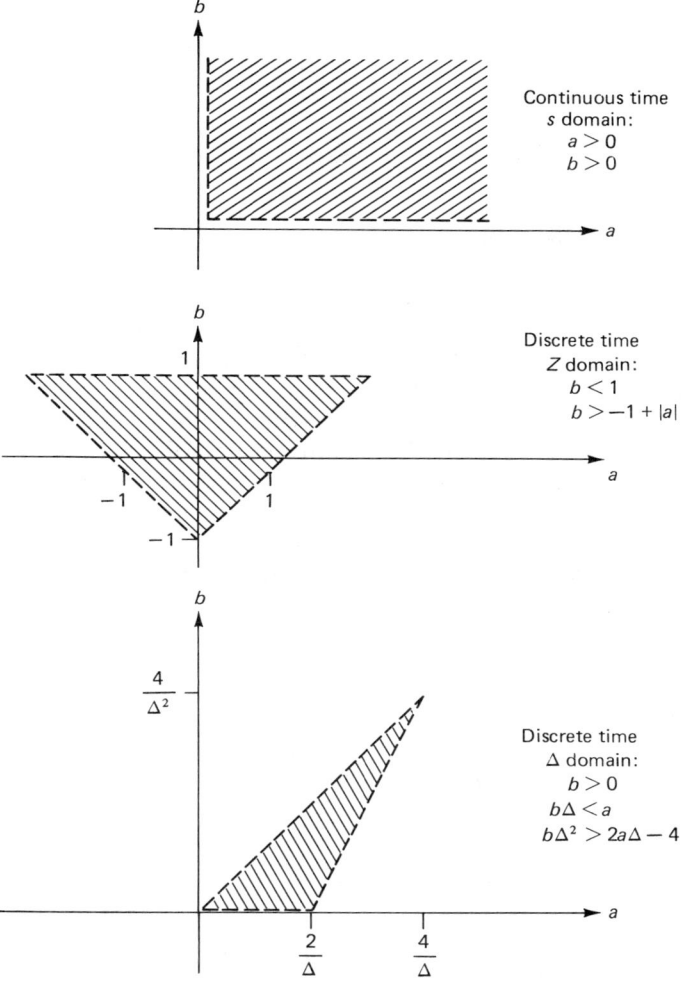

Figure 7.3.1 Second-order polynomial stability regions in the coefficient space.

As a general rule, we strongly recommend that, whenever possible, stability tests be carried out in the delta rather than the shift domain as in Example 7.3.2; otherwise, numerical problems can occur, especially when $|p_i|\Delta \ll 1$, where p_i denotes the roots of the discrete polynomial. This is illustrated in Problem 11.

Using the technique just discussed, we can show that the stability regions for a second-order polynomial, $p(r) = r^2 + ar + b$, are as shown in Figure 7.3.1. Clearly, as $\Delta \to 0$ we see that the discrete time delta domain stability region expands to cover the first quadrant, as in the continuous time case. Thus the link between continuous and discrete systems is again clarified by using the delta operator.

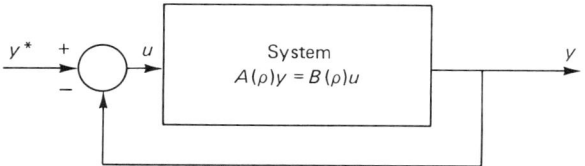

Figure 7.4.1 Unity feedback system.

7.4 ROOT LOCUS

In previous sections, stability was considered for a given linear system. In this and subsequent sections, we will be particularly interested in stability for feedback systems of the form shown in Figure 7.4.1. We will call the system linking u to y [that is, $A(\rho)y = B(\rho)u$] the open loop system, and we will call the system linking y^* to y the closed loop system. The notions of open and closed loop poles, zeros, and transfer functions follow directly from these definitions.

One method for studying closed loop stability is the root locus technique. This method gives insights into closed loop stability and performance based on open loop data.

Consider a negative unity feedback system, as shown in Figure 7.4.1, with a rational open loop transfer function

$$H_{\text{ol}}(\gamma) = \frac{k\prod_{i=1}^{m}(\gamma - z_i)}{\prod_{i=1}^{n}(\gamma - p_i)} = \frac{B(\gamma)}{A(\gamma)} \qquad (7.4.1)$$

It is readily seen from Figure 7.4.1 that the closed transfer function relating y^* to y is $H_{\text{ol}}(\gamma)/(1 + H_{\text{ol}}(\gamma))$. Hence the closed loop poles of the system satisfy the following equation:

$$H_{\text{ol}}(\gamma) = \frac{k\prod_{i=1}^{m}(\gamma - z_i)}{\prod_{i=1}^{n}(\gamma - p_i)} = -1 \qquad (7.4.2)$$

The root locus is then defined to be the graph (of n branches) in the complex plane of the closed loop poles as k varies between 0 and ∞. Thus the root locus is a graph of those values of γ which satisfy the equation (7.4.2).

Several useful properties of root loci are established in Lemma 7.4.1.

Lemma 7.4.1

The root locus of values of γ satisfying (7.4.2) for $k = 0$ to ∞ has the following properties (for $m \leq n$):

(i) The n branches of the root locus start (for $k = 0$) at the n open loop poles p_i.
(ii) m of these n branches terminate (for $k = \infty$) at the m open loop zeros z_i.
(iii) The remaining $l = n - m$ branches tend to infinity as $k \to \infty$ with asymptotes making angles $(\pi/l) + (\nu 2\pi/l)$ (for $\nu = 0, 1, \ldots, l - 1$) with the positive real

axis, and intersecting at $(\bar{\sigma} + j0)$, where

$$\bar{\sigma} = \frac{\sum_{i=1}^{n} p_i - \sum_{i=1}^{m} z_i}{n - m} \tag{7.4.3}$$

(iv) A point, $\bar{\gamma}$, will lie on the root locus if and only if

$$\sum_{i=1}^{n} \angle(\bar{\gamma} - p_i) - \sum_{i=1}^{m} \angle(\bar{\gamma} - z_i) = \pi \pm 2\nu\pi \tag{7.4.4}$$

for some integer ν.

(v) A point on the real axis, $\sigma + j0$, will lie on the root locus if and only if the number of poles plus zeros to the right of $\sigma + j0$ and on the real axis is odd.

(vi) the gain $\bar{k}$ that will place a closed loop pole at $\bar{\gamma}$ on the root locus can be found as

$$|\bar{k}| = \frac{\prod_{i=1}^{n}|\bar{\gamma} - p_i|}{\prod_{i=1}^{m}|\bar{\gamma} - z_i|} \tag{7.4.5}$$

(vii) The angle of departure from a single pole, p_i, is given by

$$\theta_{p_i} = \pi + \sum_{j=1}^{m} \angle(p_i - z_j) - \sum_{\substack{j=1 \\ j \neq i}}^{n} \angle(p_i - p_j) \tag{7.4.6}$$

(viii) The angle of approach to a single zero, z_i, is given by

$$\theta_{z_i} = \pi - \sum_{\substack{j=1 \\ j \neq i}}^{m} \angle(z_i - z_j) + \sum_{j=1}^{n} \angle(z_i - p_j) \tag{7.4.7}$$

(ix) A necessary (though *not* sufficient) condition for a νth-order breakpoint to occur at $\gamma = \bar{\gamma}$ is

$$\left(\frac{d}{d\gamma}\right)^r \left\{\frac{1}{k} H_{ol}(\gamma)\right\}_{\gamma = \bar{\gamma}} = 0, \quad \text{for } r = 1, 2, \ldots, \nu \tag{7.4.8}$$

Note: A νth-order breakpoint refers to a point $\bar{\gamma}$ that is a closed loop pole of multiplicity $(\nu + 1)$ for some k. These points have $(\nu + 1)$ branches entering and $(\nu + 1)$ branches leaving in the root locus.

Proof:

(i) Equation (7.4.2) may be rewritten as

$$k \prod_{i=1}^{m}(\gamma - z_i) + \prod_{i=1}^{n}(\gamma - p_i) = 0 \tag{7.4.9}$$

Clearly, then, for $k = 0$, the n solutions are $\gamma = p_i$.

(ii) Equation (7.4.2) can also be rewritten as

$$\prod_{i=1}^{m}(\gamma - z_i) + \frac{1}{k}\prod_{i=1}^{n}(\gamma - p_i) = 0 \tag{7.4.10}$$

and thus as $1/k \to 0$ (or $k \to \infty$) m of the solutions tend to $\gamma = z_i$.

(iii) It can be shown that, for γ large, (7.4.2) can be written as

$$\gamma^l\left\{\frac{1 - (1/\gamma)\Sigma p_i}{1 - (1/\gamma)\Sigma z_i} + 0\left(\frac{1}{\gamma^2}\right)\right\} = -k \tag{7.4.11}$$

or

$$\gamma^l\left\{1 - \frac{1}{\gamma}\left(\Sigma p_i - \Sigma z_i\right)\right\} \approx -k \tag{7.4.12}$$

(where $0(1/\gamma^2)$ denotes terms that for large γ tend to zero as $1/\gamma^2$). Thus we have

$$\gamma\left(1 - \frac{1}{\gamma}\left(\Sigma p_i - \Sigma z_i\right)^{1/l}\right) \approx (-k)^{1/l} \tag{7.4.13}$$

or (for γ large)

$$\gamma\left(1 - \frac{1}{\gamma}\frac{1}{l}\left(\Sigma p_i - \Sigma z_i\right)\right) \approx (-k)^{1/l} \tag{7.4.14}$$

and so

$$\gamma - \bar{\sigma} \approx (-k)^{1/l} \tag{7.4.15}$$

Note that the only approximations in the above derivation are terms $0(1/\gamma^2)$, and so (7.4.15) gives the asymptotes for the root locus.

(iv) This result follows by converting (7.4.2) to polar form and noting that the left side of (7.4.4) is the angle of the left side of (7.4.2).

(v) Follows directly from (iv).

(vi) Follows similarly to (iv) on taking the complex modulus of both sides of (7.4.2).

(vii) Let $\bar{\gamma}$ be a point on the root locus near p_i; that is, let $\bar{\gamma} = p_i + \epsilon$. Then (iv) gives

$$\theta_{p_i} = \angle\epsilon = \pi - \sum_{\substack{j \neq i \\ j=1}}^{n} \angle(p_i + \epsilon - p_j) + \sum_{j=1}^{m} \angle(p_i + \epsilon - z_j)$$

$$\approx \pi - \sum_{\substack{j \neq i \\ j=1}}^{n} \angle(p_i - p_j) + \sum_{j=1}^{m} \angle(p_i - z_j) \tag{7.4.16}$$

for small ϵ since the poles are distinct (and assuming no pole zero cancellations occur).

(viii) As for (vii).

(ix) Suppose $\bar{\gamma}$ is a root of multiplicity ν of the closed loop system for some k. Then $f(\gamma) = H_{ol}(\gamma) + 1$ satisfies

$$\left.\begin{aligned} f(\bar{\gamma}) &= 0 \\ f^1(\bar{\gamma}) &= 0 \\ &\vdots \\ f^{\nu}(\bar{\gamma}) &= 0 \end{aligned}\right\} \quad (7.4.17)$$

where $f^i(\bar{\gamma})$ denotes $(d/d\gamma)^i\{f(\gamma)\}|_{\gamma=\bar{\gamma}}$ for some k. The result then follows. Note that if $\bar{\gamma}$ is real the condition is necessary and sufficient if either $(1/k)H_{ol}(\bar{\gamma}) < 0$ or we allow k to be positive or negative. ▽▽▽

Corollary 7.4.1

In the case where k takes negative values, the preceding result becomes:

(i) The n branches start at the n open loop poles.
(ii) m of these branches terminate at the open loop zeros.
(iii) The remaining l branches tend to ∞ as $k \to -\infty$ with asymptotes making angles $(\nu 2\pi)/l$ with the positive real axis and intersecting at $\bar{\sigma} + j0$, where $\bar{\sigma}$ is as in (7.4.3).
(iv) The point $\bar{\gamma}$ will lie on the root locus if and only if

$$\sum_{i=1}^{n} \angle(\bar{\gamma} - p_i) - \sum_{i=1}^{m} \angle(\bar{\gamma} - z_i) = \pm 2\nu\pi \quad (7.4.18)$$

for some integer ν.

(v) A point on the real axis will lie on the root locus if and only if there is an even number of poles plus zeros to the right of the point in the real axis.
(vi) The gain $\bar{k}$, which places a closed loop pole at $\bar{\gamma}$ on the root locus, can be found using (7.4.5).
(vii) The angle of departure from a single pole, p_i, is given by

$$\theta_{p_i} = \sum_{j=1}^{m} \angle(p_i - z_j) - \sum_{\substack{j=1 \\ j \neq i}}^{n} \angle(p_i - p_j) \quad (7.4.19)$$

(viii) The angle of approach to a single zero, z_i, is given by

$$\theta_{z_i} = -\sum_{\substack{j=1 \\ j \neq i}}^{m} \angle(z_i - z_j) + \sum_{j=1}^{n} \angle(z_i - p_j) \quad (7.4.20)$$

(ix) Equation (7.4.8) is a necessary (but not sufficient) condition for νth-order breakpoint at $\gamma = \bar{\gamma}$.

Proof: See Problem 7.12. ▽▽▽

A key point of the preceding result is that the criteria given, in nearly all cases, have a simple geometric interpretation. This is of considerable value as tools for graphical construction, as we illustrate in Example 7.4.1.

Example 7.4.1

For a discrete time (delta) system with open loop transfer function,

$$G(\gamma) = k \frac{\gamma^2 + 2\gamma + 2}{\gamma^4 + \gamma^3 + 14\gamma^2 + 8\gamma} \tag{7.4.21}$$

where $\Delta = 0.25$. Find the root locus for this system for positive k. Find the largest value of k that will give closed loop stability.

First, we note that the open loop poles are at $\gamma = 0, -1, -2, -4$ and the open loop zeros are at $\gamma = -1 \pm j$. Using rules (i) to (iii), we see that the root loci start at the four open loop poles. Two of the branches terminate at the open loop zeros, and the remaining two tend to infinity along the asymptotes making an angle of $\pm \pi/2$ with intersection point $(-2.5 \pm j0)$. Using (v), we have that the following intervals on the real axis, $[-4, -2]$ and $[-1, 0]$, are part of the root locus. The branches must break from these segments to tend to the asymptotes or open loop zeros. We thus look for the breakpoints using (ix). Since there are only two possible branches entering the breakpoints, we look for first-order breakpoints.

$$\frac{d}{d\gamma}\left(\frac{1}{k}G(\gamma)\right) = \frac{-(2\gamma^5 + 13\gamma^4 + 36\gamma^3 + 62\gamma^2 + 56\gamma + 16)}{(\gamma^4 + 7\gamma^3 + 14\gamma^2 + 8\gamma)^2} \tag{7.4.22}$$

We are thus looking for solutions to

$$\bar{\gamma}^5 + 6.5\bar{\gamma}^4 + 18\bar{\gamma}^3 + 31\bar{\gamma}^2 + 28\bar{\gamma} + 8 = 0 \tag{7.4.23}$$

in the ranges $[-4, -2]$ and $[-1, 0]$. Solving numerically, the two relevant solutions are

$$\bar{\gamma} = -0.4839, \quad -3.0157 \tag{7.4.24}$$

We now use (viii) to determine the angle of arrival at the closed loop zeros:

$$\theta_{(-1+j)} = \pi - \angle[(-1+j) - (-1-j)] + \angle[(-1+j) - 0]$$
$$+ \angle[(-1+j) - (-1)] + \angle[(-1+j) - (-2)]$$
$$+ \angle[(-1+j) - (-4)]$$
$$= \pi - \frac{\pi}{2} + \frac{3\pi}{4} + \frac{\pi}{2} + \frac{\pi}{4} + 0.3218 \tag{7.4.25}$$
$$= 0.3218^c \quad \text{or} \quad 18.4349° \tag{7.4.26}$$

Using the preceding analysis, we obtain the root locus shown in Figure 7.4.2. From Figure 7.4.2 and using Equation (7.4.5), we can evaluate the gain k for which the locus leaves the stability boundary at the points marked as □. This gives that

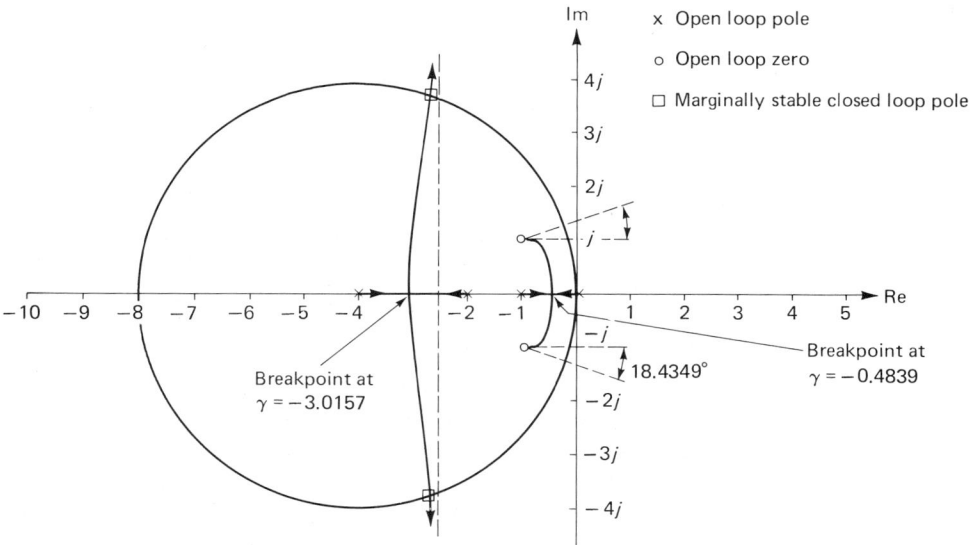

Figure 7.4.2 Root locus for Example 7.4.1.

closed loop stability is maintained for gains k such that

$$k < \frac{4 \times 3.83 \times 4.10 \times 4.63}{3.23 \times 5.05}$$
$$= 17.8 \tag{7.4.27}$$

▽▽▽

7.5 FREQUENCY DOMAIN STABILITY CRITERIA

We next turn to stability tests that are directly describable in terms of the frequency response. These tests have a particular significance in feedback systems. Thus, in the sequel we consider again the prototype unity feedback system shown in Figure 7.4.1.

The equations governing this simple feedback system are

$$A(\rho)y = B(\rho)u \tag{7.5.1}$$
$$u = y^* - y \tag{7.5.2}$$

where y^* is the external set point. Hence the closed loop is described by

$$[A(\rho) + B(\rho)]y = B(\rho)y^* \tag{7.5.3}$$

The stability of the closed loop system can thus be studied by examining the closed loop characteristic polynomial:

$$A_c(\rho) \triangleq A(\rho) + B(\rho) \tag{7.5.4}$$

This polynomial could be tested for stability using the methods of Sections (7.2) or (7.3).

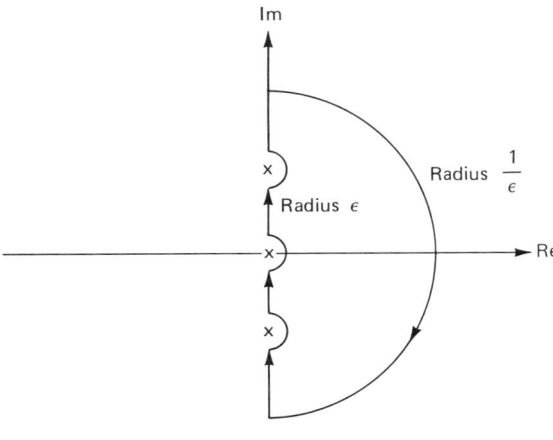

Figure 7.5.1 Continuous time Nyquist contour.

Here, however, alternative methods for analyzing closed loop stability based on the *open* loop frequency response will be examined. We will see in Section 7.6 that these tests have other attractive features, such as allowing a treatment of some nonlinearities in the system, robustness of the feedback system to imperfect modeling, and a treatment of multivariable systems with small interactions.

The frequency response of a linear system has been discussed in Chapter 6. We will need to slightly generalize this concept since technical difficulties arise if the open loop transfer function has poles on the stability boundary. (Note that this situation is reasonably frequent in practice. In particular, we often use an integrator in the controller for reasons to be discussed in Chapter 8.)

We therefore introduce the Nyquist contour shown in Figure 7.5.1 for continuous time systems and in Figure 7.5.2 for discrete time systems using shift operators and Figure 7.5.3 for discrete time systems using delta operators. Note that the Nyquist contour comprises the stability boundary with some extra contours. In each case we take the Nyquist contour as the limit $\epsilon \to 0$ of the contour shown. We then define the Nyquist *plot N* as the function formed by evaluating the open loop transfer function around the Nyquist contour. Note that, apart from the return path in continuous time and the indentations, this basically amounts to evaluation of the open loop frequency response of the system. Based on these definitions, we have the following theorem.

Theorem 7.5.1

Consider a negative feedback system with open loop transfer function $H_{ol}(\gamma)$ having n_0 unstable poles. Let n be the number of times the Nyquist plot encircles the point $-1 + j0$ in the anticlockwise direction. Then the closed loop system has $n_0 - n$ unstable poles.

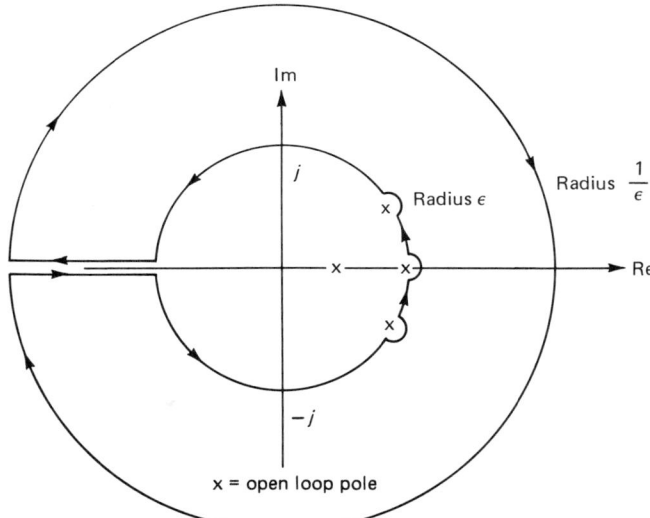

Figure 7.5.2 Discrete time Nyquist contour: Shift domain.

Proof: This result is well known in elementary treatments of control and thus will not be reproduced here. The full proof can be found in references such as Kuo (1982). Alternatively, those familiar with complex analysis will recognize the result as an elementary consequence of either conformal mapping or the principle of the argument. ▽▽▽

Corollary 7.5.1. Nyquist Criterion

A control system with open loop transfer function $kH_{ol}(\gamma)$ will give a stable closed loop system if and only if the Nyquist plot of $H_{ol}(\gamma)$ encircles the point, $-(1/k) + j0$, n_0 times in the anticlockwise sense, where n_0 is the number of unstable poles of $H_{ol}(\gamma)$.

Proof: Follows directly from Theorem 7.5.1. ▽▽▽

Since $H_{ol}(\gamma^*) = H_{ol}^*(\gamma)$, it follows that we need only evaluate the Nyquist plot over half of the Nyquist contour. The following simple example illustrates the use of the Nyquist criterion.

Example 7.5.1

Say we are interested in the set of values of k that lead to a stable closed loop system when the open loop system has continuous time transfer function

$$H_{ol}(s) = \frac{2k}{s(s+1)(s+2)} \tag{7.5.5}$$

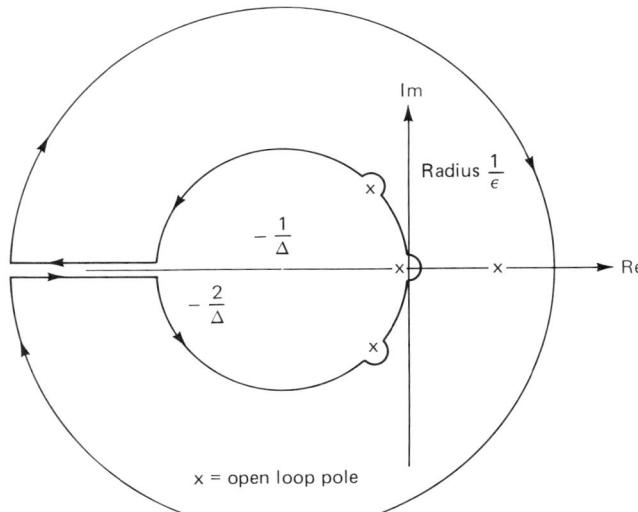

Figure 7.5.3 Discrete time Nyquist contour: Delta domain.

The Nyquist diagram for $2/[s(s + 1)(s + 2)]$ is shown in Figure 7.5.4. Using Corollary 7.5.1, we conclude that the closed loop system is stable provided $-1/k$ lies in the range $(-\infty, -\frac{1}{3})$; that is, provided $0 < k < 3$. We leave it as an exercise for the reader to verify, using the polynomial stability tests developed previously, that this is indeed the case.

The Nyquist criterion gives rise to several important concepts in the study of feedback systems. Two of these are the *gain margin* and the *phase margin*.

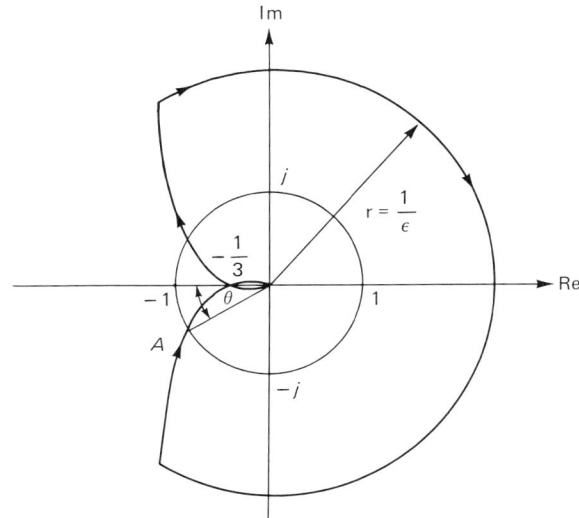

Figure 7.5.4 Nyquist diagram for Example 7.5.1.

The term gain margin describes the maximum amount the open loop gain may be increased without causing instability. For the system with open loop transfer function $2/[s(s+1)(s+2)]$, as studied in Example 7.5.1, we see from the Nyquist diagram that the gain margin is 3. As a rule of thumb, we usually aim for a gain margin of 2 or more when designing a controller for a feedback loop.

The phase margin describes the amount of extra phase lag that may be introduced, at the frequency(ies) where the open loop gain is 1, without causing instability. The point(s) where the open loop gain is 1 may be found from the intersection of a unit circle with the Nyquist plot (see, for example, point A on Figure 7.5.4). The phase margin may then be determined graphically to be the angle θ between this point and the origin with respect to the negative real axis. In Example 7.5.1, we see from Figure 7.5.4 that the phase margin is 33°. When designing a control system we usually aim for a phase margin $> 45°$.

7.6 ADVANCED FREQUENCY DOMAIN STABILITY ANALYSIS

In the previous section we gave some of the basic ideas involved in stability analysis based on open loop frequency response data. In this section we will investigate some more advanced concepts based on the frequency response: (1) robustness analysis (that is, testing for stability when we have imperfect knowledge of the system), and (2) circle criterion analysis (which gives stability conditions for systems with a nonlinearity and/or time variations).

7.6.1 Robustness of Feedback Systems

In most real situations, an accurate system model is only available over a limited range of frequencies. Let $H'_{ol}(\gamma)$ be the true open loop transfer function and $\hat{H}_{ol}(\gamma)$ be our model of the open loop system. Then we define

$$\Delta H_{ol}(\gamma) = H'_{ol}(\gamma) - \hat{H}_{ol}(\gamma) \tag{7.6.1}$$

Of course, $\Delta H_{ol}(\gamma)$ is not known. However, we may have some information about the maximum size of $\Delta H_{ol}(\gamma)$. Lemma 7.6.1 gives a sufficient condition for closed loop stability of $H'_{ol}(\gamma)$ based on $\hat{H}_{ol}(\gamma)$.

Lemma 7.6.1

Suppose that:

(i) $\hat{H}_{ol}(\gamma)$ gives a stable closed loop; that is, the Nyquist plot of $\hat{H}_{ol}(\gamma)$ encircles the -1 point anticlockwise n_0 times.
(ii) $\hat{H}_{ol}(\gamma)$ and $H'_{ol}(\gamma)$ have the same number of unstable poles.
(iii) $|\Delta H_{ol}(\gamma)| < |1 + \hat{H}_{ol}(\gamma)|$ for all γ on the stability boundary,

Then

$$H'_{ol}(r) = \hat{H}_{ol}(r) + \Delta H_{ol}(r) \quad \text{also gives a stable closed loop.}$$

Proof: The result follows directly from Nyquist's stability criterion, since condition (iii) implies $H'_{ol}(\gamma)$ encircles the -1 point the same number of times as $\hat{H}_{ol}$. ▽▽▽

Lemma 7.6.1 gives the result when the perturbation to the normal system is additive and shows that, if the perturbation is small (in a sense that depends on the nominal system), closed loop stability can be maintained. Corollary 7.6.1 gives a similar result for the case of multiplicative (cascaded) perturbations. In this case, the true open loop system is related to the nominal open loop system as follows:

$$H_{ol}(\gamma) = \hat{H}_{ol}(\gamma) G(\gamma) \qquad (7.6.2)$$

Corollary 7.6.1

Suppose that:

(i) $\hat{H}_{ol}(\gamma)$ gives a stable closed loop.
(ii) $G(\gamma)$ is open loop stable.
(iii) $|1 - G(\gamma)| < \left| \dfrac{1 + \hat{H}_{ol}(\gamma)}{\hat{H}_{ol}(\gamma)} \right|$ for all γ on the stability boundary. Then $H'_{ol}(\gamma) = \hat{H}_{ol}(\gamma) G(\gamma)$ also gives a stable closed loop.

Proof: Follows from Lemma 7.6.1, where we use $\Delta H_{ol}(\gamma) = \hat{H}_{ol}(\gamma)(G(\gamma) - 1)$. ▽▽▽

Note that the function $|(1 + \hat{H}_{ol}(\gamma)/\hat{H}_{ol}(\gamma))|$ (for multiplicative perturbations) for γ on the stability boundary may be plotted and used as a measure of the robustness of a feedback control system.

7.6.2 Systems with Nonlinearities

Another useful result based on the open loop frequency response is the circle criterion, which is useful for classes of nonlinear and/or time varying systems. Before proceeding to this result, we first need to define a sector nonlinearity.

Definition 7.6.1. Sector Nonlinearity

A nonlinear, time varying function $\phi(t, u)$ is said to belong to the sector $[\alpha, \beta]$ if

$$\alpha u^2 \leq u\phi(t, u) \leq \beta u^2, \qquad \text{for all } u \text{ and } t \qquad (7.6.3)$$

▽▽▽

One example of such a nonlinearity is the saturation nonlinearity present in many real control systems and defined by

$$\text{sat}(u) \triangleq \begin{cases} u_{max}: & u > u_{max} \\ u : & u_{min} \leq u \leq u_{max} \\ u_{min}: & u < u_{min} \end{cases} \qquad (7.6.4)$$

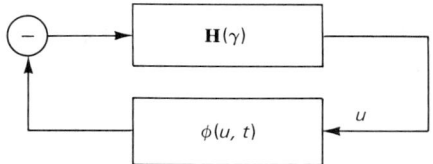

Figure 7.6.1 Feedback system with nonlinearity for circle criterion.

We can show that the saturation nonlinearity belongs to the sector $[0, 1]$ provided $u_{\min} < 0 < u_{\max}$.

Lemma 7.6.2 gives a sufficient condition for asymptotic stability of a system containing a sector nonlinearity.

Lemma 7.6.2. Circle Criterion

Consider a negative feedback system comprised of a linear, time invariant system with transfer function $H(\gamma)$ and a nonlinearity belonging to the sector $[\alpha, \beta]$ (see Figure 7.6.1). Then a sufficient condition for stability of the system is:

(i) The Nyquist plot of $H(\gamma)$ should encircle the region

$$\left| w + \frac{(1/\alpha) + (1/\beta)}{2} \right| < \left(\frac{(1/\alpha) + (1/\beta)}{2} \right)$$

(in the complex w plane) n_0 times, where n_0 is the number of unstable poles of $H(\gamma)$ and $\beta > \alpha > 0$.

(ii) For $\beta > \alpha = 0$, we require that the Nyquist plot of $H(\gamma)$ should satisfy $\text{Re}\{H(\gamma)\} > -1/\beta$ and that $H(\gamma)$ should be stable.

Proof: See Vidyasagar (1978) for a proof of this and related results. ▽▽▽

The circle criterion derives its name from the fact that the region, described by (i) in Lemma 7.6.2, is a circle in the complex w plane, with diameter the segment $(-(1/\alpha) + j0, -(1/\beta) + j0)$ (see Figure 7.6.2). Example 7.6.1 illustrates the use of the circle criterion.

Example 7.6.1

Consider two continuous time systems with open loop transfer functions:

(i) $H(s) = \dfrac{2}{s(s+1)(s+2)}$

(ii) $H(s) = \dfrac{2}{s(s+2)}$

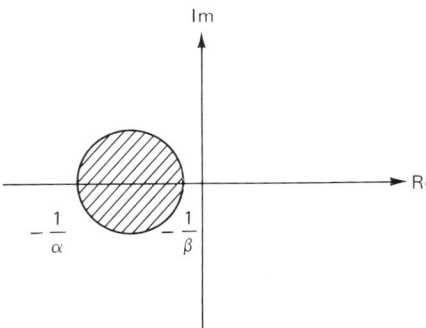

Figure 7.6.2 Region for circle criterion part (i).

If the closed loop containing these systems includes a saturation nonlinearity, which of the closed loop systems is stable?

(i) In this case, the conditions required to apply the circle criterion are *not* satisfied and so no conclusions can be drawn. For example, $H(j0.5) = -1.12941 - j1.31765$, which does *not* satisfy $\text{Re}\{H(j\omega)\} > -1/2$.

(ii) Note that in this case, by drawing the Nyquist curve (or otherwise), we may show that $\text{Re}\{H(j\omega)\} \geq -\frac{1}{2} \ \forall \ \omega$, and so by the circle criterion the system is stable. ∇∇∇

This completes our treatment of the classical approach to the question of stability for control systems. We next turn to other important issues relating to the performance of systems. We begin by examining the question of tracking performance.

7.7 TRACKING PERFORMANCE

7.7.1 Specifications

Perhaps the simplest performance specifications to give for a control system are the steady-state characteristics. The issue here is, given a class of set-point sequences, y^*, and disturbances, d, how well does the system output track y^*?

The general setup that we consider here is shown in Figure 7.7.1. Of particular interest here is the steady-state value of the tracking error e.

Let us first consider the case when $d = 0$ and the transfer function of the plant plus controller is $H_{ol}(\gamma)$. Then the closed loop transfer function relating y to y^* is

$$H_{cl}(\gamma) = \frac{H_{ol}(\gamma)}{1 + H_{ol}(\gamma)}.$$

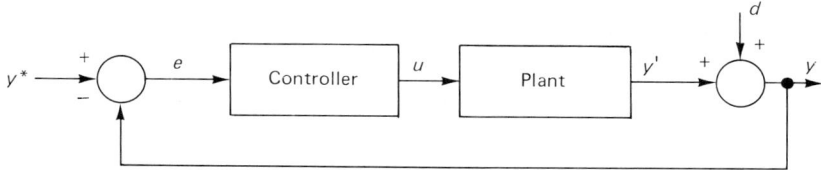

Figure 7.7.1 Closed loop system with output disturbance.

Thus a reference input, $y^*(t)$, with transform $Y^*(\gamma)$ gives an output, $y(t)$, with transform

$$Y(\gamma) = \frac{H_{ol}(\gamma)}{1 + H_{ol}(\gamma)} Y^*(\gamma) \tag{7.7.1}$$

The transform of the error e between $y(t)$ and $y^*(t)$ is

$$E(\gamma) = T\{y^*(t) - y(t)\} = \left(\frac{1}{1 + H_{ol}(\gamma)}\right) Y^*(\gamma) \tag{7.7.2}$$

If $y^*(t)$ is a unit step, then $Y^*(\gamma) = 1/\gamma$, and the final value theorem, Theorem 4.4.1(xi), gives the steady-state error as

$$e(\infty) = \lim_{\gamma \to 0} (\gamma Y(\gamma)) = \lim_{\gamma \to 0} \left(\frac{1}{1 + H_{ol}(\gamma)}\right) \tag{7.7.3}$$

$$= \frac{1}{1 + H_{ol}(0)} \tag{7.7.4}$$

provided the closed loop system is stable. Note that $H_{ol}(0)$ is simply the dc gain of the open loop, and so (7.7.4) has a simple physical interpretation.

More generally, if we have a sinusoidal input of magnitude 1 and frequency ω, then the amplitude of the steady-state error is readily seen to be (see Problem 14).

$$a(\infty) = \left|\frac{1}{1 + H_{ol}((e^{j\omega\Delta} - 1)/\Delta)}\right| \tag{7.7.5}$$

Thus the magnitude of the error is the reciprocal of the magnitude of the frequency response of $1 + H_{ol}(\gamma)$ at the frequency ω. The transfer $1 + H_{ol}(\gamma)$ is usually called the *return difference*, and its inverse is called the *sensitivity function*.

Also, with reference of sinusoidal inputs, we are frequently interested in amplitude and phase errors. The normalized amplitude error is defined to be the difference in amplitude of the output and reference; that is,

$$\bar{a}(\infty) \triangleq \|y_s\| - \|y^*\| = \left|\frac{H_{ol}[(e^{j\omega\Delta} - 1)/\Delta]}{1 + H_{ol}[(e^{j\omega\Delta} - 1)/\Delta]}\right| - 1 \tag{7.7.6}$$

where y_s is the steady-state output response and $\|\cdot\|$ denotes magnitude.

Similarly, the phase error is defined to be

$$\phi(\infty) \triangleq \text{Arg}\{y_s\} - \text{Arg}\{y^*\} \tag{7.7.7}$$

where $\text{Arg}\{y_s\}$ denotes the phase angle of y in steady state.

It is easy to see (Problem 16) that the phase error is equal to the phase angle of the open loop transfer function minus the phase angle of the return difference.

The problem of output disturbance rejection is almost the same as that of tracking. The closed loop transfer function relating the output y to the disturbance d in Figure 7.7.1 is

$$H_{d_{cl}}(\gamma) = \frac{1}{1 + H_{ol}(\gamma)} \tag{7.7.8}$$

Note that this transfer function is precisely the same as that relating $e = (y^* - y)$ to y in Equation (7.7.2). Thus the formulas (7.7.4) and (7.7.5) for steady-state errors in response to a unit step disturbance and a unit amplitude sinusoidal disturbance, respectively, are applicable also to disturbance rejection.

Thus we see that output disturbance rejection is a dual property of tracking. A feedback system that tracks well over a certain bandwidth will also reject output disturbances well over the same bandwidth.

7.7.2 System Type

A special class of set points and disturbances is those that are described as polynomial functions of time. Included in this class are steps, ramps, parabolas, and the like. It turns out that, associated with each type of set point or disturbance, in this class, there exists a corresponding type of system for which the steady-state tracking error is zero. We first establish the following result, which gives necessary and sufficient conditions for zero tracking error to a class of polynomial set points.

Lemma 7.7.1

A feedback system gives zero steady tracking error for polynomial set points of degree less than or equal to k if and only if any of the following conditions hold:

(i) $H_{cl}(0) = 1$ and $H_{cl}^{(i)}(0) = 0$, for $i = 1 \ldots k$ where $H_{cl}^{(i)}(0)$ denotes

$$\left[\left(\frac{d}{d\gamma}\right)^i H_{cl}(\gamma)\right]\bigg|_{\gamma=0} \tag{7.7.9}$$

(ii) When

$$H_{cl}(\gamma) = \frac{B_c(\gamma)}{A_c(\gamma)} = \frac{b_m \gamma^m + \cdots + b_0}{\gamma^n + a_{n-1}\gamma^{n-1} + \cdots + a_0}$$

then we require $b_i = a_i$ for $i = 0 \ldots k$

(iii) $H_{ol}(\gamma) = \frac{1}{\gamma^{k+1}} \overline{H}_{ol}(\gamma)$ \hfill (7.7.10)

where $\overline{H}_{ol}(0) \neq 0$.

Proof: See Problems 17 and 18. ▽▽▽

Part (iii) of Lemma 7.7.1 motivates the Definition 7.7.1.

Definition 7.7.1

A transfer function $H_{ol}(\gamma)$ is said to have type k if $H_{ol}(\gamma)$ can be written in the form (7.7.10); that is, it has at least k pure integrators. ▽▽▽

Thus we see that if a system has type k then the tracking error for polynomial set points of any degree less than k will be zero. Furthermore, the steady-state error to a set point of the form t^k is

$$e(\infty) = \frac{k!}{\overline{H}_{ol}(0)} \qquad (7.7.11)$$

where $\overline{H}_{ol}(\gamma)$ is defined as in (7.7.10) See Problem 7.19.

7.7.3 Internal Model Principle

One way of describing the result of the previous section is to note that the polynomial set points and disturbances t^k can be modeled as

$$D(\rho)g(t) = 0 \qquad (7.7.12)$$

where

$$g(t) = t^k, \quad D(\rho) = \rho^{k+1} \qquad (7.7.13)$$

Moreover, to achieve zero tracking error, Lemma 7.7.1(iii) tells us that the plant plus controller should contain the model for the disturbances and set point, that is, contain k pure integrators.

This result can be generalized to other disturbances and set points by introducing alternative operators, $D(\rho)$, in (7.7.12). This result is known as the internal model principle.

We have seen in Sections 2.5 and 3.5 that general deterministic disturbances can be modeled as in (7.7.12), but where $D(\rho)$ nulls the appropriate time function. For example, a sine wave of frequency ω_0 is modeled as in (7.7.12), where

$$D(\rho) = \rho^2 + \rho \left\{ \frac{2(1 - \cos \omega_0 \Delta)}{\Delta} \right\} + \frac{2(1 - \cos \omega_0 \Delta)}{\Delta^2} \qquad (7.7.14)$$

The idea of the internal model principle is to include $1/D(\rho)$ in the control law. To see how this works, consider the feedback system shown in Figure 7.7.1. Let us assume that the plant has model

$$A(\rho)y' = B(\rho)u \qquad (7.7.15)$$

The internal model principle suggests a control law of the form

$$S(\rho)L(\rho)u = P(\rho)e \qquad (7.7.16)$$

where $S(\rho)$ nulls both the disturbance and the set point. From Figure 7.7.1, we obtain the following closed loop equations:

$$Ay = Bu + Ad$$
$$e = y^* - y$$
$$SLu = Pe$$

or

$$(ASL + BP)y = BPy^* + ASLd \qquad (7.7.17)$$

Subtracting $(ASL + BP)y^*$ from both sides of (7.7.17) gives

$$(ASL + BP)(y - y^*) = ASL(d - y^*) \qquad (7.7.18)$$

Now, using the nulling property of $S(\rho)$, we see that the right side of (7.7.18) is identically zero. Thus, provided we choose L and P so that $(ASL + BP)$ is stable, then y will tend exponentially fast to y^* and there will be no steady-state error.

Note that the internal model principle is widely used in industrial control when $S(\rho)$ takes the simple form $S(\rho) = \rho$. The internal model principle then amounts to including integral action and, as shown previously, this leads to zero errors for constant set points and constant disturbances.

More sophisticated uses of the internal model principle also find application in industry. For example, $S = \rho^2$ can be used to track ramp set points. We have also used the more general form of the principle to cancel sinusoidal disturbances in rolling mills resulting from roll eccentricity (see Chapter 15).

7.8 TRANSIENT PERFORMANCE

7.8.1 Time Domain Specifications

The previous section has examined the steady-state characteristics of control systems. We now turn to the question of the dynamic or transient performance.

The usual specifications of transient performance are in terms of the step response. A typical response showing key performance measures is illustrated in Figure 7.8.1.

7.8.2 Frequency Domain Specifications

We saw in Section 7.7 that there is a very precise relationship between steady-state performance and frequency domain characteristics. However, the relationship between rise time, overshoot, and settling time (that is, the transient parameters) and frequency domain characteristics is less precise, although useful rules of thumb have been developed. Figure 7.8.2 shows features of a typical closed loop frequency response. The heuristic relationships between the frequency domain characteristics and time domain parameters are then as follows:

$$t_r \simeq \frac{2.2}{\omega_3} \qquad (7.8.1)$$

and

$$\sigma \simeq \frac{\sqrt{M_p - 1}}{2.4} \qquad (7.8.2)$$

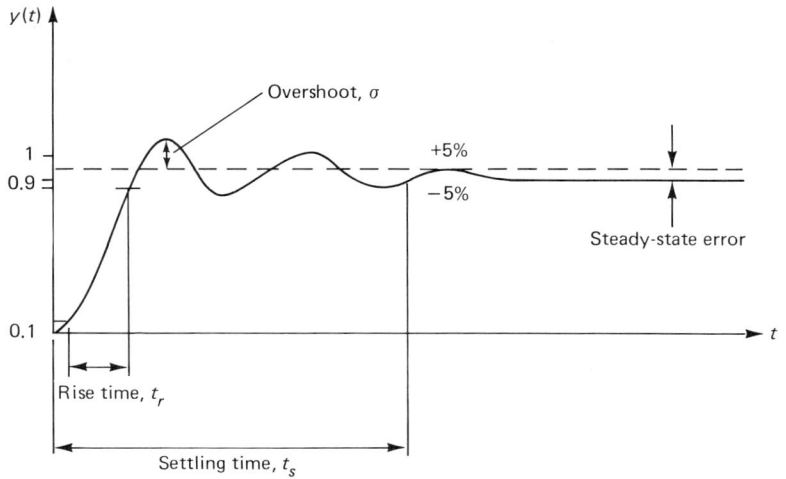

Figure 7.8.1 Typical step response.

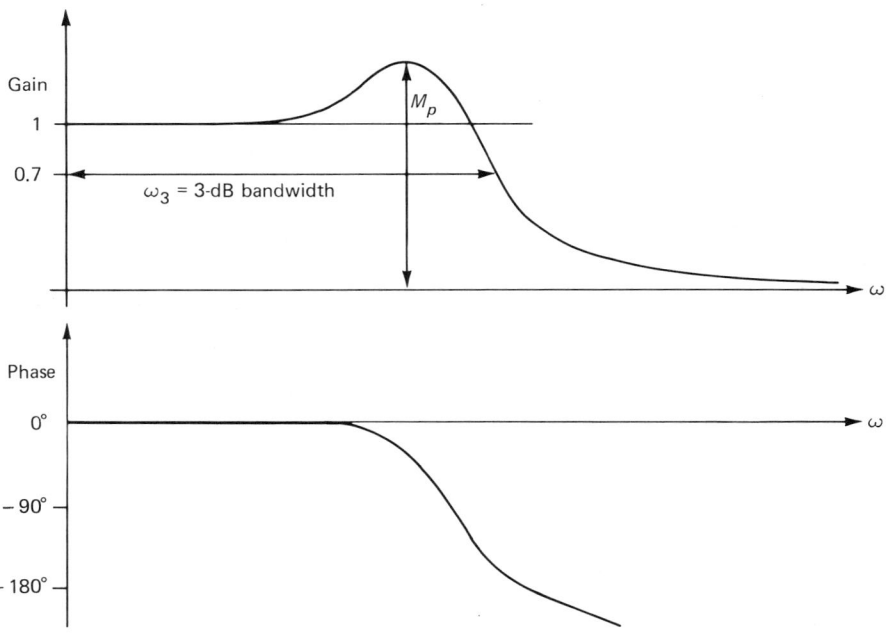

Figure 7.8.2 Bode diagram showing features of a typical closed loop frequency response.

Sec. 7.8 Transient Performance 155

Note that these relationships should be considered as rules of thumb only. However, it is still true, in general, that the bandwidth ω_3 is the dominant factor affecting the rise time, and that the peak in the magnitude response M_p is the dominant effect on the overshoot σ. Example 7.8.1 examines these relationships in a particular case.

Example 7.8.1

Consider the case where we have a second-order, continuous time, closed loop transfer function,

$$H_{cl}(s) = \frac{\omega_n^2}{s^2 + 2\xi\omega_n s + \omega_n^2} \tag{7.8.3}$$

The step response of this system, with zero initial conditions, is then

$$y_s(t) = 1 - \frac{e^{-\xi\omega_n t}}{\sqrt{1-\xi^2}} \sin(\omega t + \phi) \tag{7.8.4}$$

where

$$\phi = \tan^{-1}\left(\frac{\sqrt{1-\xi^2}}{\xi}\right) \tag{7.8.5}$$

and

$$\omega = \sqrt{1-\xi^2}\,\omega_n \tag{7.8.6}$$

In this case we can compare estimates of rise time and overshoot obtained from the frequency domain rules of thumb with the true values for various values of the damping ratio ξ. The overshoot in this case may be shown to be exactly

$$\sigma = e^{-\pi\xi/\sqrt{1-\xi^2}} \tag{7.8.7}$$

and the peak in the magnitude response may be shown to be

$$M_p = \begin{cases} \dfrac{1}{2\xi\sqrt{1-\xi^2}}, & \text{for } \xi < \dfrac{1}{\sqrt{2}} \\ 1, & \text{for } \xi \geq \dfrac{1}{\sqrt{2}} \end{cases} \tag{7.8.8}$$

Figure 7.8.3 shows graphs of the true overshoot (7.8.7) and the estimated overshoot obtained from (7.8.2) versus damping ratio ξ.

In the case of the rise time, there is no analytical solution to the precise rise time, and we must solve this problem numerically. The -3-dB frequency can,

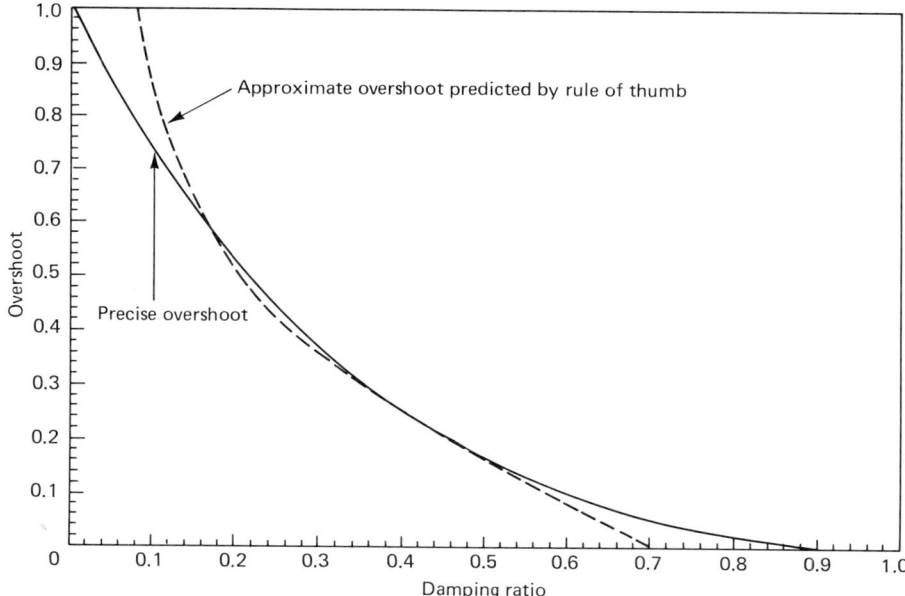

Figure 7.8.3 Precise and approximate overshoot for a second-order system.

however, be shown to be precisely

$$\omega_3 = \sqrt{(1 - \omega\xi^2) + \sqrt{(1 - 2\xi^2)^2 + 1}}\, \omega_n \qquad (7.8.9)$$

Figure 7.8.4 shows a comparison of the true rise time obtained numerically with that obtained using the rule of thumb given in (7.8.1). From the figures, we note a close correlation between the precise time domain performance and those estimated using the frequency domain rules of thumb. ▽▽▽

Thus far we have only considered continuous time systems explicitly. However, if we are *not* using slow sampling (that is, we sample at about ten times the closed loop bandwidth), then the rules of thumb given will also apply in the discrete time case.

In general, when designing a control system, we usually aim for an overshoot of less than 10% (possibly as little as 0%) in the closed loop step response. Thus, using the rule of thumb (7.8.1), we need $M_p < 1.06$ (for 10% overshoot) or $M_p = 1$ (for no overshoot). Similarly, a requirement on the maximum rise time tolerable can be translated into a requirement on the closed loop bandwidth.

We now see how the closed loop frequency domain characteristics as described previously can be translated into open loop frequency domain requirements. The main idea is to relate the open loop frequency response to the magnitude of the closed loop frequency response via M circles. The M circles are the loci of constant closed loop magnitude of frequency response plotted on the Nyquist diagram. Thus

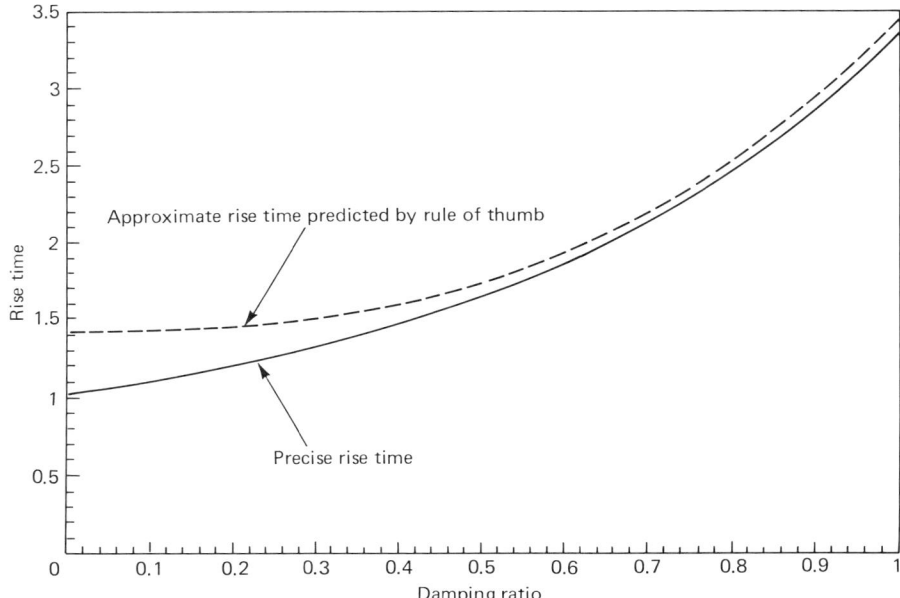

Figure 7.8.4 Precise and approximate rise times for a second-order system.

these loci are the loci of $x + jy$, which satisfy

$$M^2 = |H_{cl}(j\omega)|^2 = \left|\frac{H_{ol}(j\omega)}{1 + H_{ol}(j\omega)}\right|^2$$

$$= \left|\frac{x + jy}{1 + x + jy}\right|^2 \qquad (7.8.10)$$

That is,

$$M^2 = \frac{x^2 + y^2}{1 + 2x + x^2 + y^2} \qquad (7.8.11)$$

Equation (7.8.11) can be rearranged as

$$\left[x - \frac{M^2}{1 - M^2}\right]^2 + y^2 = \left[\frac{M}{1 - M^2}\right]^2 \qquad (7.8.12)$$

This is the equation of a set of circles (called the M circles). The M circles are shown on the Nyquist diagram for M ranging between 0.7 and 1.1 in Figure 7.8.5. (See Problem 7.21 for the case $M = 1$.) These M-circles, together with the Nyquist diagram, give us important information about the closed loop behavior. For example, the frequency corresponding to the point where the Nyquist diagram crosses the $M = 0.7$ circle gives the -3-dB frequency of the closed loop, and thus the rise time may be approximately evaluated from the rule of thumb given earlier. To keep the closed loop resonance peak small (and hence the overshoot small), we

158   Classical Control Systems Analysis   Chap. 7

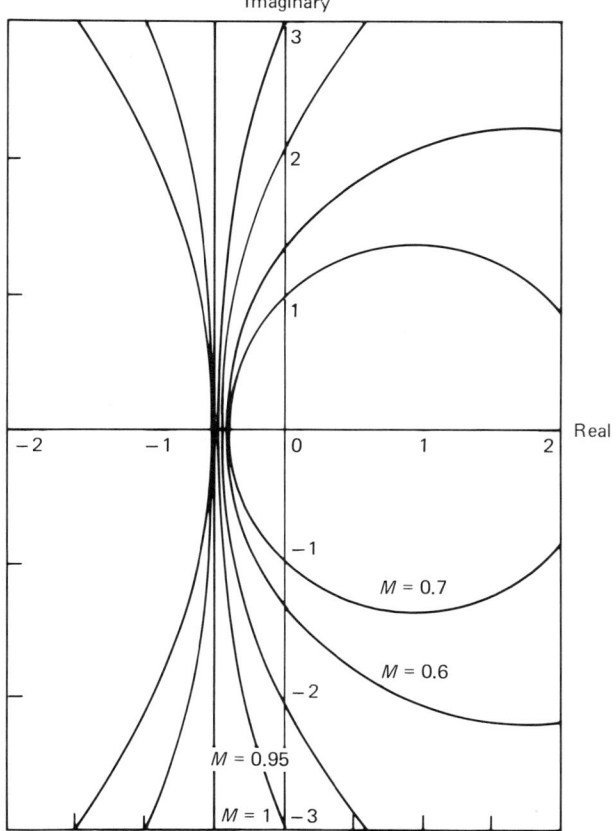

Figure 7.8.5 Constant M loci in the Nyquist diagram.

normally design the controller so that the Nyquist plot remains outside the $M = 1.1$, 1.05, or 1 contour depending on the amount of overshoot tolerable.

7.9 COMPENSATION

In many cases, it is not possible to achieve all the objectives in a control system by simply using a constant gain in the feedback controller. Usually, an upper limit to the gain is dictated by the maximum allowable overshoot. In this case, it is sometimes possible to use a dynamic controller (called a *compensator*) to achieve a better trade-off between the various design objectives, including steady-state error, rise time, overshoot, and so on.

7.9.1 Phase Advance

The basic idea of this compensator is to introduce phase advance in the appropriate region of the Nyquist diagram to allow the gain to be increased without deteriorat-

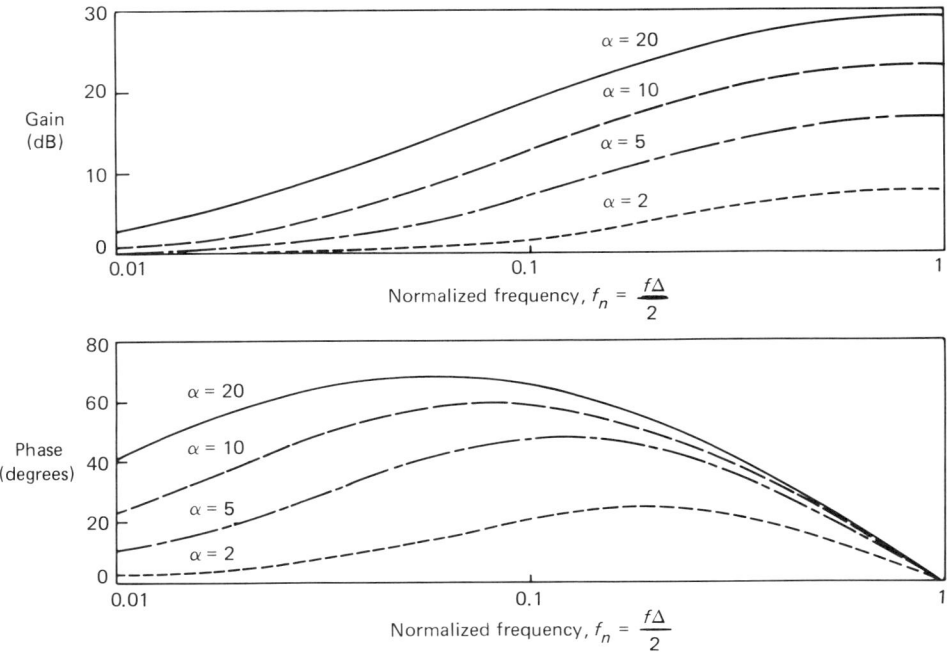

Figure 7.9.1 Bode plots for discrete lead compensator.

ing the overshoot performance. The increased gain then leads to a higher bandwidth and thus a shorter rise time.

A simple compensator that achieves the objective is

$$H_c(\gamma) = \frac{K(1 + \alpha T\gamma)}{1 + T\gamma} \quad (7.9.1)$$

where $T \geq \Delta$ and $\alpha > 1$. The discrete Bode diagrams for $T = 1.5\Delta$ and various values of α for this compensator are shown in Figure 7.9.1. From this we see that the lead compensator gives phase advance beginning at frequencies around $\omega_1 = 1/\alpha T$ and ceasing near $\omega_2 = 1/T$. Another way of viewing the phase advance compensator is as a proportional plus approximate derivative feedback law. This can be seen by reexpressing the compensator (7.9.1) in the form:

$$H_c(\gamma) = K + K(\alpha - 1)T\frac{\gamma}{1 + T\gamma} \quad (7.9.2)$$

where the term K is proportional feedback, and the remaining term is an approximation to a differentiator.

We illustrate the use of this compensator by a simple example.

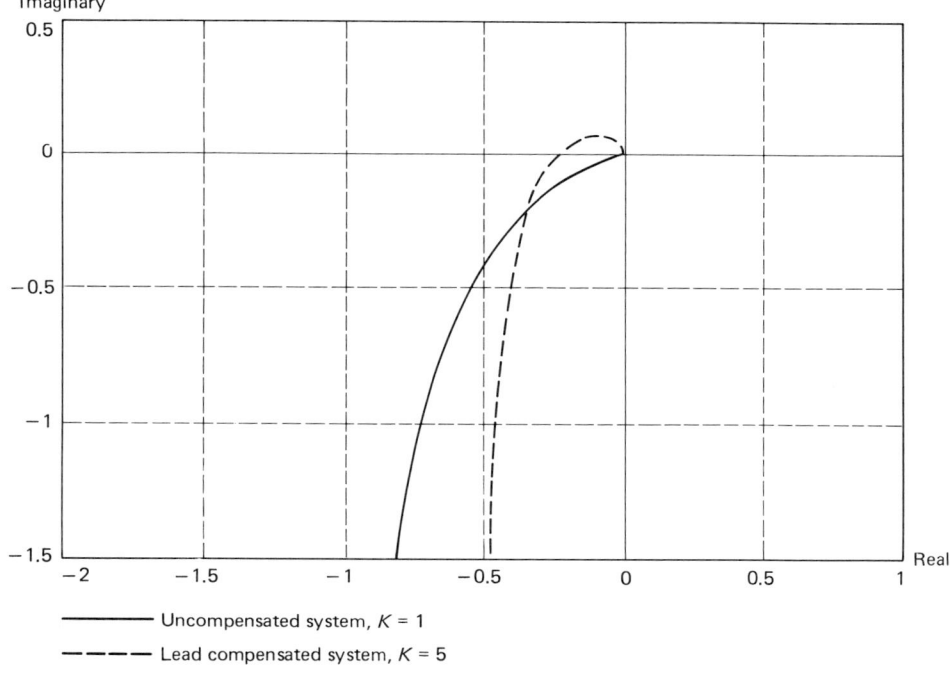

Figure 7.9.2 Nyquist plots for discrete time servosystem.

Example 7.9.1

Consider the control of an electromechanical servosystem, with transfer function

$$H(s) = \frac{1}{s(1+s)} \tag{7.9.3}$$

which gives a discrete transfer function

$$H_\Delta(\gamma) = \frac{0.048\gamma + 0.952}{\gamma(\gamma + 0.952)} \tag{7.9.4}$$

when sampled with a zero-order hold input, and $\Delta = 0.1$ s. Suppose, also, that we wish to have a closed loop bandwidth of $\omega \approx 5$ rad/s. (Note that the sampling frequency, in this case, obeys the rule of thumb, $\omega \approx \frac{1}{10}\omega_s$). The discrete time Nyquist diagram is shown in Figure 7.9.2 for this system. From this diagram we can verify that a fixed gain controller would need to have $K < 1$ to give satisfactory overshoot (that is, $< 10\%$). However, with $K = 1$, the closed loop bandwidth will be about 1 rad/s. This conflict can only be resolved by use of a more sophisticated controller. We thus suggest the use of a phase advance (lead) compensator of the form (7.9.1). From Figure 7.9.2, we see that we would like at least 45° phase advance at $\omega = 5$ rad/s^{-1} (the closed loop bandwidth), since at this frequency the

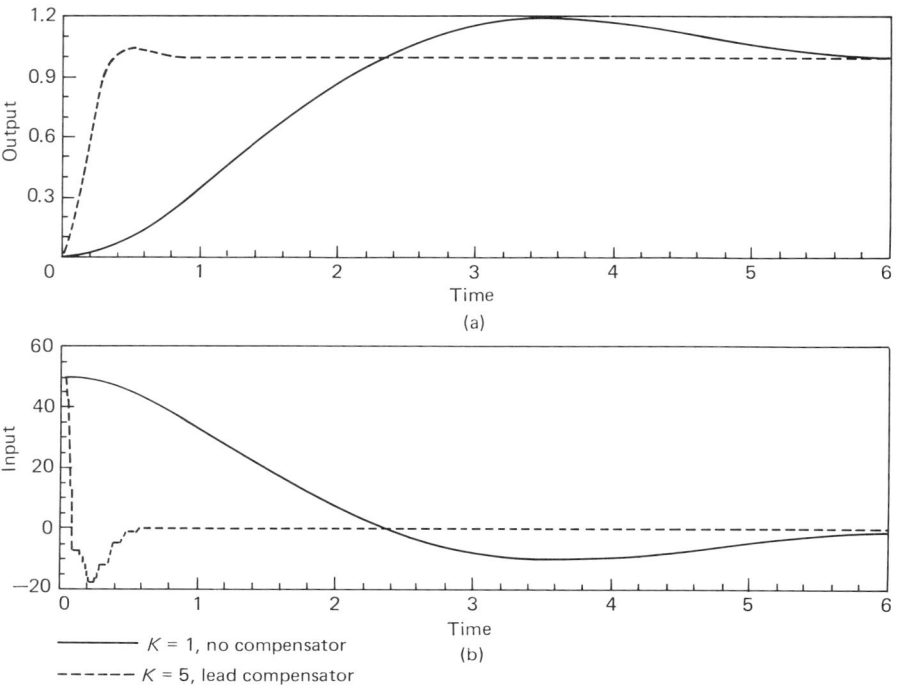

Figure 7.9.3 Plant output and input response for a step reference input. (Plant input for no compensator has been scaled up by 50).

phase is at present very close to 180°. We would also like the phase advance to commence at around $\omega = 1$ rad/s (so that all frequencies up to the desired closed loop are pushed away from the negative real axis). In this case, therefore, we might suggest a compensator of the form (7.9.1), where $T = \Delta = 0.1$ and $\alpha T = 1$ (that is, $\alpha = 10$). With this compensator, the Nyquist diagram for the modified open loop is that also shown in Figure 7.9.2, where we have used $K = 5$ to scale the plot. From the Nyquist plot of the plant plus compensator, we see that a well-damped response will occur for $K \leq 5$. Note, also, that with $K = 5$ the closed loop compensated system has a bandwidth of about 7 rad/s, which exceeds the specification.

Figure 7.9.3 shows the step response and control input for, (a) the uncompensated system with $K = 1$, and (b) the compensated system with $K = 5$. [Note that for part (a), the control input has been scaled up by 50 times.) ▽▽▽

We see from Example 7.9.1 that phase advance compensation is capable of improving the rise time of a system without deteriorating the overshoot. The cost of this improved output performance is as follows:

1. The control law must have increased gain and this may increase the likelihood of the input reaching a constraint.

2. Since the control input usually represents something with a physical cost (for example, electrical or mechanical input with associated energy cost or fuel or chemical input), the phase advance compensator will often be more expensive.
3. Since the closed loop bandwidth has been increased, the robustness to unmodeled dynamics is likely to decrease.

7.9.2 Phase Lag

Another situation where a difficult design compromise occurs is where we wish to reduce the steady-state error by increasing the gain, but this cannot be achieved without sacrificing the overshoot. In this case, a lag compensator can be used that increases the low-frequency gain but then returns to normal gain near some predetermined frequency. This can be achieved by a compensator having the following transfer function:

$$H_c(\gamma) = \frac{\gamma + \omega_1}{\gamma + \alpha\omega_1} \qquad (7.9.5)$$

with $0 \leq \alpha < 1$, $\omega_1 \ll 1/\Delta$. This compensator increases the low-frequency (that is, for $\omega < \omega_1$) gain while leaving the high-frequency gain unaltered. Note that the dc gain increases by a factor of $1/\alpha$.

As a general rule of thumb, we would like $\omega_1 \approx 0.1\omega_3$, where ω_3 is the closed loop -3-dB frequency. This will ensure that the lag compensator introduces very little phase lag at frequencies near the closed loop bandwidth, and hence the overshoot should be only slightly worse. In many cases, it suffices to set $\alpha = 0$, which corresponds to the use of integral action as described in Section 7.7.3 together with high-frequency compensation.

We illustrate the characteristics of phase lag compensation by a simple example.

Example 7.9.2

Consider an unstable plant with open loop transfer function,

$$H_p(s) = \frac{10}{(s-1)(s+5)} \qquad (7.9.6)$$

which when sampled with a rate of 10 Hz gives an open loop discrete transfer function,

$$H_\Delta(\gamma) = \frac{(0.44)\gamma + 8.28}{\gamma^2 + 2.88\gamma - 4.14} \qquad (7.9.7)$$

The Nyquist diagram for this system is shown in Figure 7.9.4. Suppose that we are satisfied with the closed loop bandwidth and overshoot obtained with $K = 1$, but the steady-state error (100%!) is considered unsatisfactory. In this case a lag compensator may be useful.

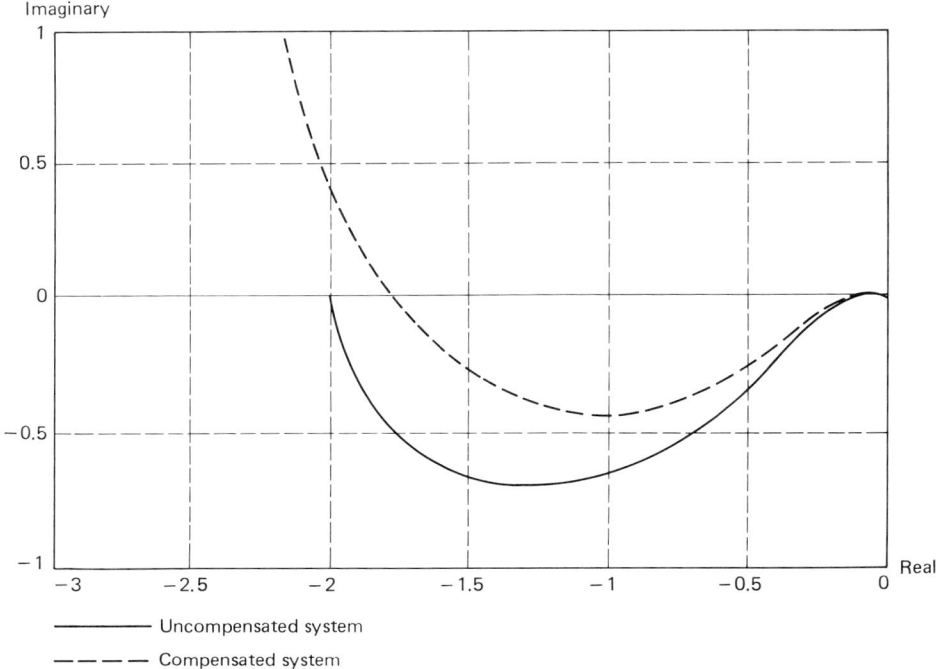

Figure 7.9.4 Nyquist plot for Example 7.9.2.

Figure 7.9.4 shows the Nyquist plot when the following lag compensator is added:

$$H_c(\gamma) = \frac{\gamma + 0.2}{\gamma} \quad (7.9.8)$$

The step responses for the closed loop system with and without compensation are shown in Figure 7.9.5. Note that the addition of the lag compensator leaves the transient performance (rise time, overshoot, and the like) virtually unchanged, while reducing the steady-state error (in this case to zero). ▽▽▽

The principal cost involved in the use of the lag compensator is the increased gain required at low frequencies. With modern technology (integrated circuit amplifiers for continuous time or microprocessors for discrete time), this cost is usually negligible, and so integral action (a special case of lag compensation) is almost invariably a feature of feedback control systems.

7.9.3 Lead – Lag

In some cases it is desirable to use both lag and lead compensation so as to benefit from the relative advantages of each method. We illustrate this by reference again to Example 7.9.2.

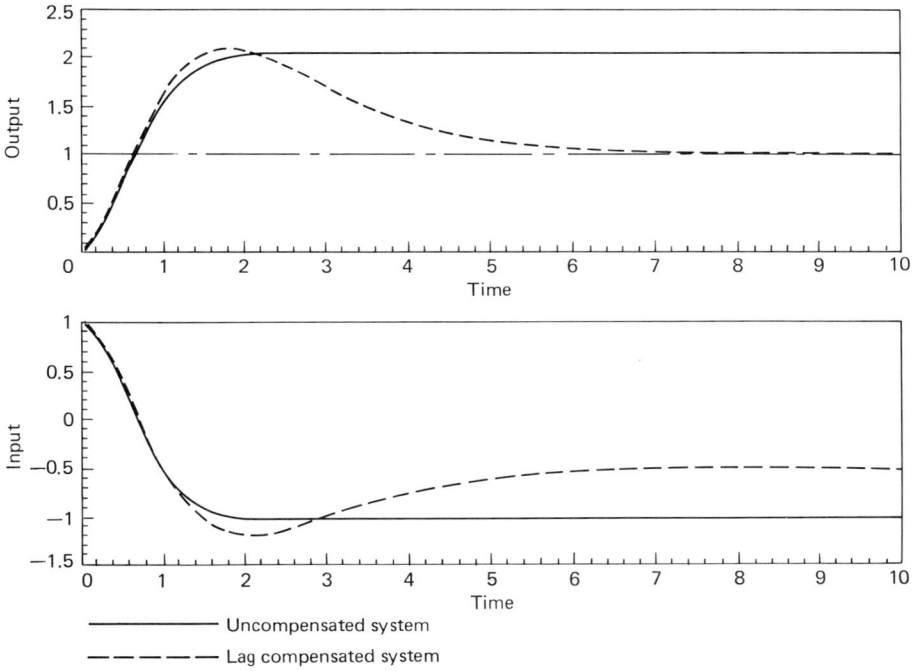

Figure 7.9.5 Closed loop step response for Example 7.9.2.

Example 7.9.2. (Continued)

Say we begin with the lag compensation of Equation (7.9.8), but we wish to improve the phase margin near $\omega = 5$ rad/s. We do this by augmenting the controller by a lead compensator as in equation (7.9.1) and choose $K = 2.4$, $T = \Delta = 0.1$, and $\alpha = 2.5$ by considering the Nyquist plot. Thus we obtain the following lead–lag compensator:

$$H_c(\gamma) = 2.4 \frac{(1 + 0.25\gamma)(\gamma + 0.2)}{\gamma(1 + 0.1\gamma)} \tag{7.9.9}$$

The step response for the closed loop system, when using this controller, is shown in Figure 7.9.6. Note that the lead–lag controller has now improved the rise time, overshoot, and steady-state error performance. ▽▽▽

We can think of a lead–lag compensator as a three-term controller having proportional, integral (lag), and derivative (lead) terms. In the case of analog control systems, this three-term (PID) controller has had widespread use.

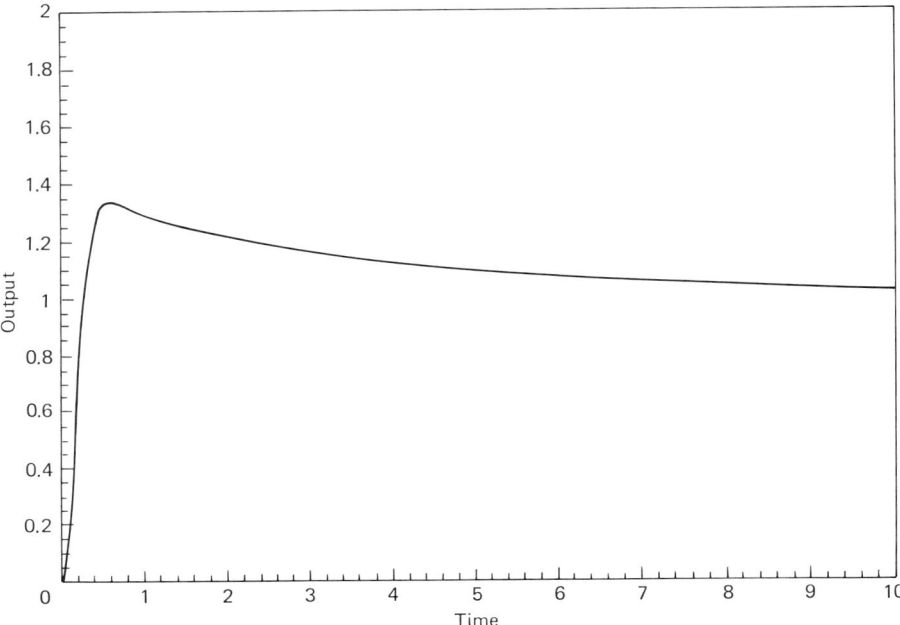

Figure 7.9.6 Closed loop step response for Example 7.9.2 with digital lead–lag controller.

7.10 MULTIVARIABLE SYSTEMS

So far in this chapter we have considered only single-input, single-output systems. We will discuss general multivariable design methods in Chapters 9 to 12. However, a common ploy frequently used in the design of practical multivariable systems is to reduce them to a single-input, single-output case by some preliminary manipulations. We will briefly discuss this idea next.

7.10.1 Multivariable Frequency Domain Stability Criterion

We can extend the idea of Nyquist's stability test to weakly coupled multivariable systems by use of the idea of diagonal dominance. Consider a multivariable feedback system with input–output transfer function matrix

$$Y(\gamma) = \begin{bmatrix} g_{11}(\gamma) & g_{12}(\gamma) & \cdots & g_{1n}(\gamma) \\ g_{21}(\gamma) & & & \vdots \\ \vdots & & & \\ g_{n1}(\gamma) & & \cdots & g_{nn}(\gamma) \end{bmatrix} U(\gamma) \qquad (7.10.1)$$

Consider, also, the case where we have a unity gain negative feedback system; that is,

$$U(\gamma) = -Y(\gamma) \qquad (7.10.2)$$

We then, introduce the following definitions:

Definition 7.10.1. Diagonal Dominance

(i) A transfer function matrix, $G(\gamma)$, is said to be *row diagonally dominant on* $\Gamma \subset \mathbb{C}$ if

$$|g_{ii}(\gamma)| > \sum_{\substack{j=1 \\ j \neq i}}^{n} |g_{ij}(\gamma)| \; \forall \; \gamma \in \Gamma, \forall \; i \qquad (7.10.3)$$

(ii) $G(\gamma)$ is said to be *column diagonally dominant on* Γ if $G^T(\gamma)$ is row diagonally dominant on Γ.

(iii) $G(\gamma)$ is said to be *diagonally dominant on* Γ if it is either row or column diagonally dominant. ▽▽▽

For the remainder of this section, we will use the term diagonally dominant and implicitly assume that Γ is the appropriate Nyquist contour. Based on the concept of diagonal dominance, we will develop stability tests for multivariable systems. Note that, in many cases, the system to be controlled might not be diagonally dominant (that is, it has strong interactions between various loops). In these cases we can often design a (possibly dynamic) precompensator that will make the open loop system diagonally dominant.

We now let

$$d_i(\gamma) = \sum_{\substack{j=1 \\ j \neq i}}^{n} |g_{ij}(\gamma)| \qquad (7.10.4)$$

and

$$d_i'(\gamma) = \sum_{\substack{j=1 \\ j \neq i}}^{n} |g_{ji}(\gamma)| \qquad (7.10.5)$$

We then define the n Gershgorin bands by

$$B_i = \{ q \in \mathbb{C} : \exists \; p \in \Gamma \text{ s.t. } |q - g_{ii}(p)| \leq d_i(p) \} \qquad (7.10.6)$$

and

$$B_i' = \{ q \in \mathbb{C} : \exists \; p \in \Gamma \text{ s.t. } |q - g_{ii}(p)| \leq d_i'(p) \} \qquad (7.10.7)$$

Note that the Gershgorin bands can be graphically constructed using a computer with little difficulty. We take a sequence of $p_i \in \Gamma$, find the appropriate radii, $d_i(p_i)$, and then draw circles centered on p_i, radius d_i. The union of the interiors of these circles then gives an approximation to the Gershgorin bands.

From the definition of the Gershgorin bands, (7.10.6) and (7.10.7), and the definition of diagonal dominance, it is clear that the following result holds.

Lemma 7.10.1

(i) $G(\gamma)$ is diagonally dominant if and only if
    (a) $0 \notin B_i$ for all $i = 1, 2, \ldots, n$, or
    (b) $0 \notin B_i'$ for all $i = 1, 2, \ldots, n$.

(ii) $(I + G(\gamma))$ is diagonally dominant if and only if
    (a) $(-1 + j0) \notin B_i$ for all $i = 1, 2, \ldots, n$, or
    (b) $(-1 + j0) \notin B_i'$ for all $i = 1, 2, \ldots, n$. ▽▽▽

The main use of this result is that (ii) may be tested graphically as a precursor to the following theorem.

Lemma 7.10.2. Multivariable Nyquist Criterion

Suppose $I + G(\gamma)$ is diagonally dominant and the corresponding Gershgorin bands, $\overline{B}_i$, encircle the point $(-1 + j0)$ n_i times in the anticlockwise sense. Then the closed loop system is stable if and only if

$$\sum_{i=1}^{n} n_i = n_0 \qquad (7.10.8)$$

where n_0 is the number of unstable modes in a minimal model for the open loop, $G(\gamma)$.

Proof: The details of the proof of this theorem are beyond the scope of this text; they may be found in references such as Rosenbrock (1974). The important concepts in the proof are diagonal dominance of a matrix, and corresponding bounds on the locations of the eigenvalue locations established by Gershgorin and the Nyquist-type argument based on conformal mappings and the principle of argument. ▽▽▽

Note that the preceding result allows, with the aid of a computer, a simple graphical stability test based on open loop frequency response data. We need only plot the Gershgorin bands and interpret them as follows:

1. If any of the Gershgorin bands contain $(-1 + j0)$, the test fails and stability/instability cannot be inferred. This means that the open loop multivariable system has strong interactions, and we should probably design a precompensator to reduce interactions.
2. Otherwise, we can directly determine stability/instability by counting the encirclements of $(-1 + j0)$.

Example 7.10.1 illustrates the use of this technique.

Example 7.10.1

Consider a multivariable feedback system with open loop transfer function matrix

$$G_{ol}(s) = \begin{bmatrix} \dfrac{1}{s+1} & \dfrac{e^{-0.5s}}{(s+1)^2} \\ \dfrac{1}{s+2} & \dfrac{2e^{-0.5s}}{(s+1)(s+2)} \end{bmatrix} \begin{bmatrix} K_1 & 0 \\ 0 & K_2 \end{bmatrix} \qquad (7.10.9)$$

where K_1 and K_2 are feedback gains. We wish to find conditions on K_1 and K_2 such that stability can be assured. By a slight extension to Lemma 7.10.2 (see Problem 7), a sufficient condition is that the row (or column) Gershgorin bands should not intersect $(-\infty, -1/K_1], (-\infty, -1/K_2]$. The Gershgorin bands for the preceding system are shown in Figure 7.10.1. From this figure we see that sufficient conditions for closed loop stability are that

$$0 < K_1 < \infty \qquad (7.10.10)$$

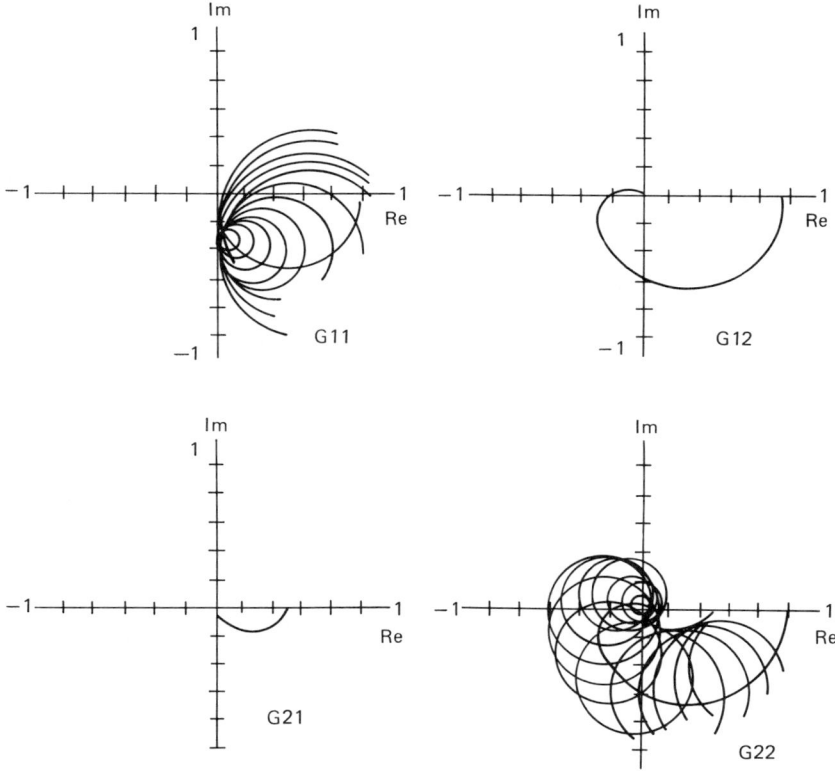

Figure 7.10.1 Gershgorin bands for Example 7.10.1.

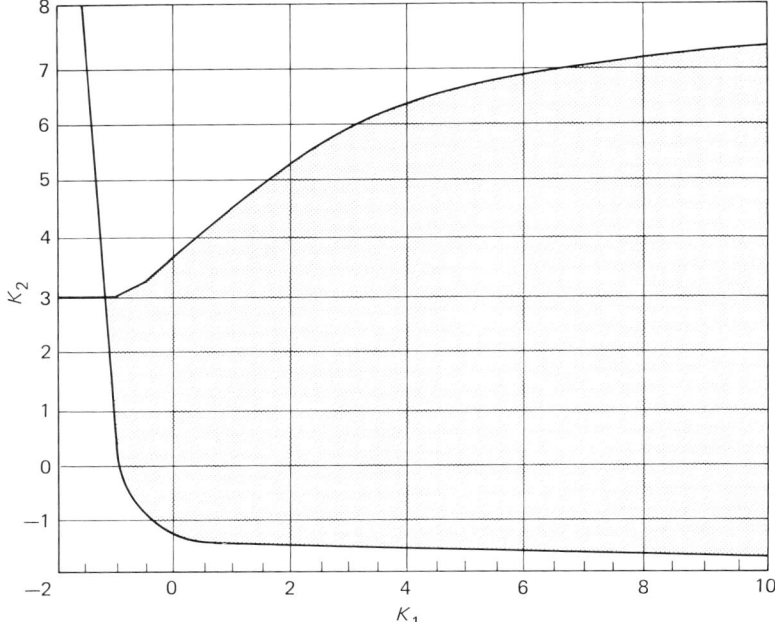

Figure 7.10.2 Precise stability boundaries for Example 7.10.1.

and

$$0 < K_2 < 1.5 \tag{7.10.11}$$

By considering $\det(I + G_{ol}(s))$, we may show that it is necessary and sufficient that K_1 and K_2 belong to the region given in Figure 7.10.2. Thus we see that the Gershgorin bands give a slightly conservative way of analyzing closed loop stability for this example. ▽▽▽

7.10.2 Precompensation

The extent of multivariable interaction can often be significantly reduced by simple precompensation. For example, the system can always be made noninteracting at dc by simply operating on the inputs by K^{-1}, where K is the plant dc gain matrix (provided that K and K^{-1} are finite).

A simple extension of this idea is to use precompensation as in Figure 7.10.3 of the form

$$\mathscr{C}(\gamma) = \frac{C_0 + C_1\gamma}{\gamma + a} \tag{7.10.12}$$

where C_0 and C_1 are matrices to be chosen and a is a fixed pole.

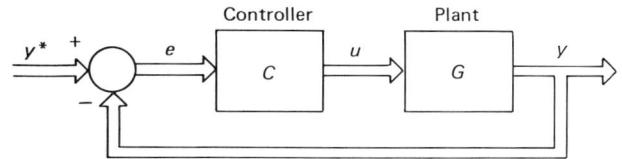

Figure 7.10.3 Precompensated plant.

If we choose any frequency ω_0, then we can evaluate the plant frequency response at that frequency as $G((e^{j\omega_0\Delta} - 1)/\Delta)$. This will, in general, be a complex number:

$$G\left(\frac{e^{j\omega_0\Delta} - 1}{\Delta}\right) = \mathscr{A} + j\mathscr{B} \qquad (7.10.13)$$

We can then choose C_0 and C_1 so that

$$\frac{C_0 + C_1[(e^{j\omega_0\Delta} - 1)/\Delta]}{[(e^{j\omega_0\Delta} - 1)/\Delta] + a} = (\mathscr{A} + j\mathrm{B})^{-1} \qquad (7.10.14)$$

The inverse in (7.10.14) is readily calculated by standard complex arithmetic as the left side is linear in C_0 and C_1. Thus C_0 and C_1 can be elevated to diagonalize the transfer function at ω_0.

Actually, by use of a high-order compensator, it is possible to diagonalize a multivariable system at every frequency. To illustrate the idea, let $a_i(\gamma)$ be the least common multiple of the denominator polynomials in the ith row of the transfer function, $G(\gamma)$. Then we can write

$$G(\gamma) = A_d(\gamma)^{-1} B(\gamma) \qquad (7.10.15)$$

where

$$A_d(\gamma) = \mathrm{diag}(a_1(\gamma) \ldots a_m(\gamma))$$

The required precompensator is then

$$\mathscr{C}(\gamma) = \mathrm{Adj}\{B(\gamma)\} E_d(\gamma)^{-1} \qquad (7.10.16)$$

where Adj{ } denotes adjoint and $E_d(\gamma)$ is a stable diagonal matrix polynomial of sufficient degree to ensure that $\mathscr{C}(\gamma)$ is proper (that is, has numerator degree less than or equal to the denominator degree).

The transfer function for e to y in Figure 7.10.3 is then diagonal:

$$G(\gamma)\mathscr{C}(\gamma) = A_d(\gamma)^{-1} E_d(\gamma)^{-1} \det\{B(\gamma)\} I \qquad (7.10.17)$$

where det{ } denotes determinant.

Example 7.10.2

Consider the system

$$G(\gamma) = \begin{bmatrix} \dfrac{\gamma - 1}{\gamma(\gamma + 1)} & \dfrac{2(\gamma - 1)}{\gamma(\gamma + 1)} \\ \dfrac{1}{\gamma + 2} & \dfrac{1}{\gamma + 1} \end{bmatrix} \qquad (7.10.18)$$

$$= \begin{bmatrix} \gamma(\gamma + 1) & 0 \\ 0 & (\gamma + 1)(\gamma + 2) \end{bmatrix}^{-1} \begin{bmatrix} (\gamma + 1) & 2(\gamma - 1) \\ (\gamma + 1) & 2(\gamma + 2) \end{bmatrix}$$

$$\triangleq A_d(\gamma)^{-1} B(\gamma) \qquad (7.10.19)$$

Then

$$\mathrm{Adj}\{B(\gamma)\} = \begin{bmatrix} 2(\gamma + 2) & -(\gamma + 1) \\ -2(\gamma - 1) & \gamma - 1 \end{bmatrix} \qquad (7.10.20)$$

If we let

$$E_d(\gamma) = \begin{bmatrix} \gamma + 1 & 0 \\ 0 & \gamma + 1 \end{bmatrix} \qquad (7.10.21)$$

then

$$G(\gamma)\mathscr{C}(\gamma) = \begin{bmatrix} \dfrac{2(\gamma - 1)}{\gamma(\gamma + 1)^2} & 0 \\ 0 & \dfrac{2(\gamma - 1)}{(\gamma + 1)^2(\gamma + 2)} \end{bmatrix} \qquad (7.10.22)$$

▽▽▽

The point about the preceding technique is that the transfer function can be diagonalized and hence the multivariable problem is reduced to a number of single-input, single-output problems.

7.11 SUMMARY

This chapter discussed classical control system analysis from a unified perspective. Key points were:

- Introduction of the concept of asymptotic stability of a system (Definition 7.2.1).
- A necessary and sufficient condition for stability in terms of the poles p_i of a linear system for both continuous and discrete delta domain cases is

$$\frac{\Delta}{2}|p_i|^2 + \mathrm{Re}\{p_i\} < 0, \quad \text{for all } i \qquad (7.2.3)$$

Equivalently, in the shift domain we require that

$$|p_i| < 1, \quad \text{for all } i \qquad (7.2.7)$$

- Tests for stability of a polynomial based on the polynomial coefficients (Routh and Jury).
- A description of the root locus technique for plotting the closed loop poles of a feedback system as a function of the controller gain.
- Frequency domain stability criteria based on the Nyquist diagram (Theorem 7.5.1).
- Extension of frequency domain stability analysis to cover uncertainty in a feedback system (Lemma 7.6.1) and/or systems with nonlinearities (Lemma 7.6.2).
- Relationship between steady-state error of a feedback system for polynomial set points and the number of open loop poles at the origin (Lemma 7.7.1).
- Extension of the steady-state error nulling idea to more general inputs using the internal model principle (section 7.7.3).
- Relationship between time domain transient characteristics and frequency domain properties, leading to the following rules of thumb:

$$t_r \simeq \frac{2.2}{\omega_3} \qquad (7.8.1)$$

where t_r is rise time and ω_3 is the -3-dB frequency and

$$\sigma \simeq \frac{\sqrt{M_p - 1}}{2.4} \qquad (7.8.2)$$

where σ is the overshoot and M_p is the peak amplitude of the frequency response.
- Introduction of the notions of phase lead and phase lag compensation (Section 7.9).
- Stability criteria for weakly coupled multivariable systems (Lemma 7.10.2).
- Techniques for diagonalization of a square multivariable system (Section 7.10.2).

7.12 REFERENCES

General background on basic stability theory is contained in:

FORTMANN, T. E., and K. L. HITZ (1977) *An Introduction to Linear Control Systems*. Marcel Dekker, New York.

GANTMACHER, F. R. (1959) *The Theory of Matrices*, Vols. I and II. Chelsea, New York.

JURY, E. I., and J. BLANCHARD (1961) "A stability test for linear discrete systems in table form." *Proc. IRE*, vol. 49 no. 12, pp. 1947–1948.

KUO, B. C. (1982) *Automatic Control Systems*. Prentice-Hall, Englewood Cliffs, N.J.

WILLEMS, J. L. (1970) *Stability Theory of Dynamical Systems*. Thomas Nelson, London.

Frequency domain stability criteria for nonlinear systems are covered in:

VIDYASAGAR, M. (1978) *Nonlinear Systems Analysis*. Prentice-Hall, Englewood Cliffs, N.J.

Multivariable frequency domain stability is discussed in:

ROSENBROCK, H. H. (1974) *Computer Aided Control System Design*. Academic Press, New York.

Classical control system analysis for continuous time systems is covered in many books on classical control including:

DOEBLIN, E. O. (1985) *Control Systems: Principles and Design*. Wiley, New York.

KUO, B. C. (1962) *Automatic Control Systems*. Prentice-Hall, Englewood Cliffs, N.J.

STEPHANOPOLIS, G. (1984) *Chemical Process Control—An Introduction to Theory and Practice*. Prentice-Hall, Englewood Cliffs, N.J.

More details on multivariable frequency domain analysis can be found in:

MACFARLANE, A. G. J., and J. J. BELLETRUTTI (1973) "The characteristic locus design method." *Automatica*, vol. 9, pp. 575–588.

MAYNE, D. Q. (1973) "The design of linear multivariable systems." *Automatica*, vol. 9, pp. 201–207.

MUNRO, N. (1972) "Multivariable design using the inverse Nyquist array." *Computer Aided Design*, vol. 4, pp. 222–227.

ROSENBROCK, H. H. (1969) "Design of multivariable control systems using the inverse Nyquist array." *Proc. IEE*, vol. 116, p. 1929.

——— (1974) *Computer Aided Control System Design*. Academic Press, New York.

7.13 PROBLEMS

1. Give the details of the proof of Lemma 7.2.1 for both the nonrepeated and repeated root cases.

2. Establish equation (7.2.8). *Hint:* Let $\bar{p}$ denote the pole that is closest to the stability boundary. For discrete time, choose λ between $-(1/\Delta) + |\bar{p} + (1/\Delta)|$ and 0. For continuous time, choose λ between $\text{Re}\{\bar{p}\}$ and 0.

3. Suppose we wish to control a continuous time system with nominal transfer function

$$H(s) = \frac{2}{s+2}$$

The control is to be performed using negative feedback with a digital control law

$$G_c(\gamma) = \frac{1}{\gamma}$$

For what values of sampling rate is the closed loop system stable? *Hint:* Show that the discrete time version of the continuous system (assuming zero-order hold input) is

$$H_\Delta(\gamma) = \frac{(1 - e^{-2\Delta})/\Delta}{\gamma + (1 - e^{-2\Delta})/\Delta}$$

Thus show that the closed loop characteristic polynomial is

$$A^*(\gamma) = \gamma\left(\gamma + \frac{1 - e^{-2\Delta}}{\Delta}\right) + \left(\frac{1 - e^{-2\Delta}}{\Delta}\right)$$

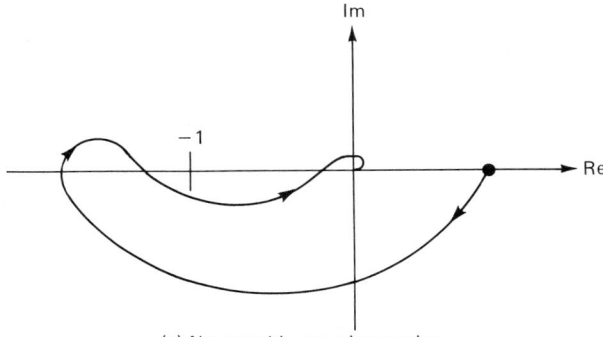

(a) No unstable open loop poles

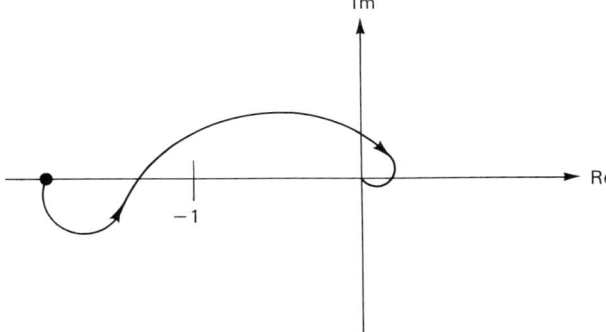

(b) One unstable open loop pole

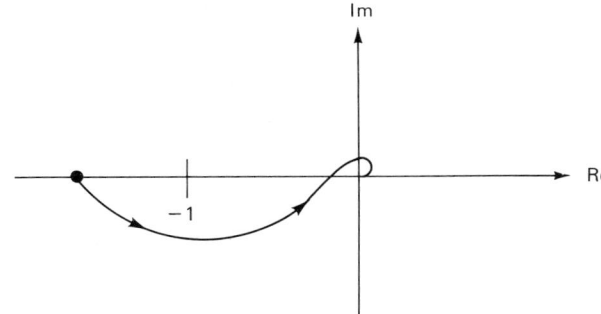

(c) Two unstable open loop poles

Figure 7.13.P5.

4. Evaluate the Nyquist diagram given in Figure 7.5.4 and compare the stability range for k with that obtained by the polynomial tests in Section 7.3.
5. (a) Which of the continuous time Nyquist diagrams in Figure 7.13.P5 (one-half, $\omega \geq 0$, only shown) will give a stable closed loop if the open loop has the number of unstable poles indicated:
   (b) What changes are needed if the system is in discrete delta form?

6. Suppose that a matrix polynomial $G(\gamma)$ is diagonally dominant. Prove that $G(\gamma)$ is nonsingular for every γ on the Nyquist contour. *Hint:* Use contradiction and assume there exists a γ and a vector $x \neq 0$ such that $G(\gamma)x = 0$. By considering the element with maximum modulus in x, show that $G(\gamma)$ cannot be diagonally dominant.

7. Consider a multivariable plant that is open loop stable. Show that a sufficient condition for closed loop stability with diagonal feedback gains, $\mathrm{diag}(\overline{K}_1, \overline{K}_2, \ldots, \overline{K}_n)$, where $0 < \overline{K}_i < \infty$, is that the ith row (or column) Gershgorin bands should not intersect
$$\left(-\infty, \overline{K}_i^{-1}\right], \quad \text{for } i = 1\ldots n$$

Hint: Let the open loop plant be $G_0(\gamma) = N(\gamma) M^{-1}(\gamma)$; the closed loop poles are the zeros of $\det(M(\gamma) + KN(\gamma))$. The closed loop poles are continuous with respect to K, and thus, since $K = 0$, gives stability; $K = \overline{K}$ can only be unstable if there exists a K^1, $0 < K^1 < \overline{K}$, such that $K = K^1$ gives poles on the stability boundary. Using Problem 6, show that this is impossible.

8. Consider a system with transfer function
$$H(\gamma) = \frac{b_1 \gamma + b_0}{\gamma + \alpha_0}$$
Show that this system has a Nyquist diagram that is a circle,
$$\text{center } \frac{1}{2}\left(\frac{b_0}{a_0} + \frac{\Delta b_0 - 2b_1}{\Delta a_0 - 2}\right) \quad \text{and} \quad \text{radius } \frac{1}{2}\left(\frac{b_0}{a_0} + \frac{\Delta b_0 - 2b_1}{\Delta a_0 - 2}\right)$$
provided $a_0 \neq 0$ and $\Delta a_0 \neq 2$. What is the Nyquist diagram if $a_0 = 0$ or if $a_0 = 2/\Delta$?

9. Show that for $\Delta > 0$ the region in the complex γ plane,
$$\frac{\Delta}{2}|\gamma|^2 + \mathrm{Re}\{\gamma\} < 0$$
is identical to the region $|1 + \Delta\gamma| < 1$.

10. Consider a third-order continuous time polynomial,
$$P_c(s) = s^3 + a_2 s^2 + a_1 s + a_0$$
Show, using Routh techniques, that this polynomial is stable if and only if
$$a_2, a_1, a_0 > 0$$
and
$$a_2 a_1 > a_0$$

11. Consider the delta domain polynomial in Example 7.3.2:
$$p(\gamma) = \gamma^3 + 4\gamma^2 + 9\gamma + 10, \qquad \Delta = 0.1$$
Test this polynomial for stability by (i) converting to shift domain using $\gamma = (z - 1)/\Delta$, and then (ii) applying Jury's stability result to the resulting polynomial. In part (ii), how many decimal places of accuracy are required to obtain the correct result?

12. Prove Corollary 7.4.1.

13. Consider a unity feedback system with open loop transfer function
$$H_{\mathrm{ol}}(\gamma) = \frac{k(-0.5\gamma + 1)}{\gamma^2 + \gamma}, \qquad \Delta = 0.1$$
Using root locus techniques, show that for any $k \leq 0$ this transfer function gives an unstable closed loop. Show that, for some $k \geq 0$, stability of the closed loop can be attained.

14. Show that for a stable closed loop system the steady-state error due to sinusoid of angular frequency ω_0 and unit amplitude is a sinusoid having amplitude as given in (7.7.5).
15. Show that the normalized amplitude error $\bar{a}(\infty)$ satisfies (7.7.6). What happens to $\bar{a}(\infty)$ as $|H_{ol}[(e^{j\omega\Delta} - 1)/\Delta]|$ becomes large?
16. Show that the phase error is equal to the phase angle of the open loop transfer function minus the phase angle of the return difference.
17. Prove Lemma 7.7.1(i). *Hint:* Show that any $y^*(t) = a_0 + a_1 t + \cdots + a_k t^k$ has a transform $Y^*(\gamma) = N(\gamma)/\gamma^{k+1}$. The properties in (7.7.9) imply that $H_{cl}(\gamma)$ can be written as $1 + \gamma^{k+1}\overline{H}(\gamma)$, where $\overline{H}(0)$ is finite. Then use the final value theorem.
18. Show that (7.7.10) is equivalent to (7.7.9). *Hint:* Show that (7.7.10) $\Rightarrow$ (7.7.9). The reverse argument follows by showing $H_{ol} = H_{cl}/(1 - H_{cl})$.
19. Show that a system of type k gives zero steady-state tracking to reference signals and/or disturbances that are polynomials of degree $< k$. *Hint:* See Problem 17. Also show that if $y^*(t)$ is a monic polynomial of degree k then the steady-state tracking error satisfies (7.7.11).
20. Consider a dc motor with transfer function $1/[s(s + 1)]$ and an input offset d, as shown:

If this system is used in a unity feedback arrangement ($u = y^* - y$). Show that although the system is of type 1 the tracking error for constant y^* and d is nonzero. Explain.
21. What is the constant M loci for $M = 1$?
22. Consider a continuous time system with transfer function,

$$F(s) = \frac{1}{s - 1}$$

  (a) If this system is to be controlled using a computer (that is, discrete time), what is the minimum sampling rate you would suggest? Briefly give a reason for your answer.
  (b) Find a discrete time model for the system when sampled with $\Delta = 0.05$ (that is, 20 Hz).
  (c) Design a discrete time controller that gives zero steady-state error to a constant reference input and/or disturbance. Use the same sampling rate as in part (b).
23. Consider the following continuous time plant

$$F(s) = \frac{s - 2}{(s + 1)(s + 2)(s + 3)}$$

  (a) Suggest a sampling rate and design a controller that gives zero steady state error to a sinusoidal disturbance of frequency 0.2 rad/s^{-1}.
  (b) Explain why it would be difficult to repeat part (a) for a sine wave with a frequency of 5 rad/s^{-1}.
24. Consider the following plant:

$$F(s) = \frac{K}{(Ts + 1)^2}$$

Sec. 7.13   Problems

where $K \in [1, 4]$

$T \in [0.5, 1]$

(a) Suggest a suitable sampling rate and design a fixed controller to give zero steady-state error to a step and a rise time of 1 s.

(b) Simulate your design for the four extreme cases.

25. Suppose that for a certain complex value of γ a multivariable system has a high open loop gain; that is, $\|H_{ol}^{-1}(\gamma)\| \leq \epsilon < 1$. If this system is used in a unity feedback arrangement, show that the closed loop is approximately the identity in the sense that

$$\|H_{cl}(\gamma) - I\| \leq \frac{\epsilon}{1 - \epsilon}$$

Hint:

$$\left\|\left(I + H_{ol}^{-1}(\gamma)\right)^{-1}\right\| \leq \frac{1}{1 - \|H_{ol}^{-1}(\gamma)\|}, \qquad \text{for } \|H_{ol}^{-1}(\gamma)\| < 1.$$

26. Let G_0 denote a given matrix transfer function. Say that the ith output is linked to the ith input by a scalar controller as shown in Figure 7.13.P26 where

$$K_i(\gamma) = \text{diag}(1, \ldots, 1, k_{ii}(\gamma), 1, \ldots, 1)$$
$$e_i^T = (0, \ldots, 0, 1, 0, \ldots, 0)$$

Show that the resulting closed loop transfer function is

$$G_c = G_0 K_i - \frac{k_{ii}^2 (G_0)_{\cdot i} (G_0)_{i \cdot}}{(1 + G_0)_{ii} k_{ii}}$$

where

$(G_0)_{\cdot i} = i$th column of G_0

$(G_0)_{i \cdot} = i$th row of G_0

$(G_0)_{ii} = i$th element of G_0

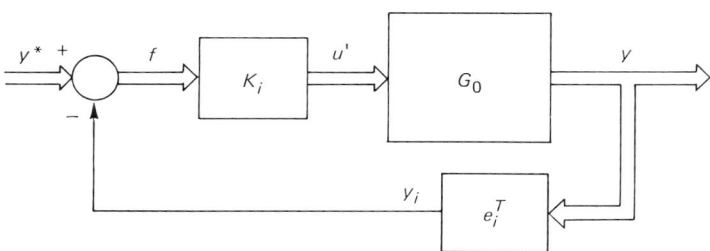

Figure 7.13.P26.

8

Time Domain Analysis of Systems

8.1 INTRODUCTION

Chapter 7 was concerned with classical analysis of control systems. In this chapter we take our study of systems theory further and study general properties of linear time invariant systems in the time domain. Our emphasis will be on state space descriptions, although connections with transfer function descriptions will also be explored.

Two key ideas to emerge in systems theory over the past 30 years are those of controllability and observability. *Controllability* refers to the issue of steering a system's response using the available inputs, whereas *observability* is concerned with inferring the current energy distribution (that is, the current state) of a system using the available output measurements. These two concepts obviously give important information about the interconnection between inputs, states, and outputs. We will see that quantitative measures of the degree of controllability and observability can also be readily derived.

Often, the appropriate set of states to use for a system description is dictated by physical insight; for example, we might choose capacitor voltages and inductor currents in electrical circuits or tank levels in a pumping system. However, it can also happen that we are only interested in the input–output behavior. In this case, the choice of the basis for the state space is somewhat arbitrary. This will lead us to consider the question of state space models, which have a special structure, but which have sufficient flexibility to describe the input–output behavior of the controllable part and/or the observable part of a system. This will further lead us to other questions, such as minimal realizations, balanced realizations, and model order reduction.

Finally, we reexamine the stability properties of linear time invariant systems and make precise the interconnections mentioned in Chapter 7. We also extend these concepts to nonlinear systems.

Thus, in summary, this chapter is aimed at gaining a deeper understanding of state space models with particular emphasis on their time domain characteristics.

8.2 STATE SPACE MODELS

The notion of linear state space models was introduced in Chapters 2 and 3. Here we will use the following general description covering both discrete and continuous time cases simultaneously:

$$\rho x(t) = Ax(t) + Bu(t); \qquad x(0) = x_0 \qquad (8.2.1)$$
$$y(t) = Cx(t) + Du(t) \qquad (8.2.2)$$

where

- ρ denotes d/dt (continuous time) or δ (discrete time)
- x denotes an $n \times 1$ state vector
- u denotes an $r \times 1$ input vector
- y denotes an $m \times 1$ output vector
- A, B, C, D are real matrices of dimension $n \times n$, $n \times r$, $m \times n$, and $m \times r$, respectively

In the following sections we explore system theory notions associated with this model. We begin by examining how to steer the state of the system using the input.

8.3 REACHABILITY, CONTROLLABILITY, STABILIZABILITY

The notions of reachability, controllability, and stabilizability are all concerned with the interconnection between the input and state of a system. Intuitively, these notions tell us to what extent we can steer the system state using the system input.

Before giving formal definitions of these notions, we will attempt to give an intuitive explanation. *Reachability* is concerned with the issue of whether or not a certain final state can be reached by driving the system using the input starting from some initial state, usually the origin. *Controllability* is concerned with whether or not a given initial state can be controlled back to the origin. Finally, a system is said to be *stabilizable* if all states can be taken to the origin in a possibly infinite time.

We now give formal definitions of these concepts.

Definition 8.3.1. Reachability from the Origin

A state x_r is said to be reachable if, given $x(0) = 0$, there exists an interval $[0, T]$ and an input $\{u(t), t \in [0, T]\}$ such that $x(T) = x_r$. If all states are reachable, then the system is said to be completely reachable.   ▽▽

We next develop a criterion for reachability. To do this, we note that the solution of (8.2.1) with zero initial conditions is

$$x(t) = \underset{0}{\overset{t}{S}} E(A, t - \tau - \Delta) Bu(\tau) \, d\tau \tag{8.3.1}$$

The following matrix will be found to play a central role in characterizing the set of possible solutions of (8.3.1), that is, in characterizing the reachability properties of the system.

Definition 8.3.2

The controllability matrix $\mathscr{C}$ for the system (8.2.1) is defined to be

$$\mathscr{C}[A, B] \triangleq [B, AB, \ldots, A^{n-1}B] \tag{8.3.2}$$

We then have the following characterization of reachability for both continuous and discrete models.

Theorem 8.3.1. Reachability via Controllability Matrix

(i) The set of all reachable states is equal to the range space of $\mathscr{C}$.
(ii) The system is completely reachable if and only if $\mathscr{C}$ has full rank n.

Proof:

(i) To find the range space of the operator in (8.3.1), we first note that the generalized exponential $E(A, t)$ can be written as follows (see Problem 27):

$$E(A, t) = \sum_{j=0}^{n-1} \varphi_j(t) A^j \tag{8.3.3}$$

where $\varphi_j(t)$, $j = 0, \ldots, n - 1$ satisfy

$$\rho \varphi_0(t) = -a_0 \varphi_{n-1}(t)$$
$$\rho \varphi_1(t) = -a_1 \varphi_{n-1}(t) + \varphi_0(t)$$
$$\vdots$$
$$\rho \varphi_{n-1}(t) = -a_{n-1} \varphi_{n-1}(t) + \varphi_{n-2}(t) \tag{8.3.4}$$

and

$$\varphi_j(0) = \begin{cases} 1, & \text{if } j = 0 \\ 0, & \text{otherwise} \end{cases} \tag{8.3.5}$$

and where

$$\det(\gamma I - A) = \gamma^n + a_{n-1}\gamma^{n-1} + \cdots + a_0 \quad (8.3.6)$$

Substituting (8.3.3) into (8.3.1) gives

$$x(t) = \sum_{j=0}^{n-1} A^j B \mathbf{S}_0^t \varphi_j(t - \tau - \Delta) u(\tau)\, d\tau \quad (8.3.7)$$

From (8.3.2) and (8.3.7), we can write

$$x(t) = \mathscr{C} \mathbf{S}_0^t V(t - \tau - \Delta) u(\tau)\, d\tau \quad (8.3.8)$$

where we have dropped the arguments from $\mathscr{C}$ in (8.3.2) and where

$$V(t) \triangleq \begin{bmatrix} \varphi_0(t) I_{r \times r} \\ \vdots \\ \varphi_{n-1}(t) I_{r \times r} \end{bmatrix} \quad (8.3.9)$$

From (8.3.8) we immediately see that $x(t)$ satisfying (8.3.7) must lie in the range space of $\mathscr{C}$.

To show the converse, let x^* be any vector in the range space of $\mathscr{C}$, that is, there exists an (nr) vector, η, such that

$$x^* = \mathscr{C}\eta \quad (8.3.10)$$

For any time $t > (n-1)\Delta$, consider the following input for $\tau \in (0, t)$:

$$u(\tau) = V(t - \tau - \Delta)^T \left[\mathbf{S}_0^t V(t - \sigma - \Delta) V(t - \sigma - \Delta)^T\, d\sigma \right]^{-1} \eta \quad (8.3.11)$$

Note that the inverse in (8.3.11) exists due to the linear independence of the functions $\varphi_0(t), \ldots, \varphi_{n-1}(t)$ (see Problem 28). Then, using (8.3.8), we have

$$x(t) = \mathscr{C} \mathbf{S}_0^t \left\{ V(t - \tau - \Delta) V(t - \tau - \Delta)^T \right.$$

$$\left. \times \left[\mathbf{S}_0^t V(t - \sigma - \Delta) V(t - \sigma - \Delta)^T\, d\sigma \right]^{-1} \eta \right\} d\tau$$

$$= \mathscr{C}\eta$$

$$= x^*, \quad \text{using (8.3.10)} \quad (8.3.12)$$

(ii) Follows immediately from part (i). ▽▽▽

An interesting observation is that if a system is not completely reachable then the nonreachable states can be isolated, leaving a completely reachable system. This is made explicit in the following result.

Theorem 8.3.2. Reachable Decomposition

If rank $\mathscr{C} = k < n$, then there exists a similarity transformation T such that with $\bar{x} = T^{-1}x$; that is,

$$\bar{A} = T^{-1}AT, \qquad \bar{B} = T^{-1}B \qquad (8.3.13)$$

$\bar{A}, \bar{B}$ have the form

$$\bar{A} = \begin{bmatrix} \bar{A}_c & \bar{A}_{12} \\ 0 & \bar{A}_{nc} \end{bmatrix}, \qquad \bar{B} = \begin{bmatrix} \bar{B}_c \\ 0 \end{bmatrix} \qquad (8.3.14)$$

where $\bar{A}_c$ has dimension k and $[\bar{A}_c, \bar{B}_c]$ is completely reachable.

Proof: Let $[T_1]$ be any basis for the range space of $\mathscr{C}$. Choose T_2 arbitrarily subject to $T \triangleq [T_1, T_2]$ being nonsingular. Also, define

$$T^{-1} \triangleq \begin{bmatrix} S_1 \\ S_2 \end{bmatrix} \qquad (8.3.15)$$

Then $T^{-1}T = I$ implies

$$S_1 T_1 = I, \qquad S_1 T_2 = 0, \qquad S_2 T_1 = 0, \qquad S_2 T_2 = I \qquad (8.3.16)$$

A suitable transformation is

$$\bar{x} = T^{-1}x \qquad (8.3.17)$$

Then it is readily seen that

$$\rho \bar{x} = \bar{A}x + \bar{B}u \qquad (8.3.18)$$

where

$$\begin{aligned} \bar{B} &= T^{-1}B \\ &= \begin{bmatrix} S_1 \\ S_2 \end{bmatrix} B \\ &= \begin{bmatrix} S_1 B \\ 0 \end{bmatrix} \end{aligned} \qquad (8.3.19)$$

since the columns of B belong to the range space of $\mathscr{C}$ and since $S_2 T_1 = 0$. Similarly,

$$\begin{aligned} \bar{A} &= T^{-1}AT \\ &= \begin{bmatrix} S_1 A T_1 & S_1 A T_2 \\ S_2 A T_1 & S_2 A T_2 \end{bmatrix} \\ &= \begin{bmatrix} S_1 A T_1 & S_1 A T_2 \\ 0 & S_2 A T_2 \end{bmatrix} \end{aligned} \qquad (8.3.20)$$

since, from the Cayley–Hamilton theorem, the columns of AT_1 belong to the range space of $\mathscr{C}$ and since $S_2 T_1 = 0$.

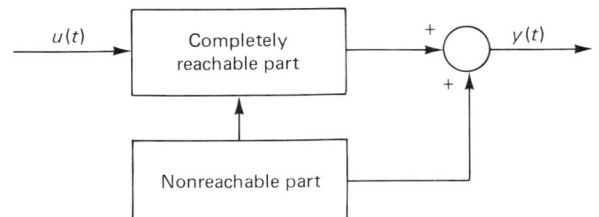

Figure 8.3.1 Decomposition into reachable and nonreachable subsystems.

Equation (8.3.14) follows immediately from (8.3.18), (8.3.19). Also, we see that

$$\begin{bmatrix} \overline{B}_c, & \overline{A}_c\overline{B}_c & ,\ldots, & \overline{A}_c^{n-1}\overline{B}_c \\ 0, & 0 & ,\ldots, & 0 \end{bmatrix} = \begin{bmatrix} \mathscr{C}[\overline{A}_c, \overline{B}_c], & \overline{A}_c^k\overline{B}_c & ,\ldots, & \overline{A}_c^{n-1}\overline{B}_c \\ 0, & 0 & ,\ldots, & 0 \end{bmatrix}$$

$$= T^{-1}\mathscr{C}[A, B] \quad (8.3.21)$$

Hence, from the Cayley–Hamilton theorem,

$$\text{rank } \mathscr{C}[\overline{A}_c, \overline{B}_c] = \text{rank } \mathscr{C}[A, B] = k \qquad \triangledown\triangledown\triangledown$$

The decomposition described in Theorem 8.3.2 is shown diagramatically in Figure 8.3.1.

Other tests for reachability can also be constructed. These tests often have advantages in analytic and/or numerical work. The following result was developed by Popov, Belevitch, and Hautus. For simplicity, we will call it the PBH test using the initial letters.

Theorem 8.3.3. PBH Reachability Test

(i) A pair (A, B) is not completely reachable if and only if there exists a row vector $q \neq 0$ and a (possibly complex) scalar λ such that

$$qA = \lambda q, \qquad qB = 0$$

that is, if and only if some left eigenvector of A is orthogonal to (all columns of) B.

(ii) The pair (A, B) is completely reachable if and only if

$$\text{rank}[B, \lambda I - A] = n, \qquad \text{for all } \lambda \qquad (8.3.22)$$

(iii) The pair (A, B) is completely reachable if and only if B and $(\lambda I - A)$ are relatively left prime.

Proof:

(i) *If*: If there exists a $q \neq 0$ such that

$$qA = \lambda q, \qquad qB = 0$$

then

$$qAB = \lambda qB = 0$$

$$qA^2B = \lambda qAB = 0$$

and so on. Hence

$$q\mathscr{C} = 0 \tag{8.3.23}$$

which implies that $\mathscr{C}$ cannot have full rank.

Only if: We begin by assuming that the system is not completely reachable and has been put into the standard form (8.3.14). A particular row vector that is orthogonal to B has the form $[0, v]$, and we choose v as a left eigenvector of $\overline{A}_{nc}$; that is,

$$v\overline{A}_{nc} = \lambda v$$

Then

$$qA = [0, v]A = [0, \lambda v] = \lambda q \tag{8.3.24}$$

(ii) If $[\lambda I - A, B]$ has rank n, there cannot be a nonzero vector q such that
$$q[\lambda I - A, B] = 0, \quad \text{for any } \lambda$$
that is, such that
$$qB = 0 \quad \text{and} \quad qA = \lambda q$$
But, then by part (i), (A, B) must be completely reachable. The converse follows easily by reversing this argument.

*(iii) *If*: For any matrices $\lambda I - A, B$, there exists a unimodular matrix $U(\lambda)$ such that

$$[\lambda I - A, B]U(\lambda) = [R(\lambda), 0]$$

where $R(\lambda)$ is the greatest common left divisor of $(\lambda I - A)$ and B. Now

$$q[\lambda I - A, B] = 0$$

implies

$$q[\lambda I - A, B]U(\lambda) = 0$$

That is

$$qR(\lambda) = 0 \tag{8.3.25}$$

Now if $\lambda I - A, B$ are relatively left prime, then $R(\lambda)$ is unimodular and hence (8.3.25) implies $q \equiv 0$. The result then follows from part (i).

Only if: From part (i), if the system is not completely reachable, then we can find a $q \neq 0$ such that $q[\lambda I - A, B] = 0$. However, from (8.3.25) this clearly implies $R(\lambda)$ cannot be unimodular. That is, $\lambda I - A, B$ cannot be relatively left prime. ▽▽▽

For linear continuous time systems, the system is completely reachable if and only if every nonzero initial state can be driven to the origin in finite time. Thus, for continuous time systems, complete reachability and complete controllability are equivalent. However, this is not quite true of discrete time systems, since in this case it is possible for some states to relax to the origin in finite time without having to be driven there.

As a trivial example, consider the following shift operator discrete time model:

$$qx(t) = \begin{bmatrix} 0 & 1 \\ 0 & 0 \end{bmatrix} x(t) \tag{8.3.26}$$

Clearly, this system is not completely reachable (since there is no input). However, it is completely controllable (to the origin) since it is easily seen that $x(t)$ always relaxes to zero in two time steps, independently of the initial conditions.

Necessary and sufficient conditions for controllability (to the origin) for both shift and delta discrete time models are given in the following.

Lemma 8.3.1

A necessary and sufficient condition for controllability (to the origin) of a state x_0 is that $E(A, t)x_0$ lie in the range space of $\mathscr{C}$ for some time t. In particular, a necessary and sufficient condition for controllability of x_0 is that $E(A, n\Delta)x_0$ lie in the range space of $\mathscr{C}$.

Proof: This follows immediately from the solution of the model equations (8.2.1).
∇∇∇

Corollary 8.3.1

A necessary and sufficient condition for complete controllability to the origin is that, when the system is transformed as in Theorem 8.3.2, then $\overline{A}_{nc}$ has all eigenvalues at $-1/\Delta$ (delta) or 0 (shift).
∇∇∇

Finally, we give a brief characterization of stabilizability. This can be readily explained in terms of the decomposition given in (8.3.14). The subsystem $(\overline{A}_c, \overline{B}_c)$ is completely reachable and hence we can drive this subsystem anywhere we like. (In fact, we will see in Chapter 9 that it is possible to determine feedback to give arbitrary closed loop poles for this subsystem.) On the other hand, the subsystem $(\overline{A}_{nc}, 0)$ is not connected to the input and thus the states of this subsystem evolve according to the matrix exponential $E(\overline{A}_{nc}, t)x_0$. We then say that a system is *stabilizable* if this matrix exponential response decays to zero, that is, if and only if $\overline{A}_{nc}$ has all its eigenvalues *strictly* inside the stability boundary.

An important class of signals whose model is neither completely reachable, completely controllable, nor stabilizable are the deterministic disturbances described in Section 2.5. These models typically are not connected to the input and have poles on the stability boundary.

8.4 CANONICAL FORMS FOR REACHABLE SYSTEMS

In the preceding discussion it was implicitly assumed that the state space model was given and fixed. Often, however, it is helpful to use a similarity transformation to convert a given state space model into a more convenient form. We will achieve this by carrying out various similarity transformations. An infinite number of such transformations are possible, but we will give special emphasis to two particular transformations that lead to state space models having a specific structure (called the controllability and controller state space forms, respectively). Throughout this section we will assume that the system is completely reachable.

8.4.1 Controllability Form

Single-input Case

We construct a similarity transformation $\bar{x} = P^{-1}x$, where P is formed from the columns of the controllability matrix as follows:

$$P = \mathscr{C} = [B, AB, \ldots, A^{n-1}B] \qquad (8.4.1)$$

Let the A matrix have characteristic polynomial $\lambda^n + \alpha_{n-1}\lambda^{n-1} + \cdots + \alpha_0$. Then it is readily seen that the resulting state space model has the following special structure:

$$\rho\bar{x}(t) = \begin{bmatrix} 0 & & & -\alpha_0 \\ 1 & & & \vdots \\ & \ddots & & \vdots \\ & & 1 & -\alpha_{n-1} \end{bmatrix} \bar{x}(t) + \begin{bmatrix} 1 \\ 0 \\ \vdots \\ 0 \end{bmatrix} u(t) \qquad (8.4.2)$$

*Multiinput Case

The preceding ideas can be extended to the multiinput case. We will only treat this briefly as we will not need the results in our subsequent work. For the multiinput case, there are many ways of searching $\mathscr{C}$ to construct P. For example, we can search the controllability matrix from left to right to isolate a set of linearly

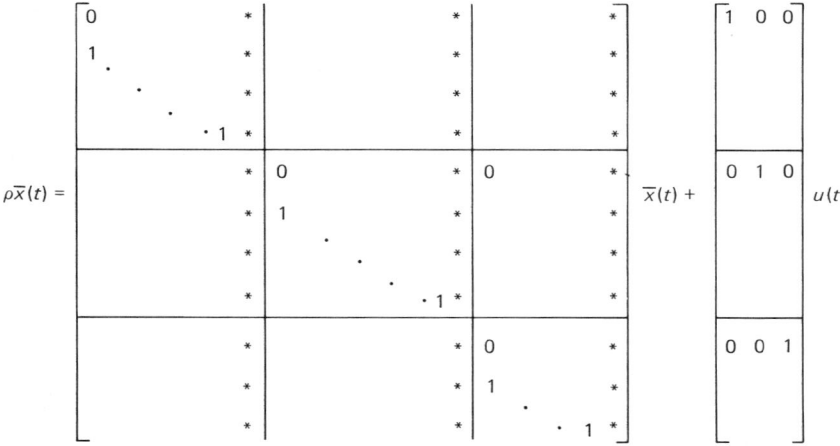

Figure 8.4.1 State space model for multiinput case.

Sec. 8.4   Canonical Forms for Reachable Systems

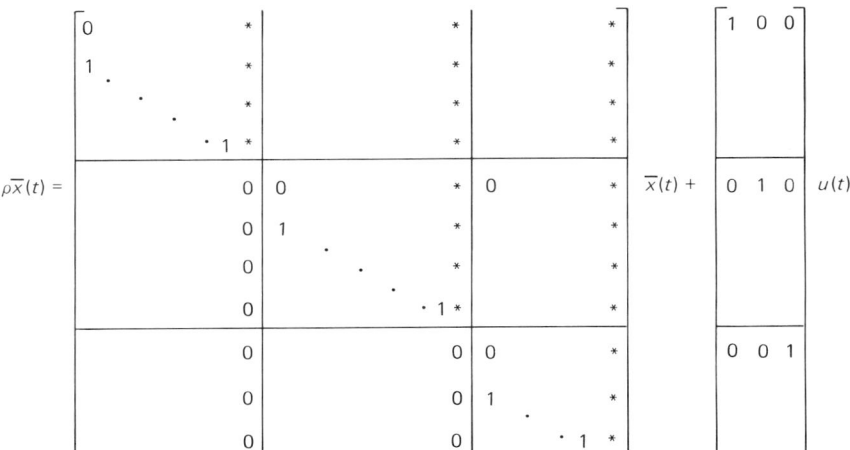

Figure 8.4.2 State space model for multiinput case, alternative method.

independent columns. The resulting columns are then rearranged into the following form:

$$P = \left[b_1 \ldots A^{k_1-1}b_1, b_2 \ldots A^{k_2-1}b_2, \ldots A^{k_r-1}b_r \right] \qquad (8.4.3)$$

where $B = [b_1 \ldots b_r]$. The indexes $k_1 \ldots k_r$ are called the controllability indexes and $k_{max} = \max k_i$ is called the controllability index.

The resulting state space model has the structure shown in Figure 8.4.1 (illustrated for $r = 3$ and where * denotes a nonzero entry).

An alternative is to search the controllability matrix starting with b_1 and then to proceed to Ab_1, A^2b_1, until the vector $A^{\nu_1}b_1$ can be expressed as a linear combination of previous vectors. Then we proceed to b_2, Ab_2, and so on. Finally, we form

$$P = \left[b_1 \ldots A^{\nu_1-1}b_1, b_2 \ldots A^{\nu_2-1}b_2, \ldots A^{\nu_r-1}b_r \right] \qquad (8.4.4)$$

Letting $\bar{x} = P^{-1}x$, the resulting state space model has the form shown in Figure 8.4.2.

8.4.2 Controller Forms

Single-input Case

Here we take the transformation found in the preceding section and postmultiply it by a nonsingular matrix M. The matrix M is formed from the characteristic polynomial of A. In the single-input case, the matrix is

$$M = \begin{bmatrix} 1 & \alpha_{n-1} & & \alpha_1 \\ 0 & 1 & & \alpha_2 \\ & & \ddots & \alpha_{n-1} \\ & & & 1 \end{bmatrix} \qquad (8.4.5)$$

where

$$\det(\lambda I - A) = \lambda^n + \alpha_{n-1}\lambda^{n-1} + \cdots + \alpha_0 \tag{8.4.6}$$

The transformation P is then given by $P = \mathscr{C}M$. The resulting state space model has the following (controller) structure:

$$\rho\bar{x}(t) = \begin{bmatrix} -\alpha_{n-1} & & & -\alpha_0 \\ 1 & & & 0 \\ & \ddots & & \vdots \\ & & 1 & 0 \end{bmatrix}\bar{x}(t) + \begin{bmatrix} 1 \\ 0 \\ \vdots \\ 0 \end{bmatrix}u(t) \tag{8.4.7}$$

$$y = \bar{c}\bar{x}(t) \tag{8.4.8}$$

This structure is shown diagrammatically in Figure 8.4.3.

*Multiinput Case

For multiinput systems, the controller state space form has the structure shown in Figure 8.4.4.

We next show that the controller state space form can be compactly written in an operator form, in which case it is termed a right matrix fraction description.

8.4.3 Right Matrix Fraction Descriptions

Let the scalar $z(t)$ denote $\bar{x}_n(t)$ in Equation (8.4.7). We then note from the latter $(n-1)$ equations in (8.4.7) that

$$\begin{bmatrix} z(t) \\ \rho z(t) \\ \vdots \\ \rho^{n-1}z(t) \end{bmatrix} = \begin{bmatrix} \bar{x}_n(t) \\ \bar{x}_{n-1}(t) \\ \vdots \\ \bar{x}_1(t) \end{bmatrix} \tag{8.4.9}$$

Then the first equation in (8.4.7) is

$$\rho\bar{x}_1(t) + \alpha_{n-1}\bar{x}_1(t) + \cdots + \alpha_0\bar{x}_n(t) = u(t) \tag{8.4.10}$$

or, using (8.4.9),

$$\alpha(\rho)z(t) = u(t) \tag{8.4.11}$$

where

$$\alpha(\rho) = \rho^n + \alpha_{n-1}\rho^{n-1} + \cdots + \alpha_0 \tag{8.4.12}$$

Also, Equation (8.4.8) can be written as

$$y(t) = \bar{c}_1\bar{x}_1(t) + \cdots + \bar{c}_n\bar{x}_n(t)$$
$$= \beta(\rho)z(t) \tag{8.4.13}$$

where

$$\beta(\rho) = \bar{c}_1\rho^{n-1} + \cdots + \bar{c}_n \tag{8.4.14}$$

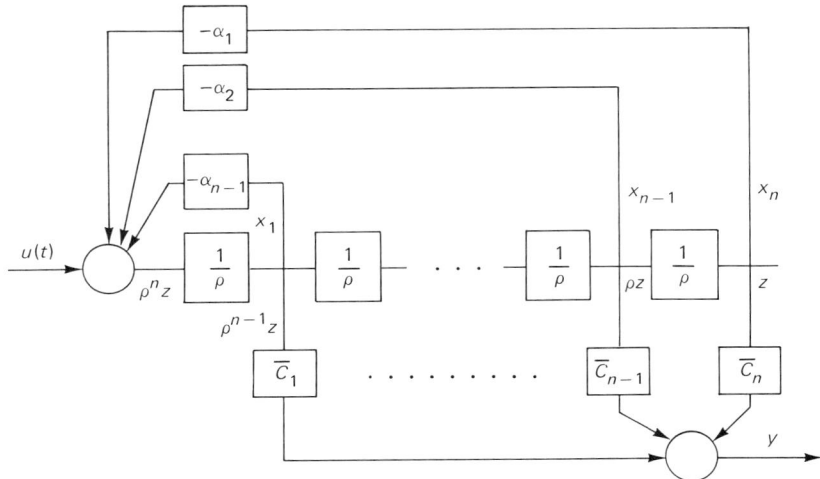

Figure 8.4.3 Structure of the controller state space form.

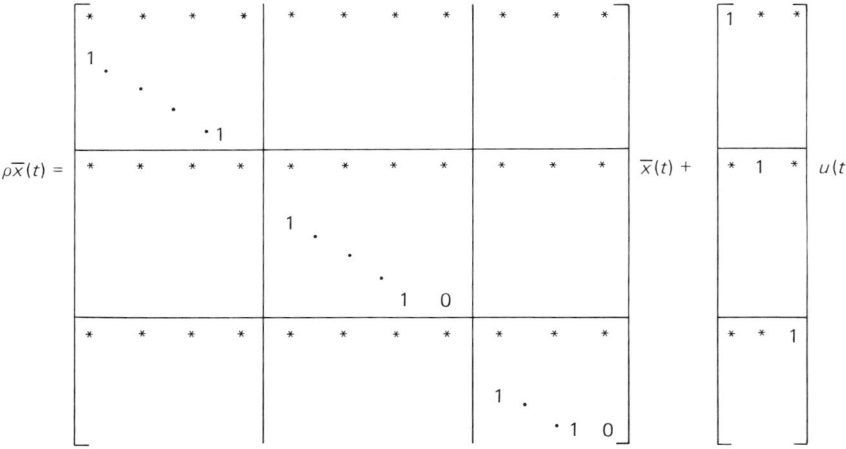

Figure 8.4.4 Controller state space form for multiinput systems.

We thus see that a right matrix fraction description of the form

$$\alpha(\rho)z(t) = u(t) \tag{8.4.15}$$
$$y(t) = \beta(\rho)z(t) \tag{8.4.16}$$

is simply a compact way of writing a controller form state space model. The quantity z is usually called the *partial state*.

A similar result holds for the multiinput case, where $\alpha(\rho)$ and $\beta(\rho)$ are now matrix polynomials of dimension $r \times r$ and $m \times r$, respectively, and $z(t)$ is an $r \times 1$-dimensional partial state vector.

8.5 CONTROLLABILITY GRAMIAN

We have seen that a completely reachable system is strongly connected to the input, in the sense that we can use the input to steer the system to any point in the state space. A natural question that arises is not just whether or not this is possible but how difficult it is to achieve for a given system. A partial answer to this question is provided by the *controllability gramian*, which we study in this section.

Say we have a completely reachable system and we wish to choose a finite energy input $u(t)$ to bring the state to x_r at time 0 starting from $x = 0$ at time $-t_0 (< 0)$. The solution of the state space model is

$$x(0) = \int_{-t_0}^{0} E(A, -\tau) B u(\tau) \, d\tau \qquad (8.5.1)$$

Since we require $x(0) = x_r$, our input must satisfy

$$\int_{-t_0}^{0} E(A, -\tau - \Delta) B u(\tau) \, d\tau = x_r \qquad (8.5.2)$$

There are an infinite number of inputs that satisfy (8.5.2). Thus, to be specific, let us choose the input from the class satisfying (8.5.2) that has least energy. This input is described in the following result.

Lemma 8.5.1

The input satisfying (8.5.2) that minimizes J, where

$$J = \int_{-t_0}^{0} u(t)^T u(t) \, dt, \qquad t_0 > (n-1)\Delta \qquad (8.5.3)$$

is

$$u^*(t) = B^T E(A, -t - \Delta)^T P(t_0)^{-1} x_r \qquad (8.5.4)$$

where $P(t_0)$ is the following matrix, which is nonsingular if and only if $\mathscr{C}$ has full rank:

$$P(t_0) = \int_{-t_0}^{0} \left[E(A, -\tau - \Delta) B B^T E(A, -\tau - \Delta)^T \right] d\tau$$

$$= \int_{0}^{t_0} \left[E(A, \tau) B B^T E(A, \tau)^T \right] d\tau \qquad (8.5.5)$$

Proof: We first establish the nonsingularity of $P(t_0)$. Following the proof of Theorem 8.3.1, we note that

$$E(A, \tau) B = \mathscr{C} V(\tau) \qquad (8.5.6)$$

where $V(\tau)$ is as defined in (8.3.9). Then, from (8.5.5),

$$P(t_0) = \mathscr{C} \int_{0}^{t_0} V(\tau) V(\tau)^T \, d\tau \, \mathscr{C}^T \qquad (8.5.7)$$

Nonsingularity of $P(t_0)$ then follows from Problem 28 and the fact that $\mathscr{C}$ has full rank.

It is easy to check that the input $u^*(t)$ satisfies (8.5.2). Next, consider an input constructed as follows:

$$u(t) = u^*(t) + \mu(t) \tag{8.5.8}$$

where $u^*(t)$ is as in (8.5.4) and $\mu(t)$ is arbitrary save that (8.5.2) is also satisfied by $u(t)$. This implies

$$\int_{-t_0}^{0} E(A, -\tau - \Delta) B \mu(\tau) \, d\tau = 0 \tag{8.5.9}$$

Thus, evaluating the cost function J, we have

$$J = \int_{-t_0}^{0} (u^*(t) + \mu(t))^T (u^*(t) + \mu(t)) \, dt$$

$$= \int_{-t_0}^{0} u^*(t)^T u^*(t) \, dt + \int_{-t_0}^{0} \mu(t)^T \mu(t) \, dt \tag{8.5.10}$$

where we have used the fact that the cross-product terms integrate to zero in view of (8.5.9). Then

$$J \geq \int_{-t_0}^{0} u^*(t)^T u^*(t) \, dt \tag{8.5.11}$$

with equality if and only if $\mu(t) = 0$. ▽▽▽

We will call the matrix in (8.5.5) the *controllability gramian* on the interval $(-t_0, 0)$. Substituting (8.5.4) into (8.5.3) shows that the minimum energy to make the transfer to the origin is

$$J^* = x_0^T P(t_0)^{-1} x_0 \tag{8.5.12}$$

Example 8.5.1

Consider the simple electrical circuit shown in Figure 8.5.1. Let x denote the voltage on the capacitor; then the equation describing this circuit is

$$\dot{x} = -\frac{1}{\tau_c} x + \frac{1}{\tau_c} u \tag{8.5.13}$$

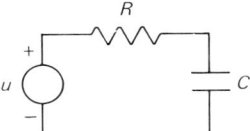

Figure 8.5.1 Electrical circuit.

where $\tau_c = RC$. For this system, we have

$$P(t_0) = \int_0^{t_0} e^{-t/\tau_c} \frac{1}{\tau_c^2} e^{-t/\tau_c} \, d\tau$$

$$= \frac{1}{2\tau_c}[1 - e^{-2t_0/\tau_c}] \tag{8.5.14}$$

As $t_0 \to \infty$, we find $P(t_0) \to 1/2\tau_c$; that is, if we allow infinite time to make the transfer, then the energy converges to

$$J^* = x_r^T P(t_0)^{-1} x_r = x_r^2 (2\tau_c)$$

As $t_0 \to 0^+$, we find $P(t_0) \to 0$; that is, if we allow an arbitrarily small time to make the transfer, the energy required becomes infinite.

We also notice that $P(t_0)^{-1}$ increases rapidly when the transfer is required to take place in times shorter than the natural time constant of the system. The graph of $P(t_0)^{-1}$ is shown in Figure 8.5.2.

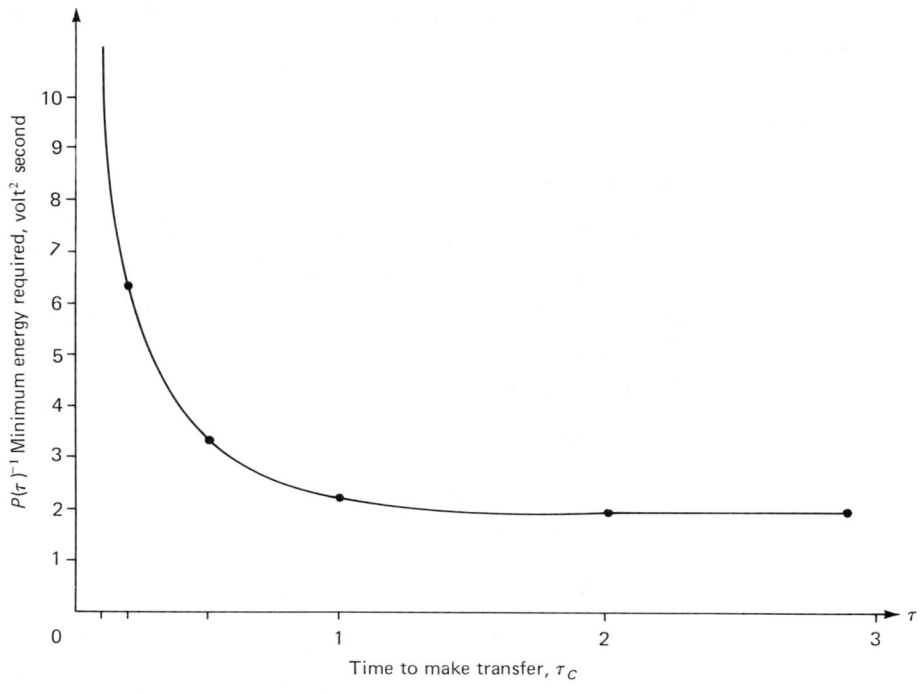

Figure 8.5.2 Energy as a function of time to make transfer (τ_c normalized to unity).

Sec. 8.5   Controllability Gramian

Example 8.5.2

Consider the problem of accelerating a car from zero velocity to v_r in T seconds. Ignoring friction, the model is

$$m\dot{v} = F \qquad (8.5.15)$$

For this problem it is readily seen that

$$P(T) = \int_0^T \frac{1}{m^2} d\tau = \frac{T}{m^2} \qquad (8.5.16)$$

Hence the integral of the force squared required to accelerate the car in T seconds to v_r is

$$J^* = v_r P(T)^{-1} v_r = \frac{m^2 v_r^2}{T} \qquad (8.5.17)$$

For this example, we see that the energy goes to zero as $T \to \infty$ as expected. Also, note that for this example the optimal input is a constant value; that is,

$$u^*(t) = \frac{m v_r}{T}, \qquad -T \leq t \leq 0 \qquad (8.5.18)$$

This suggests that $\sqrt{J^*/T}$ may give some guide as to the size of input needed to make the transfer in the allocated time. ▽▽▽

For stable systems, $\lim_{t_0 \to \infty} P(t_0)$ always exists. We will denote this limit simply by P. The matrix $P(t)$ and the limiting matrix P satisfy simple equations as follows:

Lemma 8.5.2

(i) The matrix $P(t)$ satisfies the following equation:

$$\rho P(t) = AP(t) + P(t)A^T + \Delta A P(t) A^T + BB^T; \qquad P(0) = 0 \qquad (8.5.19)$$

(ii) If all eigenvalues of A lie inside the stability boundary, then $\lim_{t \to \infty} P(t) \triangleq P$ exists and P satisfies the following Lyapunov equation:

$$AP + PA^T + \Delta A P A^T = -BB^T \qquad (8.5.20)$$

Proof:

(i) From (8.5.5), we have that

$$AP(t) + P(t)A^T + \Delta A P(t) A^T = \mathop{S}_{0}^{t} \rho_\tau \left[E(A, \tau) BB^T E(A, \tau)^T \right] d\tau$$

$$= \left[E(A, \tau) BB^T E(A, \tau)^T \right]_0^t$$

$$= \rho P(t) - BB^T \qquad (8.5.21)$$

where we have used the fact that

$$\rho P(t) = \rho \int_0^t \left[E(A, \tau) BB^T E(A, \tau)^T \right] d\tau$$

$$= E(A, t) BB^T E(A, t)^T \tag{8.5.22}$$

(ii) If the eigenvalues of A are inside the stability boundary, then for some $K > 0$, $((-1/\Delta) < \sigma < 0)$ we have

$$(\|E(A, t)\| \leq KE(\sigma, t), \qquad \text{for all } t \tag{8.5.23}$$

Equation (8.5.23) guarantees that the $\lim_{t \to \infty} P(t)$ exists.

Also, from (8.5.22) we see that $\lim_{t \to \infty} \rho P(t) = 0$. Hence, the limiting solution satisfies (8.5.20) with $\rho P(t) = 0$. ▽▽▽

Note that the inverse of the infinite time controllability gramian, P^{-1}, is a measure of the system's inherent resistance to change. In particular we see that $P^{-1} = 2\tau_c$ and 0 respectively for Examples 8.5.1, and 8.5.2.

8.6 PER UNIT VALUES

A difficulty with the controllability gramian (as discussed) is that it depends on the basis used for the state space and on the units used for the input. For example, say we let $\bar{x} = Tx$ and scale the input by defining $\bar{u} = Su$. Then the model becomes $(\bar{A}, \bar{B}) = (TAT^{-1}, TBS^{-1})$. Thus the controllability gramian becomes

$$\bar{P}(t_0) = T \left[\int_0^{t_0} SE(A, \tau) BS^{-1} S^{-T} B^T E(A, \tau)^T d\tau \right] T^T \tag{8.6.1}$$

A possible way around this scaling problem is to use per unit values for all quantities. This notion was inspired by the widespread use of per unit values in electrical engineering so that electromagnetic equipment of different sizes can be compared. Thus we normalize every variable by dividing by its rated value. In the case of input signals, this is particularly straightforward since we simply divide by the maximum available signal level. Note that this has the secondary advantage of making all quantities dimensionless. Thus, we can compare the efficacy of two very different inputs (say a voltage driving a fan speed with a fuel flow rate) without the answer depending on the units (volts, millivolts, microvolts, liters per day, megagallons per picosecond, and so on). Similarly, the states may be scaled by dividing by the expected range of variation of the state. Finally, the value of $P(t_0)$ depends on the time scale. This effect can be removed by scaling time so that the dominant poles have magnitude 1.

If these various scaling operations are carried out, then any eigenvalue of $P(1)$ much less than 1 is indicative of a state transfer that is difficult to make. The corresponding eigenvector indicates the direction in which it is difficult to steer the system.

8.7 RECONSTRUCTIBILITY, OBSERVABILITY, AND DETECTABILITY

The notions of reconstructibility, observability, and detectability are all concerned with the interaction between the system output and state. Intuitively, these notions tell us what part of the system state can be seen from the output.

Observability is concerned with our ability to detect the presence of some initial state of a system in the subsequent output response. *Reconstructibility* is concerned with our ability to establish the current state given the past output response. Finally, a system is said to be *detectable* if the only nonobservable states are those that decay to the origin. We now give formal definitions.

Definition 8.7.1

The state $x_0 \neq 0$ is said to be nonobservable if, given $x(0) = x_0$ and $u(t) = 0$ for $t \geq 0$, then $y(t) = 0$ for $t \geq 0$. The system is said to be completely observable if there exists no nonzero initial state that is nonobservable.

Without loss of generality, we can take $u(t) = 0$ since this part of the response, if present, can be readily subtracted off. The initial condition response is then

$$y(t) = CE(A, t)x_0 \qquad (8.7.1)$$

This expression immediately implies:

Theorem 8.7.1. Observability Using Observability Matrix

The set of all nonobservable states is equal to the null space of the observability matrix $\mathcal{O}$, where

$$\mathcal{O} = \begin{bmatrix} C \\ CA \\ \vdots \\ CA^{n-1} \end{bmatrix} \qquad (8.7.2)$$

The system is completely observable if and only if $\mathcal{O}$ has full column rank n.

Proof: See Problem 14.

We therefore see that reachability and observability are dual concepts in the sense that (A, B) is a completely reachable pair if and only if $(\overline{C}, \overline{A})$ is a completely observable pair, where $\overline{C} = B^T$, $\overline{A} = A^T$.

We can use this duality concept to quickly establish the corresponding results to Theorems 8.3.2, and 8.3.3.

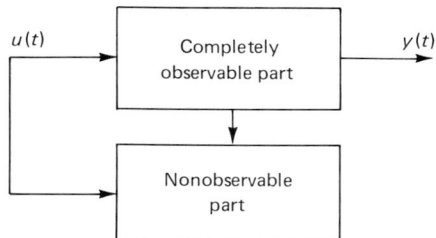

Figure 8.7.1 Decomposition into observable and nonobservable subsystems.

Theorem 8.7.2. Observable Decomposition

If rank $\mathcal{O} = k < n$, then there exists a similarity transformation T such that, with $\bar{x} = T^{-1}x$, that is,

$$\bar{A} = T^{-1}AT, \qquad \bar{C} = CT$$

$\bar{C}, \bar{A}$ have the form

$$\bar{A} = \begin{bmatrix} \bar{A}_0 & 0 \\ \bar{A}_{21} & \bar{A}_{n0} \end{bmatrix}, \qquad \bar{C} = \begin{bmatrix} \bar{C}_0 & 0 \end{bmatrix} \qquad (8.7.3)$$

where $\bar{A}_0$ has dimension k and $[\bar{C}_0, \bar{A}_0]$ is completely observable.

Proof: Dual of Theorem 8.3.2. ▽▽▽

The decomposition described is shown in Figure 8.7.1. Similarly, Theorem 8.3.3 has the following dual:

Theorem 8.7.3. PBH Observability Test

(i) A pair (C, A) will be nonobservable if and only if there exists a (column) vector $p \neq 0$ such that
$$Ap = \lambda p, \qquad Cp = 0$$
that is, if and only if some right eigenvector of A is orthogonal to (all rows of) C.

(ii) The pair (C, A) is completely observable if and only if
$$\text{rank}\begin{bmatrix} C \\ \lambda I - A \end{bmatrix} = n, \qquad \text{for all } \lambda$$

(iii) The pair (C, A) is completely observable if and only if C and $(\lambda I - A)$ are relatively right prime.

Proof: Dual of Theorem 8.3.3. ▽▽▽

Similarly, the concepts of reconstructibility and detectability are duals of controllability and stabilizability, respectively. For example, a system is detectable if the $\bar{A}_{n0}$ subsystem in Figure 8.7.1 is stable; that is, $E(\bar{A}_{n0}, t) \to 0$.

8.8 CANONICAL FORMS FOR OBSERVABLE SYSTEMS

Just as for reachable systems, there exist special canonical forms for observable systems. We will give the single-output versions and leave it to the reader to construct the multioutput generalizations.

8.8.1 Observability Form

This form has the following structure:

$$\rho \bar{x}(t) = \begin{bmatrix} 0 & 1 & & \\ & & \ddots & \\ & & & 1 \\ -\alpha_0 & \cdots & \cdots & -\alpha_{n-1} \end{bmatrix} \bar{x}(t) + \bar{B}u(t) \qquad (8.8.1)$$

$$y(t) = \begin{bmatrix} 1 & 0 & \cdots & 0 \end{bmatrix} \bar{x}(t) \qquad (8.8.2)$$

8.8.2 Observer Form

This form has the following structure:

$$\rho \bar{x}(t) = \begin{bmatrix} -\alpha_{n-1} & 1 & & \\ & & \ddots & \\ & & & 1 \\ -\alpha_0 & 0 & \cdots & 0 \end{bmatrix} \bar{x}(t) + \bar{B}u(t)$$

$$y(t) = \begin{bmatrix} 1 & 0 & \cdots & 0 \end{bmatrix} \bar{x}(t) \qquad (8.8.3)$$

This structure is shown in Figure 8.8.1 where

$$\bar{B} = [b_{n-1} \, b_{n-2} \ldots b_0]^T$$

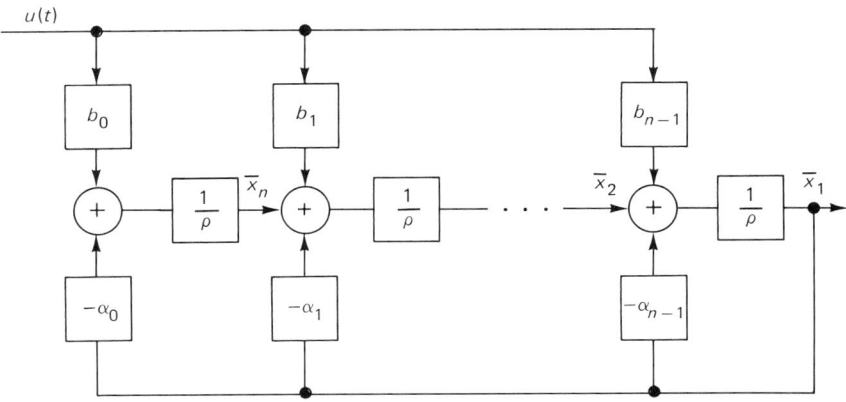

Figure 8.8.1 Structure of the observer state space form.

We next show that the observer state space form can be compactly written in operator form, in which case it is known as a left matrix fraction description.

8.8.3 Left Matrix Fraction Descriptions

We note in (8.8.3) that the state vector $\bar{x}(t)$ can be eliminated by successively differentiating $y(t)$. Thus, we have

$$\begin{aligned}
\bar{x}_1(t) &= y(t) \\
\rho \bar{x}_1(t) &= -\alpha_{n-1}\bar{x}_1(t) + \bar{x}_2(t) + b_{n-1}u(t) \\
\rho \bar{x}_2(t) &= -\alpha_{n-2}\bar{x}_1(t) + \bar{x}_3(t) + b_{n-2}u(t) \\
&\vdots \\
\rho \bar{x}_{n-1}(t) &= -\alpha_1 \bar{x}_1(t) + \bar{x}_n(t) + b_1 u(t) \\
\rho \bar{x}_n(t) &= -\alpha_0 \bar{x}_1(t) + b_0 u(t)
\end{aligned} \quad (8.8.4)$$

Hence

$$\begin{aligned}
\rho^n y(t) &= -\alpha_{n-1}\rho^{n-1}y(t) + \rho^{n-1}\bar{x}_2(t) + b_{n-1}\rho^{n-1}u(t) \\
\rho^{n-1}\bar{x}_2(t) &= -\alpha_{n-2}\rho^{n-2}y(t) + \rho^{n-2}\bar{x}_3(t) + b_{n-2}\rho^{n-2}u(t) \\
&\vdots \\
\rho^2 \bar{x}_{n-1}(t) &= -\alpha_1 \rho y(t) + \rho \bar{x}_n(t) + b_1 \rho u(t) \\
\rho \bar{x}_n(t) &= -\alpha_0 y(t) + b_0 u(t)
\end{aligned} \quad (8.8.5)$$

Finally, by successive substitution, we obtain

$$\alpha(\rho)y(t) = \beta(\rho)u(t) \quad (8.8.6)$$

where

$$\alpha(\rho) = \rho^n + \alpha_{n-1}\rho^{n-1} + \cdots + \alpha_0 \quad (8.8.7)$$

$$\beta(\rho) = b_{n-1}\rho^{n-1} + b_{n-2}\rho^{n-2} + \cdots + b_0 \quad (8.8.8)$$

Thus we see that the left matrix function description (8.8.6) is simply a compact way of representing a state space model in observer form.

8.9 OBSERVABILITY GRAMIAN

Heuristically, observability is concerned with the problem of whether or not the system state can be seen in the system output. However, it would be helpful to know the quantity of information in the output about the state. This information is provided by the observability gramian $Q(t_f)$.

Say we have an initial state of x_0 and we let the system run free with no input; then the corresponding output response is

$$y(t) = CE(A, t)x_0 \quad (8.9.1)$$

The energy recovered from the output on the interval $(0, t_f)$ is defined as

$$J_r \triangleq \int_0^{t_f} y(t)^T y(t)\, dt$$

Hence

$$J_r = x_0^T Q(t_f) x_0 \qquad (8.9.2)$$

where $Q(t_f)$ is the *observability gramian* given by

$$Q(t_f) = \int_0^{t_f} E(A,t)^T C^T C E(A,t)\, dt \qquad (8.9.3)$$

Example 8.9.1

Consider again the electrical circuit of Example 8.5.1, where $y(t) = x(t)$, the voltage across the capacitor. Then

$$Q(t_f) = \int_0^{t_f} e^{-2t/\tau_c}\, dt$$
$$= \frac{\tau_c}{2}[1 - e^{-2t_f/\tau_c}]$$

As $t_f \to \infty$, we find $Q(t_f) \to \tau_c/2$. As $t_f \to 0$, we find $Q(t_f) \to 0$.

Example 8.9.2

Consider again the car example in Example 8.5.2. For this system

$$Q(t_f) = t_f$$

Again, as for the controllability gramian, it is desirable to work with per unit values to overcome the problem of units. Thus we suggest rescaling as discussed previously.

For stable systems, $\lim_{t_f \to \infty} Q(t_f)$ exists. (Compare with Lemma 8.5.2.) We will denote the limit by Q. The matrix $Q(t)$ and the limiting matrix Q satisfy simple equations as follows:

Lemma 8.9.1

(i) The matrix $Q(t)$ satisfies

$$\rho Q(t) = A^T Q(t) + Q(t) A + \Delta A^T Q(t) A + C^T C \qquad (8.9.4)$$

(ii) For asymptotically stable systems, the (infinite time) observability gramian Q satisfies

$$A^T Q + QA + \Delta A^T Q A = -C^T C \qquad (8.9.5)$$

Proof: As for Lemma 8.5.2.

8.10 CANONICAL STRUCTURE, MINIMAL MODELS, INPUT – OUTPUT EQUIVALENCE

Further insight into the structure of linear dynamical systems is obtained by combining the results of Theorems 8.3.2 and 8.7.2. This results in the following:

Theorem 8.10.1. Canonical Structure Theorem

We can always find an invertible state transformation $\bar{x} = Tx$ such that the system, when expressed in the new coordinate system, has the structure

$$\bar{A} = \begin{bmatrix} \bar{A}_{11} & 0 & \bar{A}_{13} & 0 \\ \bar{A}_{21} & \bar{A}_{22} & \bar{A}_{23} & \bar{A}_{24} \\ 0 & 0 & \bar{A}_{33} & 0 \\ 0 & 0 & \bar{A}_{43} & \bar{A}_{44} \end{bmatrix}, \quad \bar{B} = \begin{bmatrix} \bar{B}_1 \\ \bar{B}_2 \\ 0 \\ 0 \end{bmatrix} \quad (8.10.1)$$

$$\bar{C} = \begin{bmatrix} \bar{C}_1 & 0 & \bar{C}_2 & 0 \end{bmatrix} \quad (8.10.2)$$

such that:

(i) The subsystem $R_1 = [\bar{C}_1, \bar{A}_{11}, \bar{B}_1]$ is completely observable and completely reachable.
(ii) The transfer function of the subsystem R_1 is equal to that of the original system; that is,

$$\bar{C}_1(\gamma I - \bar{A}_{11})^{-1}\bar{B}_1 = C(\gamma I - A)^{-1}B$$

(iii) There does not exist a lower-dimension model than R_1 having the same transfer function as the system.
(iv) The subsystem

$$R_{12} = \left[\begin{bmatrix} \bar{C}_1 & 0 \end{bmatrix}, \begin{bmatrix} \bar{A}_{11} & 0 \\ \bar{A}_{21} & \bar{A}_{22} \end{bmatrix}, \begin{bmatrix} \bar{B}_1 \\ \bar{B}_2 \end{bmatrix} \right]$$

is completely reachable.
(v) The subsystem

$$R_{13} = \left[\begin{bmatrix} \bar{C}_1 & \bar{C}_2 \end{bmatrix}, \begin{bmatrix} \bar{A}_{11} & \bar{A}_{13} \\ 0 & \bar{A}_{33} \end{bmatrix}, \begin{bmatrix} \bar{B}_1 \\ 0 \end{bmatrix} \right]$$

is completely observable.
(vi) The subsystem R_1 is input–output equivalent to the original system for *zero* initial state.
(vii) The subsystem R_{13} is input–output equivalent to the original system for *all* initial states (provided that the initial states are transformed consistently).

(viii) The subsystem R_{13} can be expressed compactly as a left matrix fraction description of the form

$$\alpha(\rho)y(t) = \beta(\rho)u(t)$$

Proof: Straightforward by combining Theorems 8.3.2 and 8.7.2. (See Problem 16.)
▽▽▽

Parts (ii), (iii) and (vi) of Theorem 8.10.1 are useful in that they allow a state space model of minimal dimension to be constructed to represent a given transfer function. This avoids having redundant states if our interest is in the transfer function only. Parts (v), (vii), and (viii) are also important since they show how the complete input–output behavior can be succinctly described. Often, we need to consider states that are observable but not necessarily reachable. A key application of this is when we wish to describe deterministic disturbances.

Further insights into the relationship between observability, reachability, and input–output models are given in the following result, which for simplicity is stated for the single-input, single-output case.

Theorem 8.10.2

Consider the following state space model:

$$\rho x = Ax + Bu \tag{8.10.3}$$

$$y = Cx \tag{8.10.4}$$

Then, the following two statements are equivalent

(i) (A, B) is reachable and (C, A) is observable.
(ii) The transfer function has no common poles and zeros; that is, $C \operatorname{Adj}(\gamma I - A)B$ and $\det(\gamma I - A)$ have no common factors.

Proof: See Problem 11(d).
▽▽▽

8.11 HANKEL SINGULAR VALUES AND BALANCED REALIZATIONS

Here we gain further insight into minimal models (ones that are both completely reachable and observable). We will further assume that the model is stable. Under these conditions, we will attempt to gain a deeper understanding of the relationships between observability, reachability, and the system transfer function.

We have seen earlier that the observability and controllability gramians give a measure of the connectivity between the system state and the output and input, respectively. However, these two matrices are state realization dependent. In particular, if $\bar{x} = Tx$, then it is readily seen that the gramians for the transformed system are

$$\bar{P} = TPT^T \tag{8.11.1}$$

$$\bar{Q} = T^{-T}QT^{-1} \tag{8.11.2}$$

However, the eigenvalues of the product PQ are invariant with respect to basis changes, since

$$\overline{PQ} = T(PQ)T^{-1} \tag{8.11.3}$$

We will therefore define the square root of the eigenvalues of PQ as the *Hankel singular values* of the transfer function denoted by $\sigma_1 \ldots \sigma_n$; that is,

$$\sigma_i(G(\gamma)) = \{\lambda_i(PQ)\}^{1/2} \tag{8.11.4}$$

where $G(\gamma) = C(\gamma I - A)^{-1}B$. (The name arises from the fact that the Hankel singular values are precisely the singular values of the Hankel matrix in the case of discrete shift operator models).

Now, since P and Q depend on the realization, whereas the eigenvalues of the product do not, the question arises as to whether or not we can perform a similarity transformation to share out $\sigma_1 \ldots \sigma_n$ equally between P and Q. We will call a realization in which P and Q are equal and diagonal a *balanced realization*. The following result shows how this can be achieved.

Theorem 8.11.1

Given a stable, minimal model, there exists a similarity transformation that leads to a balanced realization.

Proof: P and Q are symmetric positive semidefinite matrices. Hence we can write Q as follows (using a Cholesky factorization):

$$Q = R^T R \tag{8.11.5}$$

then $S \triangleq RPR^T$ is positive definite and hence can be expressed (using a singular value decomposition for symmetric matrices) as

$$S = U\Sigma U^T \tag{8.11.6}$$

where

$$U^T U = I$$
$$\Sigma = \text{diag}(\sigma_1^2 \ldots \sigma_n^2) \tag{8.11.7}$$

The required similarity transformation is then

$$T = \Sigma^{-1/4} U^T R \tag{8.11.8}$$

To verify that this transformation achieves a balanced realization, we see that

$$\overline{P} = TPT^T$$
$$= \Sigma^{-1/4}U^T RPR^T U \Sigma^{-1/4} = \Sigma^{1/2} \tag{8.11.9}$$

and

$$\overline{Q} = T^{-T}QT^{-1}$$
$$= \Sigma^{1/4}U^T R^{-T}R^T RR^{-1}U \Sigma^{1/4} = \Sigma^{1/2} \tag{8.11.10}$$

That is,

$$\overline{P} = \overline{Q} = \begin{bmatrix} \sigma_1 & & & \\ & \sigma_2 & & \\ & & \ddots & \\ & & & \sigma_n \end{bmatrix} \tag{8.11.11}$$

as required. We also see that $\sigma_1, \ldots, \sigma_n$ in (8.11.7) are the Hankel singular values.

The Hankel singular values give some measure of the importance of the various states in the overall system transfer function. This suggests that one way of obtaining a low-order approximation of a state space model may be to form a balanced realization and then to retain only those states corresponding to the k largest singular values by truncating the balanced realizations. This turns out to be quite a sensible procedure. The truncated realization is guaranteed to be stable (given that the original system is also stable). Moreover, the L_∞ error between the two transfer functions can be shown to be bounded as follows:

$$\|G(j\omega) - \hat{G}_k(j\omega)\|_{L_\infty} \leq 2(\sigma_{k+1} + \cdots + \sigma_n) \tag{8.11.12}$$

8.12 STABILITY FOR NONLINEAR SYSTEMS

We conclude this chapter by extending the simple notions of stability introduced in Chapter 7 to nonlinear systems.

8.12.1 Basic Definitions

Consider a dynamic system described by a nonlinear state space equation

$$\rho x(t) = f(x(t), u(t), t) \tag{8.12.1}$$

and

$$y(t) = g(x(t), u(t), t) \tag{8.12.2}$$

where

$$f(0, 0, t) \equiv 0.$$

For linear systems, we found in Chapter 7 that a very simple notion of stability could be used. However, for nonlinear systems we need to be more precise about what we mean by the term stable. We therefore introduce the following definitions:

Definition 8.12.1. Stability

The system (8.12.1) is said to be stable if, for $u(t) \equiv 0$ and any $\varepsilon > 0$, there exists $\beta(\varepsilon, T) > 0$ such that

$$\|x(T)\| < \beta(\varepsilon, T) \Rightarrow \|x(t)\| < \varepsilon, \qquad \text{for all } t > T \qquad (8.12.3)$$

▽▽▽

Stability in this context thus indicates that, if the states "start small," then they stay small.

Definition 8.12.2. Asymptotic Stability

The system (8.12.1) is said to be asymptotically stable if, for $u(t) \equiv 0$ and any $x(T)$, $x(t) \to 0$.

▽▽▽

Thus asymptotic stability implies that, with no inputs, the system states relax to zero.

Definition 8.12.3. Exponential Stability

The system (8.12.1) is said to be exponentially stable if, for $u(t) \equiv 0$, there exists $c, \lambda > 0$ such that for any T and $x(T)$ we have

$$\|x(t)\| \leq c\|x(T)\|e^{-\lambda(t-T)}, \qquad \text{for all } t > T \qquad (8.12.4)$$

▽▽▽

Exponential stability is thus a type of asymptotic stability in which $x(t) \to 0$ exponentially fast.

Definition 8.12.4. L_∞ Bounded Input, Bounded Output (BIBO) Stability

The system (8.12.1), (8.12.2) is said to be bounded input, bounded output stable if, for any finite $x(0)$ and any $u(t)$ such that $\|u(\cdot)\|_\infty$ exists, $\|y(\cdot)\|_\infty$ exists, where $\|\cdot\|_\infty$ denotes the L_∞ norm of a function; that is, $\|u(\cdot)\|_\infty = \sup\sqrt{u(t)^T u(t)}$. ▽▽▽

Note that "$\|u(\cdot)\|_\infty$ exists" means that $u(t)$ is bounded for all time; hence the term BIBO stability.

In the case of linear time invariant systems, the following relationships hold between the various definitions of stability as introduced previously.

Lemma 8.12.1

For linear time invariant systems:

(i) BIBO stability and minimality $\Rightarrow$ asymptotic stability.
(ii) Exponential stability $\Leftrightarrow$ asymptotic stability.

Proof: See Problem 17. ▽▽▽

Actually, in the case of continuous or discrete (delta) linear time invariant systems, we can find necessary and sufficient conditions for various forms of stability. These are given in the following result:

Theorem 8.12.1

The system

$$\rho x = Ax + Bu \qquad (8.12.5)$$
$$y = Cx \qquad (8.12.6)$$

is:

(i) Exponentially stable if and only if the eigenvalues λ_i of A satisfy

$$\frac{\Delta}{2}|\lambda_i|^2 + \text{Re}\{\lambda_i\} < 0 \qquad (8.12.7)$$

(ii) Stable if and only if the eigenvalues of A satisfy (8.12.7) *or*

$$\frac{\Delta}{2}|\lambda_i|^2 + \text{Re}\{\lambda_i\} = 0 \qquad (8.12.8)$$

and have Jordan block size 1.

(iii) BIBO stable if and only if:
  (a) All modes that are observable and reachable have eigenvalues that satisfy (8.12.7).
  (b) All modes that are observable and unreachable have eigenvalues that satisfy (8.12.7) or (8.12.8).

Proof:

(i) Let $A = T^{-1}\Lambda T$, where Λ is the Jordan canonical form for A (and T is nonsingular); then the unique solution to (8.12.5) is

$$x(t) = T^{-1}E(\Lambda, t)Tx(0) + \int_0^t T^{-1}E(\Lambda, t - \tau - \Delta)TBu(\tau)\,d\tau \qquad (8.12.9)$$

Thus, the system is exponentially stable if and only if $\|E(\Lambda, t)\| \to 0$ exponentially fast. Let

$$\Lambda = \begin{bmatrix} J_1 & & & 0 \\ & J_2 & & \\ & & \ddots & \\ 0 & & & J_l \end{bmatrix} \qquad (8.12.10)$$

where

$$J_i = \begin{bmatrix} \lambda_i & 1 & & & 0 \\ & \lambda_i & 1 & & \\ & & \ddots & \ddots & \\ & & & \lambda_i & 1 \end{bmatrix} \in \mathbb{R}^{\mu_i \times \mu_i} \qquad (8.12.11)$$

It then follows that

$$E(\Lambda, t) = \begin{bmatrix} E(J_1, t) & & 0 \\ & \ddots & \\ 0 & & E(J_l, t) \end{bmatrix}$$

$$= \begin{bmatrix} E(\lambda_1, t) F_{\mu_1}(t) & & 0 \\ & \ddots & \\ 0 & & E(\lambda_l, t) F_{\mu_l}(t) \end{bmatrix} \qquad (8.12.12)$$

where $F_i(t)$ denotes the matrix

$$F_i(t) = \begin{bmatrix} 1 & f_{i1}(t) & f_{i2}(t) & \cdots & \cdots & f_{i(\mu_i-1)}(t) \\ 0 & 1 & f_{i1}(t) & \cdots & \cdots & \cdots \\ 0 & 0 & 1 & \ddots & & \\ \vdots & & & \ddots & \ddots & f_{i1}(t) \\ 0 & \cdots & \cdots & \cdots & 0 & 1 \end{bmatrix} \qquad (8.12.13)$$

and

$$f_{ij}(t) \triangleq \frac{\prod_{k=0}^{j}(t - k\Delta)}{j!(1 + \Delta\lambda_i)^j}$$

Thus it is clear that $E(\Lambda, t) \to 0$ if and only if $E(\lambda_i, t) \to 0$ for all i. Condition (8.12.7) is necessary and sufficient for $E(\lambda_i, t) \to 0$ for all i and thus we have the desired result.

(ii) Stability in the case where (8.12.7) is satisfied for all λ_i follows from part (i). In the cases where some λ_i satisfy (8.12.8) and have Jordan block size 1, that is, $\mu_i = 1$, $E(J_i, t) = E(\lambda_i, t)$ and subject to (8.12.8), we have

$$|E(\lambda_i, t)| = 1, \qquad \text{for all } t \qquad (8.12.14)$$

Stability then follows. Conversely, if the Jordan block size is greater than 1,

$$\|E(J_i, t)\| = |E(\lambda_i, t)| \|F_{\mu_i}(t)\|$$
$$\geq t \qquad (8.12.15)$$

and thus the system is not stable.

(iii) If the system is not completely observable, we can use the decomposition theorem (Theorem 8.10.1) to give an equivalent input–output model that is observable. The eigenvalues of this observable system are precisely the eigenvalues of the observable part of the original system. We then need only to show that for an observable linear system BIBO ⇔ exponential stability. It is easily seen that exponential ⇒ BIBO, and we establish the converse using a negative argument. The complete details are left as an exercise, which is more easily completed using transform theory (see Problem 18).

If the system has any modes strictly outside the stability region indicated by (8.12.7), then with no input there is an initial state that diverges. Because of the observability assumption, it then follows that the output diverges.

If the system has a reachable mode that satisfies (8.12.8), then an input, $u(t) = E(\lambda_i, t)u_0$ for u_0 some nonzero vector, will affect that mode. For zero initial conditions, the state satisfies

$$x(t) = \int_0^t E(A, t - \tau - \Delta) Bu(\tau) \, d\tau \qquad (8.12.16)$$

and the integrand in (8.12.16) will contain terms of the form

$$E(\lambda_i, t - \tau) E(\lambda_i, \tau) = E(\lambda_i, t) \qquad (8.12.17)$$

When integrated from 0 to t with respect to τ, the term in (8.12.17) gives a term in $x(t)$ of the form $tE(\lambda_i, t)$, which diverges. Observability, then, implies that $y(t)$ is also unbounded. ∇∇∇

8.12.2 Lyapunov Stability

In the previous section we viewed stability from the point of view of the response of the system. An alternative way of viewing the stability of a dynamic system is in terms of the system energy. This viewpoint was originally studied by the Russian researcher Lyapunov (also spelled Liapunov or Ljapunov). This is a very powerful tool when studying nonlinear and/or time varying systems. For this reason, we will give a brief introduction to this type of analysis in this section.

We first make the following definitions:

Definition 8.12.5. Positive Definite Functions

(i) A continuous function, $W, \mathbb{R}^n \to \mathbb{R}$, is called a *positive definite function* (PDF) if:
 (a) $W(0) = 0$
 (b) $W(x) > 0$, for all $x \in \mathbb{R}^n$, $x \neq 0$
 (c) $W(x) \to \infty$, (as $\|x\| \to \infty$) uniformly in x

(ii) A continuous function, $V: \Omega \times \mathbb{R}^n \to \mathbb{R}$, is called a PDF if there exists a PDF, $W: \mathbb{R}^n \to \mathbb{R}$, such that

$$W(x) \leq V(t, x), \qquad \text{for all } t \in \Omega, \quad \text{for all } x \in \mathbb{R}^n$$

▽▽▽

$V(t, x)$ or $W(x)$ will be used to generalize the concept of system energy. The Lyapunov view of stability involves determining whether the energy in the system (with no inputs) decays to zero (corresponding to some form of stability) or increases with time (corresponding to some form of instability). We are thus interested in the behavior of the derivative (difference) of this generalized energy. We then have the following definition.

Definition 8.12.6

Suppose $V(t, x)$ is a PDF and x is a state of the system

$$\rho x(t) = f(x(t), t) \tag{8.12.18}$$

Then $V(t, x)$ is called a *Lyapunov function candidate* (LFC) if $\rho_t V(t, x(t))$ exists.

▽▽▽

In discrete time, we need not worry about the existence of the derivative since

$$\rho V(t, x(t)) = \frac{1}{\Delta} [V(t + \Delta, x(t + \Delta)) - V(t, x(t))] \tag{8.12.19}$$

In continuous time, it suffices that the partial derivatives $\partial V/\partial t$ and $\partial V/\partial x$ exist, in which case

$$\rho V(t, x(t)) = \frac{\partial V}{\partial t} + \left[\frac{\partial V}{\partial x}\right] f(x(t), t) \tag{8.12.20}$$

We then have the following theorems due to Lyapunov.

Theorem 8.12.2. Lyapunov Stability

Suppose for some LFC $V(t, x)$ and some system (8.12.18) that

$$\rho V(t, x) \leq 0, \qquad \text{for all } t, \quad \text{for all } x \tag{8.12.21}$$

Then the system (8.12.18) is stable as defined in Definition 8.12.1.

Proof: See Vidyasagar (1978). ▽▽▽

Theorem 8.12.3. Asymptotic Lyapunov Stability

Suppose for some LFC $V(t, x)$ and some system (8.12.18) that $[-\rho V(t, x(t))]$ is a PDF; then the system (8.12.18) is asymptotically stable as in Definition 8.12.2.

Proof: See Vidyasagar (1978). ▽▽▽

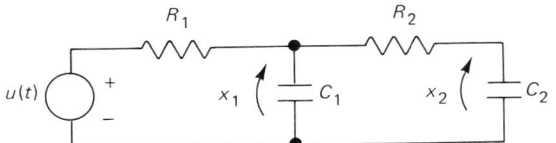

Figure 8.12.1 Simple electrical circuit.

Example 8.12.1 illustrates how we might find a Lyapunov function for a simple system.

Example 8.12.1

Consider the electrical circuit shown in Figure 8.12.1. Using the natural state variables, that is, the capacitor voltages $x = [x_1, x_2]^T$, we can show that

$$\dot{x}(t) = \begin{bmatrix} -\frac{1}{C_1}\left(\frac{1}{R_1} + \frac{1}{R_2}\right) & \frac{1}{R_2 C_1} \\ \frac{1}{R_2 C_2} & -\frac{1}{R_2 C_2} \end{bmatrix} x(t) + \begin{bmatrix} \frac{1}{R_1 C_1} \\ 0 \end{bmatrix} u(t) \qquad (8.12.22)$$

A natural energy function in this case is the total stored energy in the capacitors; that is,

$$V(t, x) = \frac{1}{2} C_1 x_1^2 + \frac{1}{2} C_2 x_2^2 = \frac{1}{2} x^T \begin{bmatrix} C_1 & 0 \\ 0 & C_2 \end{bmatrix} x \qquad (8.12.23)$$

From (8.12.23) it is clear that $V(t, x)$ is a LFC. We thus consider for $u(t) = 0$,

$$\frac{d}{dt}\{V(t, x)\} = \frac{1}{2} x^T \left\{ \begin{bmatrix} C_1 & 0 \\ 0 & C_2 \end{bmatrix} \begin{bmatrix} -\frac{1}{C_1}\left(\frac{1}{R_1} + \frac{1}{R_2}\right) & \frac{1}{R_2 C_1} \\ \frac{1}{R_2 C_2} & -\frac{1}{R_2 C_2} \end{bmatrix} \right.$$

$$\left. + \begin{bmatrix} -\frac{1}{C_1}\left(\frac{1}{R_1} + \frac{1}{R_2}\right) & \frac{1}{R_2 C_2} \\ \frac{1}{R_2 C_1} & -\frac{1}{R_2 C_2} \end{bmatrix} \begin{bmatrix} C_1 & 0 \\ 0 & C_2 \end{bmatrix} \right\} x$$

(8.12.24)

$$= -x^T \begin{bmatrix} \frac{1}{R_1} + \frac{1}{R_2} & -\frac{1}{R_2} \\ -\frac{1}{R_2} & \frac{1}{R_2} \end{bmatrix} x = -x^T Q x \qquad (8.12.25)$$

Provided that $R_1, R_2 \in (0, \infty)$, it follows that Q in (8.12.25) is positive definite, and so $-(d/dt)\{V(t, x)\}$ is a PDF. Hence, from Theorem 8.12.3, the system is asymptotically stable.

We next generalize Example 8.12.1 to show how we may construct a quadratic Lyapunov function for any stable linear system.

Theorem 8.12.4

The linear system

$$\rho x(t) = Ax(t) \tag{8.12.26}$$

is asymptotically stable if and only if, for any positive definite Q, the unique symmetric solution P to

$$A^T P + PA + \Delta A^T PA = -Q \tag{8.12.27}$$

is positive definite.

Proof:

(i) Suppose, for some Q, that the solution to (8.12.27) is positive definite. Then let

$$V(x) = x^T P x \tag{8.12.28}$$

We can then show, using (8.12.26), (8.12.27), that

$$\rho V(t, x) = -x^T Q x$$

Then by Theorem 8.12.3 it follows that the system (8.12.26) is asymptotically stable.
(ii) Conversely, suppose the system (8.12.26) is asymptotically stable. Then we have that the following integral exists:

$$P = \int_0^\infty E(A^T, \tau) Q E(A, \tau) \, d\tau \tag{8.12.29}$$

Now

$$\rho \{ E(A^T, t) Q E(A, t) \} = A^T E(A^T, t) Q E(A, t) + E(A^T, t) Q E(A, t) A$$
$$+ \Delta A^T E(A^T, t) Q E(A, t) A \tag{8.12.30}$$

and so, using (8.12.29),

$$A^T P + PA + \Delta A^T PA = \int_0^\infty \rho \{ E(A^T, t) Q E(A, t) \} \, dt$$
$$= \left[E(A^T, t) Q E(A, t) \right]_0^\infty$$
$$= -Q \tag{8.12.31}$$

The last line follows since A is an asymptotically stable matrix. Because the integrand in (8.12.29) is positive semidefinite for all τ, it follows that P is positive semidefinite. We now show that P is actually positive definite by contradiction. Suppose P is not positive definite; that is, there exists some v such that

$$Pv = 0 \tag{8.12.32}$$

Then from (8.12.31) we have that
$$v^TA^TPv + v^TPAv + \Delta v^TA^TPAv = \Delta v^TA^TPAv$$
$$= -v^TQv \tag{8.12.33}$$

From (8.12.33) it is clear that there exists a vector, $w = Av$, such that
$$w^TPw = -\frac{1}{\Delta}v^TQv < 0 \tag{8.12.34}$$

Equation (8.12.34) clearly contradicts the fact that P is positive semidefinite, and so we conclude that (8.12.34) does not hold for any $v \neq 0$ and thus P is positive definite. ▽▽▽

8.13 SUMMARY

The key points covered in this chapter were:

- The concept of reachability (Definition 8.3.1).
- Introduction of the controllability matrix (Definition 8.3.2)
$$\mathscr{C}[A, B] \triangleq [B, AB, \ldots, A^{n-1}B]$$
- A proof that the set of all reachable states is equal to the range space of $\mathscr{C}$ (Theorem 8.3.1).
- Definition of controllability and stabilizability.
- Introduction of canonical forms for reachable systems: controllability form, Equation (8.4.2); controller form, Equation 8.4.7).
- A proof that a right matrix fraction description is a compact way of writing a state space model in controller canonical form.
- The controllability gramian is defined as
$$P(t_0) \triangleq \mathbf{S}\int_0^{t_0} E(A, \tau)BB^TE(A, \tau)^T d\tau \tag{8.5.5}$$
- A proof was given that $P(t)$ satisfies
$$\rho P(t) = AP(t) + P(t)A^T + \Delta AP(t)A^T + BB^T \tag{8.5.18}$$
and for stable systems the limit P as $t \to \infty$ was shown to satisfy
$$AP + PA^T + \Delta APA^T = -BB^T \tag{8.5.20}$$
- Per unit values were introduced (Section 8.6) to normalize the controllability gramian.
- Observability was defined in Definition 8.7.1.
- Introduction of the observability matrix
$$\mathcal{O} \triangleq \begin{bmatrix} C \\ CA \\ \vdots \\ CA^{n-1} \end{bmatrix} \tag{8.7.2}$$

- A proof that the set of all unobservable states is the null space of $\mathcal{O}$ (Theorem 8.7.1).
- Introduction of canonical forms for observable systems: observability form, Equation (8.8.1); observer form, Equation (8.8.3).
- A proof that a left matrix fraction description is a compact way of writing a state space model in observer form.
- The observability gramian was defined as

$$Q(t_f) = \underset{0}{\overset{t_f}{S}} E(A, t)^T C^T C E(A, t) \, dt \tag{8.9.3}$$

- Verification that $Q(t)$ satisfies

$$\rho Q(t) = A^T Q(t) + Q(t) A + \Delta A^T Q(t) A + C^T C \tag{8.9.4}$$

and in steady state Q satisfies

$$A^T Q + QA + \Delta A^T QA = -C^T C \tag{8.9.5}$$

- A proof that a general system can be decomposed (using a similarity transformation) into reachable, unreachable, observable, and unobservable subsystems (Theorem 8.10.1).
- The notions of balanced realizations and Hankel singular values were introduced (Section 8.11).
- The introduction of various definitions of stability, including:
- Definition 8.12.1 (stability): A system is said to be stable if for $u(t) \equiv 0$ and any $\varepsilon > 0$ there exists $\beta(\varepsilon, t) > 0$ such that

$$\|x(T)\| < \beta(\varepsilon, T) \Rightarrow \|x(t)\| < \varepsilon, \quad \text{for all } t > T$$

- Definition 8.12.4 (BIBO stability): A system is said to be BIBO stable if, for any bounded input and any initial conditions, the output is bounded.
- Stability for linear time invariant systems is governed by the location of the poles. Specifically, a system is asymptotically stable if the eigenvalues λ_i of A satisfy

$$\frac{\Lambda}{2}|\lambda_i|^2 + \text{Re}\{\lambda_i\} < 0$$

- A theorem giving necessary and sufficient conditions for the various types of stability in the case of linear systems was established (Theorem 8.12.1).
- Study of the Lyapunov view of stability based on a generalized concept of energy.
- A brief discussion of Lyapunov stability for linear systems was given.

8.14 REFERENCES

Linear state space models and their properties are covered in:

CHEN, C. T. (1984) *Linear System Theory and Design*. Holt, Rinehart and Winston, New York.

FORTMANN, T. E., and K. L. HITZ (1977) *An Introduction to Linear Control Systems*. Marcel Dekker, New York.

Matrix fraction descriptions are covered in:

KAILATH, T. (1980) *Linear Systems*. Prentice-Hall, Englewood Cliffs, N.J.

WOLOVICH, W. A. (1974) *Linear Multivariable Systems*. Springer-Verlag, New York.

Balanced realizations and Hankel singular values are covered in:

FRANCIS, B. (1987) *A Course in H_∞ Control Theory*. Lecture Notes in Control and Information Sciences, Vol. 88. Springer-Verlag, New York.

Further results on nonlinear stability theory are contained in:

HILL, D. J., and P. J. MOYLAN (1988) *Dissipativeness and Stability of Nonlinear Systems*. MIT Press, Cambridge, Mass.

VIDYASAGAR, M. (1978) *Nonlinear Systems Analysis*. Prentice-Hall, Englewood Cliffs, N.J.

WILLEMS, J. L. (1970) *Stability Theory of Dynamical Systems*. Thomas Nelson, London.

8.15 PROBLEMS

1. Show that the row rank of $\mathscr{C}(N)$ defined in (8.3.2) is the same as that of $\mathscr{C}$.
2. Is the following system completely reachable?

$$\delta x = \begin{bmatrix} -1 & 2 \\ 1 & 2 \end{bmatrix} x_t + \begin{bmatrix} 1 \\ 0 \end{bmatrix} u_t, \qquad \Delta = 1$$

3. Is the following system completely observable?

$$\delta x = \begin{bmatrix} -4 & 1 \\ -4 & -1 \end{bmatrix} x_t + \begin{bmatrix} 5 \\ 6 \end{bmatrix} u_t, \qquad \Delta = 1$$
$$y_t = \begin{bmatrix} 1 & 0 \end{bmatrix} x_t$$

4. Is the following system (a) controllable (to the original) and (b) reachable?

$$\delta x = \begin{bmatrix} 0 & 1 \\ 0 & -1 \end{bmatrix} x_t + \begin{bmatrix} 1 \\ 0 \end{bmatrix} u_t, \qquad \Delta = 1$$

   (c) Specifically describe those states that can be reached from the origin.

5. Is the following system (a) reconstructible and (b) observable?

$$\delta x = \begin{bmatrix} 0 & 1 \\ 1 & 0 \end{bmatrix} x_t + \begin{bmatrix} 1 \\ 0 \end{bmatrix} u_t, \qquad \Delta = 1$$
$$y_t = \begin{bmatrix} 1 & 1 \end{bmatrix} x_t$$

   (c) Specifically describe any unobservable states.

6. Consider the following second-order continuous time system:

$$\frac{d^2 y}{dt^2} + \omega_0^2 y = 0$$

   (a) Construct a suitable continuous time state space model.
   (b) Use the results of part (a) to obtain a suitable discrete time model at sampling interval Δ.

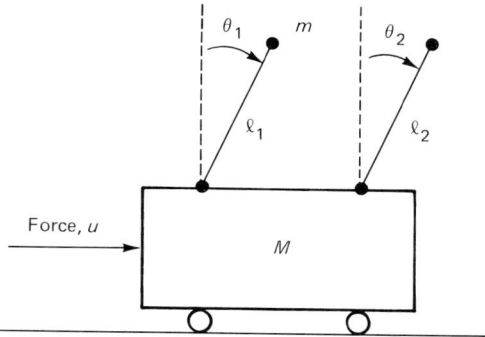

Figure 8.15.P7

(c) Show that the discrete time state space model is observable except when $\Delta = (k\pi)/\omega$.

(d) Explain the result in part (c) by sketching some sine waves.

7. A cart of mass M has two inverted pendulums having lengths l_1 and l_2, both with bobs of mass m. For small $|\theta_1|$ and $|\theta_2|$, the equations of motion can be seen to be

$$M\dot{v} = -mg\theta_1 - mg\theta_2 + u$$

$$m(\dot{v} + l_i\ddot{\theta}_i) = mg\theta_i, \quad i = 1, 2$$

where v is the velocity of the cart and u is an external force applied to the cart (see Figure 8.15.P7).

(a) Is it always possible to control both pendulums, that is, keep them both vertical, by using the input $u(\cdot)$?

(b) Is the system observable with output $y = \theta_1$?

*(c) Why do you think it might be essential to use a sixth-order model for control including state variables z and v (displacement and velocity of the cart)?

8. Consider the system

$$\dot{x} = \begin{bmatrix} -3 & 5 & 0.5 \\ -2 & 3 & 0.5 \\ 0 & 0 & 0 \end{bmatrix} x + \begin{bmatrix} 1 \\ 1 \\ 2 \end{bmatrix} u$$

$$y = \begin{bmatrix} 1 & -1 & 0 \end{bmatrix} x$$

Classify which states are reachable and which are not in this model. Decompose the system into reachable and unreachable parts, and find the eigenvalues of each part. Find the poles and zeros of the system.

9. Consider the system

$$\delta x = \begin{bmatrix} -1 & 2 \\ 0 & 1 \end{bmatrix} x_t + \begin{bmatrix} 1 \\ 1 \end{bmatrix} u, \quad \Delta = 1$$

$$y = \begin{bmatrix} 1 & 0 \end{bmatrix} x$$

Is this system controllable? reachable? observable? reconstructible?

10. Show that any discrete version (using zero-order hold input and sampled output) of a continuous time system

$$\dot{x} = Ax + Bu$$

$$y = Cx$$

has the properties

$$\text{controllability} \leftrightarrow \text{reachability}$$

$$\text{observability} \leftrightarrow \text{reconstructibility}$$

Hint: Show that these properties are true if $A_\Delta + \Delta I$ is nonsingular, where $A_\Delta =$ discrete time A matrix.

11. Suppose we have a second-order linear system with distinct eigenvalues. Then this system can always be transformed into the modal form

$$\rho x = \begin{bmatrix} \lambda_1 & 0 \\ 0 & \lambda_2 \end{bmatrix} x + \begin{bmatrix} b_1 \\ b_2 \end{bmatrix} u$$

$$y = \begin{bmatrix} c_1 & c_2 \end{bmatrix} x$$

(a) What is the transfer function $H(\gamma) = Y(\gamma)/U(\gamma)$ for the system. Write the transfer function in residue form as well; that is

$$H(\gamma) = \Sigma \frac{\gamma_i}{\gamma - \lambda_i}$$

(b) Find necessary and sufficient conditions under which the system is reachable, controllable, stabilizable, and observable.

(c) Using parts (a) and (b) (or otherwise), show that for any second-order system with distinct eigenvalues the system is both reachable and observable if and only if there are no pole zero cancellations.

(d) Generalize part (c) to a system of any order (say n) with distinct eigenvalues. *Hint:* If λ_i, $i = 1 \ldots n$ are distinct, then the following matrix is always nonsingular:

$$\begin{bmatrix} 1 & 1 & \cdots & \cdots & \cdots & 1 \\ \lambda_1 & \lambda_2 & \cdots & \cdots & \cdots & \lambda_n \\ \lambda_1^2 & \lambda_2^2 & \cdots & \cdots & \cdots & \lambda_n^2 \\ \vdots & \vdots & & & & \vdots \\ \lambda_1^{n-1} & \lambda_2^{n-1} & \cdots & \cdots & \cdots & \lambda_n^{n-1} \end{bmatrix}$$

12. Verify the form given in Equation (8.4.2).
13. Verify that the input given in Equation (8.5.4) achieves $x(0) = x_r$ while minimizing (8.5.3).
14. Establish Theorem 8.7.1. *Hint:* Define $w(t) = [\varphi_0(t)I_{m \times m} \varphi_1(t)I_{m \times m} \cdots \varphi_{n-1}(t)I_{m \times m}]$ similarly to $V(t)$ in (8.3.9). Show that $y(t) = w(t)\mathcal{O}x_0$ and consider $\int_0^t y(\tau)^T y(\tau)\, d\tau$.
15. Verify Equations (8.9.2), (8.9.3) for the energy recovered from an initial x_0 in the output.
16. Complete the proof of Theorem 8.10.1.
17. Complete the proof of Lemma 8.12.1.
18. Complete the details of the reverse of proof of Theorem 8.12.1 (iii) using transform techniques.
19. Consider the following continuous time system:

$$\frac{d}{dt}\begin{bmatrix} x_1 \\ x_2 \end{bmatrix} = \begin{bmatrix} 0 & 1 \\ -2 & -3 \end{bmatrix} \begin{bmatrix} x_1 \\ x_2 \end{bmatrix} + \begin{bmatrix} 0 \\ 1 \end{bmatrix} u(t)$$

Is the system (a) asymptotically stable? (b) stable? (c) BIBO stable?

20. Repeat Problem 19 for the following system:
$$\frac{d}{dt}x = \begin{bmatrix} -1 & 0 \\ 0 & 0 \end{bmatrix} x + \begin{bmatrix} 1 \\ 0 \end{bmatrix} u(t)$$
$$y = \begin{bmatrix} 1 & 1 \end{bmatrix} x$$

21. The output of a pure time delay is fed back to the input with a negative feedback gain K. For what results of K is the resultant closed loop system stable?

22. Consider the following system:
$$\dot{x} = \begin{bmatrix} -2 & 1 & 0 \\ -1 & 0 & 1 \\ 0 & 0 & 0 \end{bmatrix} x(t) + \begin{bmatrix} 2 \\ 3 \\ 1 \end{bmatrix} u(t)$$
$$y = \begin{bmatrix} 1 & 0 & 0 \end{bmatrix} x(t)$$

(a) Is the system (i) stable? (ii) asymptotically stable? (iii) BIBO stable?
(b) Show that the system is completely observable but not completely controllable.
(c) Find a state transformation that decomposes the system into controllable and uncontrollable subsystems; derive the transformed state equations.

23. Consider the following system:
$$\frac{d}{dt}x = \begin{bmatrix} -2 & 1 \\ 0 & 0 \end{bmatrix} x + \begin{bmatrix} 1 \\ 0 \end{bmatrix} u$$
$$y = \begin{bmatrix} 1 & 0 \end{bmatrix} x$$

(a) Is the above system (i) asymptotically stable? (ii) bounded input, bounded output stable? (iii) completely observable? (iv) completely reachable? (v) Find the corresponding input–output differential equation model.
(b) Determine a discrete time model for the system by evaluating $e^{A\Delta}$, where $\Delta = 0.1$ s.
(c) Express the discrete state space model in terms of q and δ.

24. Consider the following continuous time system:
$$G(s) = \frac{1}{(s+1)[(s/a)+1]} \quad a > 1$$

(a) Construct a suitable state space model.
(b) Convert to a balanced realization.
(c) Truncate the realization to a single state.
(d) Evaluate the L_∞ error as in Equation (8.11.12) and compare with the bound given in that equation.
(e) What happens as a goes to ∞?

25. Verify Equations (8.12.12) and (8.12.13). *Hint:* Show that
$$E(\varepsilon I + A, t) = E(\varepsilon I, t) E\left(\frac{A}{1+\varepsilon\Delta}, t\right)$$
and then evaluate $E(\bar{J}, t)$ using transform techniques, where
$$\bar{J} = \begin{bmatrix} 0 & \alpha & & & 0 & \cdots & 0 \\ \vdots & \ddots & \ddots & & \vdots & & \vdots \\ \vdots & & \ddots & \ddots & 0 & & \\ \vdots & & & & \alpha & & \\ 0 & \cdots & \cdots & & \cdots & & 0 \end{bmatrix}, \quad \alpha = \frac{1}{1+\varepsilon\Delta}$$

26. Consider the following plant:

$$G(s) = \left(\frac{1-(s/6)}{1+(s/6)}\right)^3 \frac{1}{1+10s}$$

Make a balanced realization of this plant. What are the Hankel singular values? Truncate the balanced realization to obtain approximate models or order 1 and 2. Plot the frequency response of the plant and that of the approximate models. Compare the L^∞ error achieved with the bound given in Equation (8.11.12), that is, $2(\sigma_{k+1} + \cdots + \sigma_n)$.

27. Prove Equation (8.3.3). *Hint:* Show that

$$\rho\left(\sum_{j=0}^{n-1} \varphi_j(t) A^j\right) = A\left(\sum_{j=0}^{n-1} \varphi_j(t) A^j\right)$$

using (8.3.4) and the Cayley–Hamilton theorem, and show that

$$\sum_{j=0}^{n-1} \varphi_j(0) A^j = I$$

See also method 4 of Section 3.4.3.

28. (a) Show that $\varphi_j(t)$ as defined by (8.3.4), (8.3.5) are linearly independent over the interval $[0, T]$, $T > (n-1)\Delta$. *Hint:* Let

$$f(t) = \sum_{j=0}^{n-1} \alpha_j \varphi_j(t) \equiv 0$$

Then

$$f(0) = 0 \Rightarrow \alpha_0 = 0$$
$$\rho f(0) = 0 \Rightarrow \alpha_1 = 0$$

and so on.

(b) Show that $V(t)$ defined in (8.3.9) satisfies, for $t > (n-1)\Delta$,

$$M(t) \triangleq \int_0^t V(t-\tau-\Delta)V^T(t-\tau-\Delta)\,d\tau \quad \text{is nonsingular}$$

Hint: Suppose there exists a vector w such that $w^T M w = 0$. Thus $V^T(t-\tau-\Delta)w = 0$ for $\tau \in [0, t)$. Using part (a), show that this then implies $w = 0$.

29. (a) Consider a spring–mass–damper system as shown in Figure 8.15.P29. Including the effects of gravity, let V equal the total energy in this system. Show that $dV/dt \leq 0$. Is $-dV/dt$ a PDF?

(b) For the system in part (a), let x_1 be the position, relative to the equilibrium position, and $x_2 = \dot{x}_1$. Construct a PDF W such that $d/dt(W) = -x_1^2 - x_2^2$. Thus show that the system in part (a) is asymptotically stable.

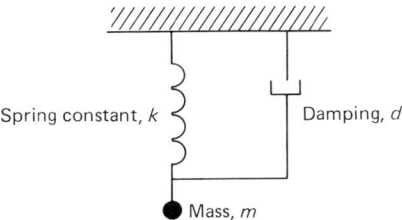

Figure 8.15.P29

9

Pole Assignment

9.1 INTRODUCTION

Chapter 8 outlined a number of basic results relating to the time domain analysis of systems. In this chapter we build on these results and show how the closed loop poles of a system can be arbitrarily assigned by appropriate choice of the feedback controller. We begin by using state space methods and later show how the same results can be obtained using input–output models. We treat the continuous and discrete time cases simultaneously using ρ to denote either d/dt (continuous) or δ (discrete).

9.2 POLE ASSIGNMENT BY STATE VARIABLE FEEDBACK

We have seen earlier that the stability properties and transient performance of control systems depend on the closed loop eigenvalues. In this section we show that, if the complete state vector of a linear system is directly measured, then, provided the system is completely reachable, it is possible to assign the closed loop eigenvalues arbitrarily.

Say we are given a completely reachable single-input, single-output state space model; then it can always be expressed in controller form as shown in (8.4.9), (8.4.10). If we then apply state variable feedback of the form

$$u(t) = -[k_{n-1}, \ldots, k_0]\bar{x}(t) + v(t) \qquad (9.2.1)$$

it is readily seen that the resultant closed loop system is

$$\rho \bar{x}(t) = \begin{bmatrix} (-k_{n-1} - \alpha_{n-1}) & ,\ldots, & (-k_0 - \alpha_0) \\ 1 & & \\ & \ddots & \\ & 1 & 0 \end{bmatrix} \bar{x}(t) + \begin{bmatrix} 1 \\ 0 \\ \vdots \\ 0 \end{bmatrix} v(t) \tag{9.2.2}$$

This closed loop system has the following characteristic polynomial:

$$\lambda^n + (\alpha_{n-1} + k_{n-1})\lambda^{n-1} + \cdots + (\alpha_0 + k_0) \tag{9.2.3}$$

Thus, by choosing $(k_{n-1}, \ldots, k_0)$, we can independently change each coefficient in the characteristic polynomial and thus arbitrarily assign the closed loop poles.

Here we have treated only the single-input case, but a similar conclusion holds for the multiinput case. We can therefore conclude:

Theorem 9.2.1

The closed loop poles of a completely reachable system can be arbitrarily assigned by state variable feedback.

9.3 STATE OBSERVERS

We next consider the situation when the complete state vector is not measured but instead only the system output $y(t)$ is available. Since we can only observe $y(t)$ and not the complete state vector $x(t)$, the question arises as to whether or not $x(t)$ can be inferred from measurements of $y(t)$. If the system is asymptotically stable, then the effect of any initial condition response will asymptotically decay to zero. This suggests that we could simply estimate the state by using a model of the system:

$$\rho \hat{x} = A\hat{x} + Bu \tag{9.3.1}$$

However, this will clearly be unsatisfactory for unstable systems, and there is a problem even for stable ones, as we have no control over the speed at which the initial condition error decays.

Elementary ideas of feedback suggests that we may be able to improve the situation by comparing the observed output with the model output and then using this error to correct the model. This suggests the following observer structure:

$$\rho \hat{x} = A\hat{x} + Bu + J(y - \hat{y}) \tag{9.3.2}$$

$$\hat{y} = C\hat{x} \tag{9.3.3}$$

Recall that the original system is described by

$$\rho x = Ax + Bu \tag{9.3.4}$$

$$y = Cx \tag{9.3.5}$$

Subtracting (9.3.4) from (9.3.2) then gives the following equation for the state error, $\tilde{x} = \hat{x} - x$:

$$\rho\tilde{x} = (A - JC)\tilde{x} \qquad (9.3.6)$$

Thus, if we can assign the eigenvalues of $(A - JC)$, then we can cause the state estimation error to decay at any desired rate. The following result is central to this issue.

Theorem 9.3.1

If a system is completely observable, then the matrix J can be found such that the poles of the state estimation error equation (9.3.6) can be arbitrarily assigned.

Proof: Since the system is completely observable, it can always be transformed into the observer canonical form as in (8.8.3). Equation (9.3.6) then has the special form

$$\rho\tilde{x}(t) = \left\{ \begin{bmatrix} -\alpha_{n-1} & 1 & & \\ & & \ddots & \\ & & & 1 \\ -\alpha_0 & 0 & \cdots & 0 \end{bmatrix} - \begin{bmatrix} j_{n-1} \\ \vdots \\ j_0 \end{bmatrix} \begin{bmatrix} 1 & 0 & \cdots & 0 \end{bmatrix} \right\} \tilde{x}$$

$$= \begin{bmatrix} -\alpha_{n-1} - j_{n-1} & 1 & & \\ & & \ddots & \\ & & & 1 \\ -\alpha_0 - j_0 & 0 & \cdots & 0 \end{bmatrix} \tilde{x} \qquad (9.3.7)$$

However, the characteristic polynomial of the A matrix in (9.3.7) is $\lambda^n + (\alpha_{n-1} + j_{n-1})\lambda^{n-1} + \cdots + (\alpha_0 + j_0)$. Hence, by choice of $j_0 \ldots j_{n-1}$, we have independent control over the coefficients of the characteristic polynomial and the result follows.
∇∇∇

Note that the result in Theorem 9.3.1 is the dual of the corresponding result in Theorem 9.2.1.

9.4 DYNAMIC OUTPUT FEEDBACK

The previous section showed that, in the case where the state is not directly measured, but the system is observable, there exists a linear dynamic system that allows the system state to be asymptotically estimated. An obvious extension of this idea is to use the state estimates in place of the states for feedback, as in Section 9.2.

This leads to the following dynamic output feedback solution. Consider a system of the form

$$\rho x = Ax + Bu \qquad (9.4.1)$$
$$y = Cx \qquad (9.4.2)$$

Then the state observer is as in (9.3.2):
$$\rho \hat{x} = A\hat{x} + Bu + J(y - C\hat{x}) \tag{9.4.3}$$
And the state estimate feedback law is as in (9.2.1) with x replaced by $\hat{x}$:
$$u = K\hat{x} + v \tag{9.4.4}$$
We then have the following result:

Theorem 9.4.1

The use of state estimates for feedback as just outlined leads to a closed loop system whose eigenvalues are the combination of the eigenvalues that would have been obtained had the true states been fed back together with the eigenvalues of the observer. Furthermore, the state estimation error $\tilde{x} = \hat{x} - x$ is unreachable from the reference input v.

Proof: Combining (9.4.3) with (9.4.1) and (9.4.2) gives
$$\rho \tilde{x} = (A - JC)\tilde{x} \tag{9.4.5}$$
Also, (9.4.4) can be rewritten as
$$u = -K[x + \tilde{x}] + v \tag{9.4.6}$$
Combining (9.4.1), (9.4.5), and (9.4.6), we see that the closed loop satisfies
$$\rho \begin{bmatrix} x \\ \tilde{x} \end{bmatrix} = \begin{bmatrix} A - BK & -BK \\ 0 & A - JC \end{bmatrix} \begin{bmatrix} x \\ \tilde{x} \end{bmatrix} + \begin{bmatrix} B \\ 0 \end{bmatrix} v \tag{9.4.7}$$
$$y = [C \quad 0] \begin{bmatrix} x \\ \tilde{x} \end{bmatrix} \tag{9.4.8}$$

The result follows immediately on noting that the eigenvalues of a block triangular matrix are equal to the union of the eigenvalues of the diagonal blocks. Thus the eigenvalues of (9.4.7) are the eigenvalues of the observer satisfying $\det(A - JC - \lambda I) = 0$, plus the eigenvalues of the closed loop (when the true states are fed back) satisfying $\det(A - BK - \lambda I) = 0$. (See also Problem 1.) ▽▽▽

We next show that the previous design procedure can be compactly expressed using operator models. This is described in the following key result (which for simplicity is stated for the single-input, single-output case only).

Theorem 9.4.2

Consider the state space pole assignment design in (9.4.3), (9.4.4). Define E_1, E_2 as the characteristic polynomials of $(A - JC)$ and $(A - BK)$, respectively.

(i) The observer (9.4.3) and the state estimate feedback (9.4.4) can be expressed succinctly in operator form as
$$\frac{L}{E_1} u = -\frac{P}{E_1} y + v \tag{9.4.9}$$
where L and P are of degree n and $n - 1$, respectively.

(ii) The operators L and P in (9.4.9) satisfy the following polynomial identity:
$$DL + NP = E_1 E_2 \tag{9.4.10}$$
where $N(\gamma)/D(\gamma)$ is the transfer function of the system; that is,
$$C(\gamma I - A)^{-1} B = \frac{N(\gamma)}{D(\gamma)} \tag{9.4.11}$$

Proof:

(i) From (9.4.3) the observer has the form
$$\rho \hat{x} = (A - JC)\hat{x} + Bu + Jy \tag{9.4.12}$$
Using the techniques of Chapter 2, this equation can be written in input–output form as
$$E_1(\rho)\hat{x} = V_1(\rho)u + V_2(\rho)y \tag{9.4.13}$$
where
$$V_1(\rho) = \text{Adj}[\rho I - A + JC]^{-1} B \tag{9.4.14}$$
$$V_2(\rho) = \text{Adj}[\rho I - A + JC]^{-1} J \tag{9.4.15}$$
$$E_1(\rho) = \det[\rho I - A + JC] \tag{9.4.16}$$
Note that $V_1(\rho)$, $V_2(\rho)$ are vectors of dimension n, of polynomials of degree $n - 1$. Recall that the state estimate feedback law is
$$u = -K\hat{x} + v \tag{9.4.17}$$
Substituting (9.4.13) into (9.4.17) gives
$$u = -K\left[\frac{1}{E_1(\rho)} V_1(\rho)u + \frac{1}{E_1(\rho)} V_2(\rho)y\right] + v \tag{9.4.18}$$
The result then follows with
$$P(\rho) = KV_2(\rho) \tag{9.4.19}$$
and
$$L(\rho) = E_1(\rho) + KV_1(\rho) \tag{9.4.20}$$

(ii) From (9.4.14) and (9.4.16) it is clear that
$$V_1(\rho) = E_1(\rho)[\rho I - A + JC]^{-1} B \tag{9.4.21}$$
$$V_2(\rho) = E_1(\rho)[\rho I - A + JC]^{-1} J \tag{9.4.22}$$
Now consider the left side of (9.4.10). Using (9.4.19), (9.4.20), we have
$$L(\rho)D(\rho) + P(\rho)N(\rho)$$
$$= [E_1(\rho) + KV_1(\rho)]D(\rho) + [KV_2(\rho)]N(\rho) \tag{9.4.23}$$

Then, using (9.4.11), we have

$$L(\rho)D(\rho) + P(\rho)N(\rho)$$
$$= D(\rho)\{E_1(\rho) + K[V_1(\rho) + V_2(\rho)C(\rho I - A)^{-1}B]\}$$
$$= D(\rho)E_1(\rho)\{1 + K(\rho I - A + JC)^{-1}B + K(\rho I - A + JC)^{-1}$$
$$\times JC(\rho I - A)^{-1}B\},$$

where we have used (9.4.21), (9.4.22)

$$= D(\rho)E_1(\rho)\{1 + K(\rho I - A + JC)^{-1}[I + JC(\rho I - A) + ^{-1}]B\}$$
$$= D(\rho)E_1(\rho)\{1 + K(\rho I - A + JC)^{-1}[(\rho I - A) + JC](\rho I - A)^{-1}B\}$$
$$= D(\rho)E_1(\rho)\{1 + K(\rho I - A)^{-1}B\}$$
$$= D(\rho)E_1(\rho)\det\{I + (\rho I - A)^{-1}BK\}, \quad \text{(see also Problem 2)}$$
$$= D(\rho)E_1(\rho)\det\{(\rho I - A)^{-1}\}\det\{\rho I - A + BK\}$$
$$= E_1(\rho)E_2(\rho) \tag{9.4.24}$$

where we have used

$$D(\rho) = \det(\rho I - A)$$
$$E_2(\rho) = \det(\rho I - A + BK)$$

Theorem 9.4.2 is illustrated in Figures 9.4.1 and 9.4.2. Figure 9.4.1 shows the original design using $\hat{x}$ as described in Equations (9.4.13) and (9.4.17). Figure 9.4.2 shows the derived form as in (9.4.17), where the scalar polynomials $P(\rho)$ and $R(\rho)$ satisfy

$$P(\rho) = KV_2(\rho) \quad \text{and} \quad R(\rho) = KV_1(\rho) \tag{9.4.25}$$

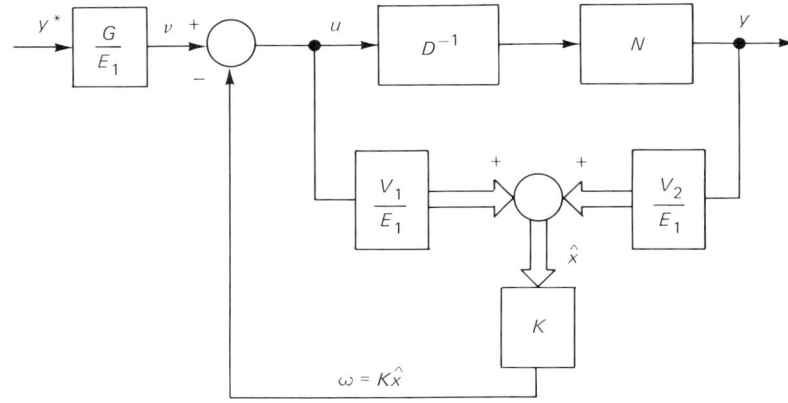

Figure 9.4.1 Feedback of estimated state.

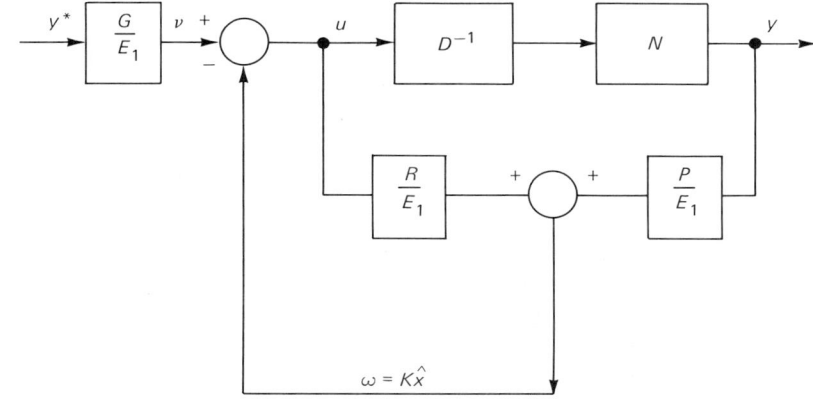

Figure 9.4.2 Operator form for feedback of estimated state.

The key point about the preceding result is that the combined problem of state estimation and feedback has been reduced to the solution of the following two equations (which are independent of state selection):

$$\frac{L}{E_1}u = -\frac{P}{E_1}y + v \tag{9.4.26}$$

where L and P satisfy

$$DL + NP = E_1 E_2 \tag{9.4.27}$$

Usually, v is chosen to have the special form

$$v = \frac{G}{E_1} y^* \tag{9.4.28}$$

in which case a minimal realization of the control law (9.4.26) is

$$Lu = -Py + Gy^* \tag{9.4.29}$$

We will next generalize the pole assignment design procedure, as outlined here, to include disturbances and set point tracking.

9.5 POLE ASSIGNMENT FOR SYSTEMS WITH DISTURBANCES

Consider the plant including disturbances as shown in Figure 9.5.1, where the symbols have the following meaning:

- u controllable input
- z measured disturbance
- d unmeasured disturbance
- y measured output.

The nominal model for the plant is assumed to be

$$y = H_0 u + H_1 z + H_2 d \tag{9.5.1}$$

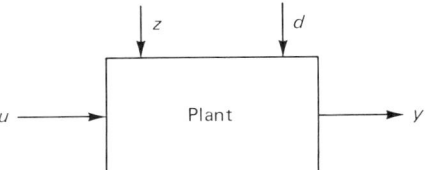

Figure 9.5.1 Plant model.

For simplicity, we will begin with the single-input, single-output case. The more general MIMO case will be treated later.

Using a common denominator for H_0, H_1, H_2, this model can be written as a left matrix fraction as follows:

$$A(\rho)y = B(\rho)u + F(\rho)z + J(\rho)d \qquad (9.5.2)$$

It is assumed that we want the output to track a given reference input y^* and to cancel the disturbance d. We further assume that $S(\rho)$ nulls both y^* and d (see the discussion of the internal model principle in Section 7.7.3).

We will base our design on the feedback law given in (9.4.29), except that we will add S to the denominator as in the internal model principle and will add an extra term to account for z. This gives the following three degrees of freedom control law:

$$u = -Hz - \frac{P}{LS}y + \frac{G}{LS}y^* \qquad (9.5.3)$$

where H may be a transfer function. The control law described in (9.5.3) is shown diagrammatically in Figure 9.5.2.

The closed loop response resulting from the feedback structure in Figure 9.5.2 is readily seen to satisfy

$$(ASL + BP)y = BGy^* + LS[F - BH]z + LSJd \qquad (9.5.4)$$

The synthesis problem now reduces to the choice of the four transfer functions L, P, G, and H.

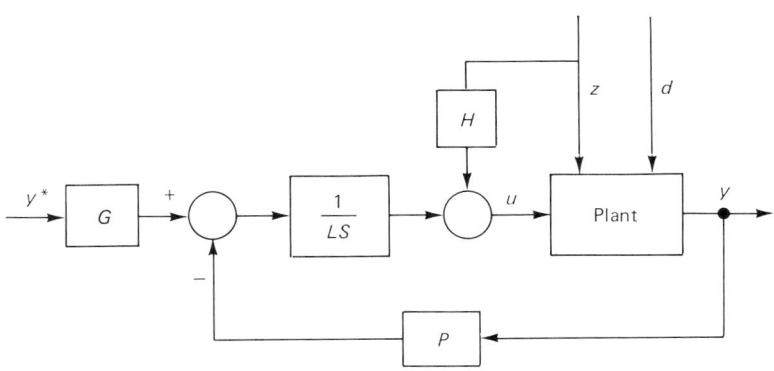

Figure 9.5.2 General feedback structure.

9.5.1 The Diophantine Equation

We saw in Section 9.4 that the feedback can be designed so as to assign the closed loop poles arbitrarily. For the simple case without disturbances, the design equations reduced to the polynomial identity (9.4.10). For the more general case with disturbances, we have seen in (9.5.4) that the closed loop characteristic polynomial is $ASL + BP$. Hence, to achieve the desired closed loop poles in this case, we need to solve the following equation for L and P:

$$ASL + BP = A^* \qquad (9.5.5)$$

where A^* is some given polynomial (the desired closed loop characteristic polynomial).

If we denote AS by $\bar{A}$, then equation (9.5.5) has the form

$$\bar{A}L + BP = A^* \qquad (9.5.6)$$

where $\bar{A}, B, A^*$ are given polynomials of the form

$$\bar{A} = \bar{a}_{\bar{n}}\rho^{\bar{n}} + \cdots + \bar{a}_0 \qquad (9.5.7)$$
$$B = b_n\rho^n + \cdots + b_0 \qquad (9.5.8)$$
$$A^* = a^*_{n^*}\rho^{n^*} + \cdots + a^*_0 \qquad (9.5.9)$$

In general, equation (9.5.6) has a unique solution, which gives a proper control law, if we choose $n^* \geq n + \bar{n} - 1$, the degree of P and L, as $\bar{n} - 1$ and $n^* - \bar{n}$ ($\geq n - 1$), respectively. To see this, let

$$L = l_{n^*-\bar{n}}\rho^{n^*-\bar{n}} + \cdots + l_0 \qquad (9.5.10)$$
$$P = p_{\bar{n}-1}\rho^{\bar{n}-1} + \cdots + p_0 \qquad (9.5.11)$$

Then equating coefficients on both sides of (9.5.6) give the result shown in Figure 9.5.3.

The $(n + \bar{n}) \times (n + \bar{n})$ submatrix on the left side of Figure 9.5.3 is called the eliminant (or Sylvester) matrix associated with the polynomial pair $(\bar{A}, B)$. The following result gives conditions for this matrix to be nonsingular.

Lemma 9.5.1. Sylvester's Theorem

Two polynomials $\bar{A}, B$ of degrees $\bar{n}, n$ are relatively prime if and only if their eliminant matrix M_e is nonsingular.

Proof:

Only if: Suppose that there is a common root γ:

$$\bar{A}(\rho) = (\rho - \gamma)(a'_{\bar{n}-1}\rho^{\bar{n}-1} + \cdots + a'_0)$$
$$B(\rho) = (\rho - \gamma)(b'_{n-1}\rho^{n-1} + \cdots + b'_0)$$

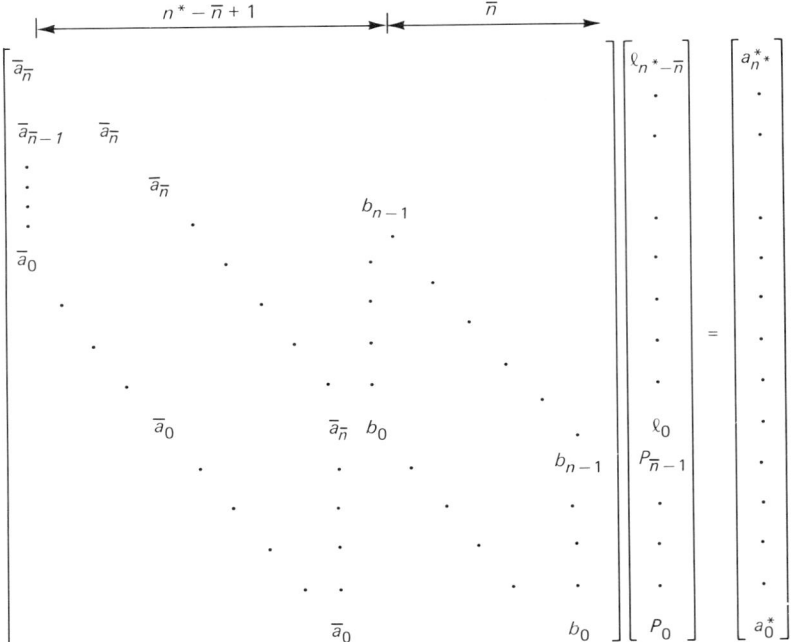

Figure 9.5.3 Linear system of equations for pole assignment.

Eliminating $(\rho - \gamma)$ gives

$$\bar{A}(\rho)[b'_{n-1}\rho^{n-1} + \cdots + b'_0] - B(\rho)[a'_{\bar{n}-1}\rho^{\bar{n}-1} + \cdots + a'_0] = 0$$

Equating coefficients on both sides gives

$$M_e \theta' = 0$$

where $\theta' = [b'_{n-1}, \ldots, b'_0, -a'_{\bar{n}-1}, \ldots, -a'_0]^T$. However, the preceding equation has a nontrivial solution for θ' if and only if $\det M_e = 0$.

If: By reversing the preceding argument. ▽▽▽

Corollary 9.5.1

Equation (9.5.6) has a unique solution for general A^* if and only if $\bar{A}, B$ are relatively prime.

Proof: Immediate from Lemma 9.5.1 on noting that the additional columns on the left of M_e in Figure 9.5.3 do not change the singularity of the matrix since $\bar{a}_n \neq 0$ by assumption. ▽▽▽

9.5.2 Solving the Diophantine Equation

In the last section we found that the pole assignment problem including use of the internal model principle could be succinctly summarized in the Diophantine equation (9.5.6). In this section we briefly discuss a number of methods for solving this equation.

(a) Simultaneous Equation Method

This method simply reduces to solving the equation in Figure 9.5.3 explicitly using any method for solving simultaneous linear equations. Note that, due to the structure in Figure 9.5.3, some of the coefficients can be obtained directly.

(b) Simplified Design for Known Plant Poles

For simplicity, we consider the case where the plant has distinct poles; that is, $\overline{A}$ in (9.5.6) can be written

$$\overline{A}(\rho) = (\rho - \alpha_1)\ldots(\rho - \alpha_{\bar{n}}) \tag{9.5.12}$$

In this case, we may evaluate the polynomial identity (9.5.6) at $\alpha_1, \ldots, \alpha_{\bar{n}}$ to obtain

$$B(\alpha_i)P(\alpha_i) = A^*(\alpha_i), \qquad i = 1, \ldots, \bar{n} \tag{9.5.13}$$

Since $P(\rho)$ is of degree less than $\bar{n}$, (9.5.13) is sufficient to completely specify $P(\rho)$. In particular, $P(\rho)$ may be evaluated as

$$P(\rho) = \sum_{i=1}^{\bar{n}} \beta_i \prod_{\substack{j=1 \\ j \neq i}}^{\bar{n}} \left(\frac{\rho - \alpha_j}{\alpha_i - \alpha_j} \right) \tag{9.5.14}$$

where

$$\beta_i \triangleq \frac{A^*(\alpha_i)}{B(\alpha_i)} \tag{9.5.15}$$

Having evaluated $P(\rho)$, we can then find $L(\rho)$ from (9.5.6) by a simple polynomial division:

$$L(\rho) = \frac{A^*(\rho) - B(\rho)P(\rho)}{\overline{A}(\rho)} \tag{9.5.16}$$

*(c) Euclid's Algorithm

Those readers who are familiar with Euclid's algorithm will be aware that, given two relatively prime polynomials $\overline{A}(\rho)$, $B(\rho)$, two other polynomials $\Lambda(\rho)$ and $\Gamma(\rho)$ can always be found such that

$$\overline{A}(\rho)\Lambda(\rho) + B(\rho)\Gamma(\rho) = 1 \tag{9.5.17}$$

Multiplying this equation by $A^*(\rho)$ gives

$$\overline{A}(\rho)\{\Lambda(\rho)A^*(\rho)\} + B(\rho)\{\Gamma(\rho)A^*(\rho)\} = A^*(\rho) \tag{9.5.18}$$

This basically solves the diophantine equation, save that we need to ensure that the degree of P is less than the degree of L. This last step can be achieved by dividing ΓA^* by A to produce a remainder of degree less than $\bar{n}$; that is, we write

$$\Gamma A^* = Q\bar{A} + P \tag{9.5.19}$$

where degree $P < \bar{n}$. Substituting (9.5.19) into (9.5.18) gives

$$\bar{A}(\Lambda A^* + QB) + BP = A^* \tag{9.5.20}$$

We then see that the required solution is

$$L = \Lambda A^* + QB \tag{9.5.21}$$

together with P, as in (9.5.19).

We illustrate these ideas by a simple example.

Example 9.5.1

Consider the following simple plant:

$$G(\gamma) = \frac{-\gamma + 1}{(\gamma + 1)(\gamma + 2)} \tag{9.5.22}$$

Say we wish to track constant set point with zero error; then we require

$$S(\rho) = \rho \tag{9.5.23}$$

The pole assignment problem is then as in (9.5.6), where $\bar{A}(\rho) = \rho^3 + 3\rho^2 + 2\rho$ and $B(\rho) = -\rho + 1$. We also choose $A^*(\rho) = (\rho + 1)^4 = \rho^4 + 4\rho^3 + 6\rho^2 + 4\rho + 1$.

The three solution methods for the diophantine equation now become:

(a) Equation (9.5.12) gives

$$\begin{bmatrix} 1 & 0 & 0 & 0 & 0 \\ 3 & 1 & -1 & 0 & 0 \\ 2 & 3 & 1 & -1 & 0 \\ 0 & 2 & 0 & 1 & -1 \\ 0 & 0 & 0 & 0 & 1 \end{bmatrix} \begin{bmatrix} l_1 \\ l_0 \\ p_2 \\ p_1 \\ p_0 \end{bmatrix} = \begin{bmatrix} 1 \\ 4 \\ 6 \\ 4 \\ 1 \end{bmatrix} \tag{9.5.24}$$

The solution of this equation is

$$(l_1, l_0, p_2, p_1, p_0) = [1, \quad 1.666, \quad 0.666, \quad 1.666, \quad 1] \tag{9.5.25}$$

(b) The poles of the augmented plant are $\alpha_1 = 0$, $\alpha_2 = -1$, $\alpha_3 = -2$. Hence

$$\beta_1 = \frac{A^*(0)}{B(0)} = \frac{1}{1} = 1 \tag{9.5.26}$$

$$\beta_2 = \frac{A^*(-1)}{B(-1)} = \frac{0}{2} = 0 \tag{9.5.27}$$

$$\beta_2 = \frac{A^*(-2)}{B(-2)} = \frac{1}{3} = \frac{1}{3} \tag{9.5.28}$$

Substituting into (9.5.14) gives

$$P(\rho) = 1\left\{\frac{(\rho+1)(\rho+2)}{(0+1)(0+2)}\right\} + 0\left\{\frac{\rho(\rho+2)}{(-1+0)(-1+2)}\right\}$$
$$+ \frac{1}{3}\left\{\frac{\rho(\rho+1)}{(-2+0)(-2+1)}\right\}$$
$$= \frac{2}{3}\rho^2 + \frac{5}{3}\rho + 1 \qquad (9.5.29)$$

Finally, substituting into (9.5.16) gives

$$L(\rho) = \frac{A^*(\rho) - B(\rho)P(\rho)}{\bar{A}(\rho)}$$
$$= \frac{(\rho^4 + 4\rho^3 + 6\rho^2 + 4\rho + 1) - (-\rho+1)(\frac{2}{3}\rho^2 + \frac{5}{3}\rho + 1)}{\rho^3 + 3\rho^2 + 2\rho}$$
$$= \frac{\rho(\rho+1)(\rho+2)(\rho+\frac{5}{3})}{\rho(\rho+1)(\rho+2)}$$
$$= \rho + \frac{5}{3} \qquad (9.5.30)$$

(c) Euclids' algorithm gives

$$\Lambda = \tfrac{1}{6}, \qquad \Gamma = \tfrac{1}{6}(\rho^2 + 4\rho + 6) \qquad (9.5.31)$$

Dividing ΓA^* by $\bar{A}$ gives

$$\Gamma A^* = \bar{A}\left(\tfrac{1}{6}\rho^3 + \tfrac{5}{6}\rho^2 + \tfrac{11}{6}\rho + \tfrac{3}{2}\right) + \left(\tfrac{2}{3}\rho^2 + \tfrac{5}{3}\rho + 1\right) \qquad (9.5.32)$$

Hence

$$Q = \tfrac{1}{6}\rho^3 + \tfrac{5}{6}\rho^2 + \tfrac{11}{6}\rho + \tfrac{3}{2} \qquad (9.5.33)$$
$$P = \tfrac{2}{3}\rho^2 + \tfrac{5}{3}\rho + 1 \qquad (9.5.34)$$

Finally,

$$L = \Lambda A^* + BQ$$
$$= \tfrac{1}{6}(\rho^4 + 4\rho^3 + 6\rho^2 + 4\rho + 1) + (-\rho+1)\left(\tfrac{1}{6}\rho^3 + \tfrac{5}{6}\rho^2 + \tfrac{11}{6}\rho + \tfrac{3}{2}\right)$$
$$= \rho + \tfrac{5}{3} \qquad (9.5.35)$$

9.5.3 Numerical Issues for the Discrete Case

In the preceding discussion we did not consider numerical issues. One point that arises in the latter context is that, in discrete time, it is usually much preferable to solve the diophthantine equation in delta form rather than shift form. This is essentially because the delta form moves the origin to the $1 + j0$ point in the Z plane and thus eliminates the offset associated with the shift operator pole zero locations. Indeed, our interest in delta models has its origins in a classroom example of pole assignment. The instructor (GCG) obtained wildly erroneous answers when using hand calculations with two decimal places. A student (RHM) obtained the correct answer by using 10 decimal places on a calculator (actually a HP15C which was, at that time, state of the art). The difference was finally traced to severe numerical difficulties associated with the Z domain form of pole assignment. Rationalizing this observation with the simplicity of the corresponding continuous time case led the authors to formulate the discrete case in delta form, and yes, this could be solved with two decimal places (otherwise we wouldn't have told you the story). The advantage of using the delta form is illustrated in Example 9.5.2.

Example 9.5.2

Consider a continuous time system having transfer function

$$H_c(s) = \frac{2(s+1)}{s(s+2)}$$

which is sampled with $\Delta = 0.1$. The discrete time δ-form Sylvester matrix is

$$M' = \begin{bmatrix} 1 & 0 & 0 & 0 \\ 1.8127 & 1 & 1.9063 & 0 \\ 0 & 1.8127 & 1.8127 & 1.9063 \\ 0 & 0 & 0 & 1.8127 \end{bmatrix}$$

which has condition number

$$\nu' = 14.8$$

In shift operator form, however, we have

$$M'' = \begin{bmatrix} 1 & 0 & 0 & 0 \\ -1.81873 & 1 & 0.19063 & 0 \\ 0.81873 & -1.81873 & -0.1725 & 0.19063 \\ 0 & 0.81873 & 0 & -0.1725 \end{bmatrix}$$

which has a condition number of

$$\nu'' = 4.43 \times 10^3$$

Thus we see that, in this example, the pole assignment equation is better conditioned (by two orders of magnitude!) in delta form than in shift form. ▽▽▽

9.5.4 Model Reference Control

A special case of pole assignment, which simplifies the solution of the diophantine equation, occurs when all zeros of the system are stable and well damped. In this case it is sensible to include the numerator polynomial B in A^*. In this case the diophantine equation becomes

$$ALS + BP = A^* = B\bar{A}^* \tag{9.5.36}$$

where $\bar{A}^*$ has degree $(n^* - m)$.

Since B is a factor of the right side and the last term on the left side of (9.5.36), we conclude that B must be a factor of ALS. If SA and B are relatively prime, then this means that L must have B as a factor; so we can write

$$A\bar{L}BS + BP = B\bar{A}^* \tag{9.5.37}$$

where $\bar{L}(\rho)$ has degree $n^* - \bar{n} - m$ (m = degree B). Equation (9.5.37) can be simplified by canceling B to give

$$A\bar{L}S + P = \bar{A}^* \tag{9.5.38}$$

Equation (9.5.38) is considerably simpler that Equation (9.5.5). In particular, Equation (9.5.38) always has a unique solution that can be found by a simple polynomial division: $\bar{L}, P$ are the quotient and remainder when $\bar{A}^*$ is divided by AS.

Example 9.5.3

Consider a simple double integrator plant having transfer function

$$G(\gamma) = \frac{1}{\gamma^2} \tag{9.5.39}$$

We see that $B = 1$, and hence (9.5.36) reduces to (9.5.38). The solution for $n^* = 3$ is obtained by the following division:

$$
\begin{array}{r}
(\gamma + 3) = \bar{L} \\
A = \gamma^2 \overline{\smash{)}(\gamma^3 + 3\gamma^2 + 3\gamma + 1)} = A^* \\
\underline{\gamma^3 + 3\gamma^2 } \\
(3\gamma + 1) = P
\end{array}
$$

▽▽▽

To show the link between the preceding control law and model reference control, let us assume that we would like the plant output y to track a signal y^*, where y^* is defined to be the output of a given reference model; that is,

$$y^* = \frac{B^*}{\bar{A}^*} r \tag{9.5.40}$$

for some external input r. That is we want to apply feedback to the system in such a way that when the closed loop system is excited by the input r it responds in exactly the same way as does the given reference model. A control law that achieves this objective is the following:

$$\bar{L}BSu = -Py + B^*r \tag{9.5.41}$$

where $\bar{L}$ and P satisfy (9.5.38).

It is readily verified that the control law (9.5.41), (9.5.38) achieves

$$y = \frac{B^*}{\bar{A}^*} r \tag{9.5.42}$$

as required.

Note that for the control law (9.5.41) to be proper we require that the relative degree (the difference between denominator and numerator degrees) of the reference model (9.5.40) should be greater than or equal to the relative degree of the plant (see Problem 15).

9.6 FRACTIONAL REPRESENTATION

In the previous section we saw that it is possible to design a feedback control law to assign the closed loop poles arbitrarily. In this section we take this idea a little further and give a characterization of *all* feedback control laws giving a closed loop plant having certain desired properties.

One way to develop this class of control laws is to model the open loop plant as a quotient of two operators each of which has a denominator in the class of desired denominator polynomials. This is commonly known as a *fractional representation*.

The advantage of being able to characterize all feedback controllers giving a closed loop plant having certain specified properties is that this pinpoints the additional degrees of freedom available to a designer to optimize the design while retaining the specified properties.

We will first take the case of single-input, single-output systems and assume that the desired property of the closed loop system is that it be stable. In the following subsections we will generalize this to multiinput, multioutput systems and to more complex desired properties (for example, closed loop poles in some prespecified region).

9.6.1 Fractional Models

Consider the elementary feedback system shown in Figure 9.6.1. Using the ideas presented in earlier chapters, we could model the plant transfer function as a quotient of coprime polynomials; that is,

$$G(\rho) = \frac{B(\rho)}{A(\rho)} \quad (9.6.1)$$

However, this model can also be rewritten in fractional form as

$$G(\rho) = \frac{N(\rho)}{M(\rho)} \quad (9.6.2)$$

where $N(\rho)$ and $M(\rho)$ are themselves stable transfer functions of the form

$$N(\rho) = \frac{B(\rho)}{E(\rho)}, \quad M(\rho) = \frac{A(\rho)}{E(\rho)} \quad (9.6.3)$$

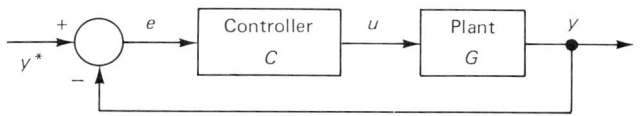

Figure 9.6.1 Elementary feedback system.

with $E(\rho)$ monic Hurwitz (that is, $1/E(\rho)$ is asymptotically stable), and with degree $\partial E(\rho) = \partial A(\rho)$, where $\partial(\cdot)$ denotes "degree of."

The model (9.6.2), (9.6.3) is called a *fractional representation*.

Example 9.6.1

As a rather trivial example of a fractional representation, consider the following model expressed in terms of the forward shift operator q.

$$G(q) = \frac{B(q)}{A(q)} \tag{9.6.4}$$

where

$$B(q) = b_m q^m + \cdots + b_0 \tag{9.6.5}$$

$$A(q) = q^n + a_{n-1} q^{n-1} + \cdots + a_0 \tag{9.6.6}$$

Say we let $E(q) = q^n$; then the representation (9.6.2) becomes

$$G(q) = \frac{B(q)/E(q)}{A(q)/E(q)} = \frac{b_m q^{m-n} + \cdots + b_0 q^{-n}}{1 + a_{n-1} q^{-1} + \cdots + a_0 q^{-n}} \tag{9.6.7}$$

Thus we see that, in this case, the fractional representation simply corresponds to using the unit delay operator, rather than the unit advance operator. ▽▽▽

Example 9.6.2

In Example 9.6.1, the polynomial $E(\rho)$ was chosen as the elementary polynomial $E(q) = q^n$. A simple extension of this is to choose $E(\rho)$ as $(\rho + \alpha)^n$, where ρ is any operator and $-\alpha$ is inside the stability boundary for the operator. Then an appropriate fractional model is

$$G(\rho) = \frac{B(\rho)/E(\rho)}{A(\rho)/E(\rho)} \tag{9.6.8}$$

where $E(\rho) = (\rho + \alpha)^n$. Equation (9.6.8) can be simplified to

$$\frac{A(\rho)}{E(\rho)} = \frac{\rho^n + a_{n-1}\rho^{n-1} + \cdots + a_0}{(\rho + \alpha)^n} \tag{9.6.9}$$

$$= \frac{(\rho + \alpha)^n + a'_{n-1}(\rho + \alpha)^{n-1} + \cdots + a'_0}{(\rho + \alpha)^n} \tag{9.6.10}$$

by rearranging the coefficients in the polynomial $A(\rho)$.

Equation (9.6.10) can now be rewritten as

$$\frac{A(\rho)}{E(\rho)} = 1 + a'_{n-1} g^{-1} + \cdots + a'_0 g^{-n} \tag{9.6.11}$$

where

$$g = \frac{1}{\rho + \alpha} \tag{9.6.12}$$

If we make a similar rearrangement to $B(\rho)/E(\rho)$, we end up with the following fractional representation:

$$G(\rho) = \frac{b'_m g^{m-n} + \cdots + b'_0 g^{-n}}{1 + a'_{n-1} g^{-1} + \cdots + a'_0 g^{-n}} \quad (9.6.13)$$

This can be viewed as a simple extension of (9.6.7) to more general operators based on $D = d/dt$ and δ.

Before leaving the topic of fractional representations, it is interesting to observe that these models can be written as

$$My = Nu \quad (9.6.14)$$

where

$$M = \{1 + a_1 F_1 + \cdots + a_n F_n\} \quad (9.6.15)$$

$$N = \{b_1 F_1 + \cdots + b_n F_n\} \quad (9.6.16)$$

and $F_1 \ldots F_n$ are a set of strictly proper (that is, having positive relative degree) and stable filters. Equation (9.6.14) can also be written in regression form as

$$y = \phi^T \theta_0 \quad (9.6.17)$$

where

$$\theta_0^T = (-a_1, \ldots, -a_n, b_1, \ldots, b_n) \quad (9.6.18)$$

$$\phi^T = (\phi_1^y, \ldots, \phi_n^y, \phi_1^u, \ldots, \phi_n^u) \quad (9.6.19)$$

$$\phi_i^y = F_i y, \quad i = 1, \ldots, n \quad (9.6.20)$$

$$\phi_i^u = F_i u, \quad i = 1, \ldots, n \quad (9.6.21)$$

Use will be made of representations of the form (9.6.17) in Chapter 12, which deals with parameter estimation.

9.6.2 Representation of All Stabilizing Controllers

We return to the fractional representation given in Equations (9.6.2), (9.6.3):

$$G(\rho) = \frac{N(\rho)}{M(\rho)} \quad (9.6.22)$$

where

$$N(\rho) = \frac{B(\rho)}{E(\rho)}, \quad M(\rho) = \frac{A(\rho)}{E(\rho)} \quad (9.6.23)$$

We assume that $\partial(B) < \partial(A)$ and that $\partial(E) = \partial(A)$. We also assume that $B(\rho)$ and $A(\rho)$ are coprime. We then have the following simple extension of Lemma 9.5.1.

Lemma 9.6.1

If $B(\rho)$, $A(\rho)$ are coprime polynomials, then $M(\rho)$, $N(\rho)$ are coprime rational functions in the sense that there exist exponentially stable rational functions $R(\rho)$, $S(\rho)$ such that

$$M(\rho)R(\rho) + N(\rho)S(\rho) = 1 \qquad (9.6.24)$$

Proof: From Lemma 9.5.1, we know that for any $A^*(\rho)$ there exist polynomials $L(\rho)$, $P(\rho)$ such that

$$A(\rho)L(\rho) + B(\rho)P(\rho) = A^*(\rho) \qquad (9.6.25)$$

The result follows by putting $A^*(\rho) = E(\rho)E'(\rho)$, where $\partial(E) = \partial(E') = \partial(A)$, and

$$R(\rho) = \frac{L(\rho)}{E'(\rho)} \qquad (9.6.26)$$

$$S(\rho) = \frac{P(\rho)}{E'(\rho)} \qquad (9.6.27)$$

▽▽▽

Note that, in the case of stable systems, we may simply choose $R(\rho) = E(\rho)/A(\rho)$ and $S(\rho) = 0$ to satisfy (9.6.24).

We next see how Lemma 9.6.1 can be exploited to characterize all stabilizing control laws.

Theorem 9.6.1

(i) Consider the system shown in Figure 9.6.1. Then the particular control law

$$C_p(\rho) = \frac{V_p(\rho)}{U_p(\rho)} \qquad (9.6.28)$$

where

$$V(\rho) = S(\rho)K(\rho) \quad \text{and} \quad U_p(\rho) = R(\rho)K(\rho) \qquad (9.6.29)$$

gives a closed loop system having denominator

$$M(\rho)U_p(\rho) + N(\rho)V_p(\rho) = K(\rho) \qquad (9.6.30)$$

(ii) A control law gives rise to the closed loop denominator $K(\rho)$ if and only if it can be expressed in the form

$$C_k(\rho) = \frac{V(\rho)}{U(\rho)} \qquad (9.6.31)$$

Sec. 9.6   Fractional Representation

where
$$V(\rho) = V_p(\rho) - \Gamma(\rho)M(\rho) \qquad (9.6.32)$$
and
$$U(\rho) = U_p(\rho) + \Gamma(\rho)N(\rho) \qquad (9.6.33)$$

where $\Gamma(\rho)$ is an exponentially stable rational proper transfer function.

(iii) A control law gives a stable loop if and only if it can be expressed as
$$C_s(\rho) = \frac{V_s(\rho)}{U_s(\rho)} \qquad (9.6.34)$$
where
$$V_s(\rho) = S(\rho) - \Omega(\rho)M(\rho)$$
and
$$U_s(\rho) = R(\rho) + \Omega(\rho)N(\rho) \qquad (9.6.35)$$

where $\Omega(\rho)$ is an exponentially stable rational proper function and $R(\rho)$, $S(\rho)$ satisfy (9.6.24).

Proof:

(i) Immediately on multiplying (9.6.24) by $K(\rho)$.

(ii) *If:* Equation (9.6.31) give the closed loop denominator as
$$\begin{aligned} &M(\rho)U(\rho) + N(\rho)V(\rho) \\ &= M(\rho)[U_p(\rho) + \Gamma(\rho)N(\rho)] + N(\rho)[V_p(\rho) - \Gamma(\rho)M(\rho)] \\ &= [M(\rho)U_p(\rho) + N(\rho)V_p(\rho)] + \Gamma(\rho)[M(\rho)N(\rho) - N(\rho)M(\rho)] \\ &= K(\rho) \end{aligned} \qquad (9.6.36)$$

This shows that any controller of the form (9.6.31) gives a closed loop denominator of $K(\rho)$.

Only if: The proof of the converse relies on the fact that the class of proper stable natural transfer functions form a Euclidean ring. Consider any control law of the form given in (9.6.31); then this gives a closed loop denominator of K if and only if
$$M(\rho)U(\rho) + N(\rho)V(\rho) = K(\rho) \qquad (9.6.37)$$

Combining this equation with the particular solution (9.6.30), we obtain
$$M(\rho)[U(\rho) - U_p(\rho)] + N(\rho)[V(\rho) - V_p(\rho)] = 0 \qquad (9.6.38)$$

Since $M(\rho)$ and $N(\rho)$ are coprime, then $N(\rho)$ must divide $U(\rho) - U_p(\rho)$; that is, there exists a $\Gamma(\rho)$ such that
$$U(\rho) - U_p(\rho) = \Gamma(\rho)N(\rho) \qquad (9.6.39)$$

Substituting (9.6.39) into (9.6.38) gives

$$V(\rho) - V_p(\rho) = -\Gamma(\rho)M(\rho) \quad (9.6.40)$$

Equations (9.6.39) and (9.6.40) are equivalent to (9.6.33) and (9.6.32), respectively.

(iii) For any $K(\rho)$ such that $K(\rho)$ and $K(\rho)^{-1}$ are stable proper rational functions, the class of all control laws that give $K(\rho)$ as the denominator is given, from part (ii), by

$$C_K(\rho) = \frac{V(\rho)}{U(\rho)}$$

$$= \frac{V_p(\rho) - \Gamma(\rho)M(\rho)}{U_p(\rho) + \Gamma(\rho)N(\rho)}, \quad \text{for any stable proper } \Gamma(\rho)$$

$$= \frac{S(\rho)K(\rho) - \Gamma(\rho)M(\rho)}{R(\rho)K(\rho) + \Gamma(\rho)N(\rho)} \quad \text{using part (i)}$$

Since $K(\rho)^{-1}$ is stable and proper, we can write

$$C_K(\rho) = \frac{S(\rho) - \Omega(\rho)M(\rho)}{R(\rho) + \Omega(\rho)N(\rho)} \quad (9.6.41)$$

where

$$\Omega(\rho) = K(\rho)^{-1}\Gamma(\rho) \quad (9.6.42)$$

From the preceding, we see that the class of functions $\Omega(\rho)$ that satisfies (9.6.42) such that $\Gamma(\rho)$, $K(\rho)$, and $K(\rho)^{-1}$ are all proper stable rational functions is equal to the set of all proper stable rational functions. ▽▽▽

The significance of Equation (9.6.35) is that it gives a simple characterization of all stabilizing controllers in terms of a rational function $\Omega(\rho)$. The function $\Omega(\rho)$ represents the degrees of freedom available to the designer subject to the constraint that the closed loop system be stable.

Another interesting feature of the use of fractional representations is that the closed loop characteristics are linear (in a sense made precise later) in the function $\Omega(\rho)$. To show this, we add input and output noise sources to Figure 9.6.1, leading to the system shown in Figure 9.6.2.

The following result characterizes the closed loop performance.

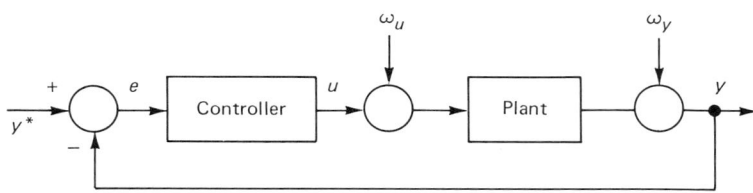

Figure 9.6.2 Feedback system with input and output noise sources.

Theorem 9.6.2

The closed loop system in Figure 9.6.2 is characterized by the following transfer functions:

$$\frac{e}{\omega_y} = -M[R + \Omega N] = -\frac{e}{y^*} \tag{9.6.43}$$

$$\frac{e}{\omega_u} = -N[R + \Omega N] \tag{9.6.44}$$

$$\frac{u}{\omega_y} = -M[S - \Omega M] = -\frac{u}{y^*} \tag{9.6.45}$$

$$\frac{u}{\omega_u} = -N[S - \Omega M] \tag{9.6.46}$$

which are all linear in the function Ω.

Proof: Straightforward, using closed loop relationships and the fact that $MR + NS = 1$. ▽▽▽

We illustrate the preceding ideas by a simple example.

Example 9.6.3

Consider the following unstable plant:

$$G(\rho) = \frac{1}{(\rho - 1)} \tag{9.6.47}$$

(a) Give a characterization of all stabilizing controllers.
(b) What is the simplest controller to give a stable closed loop system having zero tracking error to a constant input offset.

(a) We express the system in fractional form as

$$G(\rho) = \frac{N(\rho)}{M(\rho)}$$

where

$$M(\rho) = \frac{\rho - 1}{\rho + e_0}, \qquad N(\rho) = \frac{1}{\rho + e_0}$$

Equation (9.6.24) becomes

$$\frac{\rho - 1}{\rho + e_0} R(\rho) + \frac{1}{\rho + e_0} S(\rho) = 1$$

which is solved by $R(\rho) = 1$, $S(\rho) = (e_0 + 1)$. Equation (9.6.34) then gives all stabilizing control laws as

$$C_s(\rho) = \frac{V_s(\rho)}{U_s(\rho)}$$

where

$$V_s(\rho) = S - \Omega M = (e_0 + 1) - \Omega \frac{\rho - 1}{\rho + e_0}$$

$$U_s(\rho) = R + \Omega N = 1 + \Omega \frac{1}{\rho + e_0}$$

(b) We require

$$\frac{e}{\omega_u} = 0, \qquad \text{at } \rho = 0$$

From Equation (9.6.44),

$$\frac{e}{\omega_u} = -\frac{1}{\rho + e_0}\left[1 + \Omega \frac{1}{\rho + e_0}\right]$$

$$= -\frac{1}{\rho + e_0} \frac{\rho + e_0 + \Omega}{\rho + e_0}$$

Clearly, $\Omega = -e_0$ suffices. This gives

$$C_s(\rho) = \frac{(e_0 + 1) - \Omega[(\rho - 1)/(\rho + e_0)]}{1 + \Omega[1/(\rho + e_0)]}$$

$$= \frac{(2e_0 + 1)\rho + e_0^2}{\rho}$$

∇∇∇

The preceding results have a very simple interpretation when the controller is drawn in block diagram form. Combining Figure 9.6.2 with Equations (9.6.34), (9.6.35), (9.6.3), (9.6.26), and (9.6.27) leads to the characterization of all stabilizing controllers given in Figure 9.6.3. We see from this figure that the signal ν is an estimate of the combined disturbances ω_u, ω_y, and y^*. Indeed, if the plant is truly given by N/M, then ν is simply $(A/E_1)(y^* - \omega_y) - (B/E_1)\omega_u$. Thus, the upper feedback path in the figure is the normal stabilizing feedback path, while the lower feedback path through Ω injects an estimate of the plant disturbances.

The solution in Figure 9.6.3 is even easier to visualize in the case of stable open loop systems. In this case, we can take $E_1 = A$, $R = 1$, and $S = 0$. Thus, for stable open loop systems, the class of all stabilizing control laws is as shown in Figure 9.6.4, where we have assumed $\omega_u = 0$ for simplicity. In this case, we immediately see that, when the plant is truly given by B/A, then the signal ν is simply $(-y^* - \omega_y)$. The feedback structure shown in Figure 9.6.4 has a very long history in control systems and is usually called a Smith predictor structure after Otto Smith, who first suggested this structure for disturbance rejection.

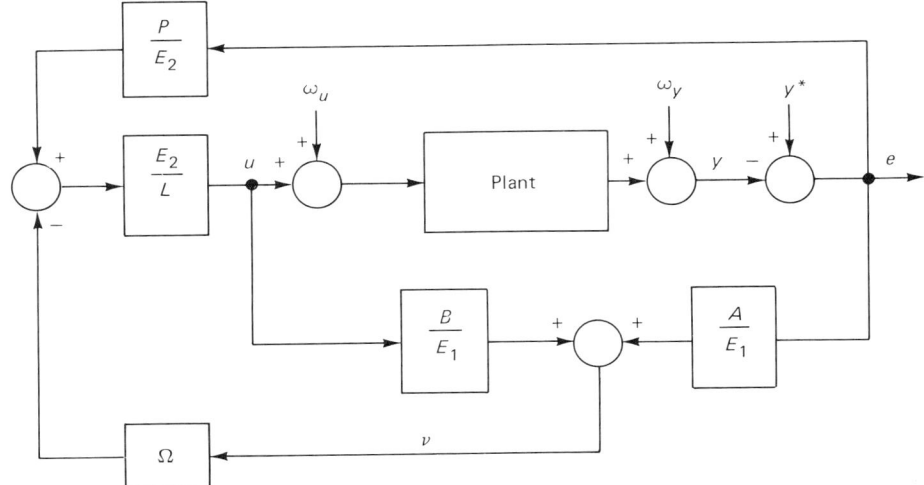

Figure 9.6.3 Block diagram representation of all stabilizing controllers.

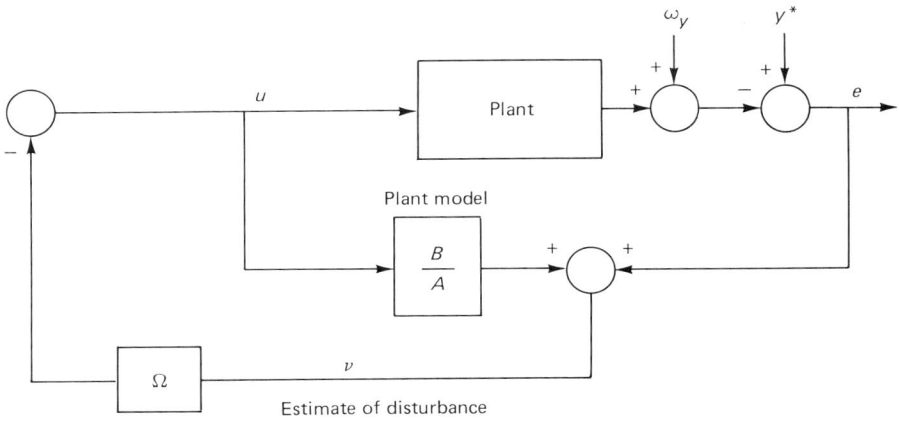

Figure 9.6.4 Block diagram representation of all stabilizing controllers for a stable plant.

The fact that the closed loop transfer functions are linear in Ω (when the true plant is N/M) is trivial to see from the block diagrams given in Figures 9.6.3 and 9.6.4. This is only approximately true when the true plant is not described by N/M. This latter situation is explored in Problem 16.

*9.6.3 Extension to Systems Defined on General Rings

The theory described can be directly extended to systems defined in an algebraic ring structure. We will treat this briefly in this section. Let $\mathscr{G}$ be a (not necessarily commutative) ring with identity. Elements of $\mathscr{G}$ will be used to describe the plant and the controllers. A simple example of $\mathscr{G}$ as used in previous sections is the class of all proper rational functions.

Let $\mathscr{H}$ be a subring of $\mathscr{G}$ that includes the identity. We will represent the plant and controllers as fractional models using elements of $\mathscr{H}$. Thus, for the previous sections, $\mathscr{H}$ would simply be the class of all *stable* proper rational functions. The general objective is to choose the compensator so that the closed loop system will be represented as an operator in the subring $\mathscr{H}$. For the simple case discussed earlier, this means that the closed loop system should be stable.

To achieve this objective, we will use the general principles of Section 9.6.2. In this development, we will need to define two subsets of $\mathscr{H}$:

$$\mathscr{J} = \{ h \in \mathscr{H} : h^{-1} \in \mathscr{G} \} \tag{9.6.48}$$

$$\mathscr{K} = \{ h \in \mathscr{H} : h^{-1} \in \mathscr{H} \} \tag{9.6.49}$$

Note that the element of $\mathscr{J}$ are those elements of $\mathscr{H}$ that are units in $\mathscr{G}$. For the simple case discussed earlier, $\mathscr{J}$ simply represents the class of proper rational stable functions that have relative degree zero. The elements of $\mathscr{K}$ are the units of $\mathscr{H}$. For the simple example, this represents those rational stable proper functions whose zeros are all strictly inside the stability boundary and that have relative degree zero.

A different example of the choice of $\mathscr{G}$, $\mathscr{H}$, $\mathscr{J}$, and $\mathscr{K}$ is the following:

$\mathscr{G}$ proper rational functions
$\mathscr{H}$ proper rational functions that have no poles outside a given area
$\mathscr{J}$ proper rational functions that have no poles outside a given area and that are nonzero at ∞
$\mathscr{K}$ proper rational functions that have no poles or zeros outside a given area and that are nonzero at ∞

Given the preceding structure, we say that $P \in \mathscr{G}$ has a right fractional representation in $(\mathscr{G}, \mathscr{H}, \mathscr{J}, \mathscr{K})$ if there exists $N \in \mathscr{H}$, $M \in \mathscr{J}$ such that

$$P = NM^{-1} \tag{9.6.50}$$

Furthermore, we say that the pair $(N, M) \in \mathscr{H} \times \mathscr{H}$ is right coprime if there exists $R \in \mathscr{H}$, $S \in \mathscr{H}$ such that

$$RM + SN = I \tag{9.6.51}$$

where I is any element in $\mathscr{K}$. Similar statements apply to left coprime representations. Given this formalism, we may then characterize the class of all controllers that give a closed loop system having denominator in $\mathscr{K}$ exactly as in Theorem 9.6.1. We illustrate this next for scalar systems having a prescribed stability margin.

Example 9.6.4

Consider the following discrete time system (sampling rate 10 Hz):

$$G(\rho) = \frac{\rho + 1}{\rho^2 + 2} \tag{9.6.52}$$

Say we wish to design a feedback control system to place all the closed loop poles inside a disc of radius 1 centered on -2 in the complex γ plane. In addition, we require that the system have zero tracking error to step inputs.

We define $\mathcal{H}$ to be those systems whose poles lie inside the specified circular area. We then write

$$G(\rho) = \frac{(\rho+1)/(\rho+2)^2}{(\rho^2+2)/(\rho+2)^2} = NM^{-1} \qquad (9.6.53)$$

We then note that (9.6.51) is satisfied by the choice

$$R = \frac{\rho + \frac{4}{3}}{\rho + 2}, \qquad S = \frac{\frac{14}{3}\rho + \frac{16}{3}}{\rho + 2} \qquad (9.6.54)$$

Now, from Equation (9.6.43), we require $(R + \Omega N)$ to be zero at $\rho = 0$. Thus let Ω be a constant Ω_0; then we require $(\rho + 2)(\rho + \frac{4}{3}) + \Omega_0(\rho + 1) = 0$ for $\rho = 0$. This requires $\Omega_0 = -\frac{8}{3}$, and thus the final control law is

$$C = \frac{S - \Omega M}{R + \Omega N}$$

$$= \frac{\dfrac{\frac{14}{3}\rho + \frac{16}{3}}{\rho + 2} + \dfrac{\frac{8}{3}(\rho^2 + 2)}{(\rho + 2)^2}}{\dfrac{\rho + \frac{4}{3}}{\rho + 2} - \dfrac{\frac{8}{3}(\rho + 1)}{(\rho + 2)^2}}$$

$$= \frac{\frac{22}{3}\rho^2 + \frac{34}{3}\rho + 16}{\rho(\rho + 2)} \qquad (9.6.55)$$

▽▽▽

We extend the characterization of all stabilizing controllers to the multiinput, multioutput case.

9.6.4 Multiinput, Multioutput Systems

In the multiinput, multioutput (MIMO) case, we have to consider both left and right matrix fraction descriptions. Thus we assume the plant transfer function G is a proper rational matrix. We then express G in fractional form as

$$G = NM^{-1} = \tilde{M}^{-1}\tilde{N} \qquad (9.6.56)$$

where $M, N, \tilde{M}, \tilde{N}$ are all proper, rational transfer functions having poles inside a prescribed region (for example, the stability boundary). We further require that N, M be right coprime and $\tilde{M}, \tilde{N}$ be left coprime. Also, it is implicit in this definition that both M and $\tilde{M}$ are square and M^{-1} and $\tilde{M}^{-1}$ exist and are proper.

The extension of Lemma 9.6.1 to the MIMO case is the following:

Lemma 9.6.2

For each proper real rational transfer function $G(\gamma)$ there exist eight proper stable rational system matrix transfer functions $M, N, \tilde{M}, \tilde{N}, R, S, \tilde{R}, \tilde{S}$ such that

$$G = NM^{-1} = \tilde{M}^{-1}\tilde{N} \qquad (9.6.57)$$

and the matrices satisfy the following Bezout identity

$$\begin{bmatrix} \tilde{R} & \tilde{S} \\ -\tilde{N} & \tilde{M} \end{bmatrix} \begin{bmatrix} M & -S \\ N & R \end{bmatrix} = I \qquad (9.6.58)$$

In particular,

$$\tilde{R}M + \tilde{S}N = I \quad \text{and} \quad \tilde{M}R + \tilde{N}S = I \qquad (9.6.59)$$

Proof: Results of this type for multiinput, multioutput systems are generally more easily visualized in state space form. Thus, let the plant be described in state space form as follows:

$$G(\rho) = C(\rho I - A)^{-1}B \qquad (9.6.60)$$

for real matrices A, B, C such that (A, B) is stabilizable and (C, A) is detectable. That is, the system can be described as

$$\rho x = Ax + Bu \qquad (9.6.61)$$
$$y = Cx \qquad (9.6.62)$$

For future use we define the following transfer functions:

$$M(\gamma) = L(\gamma I - A')^{-1}B + I \qquad (9.6.63)$$
$$N(\gamma) = C(\gamma I - A')^{-1}B \qquad (9.6.64)$$
$$\tilde{M}(\gamma) = I - C(\gamma I - A'')^{-1}J \qquad (9.6.65)$$
$$\tilde{N}(\gamma) = C(\gamma I - A'')^{-1}B \qquad (9.6.66)$$
$$R = C(\rho I - A')^{-1}J + I \qquad (9.6.67)$$
$$S = -L(\rho I - A')^{-1}J \qquad (9.6.68)$$
$$\tilde{R} = -L(\rho I - A'')^{-1}B + I \qquad (9.6.69)$$
$$\tilde{S} = -L(\rho I - A'')^{-1}J \qquad (9.6.70)$$

In view of the detectability assumption, we can choose a real matrix J so that $A'' = A - JC$ is stable (or, alternatively, to have a prescribed margin of stability). We use J to construct an observer as follows

$$\rho \hat{x} = A\hat{x} + Bu + J(y - C\hat{x})$$

or

$$\rho \hat{x} = A''\hat{x} + Bu + Jy \qquad (9.6.71)$$

Using superposition, we can write this as two models driven by u and y separately:

$$\rho \hat{x}_1 = A'' \hat{x}_1 + Bu$$
$$\rho \hat{x}_2 = A'' \hat{x}_2 + Jy$$
$$\hat{x} = \hat{x}_1 + \hat{x}_2$$

Using equation (9.6.61), (9.6.62), we obtain

$$y = C\hat{x} + \nu \qquad (9.6.72)$$

We will call ν the *output innovations* process, and we note that it is the difference between the observer output $C\hat{x}$ and what we would like it to be, the true system output, y. From (9.6.72) we also have

$$\nu \triangleq C(x - \hat{x}) = C\tilde{x}$$

and where $\tilde{x} = x - \hat{x}$ satisfies the following stable homogeneous equation:

$$\rho \tilde{x} = A'' \tilde{x}$$

Thus, among other things and in the absence of noise, ν tends to zero as $t \to \infty$. From (9.6.72) we have

$$y = C(\hat{x}) + \nu$$
$$= C(\hat{x}_1 + \hat{x}_2) + \nu$$

Expressing the equations for $\hat{x}_1$ and $\hat{x}_2$ in transfer function form and substituting into the above equation gives

$$y = C\{\rho I - A''\}^{-1} Bu + C\{\rho I - A''\}^{-1} Jy + \nu$$

or

$$\left[I - C(\rho I - A'')^{-1} J\right] y = C(\rho I - A'')^{-1} Bu + \nu$$

This equation is of the form

$$\tilde{M} y = \tilde{N} u + \nu$$

That is,

$$\nu = \tilde{M} y - \tilde{N} u = y - C\hat{x} \qquad (9.6.73)$$

Since ν tend to zero at $t \to \infty$, the input–output transfer function can be written as

$$y = \tilde{M}^{-1} \tilde{N} u$$

Next, in view of the stabilizability assumption, we can choose a real matrix L such that $A' = A + BL$ is stable (or, alternatively, to have a prescribed margin of stability). Hence adding and subtracting $BL\hat{x}$ from the observer (9.6.71) gives the following result, as illustrated in Figure 9.6.5.

$$\rho \hat{x} = (A + BL) \hat{x} + B(u - L\hat{x}) + J\nu$$

or

$$\rho \hat{x} = A' \hat{x} + Bv + J\nu \qquad (9.6.74)$$

where $v = u - L\hat{x}$.

246   Pole Assignment   Chap. 9

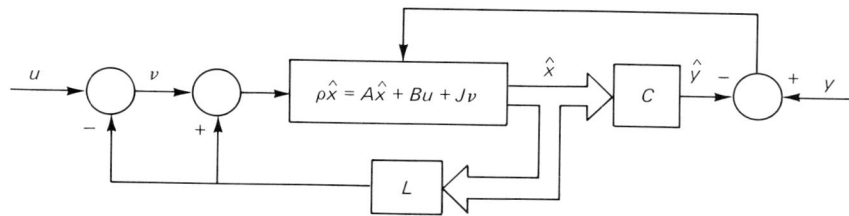

Figure 9.6.5 Introduction of canceling feedback.

We can think of v as being an *input innovations* process since it defines the difference between the true system input u and what we might like the input to be, $L\hat{x}$. Rearranging this equation, we have

$$u = L\hat{x} + v \qquad (9.6.75)$$

We also have from (9.6.72) that

$$y = C\hat{x} + \nu \qquad (9.6.76)$$

We can now write (9.6.74), (9.6.75), (9.6.76) in transfer function form as

$$\begin{bmatrix} u \\ y \end{bmatrix} = \begin{bmatrix} M & -S \\ N & R \end{bmatrix} \begin{bmatrix} v \\ \nu \end{bmatrix} \qquad (9.6.77)$$

Also, from (9.6.71) and $v = u - L\hat{x}$, we have that

$$v = \{I - L(\rho I - A'')^{-1}B\}u - L(\rho I - A'')^{-1}Jy$$

or

$$v = \tilde{R}u + \tilde{S}y \qquad (9.6.78)$$

Combining (9.6.73) and (9.6.78) gives

$$\begin{bmatrix} v \\ \nu \end{bmatrix} = \begin{bmatrix} \tilde{R} & \tilde{S} \\ -\tilde{N} & \tilde{M} \end{bmatrix} \begin{bmatrix} u \\ y \end{bmatrix} \qquad (9.6.79)$$

Finally, since v and ν can be made any signal by choice of u and by injecting noise into the system output, then it follows from (9.6.77), (9.6.79) that we must have

$$\begin{bmatrix} \tilde{R} & \tilde{S} \\ -\tilde{N} & \tilde{M} \end{bmatrix} \begin{bmatrix} M & -S \\ N & R \end{bmatrix} = [I] \qquad (9.6.80)$$

▽▽▽

Example 9.6.5 illustrates the use of the above state space formulation to determine matrices R, $\tilde{R}$, S and $\tilde{S}$ such that (9.6.58) is satisfied.

Example 9.6.5

Consider the following transfer function:

$$G(\rho) = \frac{\rho - 1}{(\rho - 2)(\rho + 1)} \quad (9.6.81)$$

A minimal realization is

$$G(\rho) = C(\rho I - A)^{-1}B \quad (9.6.82)$$

where

$$A = \begin{bmatrix} 0 & 1 \\ 2 & 1 \end{bmatrix}, \quad B = \begin{bmatrix} 0 \\ 1 \end{bmatrix} \quad (9.6.83)$$

$$C = \begin{bmatrix} -1 & 1 \end{bmatrix} \quad (9.6.84)$$

Choosing L to place the eigenvalues of A' at $-1, -1$ we get

$$L = \begin{bmatrix} -3 & -3 \end{bmatrix} \quad (9.6.85)$$

$$A' = \begin{bmatrix} 0 & 1 \\ -1 & -2 \end{bmatrix} \quad (9.6.86)$$

Then

$$N(\rho) = C(\rho I - A')^{-1}B = \frac{\rho - 1}{(\rho + 1)^2} \quad (9.6.87)$$

$$M(\rho) = L(\rho I - A')^{-1}B + I = \frac{(\rho - 2)(\rho + 1)}{(\rho + 1)^2} \quad (9.6.88)$$

Similarly, choosing J to place the eigenvalues of A'' at $-1, -1$, we get

$$J = \begin{bmatrix} 3 \\ 6 \end{bmatrix} \quad (9.6.89)$$

$$A'' = \begin{bmatrix} 3 & -2 \\ 8 & -5 \end{bmatrix} \quad (9.6.90)$$

$$\tilde{N}(\rho) = C(\rho I - A'')^{-1}B = \frac{\rho - 1}{(\rho + 1)^2} \quad (9.6.91)$$

$$\tilde{M}(\rho) = -C(\rho I - A'')^{-1}J + I = \frac{(\rho - 2)(\rho + 1)}{(\rho + 1)^2} \quad (9.6.92)$$

Finally,

$$R(\rho) = C(\rho I - A')^{-1}J + I = \frac{\rho^2 + 5\rho - 14}{(\rho + 1)^2} \quad (9.6.93)$$

$$S(\rho) = -L(\rho I - A')^{-1}J = \frac{27(\rho + 1)}{(\rho + 1)^2} \quad (9.6.94)$$

$$\tilde{R}(\rho) = -L(\rho I - A'')^{-1}B + I = R(\rho) \quad (9.6.95)$$

$$\tilde{S}(\rho) = -L(\rho I - A'')^{-1}J = S(\rho) \quad (9.6.96)$$

▽▽▽

We next extend Theorem 9.6.1 to the MIMO case and give a characterization of all stabilizing controllers. Let $C(\rho)$ denote the controller transfer function. We represent $C(\rho)$ by left and right coprime factorizations as

$$C(\rho) = V(\rho)U(\rho)^{-1} = \tilde{U}(\rho)^{-1}\tilde{V}(\rho) \tag{9.6.97}$$

The extension of Theorem 9.6.1 to the MIMO case is:

Theorem 9.6.3

1. Consider the system shown in Figure 9.6.1; then the particular control law

$$C_p = \tilde{V}_p^{-1}\tilde{U}_p = U_p V_p^{-1} \tag{9.6.98}$$

   where

$$U_p = RK, \qquad V_p = SK \tag{9.6.99}$$

   or

$$\tilde{U}_p = K\tilde{R}, \qquad \tilde{V}_p = K\tilde{S} \tag{9.6.100}$$

   gives a closed loop system having denominator K, where the plant together with R, S, $\tilde{R}$, $\tilde{S}$ is as defined in Lemma 9.6.2.

2. A controller gives rise to the closed loop denominator $K(\rho)$ if and only if it can be expressed as

$$C_k = \tilde{U}_k^{-1}\tilde{V}_k = V_k U_k^{-1} \tag{9.6.101}$$

   where

$$\tilde{U}_k = \tilde{U}_p + \Gamma\tilde{N}, \qquad \tilde{V}_k = \tilde{V}_p - \Gamma\tilde{M} \tag{9.6.102}$$

   or

$$U_k = U_p + N\Gamma, \qquad V_k = V_p - M\Gamma \tag{9.6.103}$$

   where Γ is any proper stable transfer function matrix of appropriate dimension.

3. A controller gives a stable closed loop if and only if it can be expressed as

$$C_s = \tilde{U}_s^{-1}\tilde{V}_s = V_s U_s^{-1} \tag{9.6.104}$$

   where

$$\tilde{U}_s = \tilde{R} + \Omega\tilde{N}, \qquad \tilde{V}_s = \tilde{S} - \Omega\tilde{M} \tag{9.6.105}$$

   or

$$U_s = R + N\Omega, \qquad V_s = S - M\Omega$$

   where Ω is any proper stable transfer function.

Proof:

1. Immediate.
2. *If:* Immediate as in the proof of Theorem 9.6.1.
   Only if: Consider any control law of the form

   $$\tilde{U}'_K u = -\tilde{V}'_K y$$

   This gives a closed loop denominator of K if and only if

   $$\tilde{U}'_K M + \tilde{V}'_K N = K$$

   However, from part 1 we have

   $$\tilde{U}_p M + \tilde{V}_p N = K$$

   Subtracting the last two equations gives

   $$\left(\tilde{U}_K - \tilde{U}_p\right) M + \left(\tilde{V}'_K - \tilde{V}_p\right) N = 0 \qquad (9.6.106)$$

   Now, as in the proof of Lemma 9.6.2, we have from (9.6.77), (9.7.79) that the various transfer function matrices satisfy the following reverse Bezout identity:

   $$\begin{bmatrix} M & -S \\ N & R \end{bmatrix} \begin{bmatrix} \tilde{R} & \tilde{S} \\ -\tilde{N} & \tilde{M} \end{bmatrix} = I \qquad (9.6.107)$$

   Multiplying this equation on the left by $[(\tilde{U}_K - \tilde{U}_p), (\tilde{V}_K - \tilde{V}_p)]$ and using (9.6.106) gives Equation (9.6.102), where

   $$\Gamma = \left(\tilde{U}_K - \tilde{U}_p\right) S - \left(\tilde{V}_K - \tilde{V}_p\right) R$$

3. Immediate as in Theorem 9.6.1 on multiplying the control K^{-1}. ▽▽▽

As before, Theorem 9.6.3 has a simple interpretation when the plant G is stable, since then we can take

$$N = \tilde{N} = P \qquad (9.6.108)$$

$$M = \tilde{M} = I \qquad (9.6.109)$$

and

$$\tilde{R} = R = I \qquad (9.6.110)$$

$$\tilde{S} = S = 0 \qquad (9.6.111)$$

This leads to:

Corollary 9.6.1

For stable plants, all stabilizing controllers can be expressed as

$$C_s = \tilde{U}_s^{-1} \tilde{V}_s = V_s U_s^{-1} \qquad (9.6.112)$$

where
$$\tilde{U}_s = I + \Omega P, \qquad \tilde{V}_s = -\Omega \qquad (9.6.113)$$
$$U_s = I + P\Omega, \qquad V_s = -\Omega \qquad (9.6.114)$$
where Ω is any proper stable transfer function.

Proof: Immediate from (9.6.106) to (9.6.111). ▽▽▽

*9.6.5 Multivariable Model Reference Control

We have seen in Section 9.5.4 that model reference control refers to a general class of control laws that assign the numerator as part of the closed loop denominator. We will show in this section that this idea extends in a straightforward way to the multivariable case. The essential motivation for doing this is that it leads to considerable simplifications in the design of multivariable control systems.

An important preliminary observation is that, in the case of SISO systems, feedback via a proper controller cannot decrease the relative degree (where relative degree is defined as the difference between numerator and denominator degree). Equivalently, we say that the number of plant zeros at infinity is a feedback invariant. The extension of this idea to multivariable systems requires us to characterize the nature of the zeros of a system at infinity. This is a little more subtle for multivariable systems than for scalar systems. The appropriate mechanism for characterizing the zeros at infinity of a multivariable system is via the interactor matrix $\xi(\gamma)$ defined in the following result:

Lemma 9.6.3

Given any $m \times m$ strictly proper transfer function $G(\gamma)$ whose determinant is not identically zero, then there exists a polynomial matrix $\xi(\gamma)$, known as the interactor matrix, satisfying
$$\lim_{\gamma \to \infty} \xi(\gamma) P(\gamma) = K_0 \qquad \text{(nonsingular)} \qquad (9.6.115)$$
where
$$\xi(\gamma) = U(\gamma) D(\gamma) \qquad (9.6.116)$$
$U(\gamma)$ is a unimodular matrix of the form
$$U(\gamma) = \begin{bmatrix} 1 & & & \\ h_{21}(\gamma) & \ddots & & \\ \vdots & & \ddots & \\ h_{m1}(\gamma) & \cdots & h_{mm-1}(\gamma) & 1 \end{bmatrix} \qquad (9.6.117)$$
where $h_{ij}(\gamma)$ is a polynomial and $D(\gamma)$ is a diagonal matrix of the form
$$D(\gamma) = \text{diag}[D_1(\gamma) \ldots D_m(\gamma)] \qquad (9.6.118)$$
$$D_i(\gamma) = \gamma^{f_i} + d^i_{f_i-1}\gamma^{f_i-1} + \cdots + d^i_0 \qquad (9.6.119)$$

Proof: There exist integers d_i, $i = 1, \ldots, m$, such that

$$\lim_{\gamma \to \infty} \gamma^{d_i} G_{i.}(\gamma) = \tau_i \tag{9.6.120}$$

where $G_{i.}(\rho)$ denotes the ith row of $G(\gamma)$ and τ_i is not identically zero. We define the first row of $\xi(\gamma)$ by

$$\xi_{1.}(\gamma) = \begin{bmatrix} D_1(\gamma) & 0 \ldots 0 \end{bmatrix} \tag{9.6.121}$$

where $D_1(\gamma)$ is a monic polynomial of degree d_1. Clearly,

$$\lim_{\gamma \to \infty} \xi_{1.}(\gamma) G(\gamma) = \sigma_1 = \tau_1 \neq 0 \tag{9.6.122}$$

Now consider the row vector, τ_2, defined in (9.6.120). If τ_2 is linearly independent of σ_1, then set

$$\xi_{2.}(\gamma) = (0, D_2(\gamma), 0 \ldots 0) \tag{9.6.123}$$

so that

$$\lim_{\gamma \to \infty} \xi_{2.}(\gamma) G(\gamma) = \sigma_2 = \tau_2 \tag{9.6.124}$$

On the other hand, if τ_2 and σ_1 are linearly dependent so that $\tau_2 = \alpha_1' \sigma_1$ with $\alpha_1' \neq 0$, then let

$$\xi_{2.}'(\gamma) = D_2'(\gamma) \{[0, D_2(\gamma), 0 \ldots 0] - \alpha_1' \xi_{1.}(\gamma)\} \tag{9.6.125}$$

where $D_2'(\gamma)$ has degree d_2', where d_2' is the unique integer for which $\lim_{\gamma \to \infty} \xi_{2.}' G(\gamma) = \sigma_2'$ is both finite and nonzero. If σ_2' is linearly independent of σ_1, we set

$$\xi_{2.}(\gamma) = \xi_{2.}'(\gamma) \tag{9.6.126}$$

and note that

$$\lim_{\gamma \to \infty} \xi_{2.}(\gamma) = G(\gamma) = \sigma_2' \tag{9.6.127}$$

is linearly independent of σ_1. If this is not the case, then $\sigma_2' = \alpha_1'' \sigma_1$ for some $\alpha_1'' \neq 0$. Then let

$$\xi_{2.}''(\gamma) = D_2''(\gamma) \{\xi_{2.}'(\gamma) - \alpha_1'' \xi_{1.}(\gamma)\} \tag{9.6.128}$$

where $D_2''(\gamma)$ is a monic polynomial of degree d_2'', which is the unique integer for which $\lim_{\gamma \to \infty} \xi_{2.}'' G = \sigma_2''$ is both finite and nonzero. If σ_2'' is linearly independent of τ_1, we put $\xi_{2.} = \xi_{2.}''$ and proceed. If not, we repeat the procedure until linear independence is obtained. The procedure must terminate since $\det G \neq 0$ and since $d_i \geq 1$, $i = 1, \ldots, m$. The remaining rows of $\xi(\gamma)$ are obtained in an analogous fashion. ▽▽▽

Note that $\xi(\gamma)$ has a trivial interpretation in the scalar case in terms of the plant relative degree. In the multivariable case, the situation is more subtle as illustrated in the following examples.

Example 9.6.6

Consider the following transfer function:

$$G(\gamma) = \begin{bmatrix} \dfrac{1}{\gamma(\gamma+1)} & \dfrac{2}{\gamma(\gamma+3)} \\ \dfrac{1}{\gamma^2(\gamma+1)} & \dfrac{3}{\gamma^2(\gamma+4)} \end{bmatrix} \qquad (9.6.129)$$

We can see that

$$\xi(\gamma) = \begin{bmatrix} D_1(\gamma) & 0 \\ 0 & D_2(\gamma) \end{bmatrix} \qquad (9.6.130)$$

satisfies (9.6.115), where D_1 has degree 2 and D_2 has degree 3. ▽▽▽

In Example 9.6.6, $\xi(\gamma)$ simply reduces to the relative degree of each row of $G(\gamma)$. However, this need not always be the case, as shown in Example 9.6.7.

Example 9.6.7

Consider the following transfer function:

$$G(\gamma) = \begin{bmatrix} \dfrac{1}{\gamma+1} & \dfrac{1}{\gamma+2} \\ \dfrac{1}{\gamma+4} & \dfrac{1}{\gamma+3} \end{bmatrix} \qquad (9.6.131)$$

Applying the procedure in Lemma 9.6.3, we obtain

$$\xi_{1\cdot}(\gamma) = (\gamma+1, 0)$$

giving

$$\tau_1 = (1, 1)$$

If we put $\tilde{\xi}_{2\cdot}(\gamma) = (0, \gamma+1)$, then $\tau_2 = [1\ 1]$. That is, $\tau_2 = 1[\tau_1]$. Hence try

$$\xi'_{2\cdot}(\gamma) = (\gamma+1)[\tilde{\xi}_2(\gamma) - \xi_1(\gamma)]$$

Then $\tau'_2 = [-3, -1]$, which is independent of τ_1. Hence

$$\xi(\gamma) = \begin{bmatrix} \gamma+1 & 0 \\ -(\gamma+1)^2 & (\gamma+1)^2 \end{bmatrix}$$

$$= \begin{bmatrix} 1 & 0 \\ -(\gamma+1) & 1 \end{bmatrix} \begin{bmatrix} \gamma+1 & 0 \\ 0 & (\gamma+1)^2 \end{bmatrix}$$

▽▽▽

We next establish that the zeros at infinity are preserved under feedback control.

Lemma 9.6.4

Consider the feedback system shown in Figure 9.6.1, where G is strictly proper and has interactor matrix ξ and C is proper. Then, if G_c is the closed loop transfer function,

$$\lim_{\gamma \to \infty} \xi G_c < \infty \qquad (9.6.132)$$

Proof:

$$G_c = (I + GC)^{-1} GC$$
$$= GC(I + GC)^{-1} \qquad (9.6.133)$$

Hence

$$\lim_{\gamma \to \infty} \xi G_c = \lim_{\gamma \to \infty} \xi GC(I + GC)^{-1}$$
$$= K_0 \left\{ \lim_{\gamma \to \infty} C(I + GC)^{-1} \right\}$$

where K_0 is a nonsingular real matrix.

However, $\lim_{\gamma \to \infty} C(I + GC)^{-1}$ is finite since C is proper and G is strictly proper. ▽▽▽

We are now in a position to define the model reference control law. Consider a plant parameterized in fractional form as

$$G = NM^{-1} = \tilde{M}^{-1} \tilde{N} \qquad (9.6.134)$$

where G has interactor matrix ξ. Then $[\xi N]$ is a proper matrix having a proper inverse. This is readily seen since ξ is a polynomial matrix and

$$\lim_{\gamma \to \infty} \xi G = \lim_{\gamma \to \infty} \xi N M^{-1} = \lim_{\gamma \to \infty} \xi N (M_\infty)^{-1} = K_0 \quad \text{(nonsingular)}$$

where

$$M_\infty = \lim_{\gamma \to \infty} M$$

Hence

$$\lim_{\gamma \to \infty} (\xi N) = K_0 M_\infty \quad \text{(nonsingular)}$$

and

$$\lim_{\gamma \to \infty} (\xi N)^{-1} = M_\infty^{-1} K_0^{-1} \quad \text{(nonsingular)}$$

The model reference design is now completed as in Section 9.6.4, where $K(\rho)$ has the special structure

$$K(\rho) = T(\rho) \xi(\rho) N(\rho) \qquad (9.6.135)$$

where $T(\rho), T(\rho)^{-1}$ are stable proper transfer functions. Note, in particular, that inclusion of $\xi(\rho)$ in (9.6.135) ensures that $K(\rho)$, $K(\rho)^{-1}$ are both stable and proper as required.

We then have:

Theorem 9.6.4

1. All solutions to the multivariable model reference problem satisfy
$$C = (\tilde{U}_k)^{-1} \tilde{V}_k \qquad (9.6.136)$$
where
$$\tilde{U}_k = \tilde{U}_p + \Gamma \tilde{N}, \qquad \tilde{V}_k = \tilde{V}_p - \Gamma \tilde{M} \qquad (9.6.137)$$
and
$$\tilde{U}_p = K\tilde{R}, \qquad \tilde{V}_p = K\tilde{S} \qquad (9.6.138)$$
$$\tilde{R}M + \tilde{S}N = I \qquad (9.6.139)$$
$$K = T\xi N \qquad (9.6.140)$$
and Γ is any proper stable transfer function.

2. A particular solution to the multivariable model reference problem is
$$C' = (\tilde{U}')^{-1} \tilde{V}' \qquad (9.6.141)$$
where
$$\tilde{U}' = F\tilde{N} \qquad (9.6.142)$$
$$\tilde{V}' = W \qquad (9.6.143)$$
where F and W satisfy
$$F\tilde{M} + W = T\xi \qquad (9.6.144)$$
where W is proper.

Proof:

1. Immediate from Section 9.6.4.
2. We need to check that $\tilde{U}'$ is invertible. To do this, we note that
$$\tilde{U}' = F\tilde{N} = F\tilde{M}(\tilde{M}^{-1}\tilde{N}) = F\tilde{M}G$$
Using (9.6.144),
$$F\tilde{N} = (T\xi - W)G = (T - W\xi^{-1})\xi G$$
Hence
$$\lim_{\gamma \to \infty} F\tilde{N} = \lim_{\gamma \to \infty} (T - W\xi)^{-1} \lim_{\gamma \to \infty} \xi G$$
$$= \left(\lim_{\gamma \to \infty} T \right) K_0$$
which is nonsingular since T and T^{-1} are proper and K_0 is nonsingular.

We next check that the closed loop denominator is as required. The closed loop denominator is

$$\begin{aligned}
\overline{K} &= \tilde{U}'M + \tilde{V}'N \\
&= F\tilde{N}M + WN, \quad \text{where we have used } (9.6.142), (9.6.143) \\
&= F\tilde{M}N + WN, \quad \text{by } (9.6.58) \\
&= (F\tilde{M} + W)N \\
&= T\xi N, \quad \text{by } (9.6.144)
\end{aligned}$$

▽▽▽

A key point about multivariable model reference control is that (9.6.144) is much easier to solve than (9.6.59). In fact, $W\tilde{M}^{-1}$ is simply the proper part of $T\xi\tilde{M}^{-1}$ and F is the improper part. Hence, the design identity (9.6.144) is easily solved as follows.

Let

$$T(\gamma) = \frac{\overline{T}(\gamma)}{E(\gamma)} = \frac{\overline{T}_0 + \overline{T}_1\gamma + \cdots + \gamma^n}{E(\gamma)}$$

$$\tilde{M}(\gamma) = \frac{\overline{M}(\gamma)}{E(\gamma)} = \frac{\overline{M}_0 + \overline{M}_1\gamma + \cdots + \gamma^n}{E(\gamma)}$$

and

$$W(\Gamma) = \frac{\overline{W}(\gamma)}{E(\gamma)} = \frac{\overline{\Gamma}_0 + \overline{\Gamma}_1\gamma + \cdots + \overline{\Gamma}_n\gamma^n}{E(\gamma)}$$

where $E(\gamma)$ is a stable, scalar polynomial. We then rewrite (9.6.144) as

$$\begin{aligned}
F(\gamma)\overline{M}(\gamma) + \overline{W}(\gamma) &= \overline{T}(\gamma)\xi(\gamma) \\
&\triangleq X_0 + X_1\gamma + \cdots + X_{r+n}\gamma^{r+n}
\end{aligned} \quad (9.6.145)$$

Then, if $F(\gamma) = F_1\gamma + \cdots + F_r\gamma^r$, we have, by equating coefficients of γ^{n+i} in (9.6.149),

$$\begin{aligned}
F_r &= X_{r+n} \\
F_{r-1} &= X_{r+n-1} - F_r M_{n-1} \\
&\vdots \\
F_1 &= X_{n+1} - \sum_{i=2}^{r} F_i M_{m-i+1}
\end{aligned}$$

and, by equating the remaining coefficients of γ^i, we obtain

$$\begin{aligned}
\overline{W}_0 &= X_0 \\
\overline{W}_1 &= X_1 - F_1 \overline{M}_0 \\
&\vdots \\
\overline{W}_n &= X_n - \sum_{i=1}^{r} F_i \overline{M}_{n-i}
\end{aligned}$$

(Note that these calculations can be done on the back of an envelope, even if the envelope has to be as big as Texas.)

As described above, the design law is really a zero canceling controller. We now show how this can be readily converted into model reference control.

In Section 9.5.3 we saw that for the SISO case the relative degree of the reference model had to be greater than or equal to that of the plant. In the MIMO case, we achieve this result by defining the reference model as

$$y^* = \xi^{-1} T^{-1} \tilde{N}^* r \qquad (9.6.146)$$

where T, T^{-1}, and $\tilde{N}^*$ are rational, proper, and stable.

Note that the reference model has the property

$$\lim_{\gamma \to \infty} \xi \left[\xi^{-1} T^{-1} \tilde{N}^* \right] < \infty \qquad (9.6.147)$$

That is, it preserves the zeros at ∞ as required by Lemma 9.6.4.

Model reference control, as defined above, can be achieved by using the control law

$$(F\tilde{N})u = -Wy + \tilde{N}^* r \qquad (9.6.148)$$

where F and W satisfy (9.6.144).

It is readily verified that this control law gives

$$y = \xi^{-1} T^{-1} \tilde{N}^* r \qquad (9.6.149)$$

as required.

9.7 SUMMARY

The key points of this chapter were:

- Pole assignment for reachable systems can be achieved by state variable feedback.
- The dynamics of a state observer for an observable system can be arbitrarily prescribed.
- The combined use of an observer and state estimate feedback gives a closed loop system whose poles are the sum of the poles of the observer and those that would have been achieved by true state variable feedback.
- The combined use of an observer and state estimate feedback can be succinctly described in operator form by a control law of the form

$$Lu = -Py + Gy^* \qquad (9.4.29)$$

where L and P satisfy the diophantine equation

$$AL + BP = A^* \qquad (9.5.6)$$

- Model reference control can be viewed as a special case of pole assignment in which the closed loop polynomial A^* includes the numerator polynomial B.
- Models of the following form were introduced (called fractional models):

$$G(\gamma) = \frac{N(\gamma)}{M(\gamma)} \qquad (9.6.2)$$

where $N(\gamma)$, $M(\gamma)$ are stable proper transfer functions of the form

$$N(\gamma) = \frac{B(\gamma)}{E(\gamma)}, \qquad M(\gamma) = \frac{A(\gamma)}{E(\gamma)} \tag{9.6.3}$$

- The class of all stabilizing controllers was shown to be

$$C_s(\gamma) = \frac{V_s(\gamma)}{U_s(\gamma)} \tag{9.6.34}$$

where

$$\begin{aligned} U_s(\gamma) &= R(\gamma) + \Omega(\gamma) N(\gamma) \\ V_s(\gamma) &= S(\gamma) - \Omega(\gamma) M(\gamma) \end{aligned} \tag{9.6.35}$$

where $\Omega(\gamma)$ is any rational stable function and where R and S satisfy

$$MR + NS = 1 \tag{9.6.24}$$

- The above notions were extended to multivariable systems.
- Multivariable model reference control was discussed and a particular solution exhibited.

9.8 REFERENCES

Pole assignment by the state variable feedback and state observers are treated in most books on modern control theory including:

CHEN, C. T. (1984) *Linear System Theory and Design*. Holt, Rinehart and Winston, New York.

FORTMANN, T. E., and K. L. HITZ (1977) *An Introduction to Linear Control Systems*. Marcel Dekker, New York.

KAILATH, T. (1980) *Linear Systems*. Prentice-Hall, Englewood Cliffs, N.J.

The polynomial approach to design, including model reference control, is treated in:

ÅSTRÖM, K. J., and B. WITTENMARK (1984) *Computer Controlled Systems, Theory and Design*. Prentice-Hall, Englewood Cliffs, N.J.

GOODWIN, G. C., and K. S. SIN (1984) *Adaptive Filtering Prediction and Control*. Prentice-Hall, Englewood Cliffs, N.J.

WOLOVICH, W. A. (1974) *Linear Multivariable Systems*. Springer-Verlag, New York.

Fractional representations for systems are covered in:

DESOER, C. A., R. W. LIU, J. MURRAY, and R. SAEKS (1980) "Feedback system design, fractional representation approach to analysis and synthesis." *IEEE Trans. Auto. Control*, vol. AC-25, no. 3, pp. 399–412.

FRANCIS, B. (1987) *A Course in H_∞ Control Theory*. Lecture Notes in Control and Information Services, Vol. 88. Springer-Verlag, New York.

VIDYASAGAR, M. (1985) *Control System Synthesis: A Factorization Approach*. MIT Press, Cambridge, Mass.

YOULA, D. C., H. A. JABR, and J. J. BONGIORNO (1976) "A modern Wiener Hopf design of optimal controllers." *IEEE Trans. Auto. Control*, vol. 21, pp. 3–13 and 319–338.

9.9 PROBLEMS

1. If the square matrix A is block triangular, that is,

$$A = \begin{bmatrix} A_{11} & A_{12} \\ 0 & A_{22} \end{bmatrix}$$

   where A_{11} and A_{22} are square, show that the eigenvalues of A are the eigenvalues of A_{11} together with the eigenvalues of A_{22}.

2. Show that $1 + LM = \det(I + ML)$ for any vectors L, M of appropriate dimensions. *Hint:* $\det(I + ML) = \prod_{i=1}^{n} \lambda_i(I + ML)$, where $\lambda_i(\cdot)$ denotes the ith eigenvalue. $\lambda_i(I + ML) = 1 + \lambda_i(M_L)$, and all eigenvalues of (ML) are zero except for (possibly) one, which has eigenvector M.

3. The diophantine equation $\bar{A}L + BP = A^*$ can be examined using Euclid's algorithm since the set of polynomials with real coefficients forms a Euclidean ring. Show that a solution to the diophantine equation exists if and only if the greatest common divisor of $(\bar{A}, B)$ divides A^*.

4. We have seen in Section 9.5 that there is a simple method for solving the diophantine equation when the poles are known. Extend this idea to the case when the zeros and the relative degree are known.

5. Consider a plant expressed in right matrix fraction form:

$$Dz = u$$
$$y = Nz$$

   (a) Verify that this is equivalent to a controller form state space model with state

   $$x = \Gamma(\rho)z$$

   where

   $$\Gamma^T(\rho) = (\rho^{n-1}, \rho^{n-2}, \ldots, 1)$$

   (b) Show that the system and state observer can be described in operator form as

   $$E_1(\rho)x = V_1(\rho)u + V_2(\rho)y$$
   $$E_1(\rho)\hat{x} = V_1(\rho)u + V_2(\rho)y$$

   where

   $$E_1(\rho) = \det(\rho I - A + JC)$$

   (c) Verify the following identity:

   $$E_1(\rho)\Gamma(\rho) = V_1(\rho)D(\rho) + V_2(\rho)N(\rho)$$

   (d) Consider state variable feedback

   $$u = -K\hat{x} + v$$

   Show that this can be expressed as

   $$u = -R(\rho)\frac{u}{E_1(\rho)} + P(\rho)\frac{y}{E_1(\rho)} + v$$

   where $R(\rho) = KV_1(\rho)$ and $P(\rho) = KV_2(\rho)$.

   (e) Using parts (c) and (d), show that $R(\rho)$ and $P(\rho)$ satisfy

   $$E_1(\rho)W(\rho) = R(\rho)D(\rho) + P(\rho)N(\rho)$$

   where $W(\rho) = K\Gamma(\rho)$.

6. Suppose we are designing a pole placement controller for a system with transfer function $[B_\alpha(\gamma)B_\beta(\gamma)]/[A_\alpha(\gamma)A_\beta(\gamma)]$, where $B_\alpha(\gamma)$ and $A_\alpha(\gamma)$ are stable and well damped. Show that if we include B_α and A_α in A^* the diophantine equation may be simplified. *Hint:* Let $L = B_\alpha L'$ and $P = A_\alpha P'$.

7. Give a characterization of all the compensators for the plant

$$G(s) = \frac{s+5}{s^2 - 1}$$

that will place the poles of the feedback system in the region $\text{Re}\{s\} \leq -2$.

8. Verify that (9.6.67) to (9.6.70) satisfy (9.6.58), where $\tilde{N}$, $\tilde{M}$, N, and M are defined in (9.6.63), (9.6.64), (9.6.65), and (9.6.66).

*9. In the case of a multivariable, zero canceling control law, show that if the interactor matrix is given by $\xi(\rho) = U(\rho)D(\rho)$, where $U(\rho)$ is unimodular, then the zeros of the plant can be included in the closed loop denominator without using knowledge of $U(\rho)$. *Hint:* Define F and W by $F\tilde{M} + W = D$ and set $C = (F\tilde{N})^{-1}W$. What is the resulting closed loop denominator?

10. Consider the following plant:

$$G(\rho) = \frac{\rho + 3}{\rho(\rho + 2)}$$

Design a model reference controller so that the plant output tracks the output of the following reference model:

$$y^* = \left[\frac{10}{\rho + 10}\right]r$$

*11. Consider the plant given in Equation (9.6.129). Design a model reference controller so that the plant output tracks the output of the following reference model:

$$y^* = \begin{bmatrix} \dfrac{1}{(\rho+5)^2} & 0 \\ 0 & \dfrac{1}{(\rho+6)^3} \end{bmatrix} r$$

*12. Repeat Problem 11 for the plant given in Equation (9.6.131).

13. For which of the following systems, can the closed loop poles be arbitrarily assigned:

   (a) $G(\gamma) = \dfrac{\gamma^2 + \gamma + 1}{\gamma(\gamma^2 + \gamma + 1)}$

   (b) $G(\gamma) = \dfrac{\gamma^2 - 2\gamma + 1}{\gamma(\gamma^2 + 3\gamma + 2)}$

   (c) $\rho x = \begin{bmatrix} 1 & 3 \\ 2 & -4 \end{bmatrix} x + \begin{bmatrix} 1 \\ -2 \end{bmatrix} u$

   $y = \begin{bmatrix} 1 & 0 \end{bmatrix} x$

   (d) $\rho x = \begin{bmatrix} 1 & 0 & 0 \\ 0 & 0 & 0 \\ 0 & 0 & -1 \end{bmatrix} x + \begin{bmatrix} 1 \\ 1 \\ 1 \end{bmatrix} u$

   $y = \begin{bmatrix} 1 & 1 & 1 \end{bmatrix} x$

14. Consider a system described in state space form as
$$\rho x = x + u$$
$$y = 2x$$
   (a) What is the transfer function for this system?
   (b) Design a state variable feedback law that places the closed loop eigenvalue at -1.
   (c) Design a state observer which gives an observer eigenvalue of -2. (*Note:* In this first-order example, an observer is not needed. Design it in any case.)
   (d) Show that the state observer plus state estimate feedback design is equivalent to the control law
$$(\rho + 4)u = -3y$$
   Hint: Manipulate the observer equation and state estimate feedback equations
$$\rho \hat{x} = \hat{x} + u + L(y - 2\hat{x})$$
$$u = -K\hat{x}.$$

15. Verify the statement that, in the single-input, single-output case, model reference control gives a proper control law if and only if the relative degree of the reference model is greater than or equal to the relative degree of the plant.

16. Consider the block diagram representation of all stabilizing controllers given in Figure 9.6.3. Show that, if the plant is given by
$$G = \left(\frac{N}{M} + \Delta G\right)$$
where ΔG denotes modeling errors, then the transfer function from y^* to e satisfies
$$\frac{e}{y^*} = \left[\frac{M(R + \Omega N)}{1 + M(S - \Omega M)\Delta G}\right]$$

17. Consider the same setup as in Problem 16 but assume that the open loop plant is stable. Recall that we can then use $M = 1$, $N = B/A$, $R = 1$, and $S = 0$. Show that, in this case, a sufficient condition for closed loop stability is that the product of the magnitude of Ω and the magnitude of the modeling error ΔG should be less than 1.

18. (a) Consider again the setup of Problem 16. If we require zero steady state error to an input of frequency ω, show that this requires that $1 + \Omega(B/A)$ be zero when evaluated at $\rho = (e^{j\omega\Delta} - 1)/\Delta$.
    (b) Use the results of Problem 17 to show that stability is guaranteed provided the magnitude of the relative error of the plant transfer function, $|\Delta G/G|$, is less than 1 at $\rho = (e^{j\omega\Delta} - 1)/\Delta$.
    (c) Use part (b) to comment on the maximum bandwidth over which small tracking errors can be achieved.

19. Consider an unstable plant $G = B/A$. Let P and L satisfy $AL + BP = A^*$, where A^* is stable. We can then stabilize the plant by preliminary feedback of the form
$$Lu = -Py + v$$
If we now consider an alternative set of inputs and outputs, v and y, then we have a stable plant having transfer function $G' = B/A^*$. Consider the simplified block diagram of all stabilizing controllers given in Figure 9.6.4 and apply this to the stable plant B/A^*. Show that this structure can also be obtained directly from equations (9.6.34), (9.6.35)

provided we make the choices

$$R = \frac{L}{E_2}, \qquad S = \frac{P}{E_2}$$

$$M = \frac{A}{E_1}, \qquad N = \frac{B}{E_1}$$

$$\Omega = \frac{\Omega' L}{E_1 E_2}$$

where Ω' is the value of Ω used in Figure 9.6.4 for the plant $(BL)/A^*$. *Hint:* The total control law is given by

$$Lu = -Py + v$$

$$v = \left[\frac{\Omega}{1 + \Omega(B/A^*)}\right] y$$

Eliminate v and express the control law as in (9.6.34), (9.6.35).

20. Consider the state observer plus feedback law given in (9.4.3), (9.4.4). Show that the transfer function from y to u is $-P/L$, where $L = \det(\rho I - A + BK + JC)$ and where P is given by the following equivalent expressions:

$$P(\rho) = K \operatorname{Adj}(\rho I - A + BK + JC) J$$
$$= K \operatorname{Adj}(\rho I - A + JC) J$$
$$= K \operatorname{Adj}(\rho I - A + BK) J$$
$$= K \operatorname{Adj}(\rho I - A) J$$

Hint: Show that

$$\phi_2^T \operatorname{Adj}(X + \phi_1 \phi_2^T) = \phi_2^T \operatorname{Adj}(X)$$

and

$$\operatorname{Adj}(X + \phi_1 \phi_2^T) \phi_1 = \operatorname{Adj}(X) \phi_1.$$

10

Optimal State Estimation

10.1 INTRODUCTION

In Chapter 9, it was shown that the state of a linear dynamic system can be estimated by use of an observer. We also found that the design of feedback controllers can be conveniently described in terms of a state observer together with feedback of the resultant state estimates. In this chapter, we will study in greater depth the design of observers. In particular, we will show that if certain assumptions are made about the nature of the noise then an optimal observer can be designed.

Let us first recall the simple observer design presented in Chapter 9 for the following model:

$$\rho x = Ax + Bu \qquad (10.1.1)$$

$$y = Cx \qquad (10.1.2)$$

The basic idea of the state observer was to use the model to estimate the current state vector x given the input u and measurement of the output y. The observer state equation was then given by

$$\rho \hat{x} = A\hat{x} + Bu + H(y - C\hat{x}) \qquad (10.1.3)$$

This filter is shown diagrammatically in Figure 10.1.1.

As shown in Chapter 9, the state estimation error $\tilde{x} = \hat{x} - x$ satisfies

$$\rho \tilde{x} = (A - HC)\tilde{x} \qquad (10.1.4)$$

Moreover, we have seen that if (C, A) is observable then the eigenvalues of $(A - HC)$ can be arbitrarily assigned by appropriate choice of H. The design of an

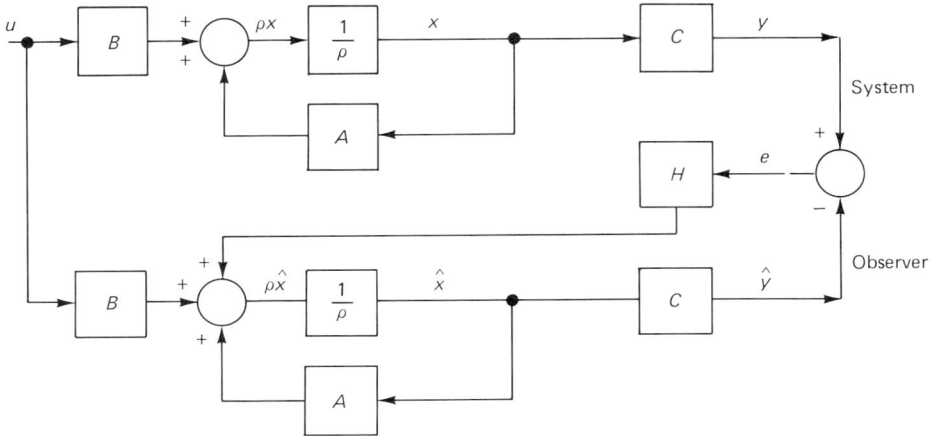

Figure 10.1.1 State observer for linear system.

optimal state estimator then reduces to the optimization of H based on some criteria. For example, if noise is added to the linear model (10.1.1), (10.1.2), then there is an optimal (in terms of mean square error) choice for the observer gain H. Moreover, this optimal gain can be precomputed by solving a special kind of nonlinear equation known as a Riccati equation. Alternatively, a frequency domain design can be carried out in which a worst-case scenario is considered for the noise leading to a min–max or H^∞ type of optimization. We will describe both of these optimization procedures in this chapter. We also investigate the robustness of optimal state estimators to errors in the system model. These robustness considerations are particularly significant in the case where the state estimates are used as part of a feedback control system.

10.2 NONRECURSIVE OPTIMAL ESTIMATION

To begin, let us consider a rather general estimation problem. Say we wish to estimate the value of one variable x at a particular time t based on measurements of another variable y at times $1, 2, \ldots, n$. If $t < n$, we call this *data smoothing*, whereas if $t > n$, we call this *prediction*. We will use the term *filtering* to describe the general case.

We restrict the class of estimates to be a linear function of the available data; that is, we assume

$$\hat{x}(t) = H^T Y \qquad (10.2.1)$$

where $Y^T = \{y(1), \ldots, y(n)\}$.

We then choose H so as to minimize the estimation error, that is, the error between $\hat{x}$ and x. A suitable minimization criterion is the mean square error defined by

$$J = \mathcal{E}\{[x(t) - \hat{x}(t)]^T[x(t) - \hat{x}(t)]\} \quad (10.2.2)$$

where $\mathcal{E}$ denotes expected value.

As might be guessed, the optimal estimate depends on the correlation between $x(t)$ and Y. Thus let us assume that $x(t)$ and Y have zero mean and known joint covariance function given by

$$\mathcal{E}\left\{\begin{pmatrix} x \\ Y \end{pmatrix}(x^T \quad Y^T)\right\} = \begin{bmatrix} \Sigma_{xx} \Sigma_{xY} \\ \Sigma_{xY}^T \Sigma_{YY} \end{bmatrix}$$

We then have the following result:

Lemma 10.2.1

1. The optimal linear estimate [in the sense of minimizing (10.2.2)] is as in (10.2.1), where H depends on the cross-correlation between $x(t)$ and Y in the following way:

$$H^T = [\mathcal{E}\{x(t)Y^T\}][\mathcal{E}YY^T]^{-1} = \Sigma_{xY}\Sigma_{YY}^{-1} \quad (10.2.3)$$

Equivalently, we can write

$$\hat{x}(t) = \tilde{\mathcal{E}}\{x(t)|Y\} \quad (10.2.4)$$

where the special symbol $\tilde{\mathcal{E}}(\cdot|\cdot)$ is called the *wide sense conditional expectation* and is given by

$$[\tilde{\mathcal{E}}\{x(t)|Y\}] = \Sigma_{xY}\Sigma_{YY}^{-1}Y \quad (10.2.5)$$

2. The estimate $\hat{x}(t)$ has the following orthogonality property:

$$\mathcal{E}\{[\hat{x}(t) - x(t)]Y^T\} = 0 \quad (10.2.6)$$

Proof:

1. From (10.2.2), (10.2.1),

$$\frac{\partial J}{\partial H} = 2\mathcal{E}[Y\{x(t)^T - Y^TH\}] \quad (10.2.7)$$

Setting this to zero gives (10.2.3).

2. Immediate from (10.2.7). ▽▽▽

The reason for the terminology wide sense conditional expectation is that, if a and b have a joint Gaussian distribution of zero mean and covariance $\begin{bmatrix} \Sigma_{aa} \Sigma_{ab} \\ \Sigma_{ab}^T \Sigma_{bb} \end{bmatrix}$, then

$$\tilde{\mathcal{E}}(a|b) = \Sigma_{ab}\Sigma_{bb}^{-1}b \quad (10.2.8)$$

which is, in this case, precisely the conditional expectation of a given b. Thus, by analogy, in the non-Gaussian case, we call $\tilde{\mathscr{E}}(a|b)$ a wide sense conditional expectation.

Some elementary properties of wide sense conditional expectations are described in the following result:

Lemma 10.2.2

Let $\hat{c}$ denote the estimate of c given b and let $\tilde{c}$ denote the corresponding estimation error; that is, let

$$\hat{c} \triangleq \tilde{\mathscr{E}}(c|b) = \Sigma_{cb}\Sigma_{bb}^{-1}b \tag{10.2.9}$$

and

$$\tilde{c} \triangleq c - \hat{c} \tag{10.2.10}$$

We then have:

1. The estimation error is orthogonal to the data; that is,

$$\mathscr{E}(b\tilde{c}^T) = 0, \qquad \mathscr{E}(\hat{c}\tilde{c}^T) = 0 \tag{10.2.11}$$

2. Let a denote a third variable. Then the wide sense conditional expectation of a given b and $\tilde{c}$ is linear in the following sense:

$$\tilde{\mathscr{E}}\{a|b, \tilde{c}\} = \tilde{\mathscr{E}}\{a|b\} + \tilde{\mathscr{E}}\{a|\tilde{c}\} \tag{10.2.12}$$

3. The estimate of a given b and c is the same as that obtained when c is replaced by $\tilde{c}$; that is,

$$\tilde{\mathscr{E}}\{a|b, c\} = \tilde{\mathscr{E}}\{a|b\} + \tilde{\mathscr{E}}\{a|\tilde{c}\} = \tilde{\mathscr{E}}\{a|b, \tilde{c}\} \tag{10.2.13}$$

Proof: Let a, b, c have joint covariance given by

$$\mathscr{E}\begin{pmatrix} a \\ b \\ c \end{pmatrix}(a^T b^T c^T) = \begin{pmatrix} \Sigma_{aa} \Sigma_{ab} \Sigma_{ac} \\ \Sigma_{ab}^T \Sigma_{bb} \Sigma_{bc} \\ \Sigma_{ac}^T \Sigma_{bc}^T \Sigma_{cc} \end{pmatrix}$$

The proof of the result is then as follows:

1. Immediate from (10.2.7) and using the fact that $\hat{c}$ is linear in b.
2.

$$\tilde{\mathscr{E}}(a|b, \tilde{c}) = (\Sigma_{ab} \Sigma_{a\tilde{c}}) \begin{pmatrix} \Sigma_{bb} & \Sigma_{b\tilde{c}} \\ \Sigma_{b\tilde{c}}^T & \Sigma_{\tilde{c}\tilde{c}} \end{pmatrix}^{-1} \begin{pmatrix} b \\ \tilde{c} \end{pmatrix}$$

$$= (\Sigma_{ab} \Sigma_{a\tilde{c}}) \begin{pmatrix} \Sigma_{bb}^{-1} & 0 \\ 0 & \Sigma_{\tilde{c}\tilde{c}}^{-1} \end{pmatrix} \begin{pmatrix} b \\ \tilde{c} \end{pmatrix}, \qquad \text{using part 1}$$

$$= \tilde{\mathscr{E}}(a|b) + \tilde{\mathscr{E}}(a|\tilde{c})$$

3. From part 2 we have that

$$\tilde{\mathcal{E}}(a|b,\tilde{c}) = (\Sigma_{ab}\Sigma_{a\tilde{c}})\begin{pmatrix} \Sigma_{bb}^{-1} & 0 \\ 0 & \Sigma_{\tilde{c}\tilde{c}}^{-1} \end{pmatrix}^{-1}\begin{pmatrix} b \\ \tilde{c} \end{pmatrix} \qquad (10.2.14)$$

Now

$$\begin{pmatrix} b \\ \tilde{c} \end{pmatrix} = \begin{pmatrix} I & 0 \\ -\Sigma_{cb}\Sigma_{bb}^{-1} & I \end{pmatrix}\begin{pmatrix} b \\ c \end{pmatrix} \qquad (10.2.15)$$

Also, from (10.2.8) we have

$$\Sigma_{ac} = \Sigma_{a\hat{c}} + \Sigma_{a\tilde{c}} = \Sigma_{ab}\Sigma_{bb}^{-1}\Sigma_{b\hat{c}} + \Sigma_{a\tilde{c}} \qquad (10.2.16)$$

Hence

$$(\Sigma_{ab}\Sigma_{a\tilde{c}}) = (\Sigma_{ab}\Sigma_{ac})\begin{pmatrix} I & -\Sigma_{bb}^{-1}\Sigma_{bc} \\ 0 & I \end{pmatrix} \qquad (10.2.17)$$

Substituting (10.2.15), (10.2.17) into (10.2.14) gives

$$\tilde{\mathcal{E}}(a|b,\tilde{c}) = (\Sigma_{ab}\Sigma_{ac})\begin{pmatrix} I & -\Sigma_{bb}^{-1}\Sigma_{bc} \\ 0 & I \end{pmatrix}\begin{pmatrix} \Sigma_{bb}^{-1} & 0 \\ 0 & \Sigma_{\tilde{c}\tilde{c}}^{-1} \end{pmatrix}$$

$$\times \begin{pmatrix} I & 0 \\ -\Sigma_{cb}\Sigma_{bb}^{-1} & I \end{pmatrix}\begin{pmatrix} b \\ c \end{pmatrix}$$

$$= (\Sigma_{ab}\Sigma_{ac})\left\{\begin{pmatrix} I & 0 \\ \Sigma_{cb}\Sigma_{bb}^{-1} & I \end{pmatrix}\begin{pmatrix} \Sigma_{bb} & 0 \\ 0 & \Sigma_{\tilde{c}\tilde{c}} \end{pmatrix}\right.$$

$$\left.\times \begin{pmatrix} I & \Sigma_{bb}^{-1}\Sigma_{bc} \\ 0 & I \end{pmatrix}\right\}^{-1}\begin{pmatrix} b \\ c \end{pmatrix} \qquad (10.2.18)$$

Thus

$$\tilde{\mathcal{E}}(a|b,\tilde{c}) = (\Sigma_{ab}\Sigma_{ac})\begin{pmatrix} \Sigma_{bb} & \Sigma_{cb} \\ \Sigma_{bc} & \Sigma_{\tilde{c}\tilde{c}} + \Sigma_{cb}\Sigma_{bb}^{-1}\Sigma_{bc} \end{pmatrix}^{-1}\begin{pmatrix} b \\ c \end{pmatrix} \qquad (10.2.19)$$

From (10.2.11) we have

$$\Sigma_{cc} = \Sigma_{\hat{c}\hat{c}} + \Sigma_{\tilde{c}\tilde{c}} \qquad (10.2.20)$$

Using (10.2.9) we obtain

$$\Sigma_{\tilde{c}\tilde{c}} = \Sigma_{cc} - \Sigma_{cb}\Sigma_{bb}^{-1}\Sigma_{bc} \qquad (10.2.21)$$

Substituting (10.2.21) into (10.2.19),

$$\tilde{\mathcal{E}}(a|b,\tilde{c}) = (\Sigma_{ab}\Sigma_{ac})\begin{pmatrix} \Sigma_{bb} & \Sigma_{cb} \\ \Sigma_{bc} & \Sigma_{cc} \end{pmatrix}^{-1}\begin{pmatrix} b \\ c \end{pmatrix}$$

$$\triangleq \tilde{\mathcal{E}}(a|b,c) \qquad (10.2.22)$$

▽▽▽

10.3 SEQUENTIAL PROCESSING

A situation that often arises in practice is that the data $y(\cdot)$ arrive sequentially in time. In such cases, it seems that it may be inefficient to repeat all the computations each time a new piece of data is received. Thus we want to see if we can simplify the computation of $\tilde{\mathscr{E}}\{x(t)|y(1),\ldots,y(n)\}$ if we have already computed $\tilde{\mathscr{E}}\{x(t)|y(1),\ldots,y(n-1)\}$.

Let Y_1^{n-1} denote $\{y(1),\ldots,y(n-1)\}^T$; then we can write

$$Y_1^n = \left\{(Y_1^{n-1})^T, y(n)\right\}^T \tag{10.3.1}$$

We also define

$$e(n) = y(n) - \tilde{\mathscr{E}}\{y(n)|Y_1^{n-1}\} \tag{10.3.2}$$

In Equation (10.3.2), $e(n)$ represents the part of $y(n)$ that cannot be predicted from $y(1),\ldots,y(n-1)$. Thus $e(n)$ represents the new information in $y(n)$ not contained in $y(1),\ldots,y(n-1)$. For this reason, $e(n)$ is given the name *innovation*.

Elementary properties of the innovations sequence are described next.

Lemma 10.3.1

1. $\tilde{\mathscr{E}}\{x(t)|Y_1^n\} = \tilde{\mathscr{E}}\{x(t)|Y_1^{n-1}, e(n)\}$ (10.3.3)
2. $\tilde{\mathscr{E}}\{x(t)|Y_1^{n-1}, e(n)\} = \tilde{\mathscr{E}}\{x(t)|Y_1^{n-1}\} + \tilde{\mathscr{E}}\{x(t)|e(n)\}$ (10.3.4)

Proof: Immediate from Lemma 10.2.2. ▽▽▽

This leads to the following result:

Lemma 10.3.2. Recursive Optimal Estimator

Let $\hat{x}(t|n)$ denote $\tilde{\mathscr{E}}(x(t)|Y_1^n)$. Then

$$\hat{x}(t|n) = \hat{x}(t|n-1) + \mathscr{E}\{x(t)e(n)^T\}\left[\mathscr{E}\{e(n)e(n)^T\}\right]^{-1}e(n) \tag{10.3.5}$$

where

$$e(n) = y(n) - \tilde{\mathscr{E}}(y(n)|Y_1^{n-1}) \tag{10.3.6}$$

Proof: Immediate from Lemmas 10.3.1 and 10.2.2. ▽▽▽

Equation (10.3.5) shows how we can go from $\hat{x}(t|n-1)$ to $\hat{x}(t|n)$ without having to recompute everything again. Note that, in the case of a scalar measurement $y(n)$, Equation (10.3.5) requires only a scalar inverse to be computed. The remaining problem is to obtain workable expressions for $\mathscr{E}\{x(t)e(n)\}$ and $\mathscr{E}\{e(n)e(n)^T\}$ since these quantities completely define the recursive optimal estimator. Unfortunately, this is often a rather difficult (if not impossible) task.

However, one case when it is straightforward to compute these quantities is when the underlying signal model is the following (shift operator) joint Markov model. This model is obtained from (10.1.1) by adding *process noise* ($v_1(t)$) and *measurement noise* ($v_2(t)$), giving

$$x(t+1) = A(t)x(t) + v_1(t) \qquad (10.3.7)$$

$$y(t) = C(t)x(t) + v_2(t) \qquad (10.3.8)$$

where $v_1(t), v_2(t)$ are noise sequences having the following covariances:

$$\mathcal{E}\left\{\begin{bmatrix} v_1(t) \\ v_2(t) \end{bmatrix}\begin{bmatrix} v_1(s)^T & v_2(s)^T \end{bmatrix}\right\} = \left\{\begin{matrix}\begin{bmatrix} Q(t) & 0 \\ 0 & R(t) \end{bmatrix}, & \text{for } t=s \\ 0, & \text{otherwise} \end{matrix}\right\} \qquad (10.3.9)$$

For simplicity, we will restrict attention to the case of one-step-ahead prediction. In this case, Equations (10.3.5) and (10.3.6) can be rewritten as

$$\hat{x}(t+1|t) = \hat{x}(t+1|t-1) + L(t)\Sigma(t)^{-1}e(t) \qquad (10.3.10)$$

where

$$L(t) = \mathcal{E}\{x(t+1)e(t)^T\} \qquad (10.3.11)$$

$$\Sigma(t) = \mathcal{E}\{e(t)e(t)^T\} \qquad (10.3.12)$$

$$e(t) = y(t) - \tilde{\mathcal{E}}(y(t)|Y_1^{t-1}) \qquad (10.3.13)$$

In the case of the joint Markov model, we can explicitly evaluate $L(t), \Sigma(t)$ as follows:

Theorem 10.3.1. The Discrete Kalman Filter

For the joint Markov model of Equations (10.3.7), (10.3.8), we have:

$$L(t) = A(t)P(t)C(t)^T \qquad (10.3.14)$$

$$\Sigma(t) = C(t)P(t)C(t)^T + R(t) \qquad (10.3.15)$$

where

$$P(t) \triangleq \mathcal{E}\{[\hat{x}(t|t-1) - x(t)][\hat{x}(t|t-1) - x(t)]^T\} \qquad (10.3.16)$$

and satisfies the following Riccati difference equation:

$$P(t+1) = A(t)P(t)A(t)^T + Q(t) - L(t)\Sigma(t)^{-1}L(t)^T \qquad (10.3.17)$$

The initial condition for the Riccati difference equation is

$$P(0) = \mathcal{E}\{[\hat{x}(0) - x(0)][\hat{x}(0) - x(0)]^T\} \qquad (10.3.18)$$

Finally, the optimal state estimator can be written as

$$\hat{x}(t+1|t) = A(t)\hat{x}(t|t-1) + L(t)P(t)^{-1}e(t)$$

$$e(t) = y(t) - C(t)\hat{x}(t|t-1) \qquad (10.3.19)$$

Proof: We first note that

$$P(t) = \mathcal{E}\{[\hat{x}(t|t-1) - x(t)][\hat{x}(t|t-1) - x(t)]^T\}$$
$$= \Pi(t) - \Gamma(t) \tag{10.3.20}$$

where

$$\Pi(t) \triangleq \mathcal{E}\{x(t)x(t)^T\} \tag{10.3.21}$$

$$\Gamma(t) \triangleq \mathcal{E}\{\hat{x}(t|t-1)\hat{x}(t|t-1)^T\} \tag{10.3.22}$$

Also, from (10.3.7), we have

$$\hat{x}(t+1|t-1) = \tilde{\mathcal{E}}\{x(t+1)|Y_1^{t-1}\}$$
$$= \tilde{\mathcal{E}}\{A(t)x(t) + v_1(t)|Y_1^{t-1}\}$$
$$= A(t)\tilde{\mathcal{E}}\{x(t)|Y_1^{t-1}\}$$
$$= A(t)\hat{x}(t|t-1) \tag{10.3.23}$$

Substituting (10.3.23) into (10.3.10) gives (10.3.19). From (10.3.19), we have

$$\Gamma(t+1) = A(t)\Gamma(t)A(t)^T + L(t)\Sigma(t)^{-1}L(t)^T \tag{10.3.24}$$

Also, from (10.3.7) we have

$$\Pi(t+1) = A(t)\Pi(t)A(t)^T + Q(t) \tag{10.3.25}$$

Subtracting (10.3.25) from (10.3.24) and using (10.3.20) gives (10.3.17).
Finally,

$$L(t) = \mathcal{E}\{x(t+1)e(t)^T\}$$
$$= \mathcal{E}\{[A(t)x(t) + v_1(t)][C(t)(x(t) - \hat{x}(t|t-1)) + v_2(t)]^T\}$$
$$= A(t)P(t)C(t)^T \tag{10.3.26}$$

since $x(t) - \hat{x}(t|t-1)$ is orthogonal to $\hat{x}(t|t-1)$.
Also,

$$\Sigma(t) = \mathcal{E}\{e(t)e(t)^T\}$$
$$= \mathcal{E}\{[C(t)(x(t) - \hat{x}(t|t-1)) + v_2(t)]$$
$$\times [C(t)(x(t) - \hat{x}(t|t-1)) + v_2(t)]^T\}$$
$$= C(t)P(t)C(t)^T + R(t) \tag{10.3.27}$$

▽▽▽

We see that the optimal estimator as given in (10.3.19) has exactly the same form as the state observer illustrated in Figure 10.1.1, where the gain matrix H is time varying and has the optimal value $L(t)\Sigma(t)^{-1}$.

The preceding ideas are illustrated in the following simple examples.

Example 10.3.1

Consider the following system with $\Delta = 1$:
$$x(t+1) = x(t) + v_1(t) \tag{10.3.28}$$
$$y(t) = x(t) + v_2(t) \tag{10.3.29}$$

where

$$\mathcal{E}\left\{\begin{bmatrix} v_1(t) \\ v_2(t) \end{bmatrix}[v_1(s) \quad v_2(s)]\right\} = \left\{\begin{matrix} \begin{bmatrix} 1 & 0 \\ 0 & 10 \end{bmatrix}, & \text{for } t = s \\ 0, & \text{otherwise} \end{matrix}\right\} \tag{10.3.30}$$

Also, we assume
$$\hat{x}(0) = 0 \tag{10.3.31}$$

and
$$\mathcal{E}\{(\hat{x}(0) - x(0))^2\} = 100 \tag{10.3.32}$$

The optimal state estimate then satisfies
$$\hat{x}(t+1) = \hat{x}(t) + H(t)[y(t) - \hat{x}(t)] \tag{10.3.33}$$

where we have used the shorthand notation $\hat{x}(t)$ to denote $\hat{x}(t|t-1)$.

In equation (10.3.33), the optimal estimator gain is given by

$$H(t) = L(t)\Sigma(t)^{-1}$$
$$= \frac{P(t)}{P(t) + 10} \tag{10.3.34}$$

and $P(t)$ satisfies the following Riccati difference equation:

$$P(t+1) = P(t) + 1 - \frac{P(t)^2}{P(t) + 10}, \quad P(0) = 100 \tag{10.3.35}$$

$$= \frac{11P(t) + 10}{P(t) + 10} \tag{10.3.36}$$

The values of P and H as a function of t are given in Table 10.3.1.

Example 10.3.2

We have seen in Section 9.6 that many fractional models for linear systems can be expressed in regression form as

$$y(t) = \phi(t)^T \theta_0 \tag{10.3.37}$$

Let us say that the measurements of the scalar quantity $y(t)$ are corrupted by noise; then we might rewrite (10.3.37) as

$$y(t) = \phi(t)^T \theta_0 + v_2(t) \tag{10.3.38}$$

TABLE 10.3.1. VALUES OF P AND H AS A FUNCTION OF t.

t	P	H
0	100	0.909
1	10.09	0.502
2	6.02	0.375
3	4.76	0.322
4	4.22	0.297
5	3.97	0.284
6	3.84	0.277
7	3.78	0.274
8	3.74	0.272
9	3.72	0.271
10	3.71	0.271
⋮	⋮	⋮
∞	3.70	0.270

We also express the fact that θ_0 is time invariant by using the following shift operator state equation:

$$\theta(t+1) = \theta(t), \qquad \theta(0) = \theta_0 \tag{10.3.39}$$

Equations (10.3.39), (10.3.38) are precisely in the form of the joint Markov model (10.3.7), (10.3.8), where we set

$$x(t) = \theta(t) \tag{10.3.40}$$
$$A(t) = I \tag{10.3.41}$$
$$C(t) = \phi(t)^T \tag{10.3.42}$$
$$v_1(t) = 0 \tag{10.3.43}$$
$$E\{v_2(t)^2\} = R \tag{10.3.44}$$

In this case, the Kalman filter equations reduce to

$$\hat{\theta}(t+1) = \hat{\theta}(t) + \frac{P(t)\phi(t)e(t)}{R + \phi(t)^T P(t)\phi(t)} \tag{10.3.45}$$

where

$$e(t) = y(t) - \phi(t)^T \hat{\theta}(t) \tag{10.3.46}$$

$$P(t+1) = P(t) - \frac{P(t)\phi(t)\phi(t)^T P(t)}{R + \phi(t)^T P(t)\phi(t)} \tag{10.3.47}$$

Equations (10.3.45) to (10.3.47) give an optimal way of estimating the unknown quantity θ_0 from measurements of $\{y(t)\}$. Equations (10.3.45) to (10.3.47) will form the basis of a more in-depth discussion of the parameter estimation problem in Chapter 12.

In the derivation of the optimal filter just given, we have not assumed any particular form for the probability distribution function for the noise. Under these conditions, we have shown that the Kalman filter is the optimal filter among the class of all *linear* estimators. If the noise distribution is Gaussian, then more can be said. In this case, the Kalman filter gives the conditional mean that is the optimal estimate among the class of *all* estimators (not necessarily linear).

10.4 STATE ESTIMATION FOR LINEAR TIME INVARIANT SYSTEMS

In Section 10.3, the matrices defining the joint Markov model (A, C, R, and Q) were allowed to depend on time. However, in practice, it is often the case that these matrices are time invariant. The shift operator joint Markov model then simplifies to

$$x(t+1) = A_q x(t) + v_1(t) \tag{10.4.1}$$

$$y(t) = C_q x(t) + v_2(t) \tag{10.4.2}$$

where A_q and C_q are constant matrices of appropriate dimensions and $\{v_1(t)\}, \{v_2(t)\}$ are uncorrelated zero mean "white" sequences having covariance matrices Q_q and R_q, respectively, with $Q_q \geq 0$ and $R_q > 0$. [Note we have introduced the notation A_q and the like to signify that the model (10.4.1), (10.4.2) is expressed in terms of the shift operator.] We also assume that the initial state $x(0)$ is a random variable uncorrelated with $\{v_1(t)\}$ and $\{v_2(t)\}$, having mean $\bar{x}_0$ and covariance matrix Σ_0. Thus

$$\mathcal{E}\left\{\begin{bmatrix} v_1(t) \\ v_2(t) \end{bmatrix} \begin{bmatrix} v_1(s)^T & v_2(s)^T \end{bmatrix}\right\} = \left\{\begin{array}{ll} \begin{bmatrix} Q_q & 0 \\ 0 & R_q \end{bmatrix}, & \text{for all } t = s \\ 0, & \text{otherwise} \end{array}\right\} \tag{10.4.3}$$

For this time invariant joint Markov model, the Kalman filter simplifies to

$$\hat{x}(t+1) = A_q \hat{x}(t) + H_q(t)[y(t) - C_q \hat{x}(t)] \tag{10.4.4}$$

where $H_q(t)$ is given by

$$H_q(t) = A_q P(t) C_q^T [C_q P(t) C_q^T + R_q]^{-1} \tag{10.4.5}$$

and $P(t)$ satisfies the following Riccati difference equation (RDE) [compare with (10.3.17)]:

$$P(t+1) = A_q P(t) A_q^T - A_q P(t) C_q^T [C_q P(t) C_q^T + R_q]^{-1} C_q P(t) A_q^T + Q_q$$
$$P(0) = P_0 \tag{10.4.6}$$

Using Equation (10.4.5), Equation (10.4.6) can be reexpressed in the following form:

$$P(t+1) = A_q P(t) A_q^T - H_q [C_q P(t) C_q^T + R_q] H_q^T + Q_q \tag{10.4.7}$$

Equation (10.4.4) can also be expressed in the following form:

$$\hat{x}(t+1) = \bar{A}_q(t) \hat{x}(t) + H_q(t) y(t) \tag{10.4.8}$$

$$\hat{x}(0) = \tilde{x}_0 \tag{10.4.9}$$

where
$$\overline{A}_q(t) \triangleq A_q - H_q(t)C_q \tag{10.4.10}$$
$$H_q(t) \triangleq A_q P(t) C_q^T \left(C_q P(t) C_q^T + R_q \right)^{-1} \tag{10.4.11}$$
and $P(t)$ satisfies the matrix Riccati difference equation (10.4.6) or (10.4.7).

Remark 10.4.1

A straightforward extension of the model (10.4.1) to (10.4.3) can be made to include cases when v_1 and v_2 are mutually correlated (but still white); that is, (10.4.3) becomes

$$\mathscr{E}\left\{\begin{bmatrix} v_1(t) \\ v_2(t) \end{bmatrix} \begin{bmatrix} v_1(s)^T & v_2(s)^T \end{bmatrix}\right\} = \left\{\begin{matrix} \begin{bmatrix} Q_q & S_q \\ S_q^T & R_q \end{bmatrix}, & \text{for } t = s \\ 0, & \text{otherwise} \end{matrix}\right\} \tag{10.4.12}$$

Equation (10.4.1) can be written as
$$\begin{aligned} x(t+1) &= A_q x(t) + v_1(t) - S_q R_q^{-1}(y(t) - y(t)) \\ &= A_q x(t) + v_1(t) - S_q R_q^{-1}(C_q x(t) + v_2(t) - y(t)) \\ &= A_q' x(t) + S_q R_q^{-1} y(t) + v_1'(t) \end{aligned} \tag{10.4.13}$$

where
$$A_q' = A_q - S_q R_q^{-1} C_q \tag{10.4.14}$$
$$v_1'(t) = v_1(t) - S_q R_q^{-1} v_2(t) \tag{10.4.15}$$

Then
$$\mathscr{E}\left\{\begin{bmatrix} v_1'(t) \\ v_2(t) \end{bmatrix} \begin{bmatrix} v_1'(s)^T & v_2(s)^T \end{bmatrix}\right\} = \begin{bmatrix} Q_q' & 0 \\ 0 & R_q \end{bmatrix} \tag{10.4.16}$$

$$Q_q' = Q_q - S_q R_q^{-1} S_q^T \tag{10.4.17}$$

Since $\{y(t)\}$ is assumed to be known, then the optimal filter for the system (10.4.13) to (10.4.17) is as in (10.4.8) to (10.4.11) with A_q, Q_q replaced by A_q', Q_q', respectively. ∇∇∇

10.5 DELTA FORM OF THE OPTIMAL STATE ESTIMATOR

The form of the optimal state estimator just presented is the one usually presented using the standard shift operator notation. However, this form is *not* consistent with the corresponding continuous time results. As in the previous chapters, to obtain a unified treatment of continuous and discrete systems, it is helpful to change the discrete time operator from shift to delta.

Using the delta operator, the discrete system description (10.4.1), (10.4.2), (10.4.12) becomes

$$\delta x(k) = A_\delta x(k) + v_\delta(k) \tag{10.5.1}$$

$$y(k) = C_\delta x(k) + w_\delta(k) \tag{10.5.2}$$

where

$$A_\delta = \frac{A_q - I}{\Delta}, \qquad C_\delta = C_q \tag{10.5.3}$$

and

$$v_\delta = \frac{1}{\Delta} v_q, \qquad w_\delta = w_q$$

Hence

$$\text{Cov}\{v_\delta\} = \frac{1}{\Delta^2} Q_q \triangleq Q_\delta \tag{10.5.4}$$

$$\text{Cov}\{w_\delta\} = R_q \triangleq R_\delta \tag{10.5.5}$$

$$\text{Cov}\{v_\delta, w_\delta\} = \frac{1}{\Delta} S_q \triangleq S_\delta'; \qquad S_\delta \triangleq \Delta S_\delta' \tag{10.5.6}$$

We will see later that for fast sampling Q_q, R_q are typically of order Δ and $1/\Delta$, respectively. Thus Q_δ, R_δ are typically both of order $1/\Delta$. Hence, to achieve sampling independence, it is desirable to define *spectral densities* Ω_δ, Γ_δ by dividing Q_δ, R_δ by the sampling frequency $f_s = 1/\Delta$:

$$\Omega_\delta \triangleq \frac{Q_\delta}{f_s} = \Delta Q_\delta = \frac{1}{\Delta} Q_q \tag{10.5.7}$$

$$\Gamma_\delta \triangleq \frac{R_\delta}{f_s} = \Delta R_\delta = \Delta R_q \tag{10.5.8}$$

We then have the following result:

Lemma 10.5.1

When expressed in terms of the delta operator, the optimal filter for the system (10.5.1), (10.5.2) has the form

$$\delta \hat{x}(k) = A_\delta \hat{x}(k) + H_\delta(k)[y(k) - C_\delta \hat{x}(k)] \tag{10.5.9}$$

where

$$H_\delta(k) = \frac{1}{\Delta} H_q(k) \tag{10.5.10}$$

$$= [(\Delta A_\delta + I) P_\delta(k) C_\delta^T + S_\delta][\Delta C_\delta P_\delta(k) C_\delta^T + \Gamma_\delta]^{-1} \tag{10.5.11}$$

The delta form of (10.4.6) is

$$\delta P_\delta(k) = \Omega_\delta + P_\delta(k)A_\delta^T + A_\delta P_\delta(k)$$
$$+ \Delta A_\delta P_\delta(k)A_\delta^T - H_\delta(k)\left[\Gamma_\delta + \Delta C_\delta P_\delta(k)C_\delta^T\right]H_\delta(k)^T \quad (10.5.12)$$

Proof: Immediate on substituting (10.5.1) to (10.5.8) into the results of Section 10.4.
∇∇∇

10.6 UNIFICATION OF CONTINUOUS AND DISCRETE RESULTS

In this section we explore the connection between the continuous time Riccati equation and the corresponding discrete results developed previously. This connection is made particularly transparent by using the delta operator form of the discrete results.

Consider a continuous time, lumped, linear, time invariant stochastic system given by

$$dx = Ax\,dt + dv \quad (10.6.1)$$
$$dz = Cx\,dt + dw, \quad \text{with } x(0) = x_0 \quad (10.6.2)$$

where $x \in \mathbb{R}^n$, $z \in \mathbb{R}^l$, $A \in \mathbb{R}^{n \times n}$, $C \in \mathbb{R}^{l \times n}$, $\mathscr{E}[x_0] = \bar{x}_0$, $\text{Cov}[x_0] = P_0$; $v(t) \in \mathbb{R}^n$ is a vector Wiener process with incremental covariance $\Omega\,dt$; $w(t) \in \mathbb{R}^l$ is a vector Wiener process with incremental covariance $\Gamma\,dt$; $x(0)$, $v(t)$, and $w(t)$ are independent; Ω and P_0 are symmetric positive semidefinite matrices; and Γ is a symmetric positive definite matrix.

Remark 10.6.1

For those readers not familiar with Wiener processes, Equations (10.6.1), (10.6.2) can be thought of in the following differential form:

$$\dot{x} = Ax + \dot{v} \quad (10.6.3)$$
$$y' = \dot{z} = cx + \dot{w} \quad (10.6.4)$$

where $\dot{v}$ and $\dot{w}$ denote continuous time white noise processes. These processes are assumed to have spectral density Ω and Γ, respectively. The corresponding covariance functions are easily obtained from the inverse transform of the spectral density as

$$\mathscr{E}\left\{\dot{v}(t)\dot{v}(s)^T\right\} = \Omega i(t-s) \quad (10.6.5)$$
$$\mathscr{E}\left\{\dot{w}(t)\dot{w}(s)^T\right\} = \Gamma i(t-s) \quad (10.6.6)$$

where $i(0)$ denotes the continuous time Dirac delta function.

The variances given in (10.6.5), (10.6.6) are infinite, and thus the processes $\dot{v}$ and $\dot{w}$ do not, strictly speaking, exist. This is why the incremental form given in (10.6.1), (10.6.2) is generally preferred.

The incremental covariance of $\dot{v}$ can be evaluated as follows:

$$\mathcal{E}\{dv\, dv^T\} \triangleq \mathcal{E}\{v(t) - v(s)][v(t) - v(s)]^T\}$$

$$= \mathcal{E}\int_s^t \int_s^t \dot{v}(\tau)\dot{v}(\sigma)\, d\tau\, d\sigma$$

$$= \int_s^t \int_s^t \Omega i(t-s)\, d\tau\, d\sigma$$

$$= \int_s^t \Omega\, dt$$

$$= \Omega(t-s)$$

$$\triangleq \Omega\, dt \tag{10.6.7}$$

Similarly,

$$\mathcal{E}\{dw\, dw^T\} = \Gamma\, dt \tag{10.6.8}$$

We see from the preceding discussion that Ω and Γ are actually spectral densities and that use of the incremental equations avoids explicitly mentioning $\dot{v}$ and $\dot{w}$.

We will consider (10.6.1), (10.6.2) as the underlying system. We will next rederive the discrete Kalman filter by sampling (10.6.1), (10.6.2).

A key aspect of this derivation is the specification of the sampling process. For example, it makes no sense to directly sample the output dz/dt in (10.6.2) since this would lead to a discrete system having output noise of infinite variance! This problem is resolved by replacing the impractical ideal sampler by a practical form of sampling in which the signal is passed through an antialiasing low-pass filter prior to sampling (see Chapter 3). We will denote the impulse response of this low-pass filter by $h(\tau)$. For simplicity, we assume that $h(\tau)$ is zero for $\tau \geq \Delta$. More general cases are explored in the problems.

We then have the following result:

Lemma 10.6.1

If the system (10.6.1), (10.6.2) is sampled at interval Δ using a presampling filter of impulse response $h(\tau)$, then the sampled output response is described by the following stationary discrete time linear system (expressed in shift operator form):

$$\bar{x}(k+1) = A_q \bar{x}(k) + v_q(k) \tag{10.6.9}$$

$$y(k) = C_q \bar{x}(k) + w_q(k), \quad \text{with } x(0) = x_0 \tag{10.6.10}$$

where $\bar{x}(k) \triangleq x(k\Delta)$.

$$A_q = \phi(k\Delta + \Delta, \Delta) = e^{A\Delta} \tag{10.6.11}$$

$$C_q = \int_{-\infty}^{\infty} h(k\Delta + \Delta - \tau)C\phi(\tau, k\Delta)\, d\tau \tag{10.6.12}$$

$$\text{Cov}\{v_q\} = \int_0^{\Delta} \phi(\Delta, \tau)\Omega\phi(\Delta, \tau)^T\, d\tau \triangleq Q_q \tag{10.6.13}$$

$$\text{Cov}\{w_q\} = R_1(\Delta) + R_2(\Delta) \triangleq R_q \tag{10.6.14}$$

with

$$R_1(t) \triangleq \int_{-\infty}^{\infty}\int_{-\infty}^{\infty} h(t-\tau)R_3(\tau, T)h(t-T)^T\, d\tau\, dT \tag{10.6.15}$$

$$R_2(t) \triangleq \int_{-\infty}^{\infty} h(t-\tau)\Gamma h(t-\tau)^T\, d\tau \tag{10.6.16}$$

and

$$R_3(t_1, t_2) \triangleq \int_{k\Delta}^{\min(t_1, t_2)} C\phi(t_1, \tau)\Omega\phi(t_2, \tau)^T C^T\, d\tau \tag{10.6.17}$$

Also,

$$\text{Cov}(v_q, w_q) \triangleq E\{v_q w_q^T\}$$
$$= \int_0^{\Delta}\int_{-\infty}^{\infty} \phi(\Delta, \tau)\Omega\phi(\alpha, \tau)^T C^T h(\Delta - \alpha)^T\, d\alpha\, d\tau \triangleq S_q$$
$$\tag{10.6.18}$$

Proof: The output of the presampling filter for $t \geq k\Delta$ is

$$y_f(t) = \int_{-\infty}^{\infty} h(t-\tau)\, dz(\tau)$$
$$= \int_{-\infty}^{\infty} h(t-\tau)Cx(\tau)\, d\tau + \int_{-\infty}^{\infty} h(t-\tau)\, dw(\tau)$$
$$= \left[\int_{-\infty}^{\infty} h(t-\tau)C\phi(\tau, k\Delta)\, d\tau\right]x(k\Delta) + \omega_1(t) + \omega_2(t) \tag{10.6.19}$$

where

$$\omega_1(t) \triangleq \int_{-\infty}^{\infty} h(t-\tau)\omega_3(\tau)\, d\tau \tag{10.6.20}$$

$$\omega_2(t) \triangleq \int_{-\infty}^{\infty} h(t-\tau)\, dw(\tau) \tag{10.6.21}$$

$$\omega_3(t) \triangleq C\int_{k\Delta}^{t} \phi(t, \tau)\, dv(\tau) \tag{10.6.22}$$

$$\phi(t, \tau) \triangleq e^{A(t-\tau)} \tag{10.6.23}$$

Using the properties of $dv(t)$ and $dw(t)$, we have

$$\mathcal{E}\{\omega_1(t)\} = \mathcal{E}\{\omega_2(t)\} = \mathcal{E}\{\omega_3(t)\} = 0 \tag{10.6.24}$$

$$\mathcal{E}\{\omega_3(t_1)\omega_3(t_2)^T\} = R_3(t_1, t_2)\} \tag{10.6.25}$$

$$\text{Covar}\{\omega_2(t)\} = R_2(t) \tag{10.6.26}$$

$$\text{Covar}\{\omega_1(t)\} = R_1(t) \tag{10.6.27}$$

Finally, the appropriate discrete time system is as in (10.6.9), (10.6.10), where $y(k) \triangleq y_f[(k+1)\Delta]$ and ω_q are serially uncorrelated due to the assumption $h(\tau) = 0$ for $\tau \geq \Delta$.

The optimal discrete time filter for the system (10.6.9), (10.6.10) is then given by the appropriate results from Section 10.4:

$$\hat{x}(k+1) = A_q \hat{x}(k) + H_q(k)[y(k) - C_q \hat{x}(k)] \tag{10.6.28}$$

$$H_q(k) = [A_q P_q(k) C_q^T + S_q](C_q P_q(k) C_q^T + R_q)^{-1},$$

$$\text{with } \hat{x}(0) = \bar{x}_0 \tag{10.6.29}$$

where $P(k)$ satisfies the following discrete Riccati difference equation (DRDE):

$$P_q(k+1) = Q_q + A_q P_q(k) A_q^T$$

$$- [A_q P_q(k) C_q^T + S_q][R_q + C_q P_q(k) C_q^T]^{-1}[C_q P_q(k) A_q^T + S_q^T] \tag{10.6.30}$$

with

$$P_q(0) = P_0 \tag{10.6.31}$$

A simple antialiasing filter that captures the essence of the required presample filtering action is an average filter obtained from a reset and integrate circuit; specifically,

$$y(k\Delta + \Delta) = \frac{1}{\Delta} \int_{k\Delta}^{(k+1)\Delta} dz(t) \tag{10.6.32}$$

The corresponding impulse response $h(t)$ for this filter is given by

$$h(t) = \frac{1}{\Delta}, \quad 0 \leq t \leq \Delta$$

$$= 0, \quad \text{otherwise} \tag{10.6.33}$$

From equation (10.6.13), it can be seen that for small Δ

$$Q_q \simeq \Omega \Delta \tag{10.6.34}$$

Also, for the preceding typical presampling filter, it is readily seen that for small Δ

$$R_q \simeq \frac{1}{\Delta}\Gamma \tag{10.6.35}$$

$$C_q \simeq C \tag{10.6.36}$$

$$S_q \simeq 0 \tag{10.6.37}$$

Using these results, we are able to see what happens to the discrete Kalman filter in the limit as $\Delta \to 0$. A major conclusion is that the shift form of the filter, as given in Equations (10.6.34) to (10.6.37), is not appropriate when fast sampling is used. For example, as the sampling rate increases, we obtain the following misleading results, which hold *irrespective* of the underlying continuous time system:

$$\lim_{\Delta \to 0} Q_q = 0 \tag{10.6.38}$$

$$\lim_{\Delta \to 0} R_q = \infty \tag{10.6.39}$$

$$\lim_{\Delta \to 0} A_q = I \tag{10.6.40}$$

$$\lim_{\Delta \to 0} H_q = 0 \tag{10.6.41}$$

As in earlier sections of the book, these anomalous results are a direct consequence of the use of the shift operator to describe the discrete time model. The situation is remedied if we reformulate the discrete model using the delta operator as in Lemma 10.5.1. In this case the optimal filter becomes

$$\delta \hat{x}(k) = A_\delta \hat{x}(k) + H_\delta(k)[y(k) - C_\delta \hat{x}(k)] \tag{10.6.42}$$

$$H_\delta(k) = \left[(\Delta A_\delta + I)P_\delta(k)C_\delta^T + S_\delta\right]\left[\Delta C_\delta P_\delta(k)C_\delta^T + \Gamma_\delta\right]^{-1} \tag{10.6.43}$$

where P_δ satisfies (10.5.12) and where

$$A_\delta = \frac{A_q - I}{\Delta}, \quad \text{with } A_q \text{ as in (10.6.11)} \tag{10.6.44}$$

$$C_\delta = C_q, \quad \text{with } C_q \text{ as in (10.6.12)} \tag{10.6.45}$$

$$\Omega_\delta = \frac{1}{\Delta} Q_q, \quad \text{with } Q_q \text{ as in (10.6.13), (10.6.34)} \tag{10.6.46}$$

$$\Gamma_\delta = \Delta R_q, \quad \text{with } R_q \text{ as in (10.6.14), (10.6.35)} \tag{10.6.47}$$

$$S_\delta = S_q, \quad \text{with } S_q \text{ as in (10.6.18), (10.6.37)} \tag{10.6.48}$$

$$H_\delta = \frac{1}{\Delta} H_q, \quad \text{with } H_q \text{ as in (10.6.29)} \tag{10.6.49}$$

We then have the following result connecting continuous and discrete cases:

Lemma 10.6.2

(i) The quantities appearing in the delta form of the discrete model converge to the corresponding continuous quantities; that is,

(a) $\lim_{\Delta \to 0} A_\delta = A$ (10.6.50)

(b) $\lim_{\Delta \to 0} C_\delta = C$ (10.6.51)

(c) $\lim_{\Delta \to 0} \Omega_\delta = \Omega$ (10.6.52)

(d) $\lim_{\Delta \to 0} \Gamma_\delta = \Gamma$ (10.6.53)

(e) $\lim_{\Delta \to 0} S_\delta = 0$ (10.6.54)

(ii) The delta form of the discrete filter converges to the following continuous time filter:

$$d\hat{x}(t) = A\hat{x}(t)\,dt + H(t)[dz(t) - C\hat{x}(t)\,dt], \qquad \hat{x}(0) = \tilde{x}(0) \tag{10.6.55}$$

where

$$H = \lim_{\Delta \to 0} H_\delta \tag{10.6.56}$$

and H satisfies

$$H(t) = P(t)C^T\Gamma^{-1} \tag{10.6.57}$$

where $P(t)$ satisfies the following continuous Riccati differential equation (CRDE):

$$\frac{d}{dt}P(t) = P(t)A^T + AP(t) - P(t)G^T\Gamma^{-1}CP(t) + \Omega \tag{10.6.58}$$

with $P(0) = P_0$.

Proof: Immediate from (10.6.42) to (10.6.49) on letting $\Delta \to 0$. ▽▽▽

In summary, the Kalman filter for both continuous and discrete time systems can be described in our usual unified notation as

$$\rho\hat{x} = A\hat{x} + H[y - C\hat{x}] \tag{10.6.59}$$

where

$$H = [(\Delta A + I)PC^T + S][\Delta CPC^T + \Gamma]^{-1} \tag{10.6.60}$$

where P satisfies

$$\rho P = \Omega + PA^T + AP + \Delta APA^T - H[\Gamma + \Delta CPC^T]H^T \tag{10.6.61}$$

We conclude that the delta form of the optimal filter has the following key properties:

(i) The corresponding continuous time result is simply obtained by setting $\rho = d/dt$, $\Delta = 0$. [Note that for the model (10.6.1), (10.6.2), $S \to 0$ as $\Delta \to 0$.]

(ii) The delta form clearly shows that the design and performance of the optimal filter are best expressed as functions of the noise *spectral densities* rather than the noise variances, as seems to be the case in the alternative shift formulation. The dependence of the continuous formulation on spectral density has been discussed in Remark 10.6.1. Also, in Equations (10.5.7), (10.5.8) we see that Ω_δ and Γ_δ are noise variances divided by the sampling frequency. Hence they represent discrete time spectral densities. These quantities are roughly invariant as the sampling rate changes, whereas the noise variances increase proportionally to the sampling frequency.

The conclusion that filter design and performance depend on the noise spectral density is intuitively appealing since it becomes clear that estimation accuracy is only a function of the ratio of signal power to noise power at each frequency of interest.

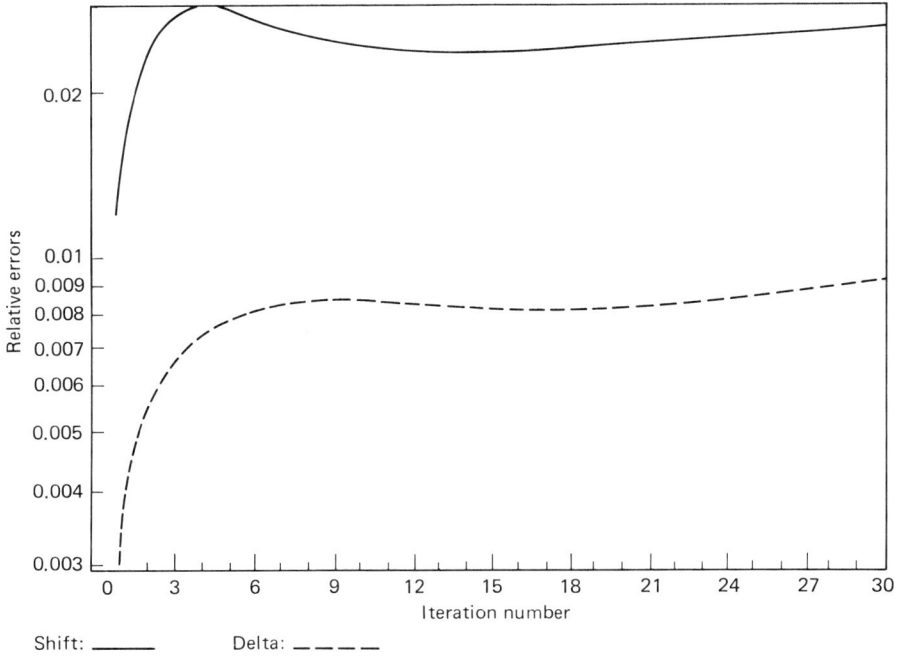

Figure 10.6.1 Comparison of errors for dynamic Riccati equation using delta (δ) and shift (q) operators.

We can draw a further inference from this discussion. In particular, say the noise is actually colored with a finite bandwidth, say f_B. Provided this bandwidth, exceeds the signal bandwidth, then the design and performance of the optimal filter are virtually unaltered provided they are expressed in terms of noise spectral densities. Note that, if we are given the colored noise variance as v with bandwidth f_B, then the appropriate spectral density is $v/2f_B$. For white noise this still gives the correct result, as the bandwidth is then $f_s/2$.

(iii) A further bonus arising from the use of the delta formulation is that the numerical properties of this form generally turn out to be much superior to the shift form. This is verified in Example 10.6.1.

Example 10.6.1

Consider a system having transfer function $H(s) = 1/[s(s + 1)]$ with continuous state space model in observer form. The Riccati equation was solved using five mantissa bits to represent A_q (shift form) and A_δ (delta form). The following constants were also used: $R = 1$, $Q = 1$, $\Delta = 0.075$, $P_0 = I$.

Figure 10.6.1 shows the propagation of the relative error defined as

$$\eta_l = \frac{\|[P(l)]_{FP} - P(l)\|_F}{\|P(l)\|_F} \tag{15.6.62}$$

where $[P(l)]_{FP}$, $P(l)$ denote the floating point and infinite precision solution of the Riccati equation, and $\|\cdot\|_F$ denotes the Fröbenius norm. It can be seen from the figure that the delta formulation leads to a significant improvement in the relative error in the computation of the DRDE as predicted.

A complete numerical analysis is beyond the scope of this text. However, a brief treatment of the reasons for the numerical advantages of delta from over shift are given in Appendix C.

10.7 PROPERTIES OF THE OPTIMAL FILTER

The previous sections have derived the optimal mean square filter for both continuous and discrete time systems. These optimal filters have a number of interesting properties, which hold in the limit as $t \to \infty$. Let us begin by briefly reviewing Example 10.3.1.

Example 10.7.1

In Example 10.3.1 we found, by numerical evaluation, that $P(t)$ converged to the limit 3.70 as $t \to \infty$. This answer can also be obtained by examining all possible fixed points of the Riccati difference equation (10.3.36). Setting $P(t + 1) = P(t) = P$

in this equation leads to
$$P = \frac{11P + 10}{P + 10} \tag{10.7.1}$$
or
$$P^2 - P - 10 = 0 \tag{10.7.2}$$
This quadratic equation has two possible solutions:
$$P = 3.70 \quad \text{and} \quad -2.70 \tag{10.7.3}$$

We will call the general form of (10.7.2) the algebraic Riccati equation (ARE). We will be interested in such questions as:

- How many solutions are there of the ARE?
- Do any of these solutions have special properties; that is, how many are positive definite?
- Which solutions lead to a stable state estimator?
- Under what conditions do the solutions of the Riccati difference equation (RDE) converge to particular solutions of the ARE?

For Example 10.7.1, we have found that:

- There are two solutions of the ARE.
- There is one positive solution.
- $P = 3.7$ (the positive solution) gives a filter having state transition matrix
$$A - HC = 1 - 0.27 = 0.73$$
which is stable for a shift model. On the other hand, $P = -2.7$ gives
$$A - HC = 1 - (-0.37) = 1.37$$
which is unstable for a shift operator model.
- In the example, we found that the solution of the RDE converged to the positive solution of the ARE. In fact, we can show that for this example convergence to the positive solution occurs for all $P(0) \geq 0$.

In the sequel, we will generalize the properties mentioned in Example 10.7.1. We will use the unified form throughout, using ρ to denote d/dt in continuous time and δ in discrete time, as usual.

Most of these asymptotic properties are associated with the fact that the solution P of the Riccati equation converges to a well-defined limit as $t \to \infty$. We will presently discuss under what condition this occurs and what characterizes the limiting solution. Before doing this, however, we note that the fixed points of the RDE satisfy an equation obtained by setting ρP to zero. Hence, from equation (10.6.61) the limiting solution for P should satisfy the following equation:

$$0 = \Omega + PA^T + AP + \Delta APA^T - H[\Gamma + \Delta CPC^T]H^T \tag{10.7.4}$$

Equation (10.7.4) is generally known as an algebraic Riccati equation (ARE). Since it is a matrix nonlinear equation, it has many solutions. Among these solutions we will be particularly interested in those solutions that give rise to a corresponding filter having poles inside or on the stability boundary. We will call these solutions the *strong solutions* of the algebraic Riccati equation. If the corresponding filter poles are all strictly inside the stability boundary, then we will call this a *stabilizing solution* of the algebraic Riccati equation. Clearly, from these definitions, the stabilizing solutions of the ARE are a subset of the strong solutions. The following result clarifies the existence and uniqueness of these particular solutions of the ARE.

Theorem 10.7.1

(i)   The ARE has a unique strong solution if and only if (C, A) is detectable.
(ii) We factor Ω as DD^T. Then the strong solution is the only nonnegative definite solution of the ARE if and only if (C, A) is detectable and (A, D) has no unreachable mode outside the stability boundary.
(iii) The strong solution is stabilizing if and only if (C, A) is detectable and (A, D) has no unreachable mode on the stability boundary. If, in addition, (A, D) has no unreachable mode *inside* the stability boundary, then the strong solution is also positive definite.

Proof: The proof of this result is beyond the scope of this text. However, a full discussion is given in the references at the end of the chapter. ▽▽▽

Example 10.7.1. (Continued)

For this example, we have

(a) (C, A) is detectable and therefore the strong solution ($P = 3.70$) of the ARE exists and is unique.
(b) $\Omega = 1 = 1*1 = DD^T$. Thus (A, D) is completely reachable and thus the strong solution is the only nonnegative solution of the ARE.
(c) The strong solution is stabilizing and is positive definite. ▽▽▽

There exists many ways of explicitly evaluating the strong solution of the ARE. A common method is outlined below.

Lemma 10.7.1

Consider the following generalized Hamiltonian matrix:

$$M \triangleq \begin{bmatrix} A^T + \Delta C^T \Gamma^{-1} C(I + A\Delta)^{-1}\Omega & -C^T \Gamma^{-1} C(I + A\Delta)^{-1} \\ -(I + A\Delta)^{-1}\Omega & -(I + A\Delta)^{-1}A \end{bmatrix} \quad (10.7.5)$$

(i) This matrix has the property that the eigenvalues of M can be grouped into two disjoint subsets Γ_1 and Γ_2 such that for every $\lambda_c \in \Gamma_1$ there exist a $\lambda_d \in \Gamma_2$ such that $\lambda_c + \lambda_d + \Delta\lambda_c\lambda_d = 0$. We can thus choose either Γ_1 or Γ_2 to contain only those eigenvalues that are inside or on the stability boundary.

(ii) The strong solution of the ARE can be obtained by choosing $\begin{bmatrix} X_{11} \\ X_{21} \end{bmatrix} \in \mathbb{R}^{2n \times n}$ to span the nth-order stable invariant subspace of M. Then the strong solution for (10.7.4) is

$$P_s = X_{21}X_{11}^{-1} \tag{10.7.6}$$

(iii) The eigenvalues of the corresponding filter are equal to the eigenvalues of M corresponding to the choice of (X_{11}, X_{21}).

(iv) For continuous time systems, the preceding result holds with $\Delta = 0$ on noting that the generalized Hamiltonian matrix can be written as

$$M = \begin{bmatrix} I & \Delta C^T \Gamma^{-1} C \\ 0 & I + \Delta A \end{bmatrix}^{-1} \begin{bmatrix} A^T & -C^T \Gamma^{-1} C \\ -\Omega & -A \end{bmatrix}$$

and hence it is clear that

$$\lim_{\Delta \to 0} M = \begin{bmatrix} A^T & -C^T \Gamma^{-1} C \\ -\Omega & -A \end{bmatrix} \tag{10.7.7}$$

Proof: We will defer the proof of this result until Chapter 11, where we will present it for the dual optimal control problem—see Lemma 11.3.1. ▽▽▽

Example 10.7.1. (Continued)

When expressed in delta form for $\Delta = 1$, the model (10.3.28), (10.3.29) becomes

$$\delta x = v_1 \tag{10.7.8}$$
$$y = x + v_2 \tag{10.7.9}$$

where v_1, v_2 have spectral density $\Omega = 1$ and $\Gamma = 10$, respectively. The generalized Hamiltonian matrix is

$$M = \begin{bmatrix} 0.1 & -0.1 \\ -1 & 0 \end{bmatrix} \tag{10.7.10}$$

The eigenvalues of this matrix are -0.27 and 0.37. The eigenvector corresponding to the stable eigenvalue is

$$\begin{pmatrix} X_{11} \\ X_{21} \end{pmatrix} = \begin{pmatrix} 0.27 \\ 1 \end{pmatrix} \tag{10.7.11}$$

Hence, using Lemma 10.7.1, the strong solution is

$$P_s = \frac{1}{0.27} = 3.7 \tag{10.7.12}$$

▽▽▽

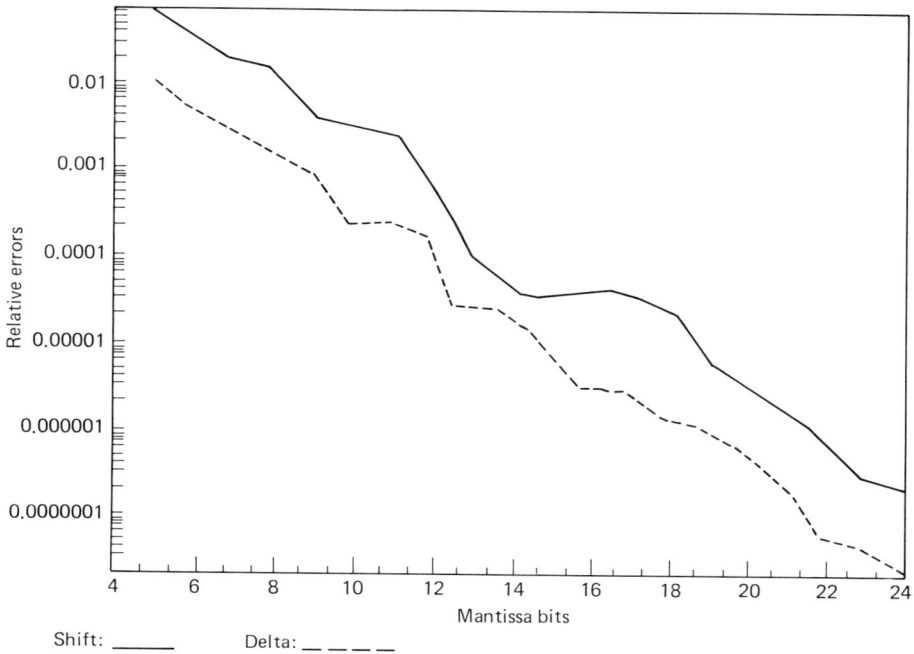

Figure 10.7.1 Comparison of errors for algebraic Riccati equations using delta (δ) and shift (q) operators with $\Delta = 0.075$.

A further question that arises is whether it is preferable to use shift or delta form in solving the ARE. Arguments establishing the superiority of the delta form are given in Appendix C. The conclusion is illustrated in Example 10.7.2.

Example 10.7.2

Consider the same system as in Example 10.6.1 with the following parameters: $R = 1$, $Q = 1$. The steady-state solution of the Riccati equation was found using the eigenvector method for two different sampling periods ($\Delta = 0.075$, $\Delta = 0.05$) and for mantissa lengths ranging from 5 to 24 bits. Figure 10.7.1 compares the relative error between delta and shift for $\Delta = 0.075$, while Figure 10.7.2 gives the corresponding results for $\Delta = 0.05$. It is clear from the figures that the delta performance is better.

We see from the figures that, for a fixed number of mantissa bits, the delta form gives errors that are roughly an order of magnitude smaller than those obtained using the shift form. ▽▽▽

Next, we want to bring all these trails together and study under what conditions the solutions of the RDE converge to the strong solution of the ARE. We

Sec. 10.7 Properties of the Optimal Filter

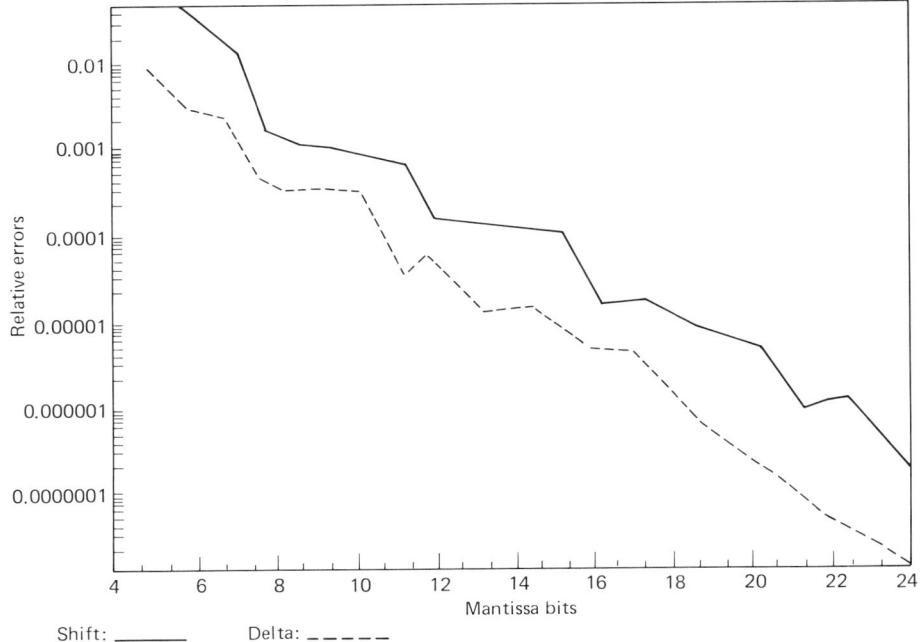

Figure 10.7.2 Comparison of errors for algebraic Riccati equations using delta (δ) and shift (q) operators with $\Delta = 0.05$.

found in Example 10.3.4 that the solution of the RDE converged to the strong solution of the ARE for any nonnegative initial conditions $P(0)$. This is no coincidence. Indeed the following result, which we quote without proof, gives general conditions under which the solutions of the RDE converge to the strong solution of the ARE.

Theorem 10.7.2

(i) Detectability only

Subject to $P_0 - P_s \geq 0$, then $\lim_{t \to \infty} P(t) = P_s$ if and only if (C, A) is detectable, where $P(t)$ is the solution of the RDE with initial condition P_0, and P_s is the unique *strong* solution of the ARE.

(ii) Detectability and no unreachable modes on stability boundary

Subject to $P_0 > 0$, then the detectability of (C, A) and the nonexistence of unreachable modes of (A, D) on the stability boundary are necessary and sufficient conditions for $\lim_{t \to \infty} P(t) = P_s$ (exponentially fast), where $P(t)$ is the solution of the RDE with initial condition P_0, and P_s is the unique *stabilizing* solution of the ARE.

Proof: See the references given at the end of the chapter. ▽▽▽

Part (ii) of Theorem 10.7.2 generalizes the results obtained in Example 10.3.1. Example 10.7.3 illustrates part (i) for the case where there are unreachable modes on the stability boundary.

Example 10.7.3

Consider the problem of estimating a sine wave buried in noise. A model for this was developed in Chapter 3 [see Equations (3.4.32) to (3.4.36)]. Adding measurement noise to these equations leads to

$$\rho \begin{bmatrix} x_1 \\ x_2 \end{bmatrix} = \begin{bmatrix} \dfrac{\cos \omega_0 \Delta - 1}{\Delta} & \dfrac{1}{\omega_0 \Delta} \sin \omega_0 \Delta \\ \dfrac{-\omega_0}{\Delta} \sin \omega_0 \Delta & \dfrac{\cos \omega_0 \Delta - 1}{\Delta} \end{bmatrix} \begin{bmatrix} x_1 \\ x_2 \end{bmatrix}$$

$$y = \begin{bmatrix} 1 & 0 \end{bmatrix} x + v_1 \qquad (10.7.13)$$

Using (10.6.59) to (10.6.61), the optimal filter for estimating the state is

$$\rho \hat{x}(t) = \begin{bmatrix} \dfrac{\cos \omega_0 \Delta - 1}{\Delta} & \dfrac{1}{\omega_0 \Delta} \sin \omega_0 \Delta \\ \dfrac{-\omega_0}{\Delta} \sin \omega_0 \Delta & \dfrac{\cos \omega_0 \Delta - 1}{\Delta} \end{bmatrix} \hat{x} + H(t)[y(t) - \hat{x}_1(t)]$$

$$(10.7.14)$$

To put this problem into the context of the theory presented in this section, we see that the model (10.7.5) is observable; but since $\Omega = 0$, then $D = 0$ and hence (A, D) is uncontrollable. Also, the roots of (10.7.13) are on the stability boundary. Hence (A, D) has unreachable roots on the stability boundary.

Part (i) of Theorem 10.7.1 predicts that the strong solution of the ARE exists and that it is unique. However part (iii) predicts that the strong solution is *not* stabilizing. Furthermore, part (ii) of Theorem 10.7.1 tells us that the strong solution is the only nonnegative definite solution of the ARE.

A simple calculation shows that $P = 0$ is a nonnegative solution of the ARE in this case and hence this must correspond to the strong solution. Since $P_s = 0$, then the corresponding steady-state filter gain is

$$H(t) = 0 \qquad (10.7.15)$$

Hence the optimal steady-state filter is

$$\rho \hat{x}(t) = \begin{bmatrix} \dfrac{\cos \omega_0 \Delta - 1}{\Delta} & \dfrac{1}{\omega_0 \Delta} \sin \omega_0 \Delta \\ \dfrac{-\omega_0}{\Delta} \sin \omega_0 \Delta & \dfrac{\cos \omega_0 \Delta - 1}{\Delta} \end{bmatrix} \hat{x}(t) \qquad (10.7.16)$$

This steady-state filter is strong; that is, it has its roots on the stability boundary. However, since we have also argued that the filter is not stabilizing, we conclude that the poles must actually be on the stability boundary, as is indeed the case.

Theorem 10.7.2 predicts that $\Sigma(t)$ will converge to the strong solution (zero) but at a rate that is less than exponential. It is clearly undesirable to go to the limit as $t \to \infty$ since in the limit the filter output becomes disconnected from the data. This suggests that we should stop the RDE before the limit is reached and use a small (but nonzero) filter gain. A possible choice is

$$H = 2\xi\omega_0 \begin{bmatrix} \cos \omega_0 \Delta \\ \sin \omega_0 \Delta \end{bmatrix} \tag{10.7.17}$$

where ξ is a small constant. This leads to a suboptimal steady-state filter, which is described by

$$\rho \hat{x} = \begin{bmatrix} \dfrac{\cos \omega_0 \Delta - 1}{\Delta} & \dfrac{1}{\omega_0 \Delta} \sin \omega_0 \Delta \\ \dfrac{-\omega_0}{\Delta} \sin \omega_0 \Delta & \dfrac{\cos \omega_0 \Delta - 1}{\Delta} \end{bmatrix} \hat{x} + 2\xi\omega_0 \begin{bmatrix} \cos \omega_0 \Delta \\ \sin \omega_0 \Delta \end{bmatrix}(y - \hat{x}_1)$$

$$= \begin{bmatrix} \dfrac{(1 - 2\xi\omega_0\Delta)\cos \omega_0 \Delta - 1}{\Delta} & \dfrac{1}{\omega_0 \Delta} \sin \omega_0 \Delta \\ -\left(\dfrac{\omega_0}{\Delta} + 2\xi\omega_0\right)\sin \omega_0 \Delta & \dfrac{\cos \omega_0 \Delta - 1}{\Delta} \end{bmatrix} \hat{x}$$

$$+ 2\xi\omega_0 \begin{bmatrix} \cos \omega_0 \Delta \\ \sin \omega_0 \Delta \end{bmatrix} y \tag{10.7.18}$$

In continuous time ($\Delta \to 0$), this filter has the simple form

$$\frac{d}{dt}\hat{x} = \begin{bmatrix} -2\xi\omega_0 & 1 \\ -\omega_0^2 & 0 \end{bmatrix} \hat{x} + 2\xi\omega_0 \begin{bmatrix} 1 \\ 0 \end{bmatrix} y \tag{10.7.19}$$

Note that the characteristic equation for this filter is $s^2 + 2\xi\omega_0 s + \omega_0^2$. Thus ξ is the damping ratio for a stable second-order system.

The continuous time transfer function from y to $\hat{x}$ is simply given by

$$F(s) = \begin{bmatrix} \dfrac{2\xi\omega_0 s}{s^2 + 2\xi\omega_0 s + \omega_0^2} \\ \dfrac{-2\xi\omega_0^2}{s^2 + 2\xi\omega_0 s + \omega_0^2} \end{bmatrix} \tag{10.7.20}$$

It is readily verified that *both* the discrete and continuous suboptimal filters given here are exponentially stable. At frequency $\omega = \omega_0$, the frequency response for y to $\hat{x}$ is precisely $\begin{bmatrix} 1 \\ j \end{bmatrix}$; that is, the filters give the in-phase and quadrature components of a single sine wave of frequency ω_0. The estimator is actually a narrow-band filter that has a gain of 1 at ω_0, but that filters off the noise at frequencies away from ω_0. This appears to be an entirely reasonable result. A practical application of this state estimator will be discussed in Chapter 15.

10.8 SPECTRAL FACTORIZATION

The optimal state estimation problem, as described, can be linked with a problem known as the spectral factorization problem. This latter problem turns out to be the frequency domain dual of the time domain Kalman filter.

The spectral factorization problem is as follows: given any spectral density matrix $\Phi(\gamma)$, find a transfer function $T(\gamma)$ and a real matrix $\bar{\Gamma}$ such that

(i) $\Phi(\gamma) = T(\gamma)\bar{\Gamma}T\left(-\dfrac{\gamma}{1+\gamma\Delta}\right)^T$ (10.8.1)

(ii) $\bar{\Gamma} = \bar{\Gamma}^T > 0$ (10.8.2)

(iii) $T(\gamma)$ and $T(\gamma)^{-1}$ are stable (10.8.3)

(iv) $\lim_{\gamma \to \infty} T(\gamma) = I$ (10.8.4)

The motivation for this problem is that, if y is a stochastic process having spectral density $\Phi(\gamma)$, then the representation (10.8.1) is equivalent to describing the process y as the output of a linear dynamic system. This system has transfer function $T(\gamma)$ and is driven by a white noise process having spectral density $\bar{\Gamma}$.

For the case where the process y is described in state space form, we have the following result.

Lemma 10.8.1

Consider a process y described by a stable state space model of the form

$$\rho x = Ax + v_1 \qquad (10.8.5)$$
$$y = Cx + v_2 \qquad (10.8.6)$$

where v_1 and v_2 are white processes that are mutually uncorrelated and have spectral densities Ω and Γ, respectively.

The corresponding spectral density function for y is

$$\Phi_y(\gamma) = C(\gamma I - A)^{-1}\Omega\left(-\dfrac{\gamma}{1+\gamma\Delta}I - A^T\right)^{-1}C^T + \Gamma \qquad (10.8.7)$$

This can be written as

$$\Phi_y(\gamma) = T(\gamma)\bar{\Gamma}T\left(-\dfrac{\gamma}{1+\gamma\Delta}\right)^T \qquad (10.8.8)$$

where

$$T(\gamma) = I + C(\gamma I - A)^{-1}H \qquad (10.8.9)$$
$$\bar{\Gamma} = \Gamma + \Delta CPC^T \qquad (10.8.10)$$

and where H, P satisfy the algebraic Riccati equation

$$0 = \Omega + PA^T + AP + \Delta A P A^T - H(\Gamma + \Delta C P C^T) H^T \qquad (10.8.11)$$

$$H = [\Delta A P C^T + P C^T][\Gamma + \Delta C P C^T]^{-1} \qquad (10.8.12)$$

Proof: Adding and subtracting terms to (10.8.11) lead to the following identity:

$$\Omega = (\gamma I - A) P (I + A^T \Delta) + (I + A\Delta) P \left(-\frac{\gamma}{1 + \gamma \Delta} I - A^T \right)$$

$$+ \Delta (\gamma I - A) P \left(-\frac{\gamma}{1 + \gamma \Delta} I - A^T \right) + H(\Gamma + \Delta C P C^T) H^T \qquad (10.8.13)$$

Multiplying on the left by $C(\gamma I - A)^{-1}$ and on the right by $(-\gamma/(1 + \gamma\Delta) - A^T)^{-1} C^T$ gives

$$C(\gamma I - A)^{-1} \Omega \left(-\frac{\gamma}{1 + \gamma \Delta} I - A^T \right)^{-1} C^T$$

$$= CP(I + A^T \Delta) \left(-\frac{\gamma}{1 + \gamma \Delta} I - A^T \right)^{-1} C^T + C(\gamma I - A)^{-1} (I + A\Delta) P C^T$$

$$+ \Delta C P C^T$$

$$+ C(\gamma I - A)^{-1} H(\Gamma + \Delta C P C^T) H^T \left(-\frac{\gamma}{1 + \gamma \Delta} I - A^T \right)^{-1} C^T$$

$$= \left(I + C(\gamma I - A)^{-1} H \right) (\Gamma + \Delta C P C^T)$$

$$\times \left(I + H^T \left(-\frac{\gamma I}{1 + \gamma \Delta} - A^T \right) C^T \right) - \Gamma \qquad (10.8.14)$$

Adding Γ to both sides of (10.8.14) immediately gives (10.8.8). ▽▽▽

Example 10.8.1

Consider the following spectral density:

$$\Phi(\gamma) = \frac{10\gamma^2 - 1.222\gamma - 12.222}{\gamma^2 - 0.111\gamma - 1.111} \qquad (10.8.15)$$

for a discrete time system with $\Delta = 0.1$. Note that $\Phi(\gamma)$ is a well-defined spectral density, since for $\gamma = (e^{j\omega\Delta} - 1)/\Delta$ (that is, for γ on the stability boundary) $\Phi(\gamma)$ is real and nonnegative.

By taking out the direct term, $\Phi(\gamma)$ can be written as

$$\Phi(\gamma) = 10 - \frac{1.1111(0.1\gamma + 1)}{(\gamma + 1)(\gamma - 1.1111)} \qquad (10.8.16)$$

It is readily checked that this spectral density may be generated by the following state space model:

$$\rho x = -x + v_1 \qquad (10.8.17)$$

$$y = x + v_2 \qquad (10.8.18)$$

where $\Delta = 0.1$ and v_1, v_2 are mutually uncorrelated white noise sequences with spectral densities 1 and 10, respectively. The corresponding algebraic Riccati equation is

$$0 = 1 - 2P + 0.1P - \frac{(0.9P)^2}{10 + 0.1P} \qquad (10.8.19)$$

The stabilizing solution to this equation is

$$P = 0.515 \qquad (10.8.20)$$

which yields

$$H = 0.0461 \qquad (10.8.21)$$

Hence the spectral factor is obtained from (10.8.9), (10.8.10) as

$$T(\gamma) = 1 + \frac{0.0461}{\gamma + 1} = \frac{\gamma + 1.0461}{\gamma + 1} \qquad (10.8.22)$$

$$\overline{\Gamma} = 10.0515 \qquad (10.8.23)$$

For this simple problem, the preceding result is easily obtained from (10.8.15) by splitting the poles and zeros into stable and unstable groups. ▽▽▽

10.9 CHARACTERIZATION OF ALL STATE ESTIMATORS

In this section, we show how if we are given one particular state estimator then all possible state estimators can be generated in an affine fashion. This is the dual of the results given in Section 9.6 for the control problem.

We will constrain the class of allowable state estimators in the following way. We require that:

1. The state estimate should be a stable proper linear time invariant function of the plant input and output (that is, be *stable*).
2. In the absence of modeling errors and noise, and for any input signal, the state estimation error should decay to zero (that is, be *unbiased*). (The term unbiased can be justified by noting that, if zero mean noise is added to the process and observations, then the mean value of the state estimation error converges to zero.) Let the nominal plant without noise, be described by

$$\rho x = Ax + Bu \qquad (10.9.1)$$

$$y = Cx \qquad (10.9.2)$$

Let J be any matrix such that $A'' = A - JC$ is stable and define the corresponding state estimator as

$$\rho \hat{x} = A\hat{x} + Bu + J(y - C\hat{x}) \qquad (10.9.3)$$

The corresponding output estimation error is

$$\nu = y - C\hat{x} \qquad (10.9.4)$$

We then have the following result.

Theorem 10.9.1

A necessary and sufficient condition for $\hat{x}'$ to be a stable, unbiased state estimate of x is that $\hat{x}'$ can be expressed as

$$\hat{x}' = \hat{x} + Q\nu \qquad (10.9.5)$$

where Q is a stable proper linear vector transfer function and $\hat{x}$ and ν are as in (10.9.3), (10.9.4).

Proof: Sufficiency. Clearly, $\hat{x}'$ is a stable estimate since $\hat{x}$ is. Also, since $\hat{x}$ is unbiased and ν tends to zero, it is clear that $\hat{x}'$ is also unbiased.

Necessity. Stability implies that $\hat{x}'$ can always be expressed as

$$\hat{x}' = \Gamma' u + X' y \qquad (10.9.6)$$

for stable proper linear vector transfer functions Γ' and X'.

Similarly, since $\hat{x}$ is stable, we can write

$$\hat{x} = \Gamma u + X y \qquad (10.9.7)$$

Using the above expressions, the transfer functions from u to $\hat{x}' - \hat{x}$ is

$$T_e = (\Gamma' - \Gamma) + (X' - X)NM^{-1} \qquad (10.9.8)$$

where NM^{-1} is a right matrix fraction description of the plant.

For $\hat{x}'$ and $\hat{x}$ to be unbiased, we require that the transfer function from u to $(\hat{x}' - \hat{x})$ be zero. Thus from (10.9.8) we require

$$(\Gamma' - \Gamma)M + (X' - X)N = 0 \qquad (10.9.9)$$

Provided M and N are relatively prime, then it follows from Section 9.6 that all solutions to (10.9.9) can be written in the form

$$\Gamma' = \Gamma - Q\tilde{N} \qquad (10.9.10)$$

$$X' = X + Q\tilde{M} \qquad (10.9.11)$$

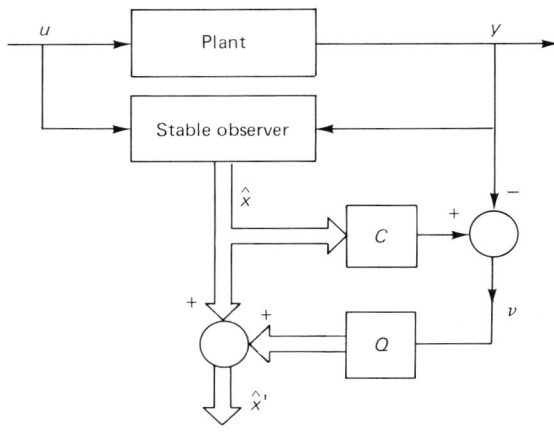

Figure 10.9.1 Classification of all stable, unbiased observers.

where $\tilde{M}$, $\tilde{N}$ are the matrices in a left matrix fraction description of the plant as defined in Equations (9.6.73), (9.6.74).

From (10.9.6), (10.9.10), and (10.9.11), we have

$$\hat{x}' = \hat{x} + Q[\tilde{M}y - \tilde{N}u] \qquad (10.9.12)$$

Finally, using (9.6.75), we obtain

$$\hat{x}' = \hat{x} + Qv \qquad (10.9.13)$$

▽▽▽

This result is depicted diagramatically in Figure 10.9.1.

10.10 ALTERNATIVE OPTIMIZATION CRITERIA

The formulation of all stable, unbiased state estimators in the previous section has the principal advantage that the filter transfer function from noise to the state estimation error is affine in the free transfer function Q. This fact is demonstrated next.

We begin by adding process noise v_1 and measurement noise v_2 to the model (10.9.1), (10.9.2) leading to

$$\rho x = Ax + Bu + v_1 \qquad (10.10.1)$$
$$y = Cx + v_2 \qquad (10.10.2)$$

Consider *any* stable, unbiased state estimator of the form

$$\rho\hat{x} = A\hat{x} + Bu + J(y - C\hat{x}) \qquad (10.10.3)$$

Then the state estimation error $\tilde{x} = \hat{x} - x$ satisfies

$$\rho\tilde{x} = (A - JC)\tilde{x} + Jv_2 - v_1 \qquad (10.10.4)$$

Let $T(\rho)$ be defined as

$$T(\rho) = (\rho I - A + JC)^{-1} \qquad (10.10.5)$$

Then $\tilde{x}$ can be expressed in transfer function form as

$$\tilde{x} = T(\rho)[Jv_2 - v_1] \qquad (10.10.6)$$

Let $\hat{x}'$ denote any other stable, unbiased state estimate generated as in (10.9.13). The corresponding estimation error $\tilde{x}'$ satisfies

$$\tilde{x}' \triangleq \hat{x}' - x$$
$$= \tilde{x} + Q(y - C\hat{x})$$
$$= \tilde{x} + Q[-C\tilde{x} + v_2]$$
$$= (I - QC)\tilde{x} + Qv_2$$
$$= [(I - QC)TJ + Q]v_2 - (I - QC)Tv_1 \qquad (10.10.7)$$
$$\triangleq T_2v_2 + T_1v_1 \qquad (10.10.8)$$

Thus, as anticipated, $\tilde{x}'$ is an affine function of Q.

This property can be exploited to design filters that optimize alternative criteria. Several possibilities are discussed next.

10.10.1 Mean Square Criterion Revisited

If we assume that v_1 and v_2 are white noise processes having spectral density Ω and Γ, respectively, then the variance R of the state estimation error $\tilde{x}'$ is given by

$$R = \frac{1}{2\pi j} \oint_C [T_2(\gamma)\Gamma T_2^*(\gamma) + T_1(\gamma)\Omega T_1^*(\gamma)]\, d\gamma \qquad (10.10.9)$$

where the contour C denotes the stability boundary and * denotes conjugate transpose.

Since T_1 and T_2 are affine in Q, (10.10.9) is a quadratic function of Q. In principle, it is easy to optimize R, thereby rederiving the Kalman filter. This is illustrated in Example 10.10.1.

Example 10.10.1

Consider the following continuous time stable system:

$$\frac{d}{dt}x = -x + u + v_1 \qquad (10.10.10)$$

$$y = x + v_2 \qquad (10.10.11)$$

where v_1 and v_2 have unit spectral density.

Since the system is stable, we can design a stable unbiased observer by simply putting $J = 0$. This leads to the following stable unbiased estimator:

$$\frac{d}{dt}\hat{x} = -\hat{x} + u \qquad (10.10.12)$$

In this case, T_2 and T_1 are given by

$$T_2(s) = Q(s), \qquad T_1(s) = -(1-Q)\frac{1}{s+1} \qquad (10.10.13)$$

Hence

$$R = \frac{1}{2\pi}\int_{-\infty}^{\infty} |Q(j\omega)|^2 + \left|(1-Q(j\omega))\frac{1}{1+j\omega}\right|^2 d\omega \qquad (10.10.14)$$

We take $Q(s)$ to have the following form (other considerations tell us that a first-order Q suffices):

$$Q = \frac{b}{s+a} \qquad (10.10.15)$$

Substituting (10.10.15) into (10.10.14) gives

$$R = \frac{b^2}{2a} + \frac{1}{2} - \frac{b}{a+1} + \frac{b^2}{2a(a+1)} \qquad (10.10.16)$$

Differentiating with respect to a and b gives

$$\frac{\partial R}{\partial a} = -\frac{b^2}{2a^2} + \frac{b}{(a+1)^2} - \frac{b^2(a+\frac{1}{2})}{a^2(a+1)^2} \qquad (10.10.17)$$

$$\frac{\partial R}{\partial b} = \frac{b}{a} - \frac{1}{a+1} + \frac{b}{a(a+1)} \qquad (10.10.18)$$

Setting $\partial R/\partial b = 0$ gives

$$b = \frac{a}{a+2} \qquad (10.10.19)$$

Similarly, $\partial R/\partial a = 0$ gives

$$b = \frac{a^2}{\frac{a^2}{2} + 2a + 1} \qquad (10.10.20)$$

Substituting (10.10.19) into (10.10.20) gives

$$\frac{a^2}{2} = 1 \qquad (10.10.21)$$

For stability, we require $a > 0$; hence

$$a = \sqrt{2} \quad \text{and} \quad b = \sqrt{2} - 1 \qquad (10.10.22)$$

Thus the optimal Q is

$$Q = \frac{\sqrt{2} - 1}{s + \sqrt{2}} \qquad (10.10.23)$$

which gives the optimal value of R as

$$R = \sqrt{2} - 1 \qquad (10.10.24)$$

Substituting into (10.9.13) gives

$$\begin{aligned}
\hat{x}' &= \hat{x} + Q(y - \hat{x}) \\
&= (1 - Q)\hat{x} + Qy \\
&= \frac{1-Q}{s+1} u + Qy \\
&= \frac{1}{s + \sqrt{2}} \{(\sqrt{2} - 1)y + u\}
\end{aligned} \qquad (10.10.25)$$

The reader should check that this result can be obtained directly via the Kalman filter.

10.10.2 Min – Max Criteria

It could be argued that the Kalman filter solution given in Section 10.10.1 requires excessive prior knowledge of the nature of the disturbances. Basically, the assumption is equivalent to knowing the noise spectral density function so that it can be prewhitened.

An alternative starting point would be to consider all possible noise sources of restricted energy; that is,

$$\|v(\cdot)\|_{\mathscr{L}_2} \triangleq \int_0^\infty \|v\|^2 \, dt \leq 1 \qquad (10.10.26)$$

where

$$v^T = \begin{pmatrix} v_1^T, v_2^T \end{pmatrix} \qquad (10.10.27)$$

We will aim to design an alternative optimal filter by minimizing the cost function R_2, where

$$R_2 = \max_{\|v(\cdot)\|_{\mathscr{L}_2} \leq 1} \left\{ \|\tilde{x}'(\cdot)\|_{\mathscr{L}_2} \right\} \qquad (10.10.28)$$

Thus the optimal Q is given by

$$Q^* = \arg\min_{Q \text{ stable}} \left\{ \max_{\|v(\cdot)\|_{\mathscr{L}_2} \leq 1} \left[\|\tilde{x}'(\cdot)\|_{\mathscr{L}_2} \right] \right\} \qquad (10.10.29)$$

From (10.10.9) the transfer function from v to $\tilde{x}'$ is G, where

$$G(\rho) = \begin{bmatrix} T_1(\rho) & T_2(\rho) \end{bmatrix} \qquad (10.10.30)$$

Using Parseval's theorem, we have

$$\|\tilde{x}'(\cdot)\|_{\mathscr{L}_2}^2 = \frac{1}{2\pi j} \oint_C \tilde{X}'(\gamma)^* \tilde{X}'(\gamma) \, d\gamma \qquad (10.10.31)$$

where C denotes the stability boundary and $\tilde{X}'(\gamma)$ is the (generalized) transform of $\tilde{x}'$. Hence

$$\|\tilde{x}'(\cdot)\|_{\mathscr{L}_2}^2 = \frac{1}{2\pi j} \oint_C V(\gamma)^* G(\gamma)^* G(\gamma) V(\gamma) \, d\gamma \qquad (10.10.32)$$

$$\leq \left\{ \sup_{\gamma \in C} \|G(\gamma)\|_2^2 \right\} \frac{1}{2\pi j} \oint_C V(\gamma)^* V(\gamma) \, d\gamma \qquad (10.10.33)$$

where $\|G(\gamma)\|_2^2$ is the maximum eigenvalue of $G(\gamma)^* G(\gamma)$ [that is, the square of the largest singular value of $G(\gamma)$].

Since the inequality in (10.10.33) is tight, and again using Parseval's theorem,

$$\|\tilde{x}'(\cdot)\|_{\mathscr{L}_2} \leq \|G(\cdot)\|_{H_\infty} \qquad (10.10.34)$$

where

$$\|G(\cdot)\|_{H_\infty} \geq \left\{ \sup_{\gamma \in C} \|G(\gamma)\|_2 \right\} \qquad (10.10.35)$$

Since the inequality in (10.10.34) is tight, it follows that

$$\max_{\|v(\cdot)\|_{\mathscr{L}_2} \leq 1} \left[\|\tilde{x}'(\cdot)\|_{\mathscr{L}_2} \right] = \|G(\cdot)\|_{H_\infty} \qquad (10.10.36)$$

Thus, combining (10.10.29) and (10.10.36), we have

$$Q^* = \arg\min_{Q \text{ stable}} \|G(\cdot)\|_{H_\infty} \qquad (10.10.37)$$

This optimization problem is known as an H^∞ optimization. This problem has been well studied and the solution is described in the references given at the end of the chapter. More details are given in Section 11.11 of Chapter 11.

A simple example is given next.

Example 10.10.2

Consider the same system as in Example 10.10.1 save that v_1 and v_2 are considered as members of the class defined by (10.10.26). For this example, the transfer function from v to $\tilde{x}'$ is

$$G = \left[-(1-Q)\frac{1}{s+1}, Q \right] \qquad (10.10.38)$$

Using the H^∞ design criterion, we have

$$Q^* = \arg\min_{Q \text{ stable}} \|G(\cdot)\|_{H^\infty}^2$$

$$= \min_{Q \text{ stable}} \max_\omega \left\{ |Q(j\omega)|^2 + \frac{1}{1+\omega^2}|1 - Q(j\omega)|^2 \right\} \qquad (10.10.39)$$

This particular H^∞ design problem is of a nonstandard type, and there turns out to be a simple solution. Let $q_0 = Q(j0)$. Then clearly

$$\|G(\cdot)\|_{H^\infty}^2 = \max_\omega \left\{ |Q(j\omega)|^2 + \frac{1}{1+\omega^2}|1 - Q(j\omega)|^2 \right\}$$

$$\geq |Q(j0)|^2 + |1 - Q(j0)|^2$$

$$= q_0^2 + (1 - q_0)^2 \qquad (10.10.40)$$

From (10.10.40) it is easy to show that

$$\|G(\cdot)\|_{H^\infty}^2 \geq 0.5 \qquad (10.10.41)$$

However, the selection $Q(j\omega) = 0.5$ gives

$$\|G(\cdot)\|_{H^\infty}^2 = \max_\omega \left\{ 0.25 + \frac{0.25}{1+\omega^2} \right\}$$

$$= 0.5 \qquad (10.10.42)$$

From (10.10.42) it is clear that the selection $Q(s) = 0.5$ is an optimal (in the H^∞ sense) choice. In fact, there is a whole family of $Q(s)$ that achieves optimality, in the H^∞ sense, and so additional performance objectives may be specified (see Problem 10.17). ▽▽▽

10.11 ROBUSTNESS CONSIDERATIONS IN OBSERVER DESIGN

So far in this chapter, we have assumed that the model of the system gives an exact description of the system behavior. In practice, this will never be true since all real models are approximations. To capture this modeling error, the model of Equations

(10.10.1) and (10.10.2) can be augmented as follows:

$$\rho x = Ax + Bu + v_1' \tag{10.11.1}$$

$$y = Cx + v_2' \tag{10.11.2}$$

where v_1' and v_2' represent composite uncertainties; that is,

$$v_1' = v_1 + G_\Delta^1 u \tag{10.11.3}$$

$$v_2' = v_2 + G_\Delta^2 u \tag{10.11.4}$$

v_1, v_2 are noise sources as before, and G_Δ^1 and G_Δ^2 are transfer functions describing model uncertainty.

If we use the same observer structure as in Figure 10.9.1, then the state estimation error becomes the same as in (10.10.7), (10.10.8), but with v_1, v_2 replaced by v_1', v_2', respectively; that is,

$$\tilde{x}' = T_2 v_2' + T_1 v_1' \tag{10.11.5}$$

where

$$T_1 \triangleq (QC - I)T \tag{10.11.6}$$

$$T_2 \triangleq (I - QC)TJ + Q \tag{10.11.7}$$

We see from (10.11.5) that $\tilde{x}'$ is still affine in Q, and this suggests that the optimization of Q can also include the modeling errors. To carry this out, we need to make assumptions about G_Δ^1 and G_Δ^2. For example, we might assume that G_Δ^1, G_Δ^2 belong to some class $\mathbf{G}$ having certain properties. In practice, it is probably not worthwhile to carry out this kind of optimization. However, the expression (10.11.5) is still useful insofar as it reminds us that the fidelity of model-based state estimators depend on the fidelity of the model, and the performance will be a function of the input signal.

10.12 BANDWIDTH CONSIDERATIONS IN OBSERVER DESIGN

In the previous sections, various optimization procedures were outlined for designing state estimators. However, irrespective of which procedure is followed, the optimal design will lie between two extremes. On the one hand, if v_1' is small (that is, the state evolution model is known accurately), then, provided the system is stable, the optimal observer is the open loop observer:

$$\rho \hat{x} = A\hat{x} + Bu \tag{10.12.1}$$

On the other hand, if v' is large (that is, the model is unreliable), then it becomes dangerous to rely upon the model when constructing $\hat{x}$. In this case, a useful strategy is to base the observer on the output measurements only.

To see the consequences of this, consider

$$\hat{x} = D(p)y \tag{10.12.2}$$

A possible choice for the transfer function $D(p)$ is:

$$D(p) = \text{Adj}(pI - A)B / N^*(p)E'(p) \qquad (10.12.3)$$

where $N^*(p)$ is the plant numerator with all roots reflected inside the stability boundary. $E'(p) = 1 + e_1 p + \cdots + e_{n-m} p^{n-m}$ is any stable polynomial with fast modes.

Substituting into (10.12.2), we see that the transfer function from u to $\hat{x}$ is given by (10.12.4) where $M(p)$ is the plant denominator.

$$\hat{x} = \frac{\text{Adj}(pI - A)B}{N^*(p)E'(p)} \cdot \frac{N(p)}{M(p)} u \qquad (10.12.4)$$

Since $x = (pI - A)^{-1} B u$, we see that $\hat{x} = xN(p)/N^*(p)E'(p)$; that is, $\hat{x}$ differs from x by an approximate all-pass network, $F(p)$, where

$$F(p) = N(p)/N^*(p)E'(p) \qquad (10.12.5)$$

Without loss of generality, we can assume that (A, B) is in controller form:

$$A = \begin{bmatrix} -a_{n-1} & \cdots & -a_0 \\ 1 & & \\ & \ddots & \\ & & 1 & 0 \end{bmatrix}, \quad B = \begin{bmatrix} 1 \\ 0 \\ \vdots \\ 0 \end{bmatrix} \qquad (10.12.6)$$

In this case

$$\text{Adj}(pI - A)B = (p^{n-1}, p^{n-2}, \ldots, 1)^T \qquad (10.12.7)$$

Equation (10.12.2) can then be written in transfer function form as

$$\hat{x}' = \frac{1}{N^*(p)E'(p)} \begin{bmatrix} p^{n-1} \\ p^{n-2} \\ \vdots \\ 1 \end{bmatrix} y \qquad (10.12.8)$$

We thus see that in this limiting case the state estimate is simply a filtered version of the derivatives of the system output. The interesting thing about (10.12.8) is that this state estimator is trivial to realize in practice and gives good robustness to model uncertainty. This procedure is generally known as Loop Transfer Recovery.

Example 10.12.1

Consider again the system of Example 10.10.1. We will design the state estimator using the Kalman filter approach in which we take the spectral density of v_1 and v_2 to be Ω and Γ, respectively.

The optimal observer gain is then given by

$$H = -1 + \sqrt{1 + \Gamma^{-1}\Omega} \qquad (10.12.9)$$

We see that for $\Gamma^{-1}\Omega$ small (that is, the state evolution model is much more reliable than the measurements) $H \to 0$ and the optimal observer becomes the open loop observer (10.12.1):

$$\frac{d}{dt}\hat{x} = -\hat{x} + u \qquad (10.12.10)$$

On the other hand, for $\Gamma^{-1}\Omega$ large (that is, the state evolution model is less reliable than the measurements) H becomes large, and the state observer becomes

$$\frac{d}{dt}\hat{x} = -\xi\hat{x} + u + (\xi - 1)y \qquad (10.12.11)$$

where

$$\xi = \sqrt{1 + \Gamma^{-1}\Omega} \qquad (10.12.12)$$

From (10.12.11), the transfer function form of the state estimator is

$$\hat{x} = \frac{1}{s + \xi}u + \frac{\xi - 1}{s + \xi}y \qquad (10.12.13)$$

For ξ large, this can be approximated by

$$\hat{x} \simeq \frac{\xi}{s + \xi}y \qquad (10.12.14)$$

which is simply a low-pass filtered version of y. ▽▽▽

10.13 SUMMARY

The main points covered in this chapter were:

- The linear least mean square estimate of x given Y is given by the wide sense conditional expectations

$$\hat{x} = \hat{\mathcal{E}}[x|Y] \qquad (10.2.4)$$

$$= \Sigma_{xY}\Sigma_{YY}^{-1}Y \qquad (10.2.5)$$

- When data arrives sequentially, then $\hat{x}(t|n)$ can be computed recursively as

$$\hat{x}(t|n) = \hat{x}(t|n-1) + \mathcal{E}\{x(t)e(n)\}\left[\mathcal{E}\{e(n)e(n)^T\}\right]^{-1}e(n) \qquad (10.3.5)$$

where

$$e(n) \triangleq y(n) - \tilde{\mathcal{E}}(y(n)|Y_1^{n-1}) \qquad (10.3.6)$$

- For the shift operator joint Markov model, the optimal recursive estimate can be expressed as

$$\hat{x}(t+1|t) = A(t)\hat{x}(t|t-1) + L(t)\Sigma(t)^{-1}e(t)$$
$$e(t) = y(t) - C(t)\hat{x}(t|t-1) \qquad (10.3.19)$$
$$L(t) = A(t)P(t)C(t)^T \qquad (10.3.14)$$
$$\Sigma(t) = C(t)P(t)C(t)^T + R(t) \qquad (10.3.15)$$

where $P(t)$ satisfies

$$P(t+1) = A(t)P(t)A(t)^T + Q(t) - L(t)\Sigma(t)^{-1}L(t)^T \qquad (10.3.17)$$

- In unified form, the optimal filter can be expressed as

$$\rho\hat{x} = A\hat{x} + H[y - C\hat{x}] \qquad (10.6.59)$$

$$H = [(\Delta A + I)PC^T][\Delta CPC^T + \Gamma]^{-1} \qquad (10.6.60)$$

where P satisfies a Riccati equation of the form

$$\rho P = \Omega + PA^T + AP + \Delta APA^T - H[\Gamma + \Delta CPC^T]H^T \qquad (10.6.61)$$

- Under weak assumptions, the solution of the Riccati equation converges and this limiting solution corresponds to a stable state estimator.
- The Kalman filter problem is equivalent to expressing a given spectral density $\Phi(\gamma)$ in the form

$$\Phi(\gamma) = T(\gamma)\overline{\Gamma}T\left(-\frac{\gamma}{1+\gamma\Delta}\right)^T \qquad (10.8.8)$$

where

$$T(\gamma) = I + C(\gamma I - A)^{-1}H \qquad (10.8.9)$$

$$\overline{\Gamma} = \Gamma + \Delta CPC^T \qquad (10.8.10)$$

- Given any stable unbiased observer

$$\rho\hat{x} = A\hat{x} + Bu + J(y - C\hat{x}) \qquad (10.9.3)$$

then the class of all stable unbiased state estimators can be expressed as

$$\hat{x}' = \hat{x} + Qv \qquad (10.9.5)$$

where v is the estimator error

$$v = y - C\hat{x} \qquad (10.9.4)$$

and Q is any stable, proper transfer function.

- It was emphasized that the fidelity of model-based state estimates are a function of the fidelity of the model.
- Irrespective of the optimization procedure used, the optimal estimator will lie between two extremes:

1. When the state evolution model is much more reliable than the measurements, then for stable plants the observer is open loop:

$$\rho\hat{x} = A\hat{x} + Bu \qquad (10.12.1)$$

2. When the state evolution model is much less reliable than the measurements, then the observer can be considered as a combination of band limited output

derivatives; that is, $\hat{x} = T^{-1}\hat{x}'$, where

$$\hat{x}' = \frac{1}{N^*(p)E'(p)} \begin{bmatrix} \rho^{n-1} \\ \rho^{n-2} \\ \vdots \\ 1 \end{bmatrix} y \qquad (10.12.8)$$

- When the observer (10.12.8) is used, then the observer is insensitive to modeling errors.

10.14 REFERENCES

Introductions to optimal filtering are contained in:

ANDERSON, B. D. O., and J. B. MOORE (1979) *Optimal Filtering*. Prentice-Hall, Englewood Cliffs, N.J.

ÅSTRÖM, K. J. (1970) *Introduction to Stochastic Control Theory*. Academic Press, New York.

More advanced ideas on filtering are contained in:

SOLO, V. (1988) *Time Series Analysis*. Springer-Verlag, New York.

WONG, E. (1970) *Stochastic Processes in Information and Dynamical Systems*. McGraw-Hill, New York.

Results on asymptotic properties of the Riccati equation are contained in:

CHAN, S. W., G. C. GOODWIN, and K. S. SIN (1984) "Convergence properties of the Riccati difference equation in optimal filtering of nonstabilizable systems." *IEEE Trans. Auto. Control*, vol. AC-29, no. 2, pp. 110–118.

DE SOUZA, C. E., M. GEVERS, and G. C. GOODWIN (1986) "Riccati equations in optimal filtering of nonstabilizable systems having singular state transition matrices." *IEEE Trans. Auto. Control*, vol. AC-31, pp. 831–839.

KUCERA, V (1972) "The discrete Riccati equation of optimal control." *Kybernetika*, vol. 8, no. 5, pp. 430–447.

WILLEMS, J. C. (1971) "Least squares stationary optimal control and the algebra Riccati equation." *IEEE Trans. Auto. Control*, vol. AC-16, pp. 621–634.

The role of spectral densities, as opposed to signal variances, in estimation is discussed in:

LJUNG, L. (1987) *System Identification: Theory for the User*. Prentice-Hall, Englewood Cliffs, N.J.

A discussion of unified optimal filtering is contained in:

GOODWIN, G. C. (1988) "Some observations on robust stochastic estimation." Plenary Address at 1988 IFAC Symposium on Identification and System Parameter Estimation, Beijing.

SALGADO, M. E., R. H. MIDDLETON, and G. C. GOODWIN (1987) "Connection between continuous and discrete Riccati equations with applications to Kalman filtering." *Proc. IEE*, Part D, vol. 135, no. 1, pp. 28–34..

Techniques for solving the H^∞ optimization problems are discussed in:

DOYLE, J., K. GLOVER, P. KHARGONEKAR, and B. FRANCIS (1989) "State space solutions to standard H_2 and H_∞ control problems." *IEEE Transactions on Automatic Control*, vol. 34, no. 8, pp. 831–847.

FRANCIS, B. (1987) *A Course in H^∞ Control Theory*. Springer-Verlag, New York.

KWAKERNAK, H. (1986) "A polynomial approach to minimax frequency domain optimization of multivariable feedback systems." *Int. J. Control*, pp. 117–156.

10.15 PROBLEMS

1. Consider the parameter estimation problem in Example 10.3.2, but modify the parameter variations such that (10.3.39) is replaced by

    $$\theta(t+1) = \theta(t) + v_1(t)$$

    where

    $$\mathscr{E} v_1 = 0$$
    $$\mathscr{E}(v_1 v_1^T) = Q$$

    Derive the optimal estimator for θ in both shift and delta form.

2. Verify equation (10.4.16).

3. (a) Consider a first-order antialiasing filter with transfer function $1/(1 + \tau s)$, where $\tau = \Delta/\pi$. What is the impulse response of this filter at $t = \Delta$?
    (b) Derive a joint Markov model when the filter in part (a) is used on the system (10.6.1), (10.6.2) and when the output is sampled with period Δ. *Hint:* Add additional states to the discrete process to account for the correlation in the discrete output noise process.
    *(c) Rewrite the optimal filter of part (b) in δ operator form and show that it converges to the continuous time filter.

4. Modify Example 10.7.2 by adding a small amount of process noise to the model (10.7.13). Note that this is one way of describing the fact that the model is inexact. What is the optimal steady-state filter in this case? Show that it is stable.

5. Consider the problem of estimating a constant in white noise; that is,

    $$y(t) = \theta + \varepsilon(t)$$

    where $\varepsilon(t)$ is noise and θ is a constant.
    (a) Set this up as a Kalman filtering problem.
    (b) Modify the filter to account for parameter variations of the form

    $$\rho\theta = -a\theta + a\varepsilon_1(t)$$

    where $\varepsilon_1(t)$ is an uncorrelated noise source and $a \geq 0$.
    (c) Evaluate the steady-state variance of the estimation error.
    (d) Modify the filter again to account for the fact that the measurement noise $\varepsilon(t)$ is colored by adjoining the model

    $$\rho\varepsilon = -a_2\varepsilon + a_2\omega$$

    where ω is another uncorrelated white noise source.

(e) Evaluate the steady-state variance of the estimation error for θ and show that the filter and its performance are insensitive to the noise coloring, provided the bandwidth of the noise (a_2) exceeds the bandwidth of the parameter variations (a).

6. Establish equation (10.2.16).

7. Consider the system of Example 10.3.1 save that the noise processes have variances

$$E\left\{\begin{pmatrix} v_1 \\ v_2 \end{pmatrix}(v_1 \quad v_2)\right\} = \begin{bmatrix} 10 & 0 \\ 0 & 1 \end{bmatrix}$$

Rederive the spectral gain H as a function of time. Compare with Example 10.3.1 and discuss in the light of the results of Section 10.12.

8. Why does it not make sense to describe simulations of a continuous time plant including discrete noise variances if the sampling rate is not given?

9. Consider the following plant:

$$\frac{d}{dt}x = -x + v_1$$
$$y = x + v_2$$

where v_1, v_2 have spectral density 1, 1, respectively.
(a) Solve the continuous Riccati equation starting from $P = 10$.
(b) Sample the system with an integrate and reset antialiasing filter with sampling period $\Delta = 1/10$. Solve the corresponding discrete time Riccati equation beginning from $P = 10$.
(c) Approximate the discrete time Riccati equation by deleting all terms of order Δ. Solve the corresponding approximate Riccati equation starting from $P = 10$.
(d) Compare the results in parts (a), (b), and (c).

10. For the following three examples, (i) does the strong solution exist? (ii) is it positive definite?, and (iii) is it stabilizing?

(a) $\frac{d}{dt}x = \begin{bmatrix} 10 & 0 \\ 0 & -1 \end{bmatrix}x + v_1$
$y = \begin{bmatrix} 0 & 1 \end{bmatrix}x + v_2$

(b) $\frac{d}{dt}x = \begin{bmatrix} -1 & 0 \\ 0 & -1 \end{bmatrix}x + v_1$
$y = \begin{bmatrix} 0 & 1 \end{bmatrix}x + v_2$

(c) $\frac{d}{dt}x = \begin{bmatrix} -1 & 0 \\ 0 & 0 \end{bmatrix}x$
$y = \begin{bmatrix} 1 & 1 \end{bmatrix}x + v_2$

11. Use the Hamiltonian method to find the steady-state solution for the Kalman filtering problem described in Problem 9.

12. Find a spectral factorization as in (10.8.8) for the problem of Problem 9.

13. Consider the following plant,

$$\frac{d}{dt}x = -2x + u$$
$$y = x$$

together with the two state estimators:

(i) $\frac{d}{dt}\hat{x}_1 = -2\hat{x}_1 + u$

(ii) $\frac{d}{dt}\hat{x}_2 = -2\hat{x}_2 + u + 5(y - \hat{x})$

Express $\hat{x}_2$ in the form
$$\hat{x}_2 = \hat{x}_1 + Q(y - \hat{x}_1)$$
giving an explicit expression for Q.

14. Complete the details of Example 10.10.1.
15. Complete the details of Example 10.10.2.
16. Consider the following system:
$$\frac{d}{dt}x = -x + u + G'_\Delta u$$
$$y = x + v_2$$
where the model error G'_Δ is a constant ε and the spectral density of v_2 is Γ.
   (a) Design an optimal state estimator ignoring G'_Δ.
   (b) Show that the state estimation error for part (a) satisfies
$$\tilde{x}_1 = \frac{1}{s+1}\varepsilon u$$
   (c) Consider another estimator of the type given in (10.12.7):
$$\hat{x}_2 = \frac{\alpha}{s+\alpha}y$$
   Find an expression for $\tilde{x}_2$ as a function of u, ε, and v_2.
   (d) Discuss the relative merits of the solutions obtained in parts (a) and (c).

*17. Consider again Example 10.10.2. In that example it was shown that one H^∞ optimal observer is generated by $Q(s) = \frac{1}{2}$.
   (a) Show that *any* H^∞ optimal observer must satisfy $Q(0) = \frac{1}{2}$.
   (b) Show that if $Q(s) = b/(s+a)$ (where b and a are free design parameters) then $Q(s)$ is H^∞ optimal if and only if:
      (i) $b = \frac{1}{2}a$, and
      (ii) $a \geq \sqrt{2}$.
   (c) Of all the $Q(s)$ having structure as in part (b), that is, that are H^∞ optimal, which $Q(s)$ minimizes the H^2 norm of $G(\cdot)$? That is, find
$$Q^*(s) = \mathop{\arg\min}_{\substack{Q(s)\text{ stable} \\ Q(s) = \frac{b}{s+a}}} \left\{ \frac{1}{2\pi}\int_{-\infty}^{\infty}\left[|Q(j\omega)|^2 + |1 - Q(j\omega)|^2 \frac{1}{1+\omega^2}\right]d\omega \right\}$$
   $Q(s)$ such that $\|G(\cdot)\|^2_{H^\infty} = 0.5$.

*18. Rederive the delta form of the Kalman filter as follows:
   (i) Let
$$\delta x = Ax + v, \qquad \mathcal{E}[vv^T] = \frac{\Omega}{\Delta}$$
$$y = Cx + w, \qquad \mathcal{E}[ww^T] = \frac{\Gamma}{\Delta}$$
   and
$$\delta\hat{x} = A\hat{x} + L(y - C\hat{x})$$
   (ii) With $\tilde{x} = \hat{x} - x$, note that $\delta\tilde{x} = (A - LC)\tilde{x} + Lw + v$ and let $P(t) = \mathcal{E}\tilde{x}(t)\tilde{x}(t)^T]$.

(iii) Thus show that since $\tilde{x}(t)$, $v(t)$, and $w(t)$ are all mutually uncorrelated, then

$$\delta P(t) = \mathcal{E}\left[\delta\left(\tilde{x}(t)\tilde{x}(t)^T\right)\right]$$
$$= AP + PA^T + \Delta A P A^T + \Omega - LCP(I + \Delta A^T) - (I + A\Delta)PC^T L^T$$
$$+ L(\Gamma + \Delta CPC^T)L^T$$
$$= AP + PA^T + \Delta A P A^T + \Omega + (L - H)(\Gamma + \Delta CPC^T)(L - H)^T$$
$$- H(\Gamma + \Delta CPC^T)H^T$$

where H is defined as

$$H \triangleq (I + A\Delta)PC^T(\Gamma + \Delta CPC^T)^{-1}$$

(iv) Thus show that at each time step, t, the gain $L(t)$ that minimizes $\delta P(t)$, and thus $P(t + \Delta)$ for given $P(t)$, is $L(t) = H(t)$.

11

Optimal Control

11.1 INTRODUCTION

In Chapter 10 we saw how the state estimation problem could be formulated as an optimization problem. In this chapter we will build on this idea and show that control problems can be similarly formulated. However, a distinction is that, whereas in the state estimation case it is clear that we wish to minimize the state estimation error, in the control context it is less clear what is an appropriate cost function. Typical criteria penalize the tracking error, as well as penalizing control energy.

The simplest form of criterion for the control problem is a quadratic performance index. If this is used and the state equations are linear, then the corresponding optimization has an analytic solution. It turns out that this solution is the dual of the minimum variance state estimation problem studied in Chapter 10. In particular, we show how the solution of the linear quadratic optimal control problem can be obtained by solving a matrix Riccati equation. We will outline the solution to this problem and its properties.

The control design outlined above turns out to require state variable feedback. In the case where the states are not measured, it can be shown that the optimal strategy is to combine state variable feedback with an optimal state estimator. This idea will be briefly explored.

Finally, all the preceding discussion has focused on the linear quadratic optimal control problem. We will conclude the chapter with a discussion of alternative optimization criteria including H^∞ designs.

11.2 FINITE HORIZON LQ OPTIMAL CONTROL

We begin by considering a linear system of the form

$$\rho x = Ax + Bu, \qquad x(t_0) = x_0 \tag{11.2.1}$$

We assume that we wish to find the input u that minimizes a quadratic criterion of the form

$$J = x(t_f)^T \Sigma_f x(t_f) + \int_{t_0}^{t_f} \left(x(t)^T Q x(t) + u(t)^T R u(t) \right) dt \tag{11.2.2}$$

with Σ_f, Q, and R symmetric, positive semidefinite matrices.

We call the problem of optimizing (11.2.2), subject to (11.2.1), the linear quadratic (LQ) optimal control problem. In the case where t_f is finite, we will call the problem a finite horizon problem.

We will follow the strategy adopted for optimal filtering and obtain the solution to the problem in two stages. We first consider the traditional shift operator form of the problem. We will then convert this to the unified delta form, which allows the continuous time results to be seen easily as a limiting case.

Thus consider the shift form of (11.2.1), (11.2.2):

$$x_{t+1} = A_q x_t + B_q u_t, \qquad x_0 \text{ given} \tag{11.2.3}$$

$$J = x_N^T \Sigma_N x_N + \sum_{i=0}^{N-1} x_i^T Q_q x_i + u_i^T R_q u_i \tag{11.2.4}$$

Let $V_j(x_j)$ denote the optimal cost accumulated from $k = j, \ldots, N$ if the system is in state x_j at $k = j$. Clearly, we have

$$V_N(x_N) = x_N^T \Sigma_N x_N \tag{11.2.5}$$

A simple argument shows that the following principle of optimality holds:

$$V_j(x_j) = \min_{u_j} \left\{ V_{j+1}(x_{j+1}) + x_j^T Q_q x_j + u_j^T R_q u_j \right\} \tag{11.2.6}$$

(This equation simply says that the optimal cost from time j can be obtained by optimizing the incremental cost from time j to $j + 1$, together with the optimal cost from time $j + 1$ onwards.) Using this principle, we establish:

Lemma 11.2.1

The LQ optimal control in shift form is given by

$$u_j = L_{j+1} x_j \tag{11.2.7}$$

where

$$L_{j+1} = \left(R + B_q^T \Sigma_{j+1} B_q \right)^{-1} B_q^T \Sigma_{j+1} A_q \tag{11.2.8}$$

and where Σ satisfies the following matrix Riccati equation:

$$\Sigma_j = Q_q + L_{j+1}^T R_q L_{j+1} + (A_q - B_q L_{j+1})^T \Sigma_{j+1}(A_q - B_q L_{j+1}) \qquad (11.2.9)$$

with Σ_N given.

Proof: We proceed by induction. We note that $V_N(x_N)$ is a quadratic function of x_N [see (11.2.5)]. We assume this is true at time $j+1$ and prove it holds for time j. Thus, assume

$$V_{j+1}(x_{j+1}) = x_{j+1}^T \Sigma_{j+1} x_{j+1} \qquad (11.2.10)$$

Substituting into (11.2.6) gives

$$V_j(x_j) = \min_{u_j} \left\{ x_{j+1}^T \Sigma_{j+1} x_{j+1} + x_j^T Q_q x_j + u_j^T R_q u_j \right\}$$

$$= \min_{u_j} \left\{ (A_q x_j + B_q u_j)^T \Sigma_{j+1}(A_q x_j + B_q u_j) + x_j^T Q_q x_j + u_j^T R_q u_j \right\}$$

$$(11.2.11)$$

This is a quadratic function of u. Hence, differentiating with respect to u_j and setting the result to zero gives

$$\frac{\partial V_j(x_j)}{\partial u_j} = 2 B_q^T \Sigma_{j+1}(A_q x_j + B_q u_j) + 2 R_q u_j = 0 \qquad (11.2.12)$$

That is,

$$u_j = -L_{j+1} x_j \qquad (11.2.13)$$

where

$$L_{j+1} = \left(R_q + B_q^T \Sigma_{j+1} B_q \right)^{-1} B_q^T \Sigma_{j+1} A_q \qquad (11.2.14)$$

Substituting back into $V_j(x_j)$ gives

$$V_j(x_j) = x_j^T \Sigma_j x_j \qquad (11.2.15)$$

where

$$\Sigma_j = Q_q + L_{j+1}^T R_q L_{j+1} + (A_q - B_q L_{j+1})^T \Sigma_{j+1}(A_q - B_q L_{j+1}) \qquad (11.2.16)$$

Induction completes the proof. ▽▽▽

This result can be converted to unified form by simply replacing the operator q by the operator $1 + \Delta\delta$. This leads to the following result.

Theorem 11.2.1

The LQ optimal control for (11.2.1), (11.2.2) in delta form is

$$u(t) = -L(t + \Delta)x(t) \qquad (11.2.17)$$

where
$$L(t) = (R + \Delta B^T \Sigma(t) B)^{-1} B^T \Sigma(t)(I + A\Delta) \quad (11.2.18)$$
and $\Sigma(t)$ satisfies
$$\bar{\rho}\Sigma(t) = Q + A^T \Sigma(t) + \Sigma(t) A + \Delta A^T \Sigma(t) A$$
$$- L(t)^T [R + \Delta B^T \Sigma(t) B] L(t) \quad (11.2.19)$$
$$\Sigma(t_f) = \Sigma_f \quad (11.2.20)$$

and $\bar{\rho}$ is the reverse time generalized derivative; that is,

$$\bar{\rho}\Sigma(t) \triangleq \begin{cases} -\left(\dfrac{\Sigma(t) - \Sigma(t-\Delta)}{\Delta}\right) & \text{for discrete time} \\ -\dfrac{d}{dt}\Sigma(t), & \text{for continuous time} \end{cases} \quad (11.2.21)$$

Proof: Immediate on using the following substitution in Lemma 11.2.1.
$$A = \frac{A_q - I}{\Delta} \quad (11.2.22)$$
$$B = \frac{B_q}{\Delta} \quad (11.2.23)$$
$$Q = \frac{Q_q}{\Delta} \quad (11.2.24)$$
$$R = \frac{R_q}{\Delta} \quad (11.2.25)$$
▽▽▽

It is interesting to note that the preceding result is the dual of the result given in Equations (10.6.59) to (10.6.61) for the optimal filtering problem. The equivalences are set out in Table 11.2.1.

TABLE 11.2.1 OPTIMAL FILTERING AND OPTIMAL CONTROL DUALS

	Optimal State Estimator	LQ Controller	
System matrix	A	A^T	System matrix
Output matrix	C	B^T	Input matrix
Optimal covariance matrix	P	Σ	Optimal cost matrix
Derivative	ρ	$\bar{\rho}$	Reverse time derivative
Process noise spectral density	Ω	Q	State deviation weighting
Measurement noise spectral density	Γ	R	Control weighting
Filter gain	H	L^T	Feedback gain

TABLE 11.2.2 SOLUTION TO EXAMPLE 11.2.1

t	$\Sigma(t)$	$L(t)$
t_f	100	0.909
$t_f - \Delta$	10.09	0.502
$t_f - 2\Delta$	6.02	0.376
$t_f - 3\Delta$	4.76	0.322
$t_f - 4\Delta$	4.22	0.297
$t_f - 5\Delta$	3.97	0.284
$\vdots$		
$t_f - \infty$	3.70	0.270

Example 11.2.1

Consider the following problem:

$$\rho x = u, \quad \Delta = 1 \tag{11.2.26}$$

$$J = 100x(t_f)^2 + \overset{t_f}{\underset{t_0}{S}} x(t)^2 + 10u(t)^2 \tag{11.2.27}$$

Using Theorem 11.2.1, the optimal control is

$$u(t) = -L(t + \Delta)x(t) \tag{11.2.28}$$

where

$$L(t) = \frac{\Sigma(t)}{\Sigma(t) + 10} \tag{11.2.29}$$

and where

$$\bar{\rho}\Sigma(t) = 1 - \frac{\Sigma(t)^2}{\Sigma(t) + 10}, \quad \Sigma(t_f) = 100 \tag{11.2.30}$$

The solution to the equation is given in Table 11.2.2. This solution should be compared with the dual result obtained in Example 10.3.1. ▽▽▽

11.3 INFINITE HORIZON LQ OPTIMAL CONTROL

A limiting form of the LQ optimal control problem is obtained for the case $t_f - t_0$ tends to ∞. This leads to the following cost function:

$$J = \lim_{t_f \to \infty} \left\{ x(t_f)^T \Sigma_f x(t_f) + \overset{t_f}{\underset{t_0}{S}} x(t)^T Q x(t) + u(t)^T R u(t) \, dt \right\} \tag{11.3.1}$$

Under suitable conditions, described later, this limit exists and the solution of the Riccati equation converges to a finite limit. Any limiting solution must satisfy the following equation obtained by setting $\bar{\rho}\Sigma(t) = 0$ in (11.2.19):

$$0 = Q + A^T\Sigma + \Sigma A + \Delta A^T \Sigma A - L^T(R + \Delta B^T \Sigma B)L \tag{11.3.2}$$

$$L = (R + \Delta B^T \Sigma B)^{-1} B^T \Sigma (I + A\Delta) \tag{11.3.3}$$

These equations have many solutions. We will be particularly interested in those solutions that give a closed loop control system having roots inside or on the stability boundary. We call these solutions the strong solutions of the algebraic Riccati equation. If the corresponding closed loop system has roots strictly inside the stability boundary, we call this a stabilizing solution of the algebraic Riccati equation. The following result clarifies the existence and uniqueness of these particular solutions of the ARE.

Theorem 11.3.1

(i) The ARE has a unique strong solution if and only if (A, B) is stabilizable.

(ii) We factor Q as $C^T C$. Then the strong solution is the only nonnegative definite solution of the ARE if and only if (A, B) is stabilizable and (C, A) has no unobservable mode outside the stability boundary.

(iii) The strong solution is stabilizing if and only if (A, B) is stabilizable and (C, A) has no unobservable mode on the stability boundary. If, in addition, (C, A) has no unobservable mode *inside* the stability boundary, then the strong solution is also positive definite.

Proof: The proof of this result is beyond the scope of this text. Note, however, that Theorem 11.3.1 is the dual of the result given for optimal filtering in Theorem 10.7.1. ▽▽▽

We will next describe a method for evaluating the strong solution of the ARE using the Hamiltonian matrix. This will be the dual of the result quoted in Lemma 10.7.1 for the optimal filtering problem.

We first note that from Equation (11.2.17), (11.2.18) we have

$$u(t) = -(R + \Delta B^T \Sigma(t + \Delta) B)^{-1} B^T \Sigma(t + \Delta)(I + A\Delta) x(t) \quad (11.3.4)$$

However, from Equation (11.2.1) we have

$$(I + A\Delta) x(t) = x(t + \Delta) - \Delta B u(t) \quad (11.3.5)$$

Substituting (11.3.5) into (11.3.4) gives

$$u(t) = -(R + \Delta B^T \Sigma(t + \Delta) B)^{-1} B^T \Sigma(t + \Delta)[x(t + \Delta) - \Delta B u(t)] \quad (11.3.6)$$

Moving $u(t)$ to the left-hand side gives the following equivalent (but non-implementable) form of the control law:

$$u(t) = -R^{-1} B^T \Sigma(t + \Delta) x(t + \Delta) \quad (11.3.7)$$

We now define $\lambda(t + \Delta)$ as the time varying quantity on the right-hand side of (11.3.7), that is

$$\lambda(t) \triangleq \Sigma(t) x(t) \quad (11.3.8)$$

(Note that in alternative derivations of the optimal control law based on Pontryagin's minimum principle, $\lambda(t)$ plays the role of a co-state or adjoint variable.) Substituting (11.3.7), (11.3.8) into (11.2.1) gives

$$\rho x(t) = Ax(t) - BR^{-1}B^T\lambda(t + \Delta) \tag{11.3.9}$$

Also, from (11.3.8) we have

$$\bar{\rho}\lambda(t + \Delta) = [\bar{\rho}\Sigma(t + \Delta)]x(t) + \Sigma(t + \Delta)[\bar{\rho}x(t + \Delta)] \tag{11.3.10}$$

where $\bar{\rho}$ is the reverse time generalized derivative given in (11.2.21).

Substituting (11.2.19), (11.2.1) into (11.3.10) gives

$$\bar{\rho}\lambda(t + \Delta) = \big[Q + A^T\Sigma(t + \Delta) + \Sigma(t + \Delta)A + \Delta A^T\Sigma(t + \Delta)A$$
$$- L(t + \Delta)^T[R + \Delta B^T\Sigma(t + \Delta)B]L(t + \Delta)\big]x(t)$$
$$- \Sigma(t + \Delta)\left[\frac{x(t + \Delta) - x(t)}{\Delta}\right]$$

$$= \big[Q + A^T\Sigma(t + \Delta) + \Sigma(t + \Delta)A + \Delta A^T\Sigma(t + \Delta)A$$
$$- L(t + \Delta)^T[R + \Delta B^T\Sigma(t + \Delta)B]L(t + \Delta)\big]x(t)$$
$$- \Sigma(t + \Delta)[A - BL(t + \Delta)]x(t)$$

$$= \big[Q + A^T\Sigma(t + \Delta) + \Delta A^T\Sigma(t + \Delta)A - (I + A\Delta)^T\Sigma(t + \Delta)^T$$
$$\times B(R + \Delta B^T\Sigma(t + \Delta)B)^{-1}B^T\Sigma(t + \Delta)(I + A\Delta)$$
$$+ \Sigma(t + \Delta)B(R + \Delta B^T\Sigma(t + \Delta)B)^{-1}B^T$$
$$\times \Sigma(t + \Delta)(I + A\Delta)\big]x(t)$$

$$= \{Q + A^T\Sigma(t + \Delta)[I + \Delta A - \Delta BL(t + \Delta)]\}x(t)$$

$$= Qx(t) + A^T\Sigma(t + \Delta)x(t + \Delta) \tag{11.3.11}$$

Finally, using (11.3.8) we have

$$\bar{\rho}\lambda(t + \Delta) = Qx(t) + A^T\lambda(t + \Delta) \tag{11.3.12}$$

Using the fact that $\bar{\rho}\lambda(t + \Delta) = -\rho\lambda(t)$ and that $\lambda(t + \Delta) = (1 + \Delta\rho)\lambda(t)$ in (11.3.12) gives

$$\rho\lambda(t) = -Qx(t) - A^T\lambda(t) - \Delta\rho A^T\lambda(t) \tag{11.3.13}$$

Then, combining (11.3.13) with (11.3.9) and again using the fact that $\lambda(t + \Delta) = (1 + \Delta\rho)\lambda(t)$ gives

$$\begin{bmatrix} I & \Delta BR^{-1}B^T \\ 0 & I + \Delta A^T \end{bmatrix}\begin{bmatrix} \rho x(t) \\ \rho\lambda(t) \end{bmatrix} = \begin{bmatrix} A & -BR^{-1}B^T \\ -Q & -A^T \end{bmatrix}\begin{bmatrix} x(t) \\ \lambda(t) \end{bmatrix} \tag{11.3.14}$$

A remarkable fact is that the solutions of this equation are intimately associated with the solutions of the Riccati equation. The development can proceed without assuming that $(I + \Delta A^T)$ is invertible. However, so as to give a clearer presentation, we will assume $(I + \Delta A^T)^{-1}$ exists. In this case, (11.3.14) can be rewritten as

$$\rho \begin{bmatrix} x(t) \\ \lambda(t) \end{bmatrix} = M \begin{bmatrix} x(t) \\ \lambda(t) \end{bmatrix} \quad (11.3.15)$$

where M is the following generalized Hamiltonian matrix

$$M = \begin{bmatrix} I & \Delta B R^{-1} B^T \\ 0 & I + \Delta A^T \end{bmatrix}^{-1} \begin{bmatrix} A & -BR^{-1}B^T \\ -Q & -A^T \end{bmatrix} \quad (11.3.16)$$

Obviously, as $\Delta \to 0$, we obtain the following continuous time Hamiltonian matrix

$$M_c = \begin{bmatrix} A & -BR^{-1}B^T \\ -Q & -A^T \end{bmatrix} \quad (11.3.17)$$

If we define the $2n \times 2n$ matrix J as

$$J = \begin{bmatrix} 0 & I \\ -I & 0 \end{bmatrix} \quad (11.3.18)$$

then it is readily seen that $J^T = J^{-1} = -J$ and

$$J^{-1} M_c^T J = -M_c \quad (11.3.19)$$

Matrices having the property (11.3.19) are said to be Hamiltonian matrices.

The shift operator form of (11.3.16) is readily shown to be

$$M_q = \begin{bmatrix} A_q + B_q R_q^{-1} B_q^T A_q^{-T} Q_q & -B_q R_q^{-1} B_q^T A_q^{-T} \\ -A_q^{-T} Q_q & A_q^{-T} \end{bmatrix} \quad (11.3.20)$$

The matrix M_q in (11.3.20) is a symplectic matrix since [analogously to (11.3.19)] we have

$$J^{-1} M_q^T J = M_q^{-1} \quad (11.3.21)$$

We find the generalized Hamiltonian matrix given in (11.3.15) more insightful than the shift form of (11.3.20). The generalized matrix M has the following generalized Hamiltonian property:

$$J^{-1} M^T J = -(I + M\Delta)^{-1} M \quad (11.3.22)$$

which includes both the continuous time result in (11.3.19) and the discrete time result in (11.3.21) as special cases.

We next relate the solutions of the ARE to the generalized Hamiltonian matrix.

Lemma 11.3.1

Consider the generalized Hamiltonian matrix

$$M = \begin{bmatrix} I & \Delta BR^{-1}B^T \\ 0 & I + \Delta A^T \end{bmatrix}^{-1} \begin{bmatrix} A & -BR^{-1}B^T \\ -Q & -A^T \end{bmatrix}$$

$$= \begin{bmatrix} A + \Delta BR^{-1}B^T(I + \Delta A^T)^{-1}Q & -BR^{-1}B^T(I + A^T\Delta)^{-1} \\ -(I + \Delta A^T)^{-1}Q & -(I + \Delta A^T)^{-1}A^T \end{bmatrix} \quad (11.3.23)$$

Then the following properties hold:

(i) The matrix has the property that the eigenvalues of M can be grouped into two disjoint sets Γ_1 and Γ_2 such that for every $\lambda_c \in \Gamma_1$ there exists a $\lambda_d \in \Gamma_2$ such that $\lambda_c + \lambda_d + \Delta\lambda_c\lambda_d = 0$. We can thus choose either Γ_1 or Γ_2 to contain only those eigenvalues which are inside or on the stability boundary.

(ii) The Algebraic Riccati Equation (ARE) can be succinctly expressed using the matrix M as

$$\begin{bmatrix} -\Sigma & I \end{bmatrix} M \begin{bmatrix} I \\ \Sigma \end{bmatrix} = 0 \quad (11.3.24)$$

(iii) The strong solution of the ARE can be obtained by choosing $\begin{bmatrix} X_{11} \\ X_{21} \end{bmatrix} \in \mathbb{R}^{2n \times n}$ to span the nth order stable invariant subspace of M. The strong solution for (11.3.2) [or (11.3.24)] is then

$$\Sigma_s = X_{21} X_{11}^{-1} \quad (11.3.25)$$

(iv) The eigenvalues of the corresponding closed-loop control system are equal to the eigenvalues of M, corresponding to the choice of (X_{11}, X_{21}).

(v) For continuous time systems, the above result holds with $\Delta = 0$ on noting that

$$\lim_{\Delta \to 0} M = \begin{bmatrix} A & -BR^{-1}B^T \\ -Q & -A^T \end{bmatrix} = M_c \quad (11.3.26)$$

Proof:

(i) Follows from (11.3.22)

(ii)

$$[-\Sigma \quad I] M \begin{bmatrix} I \\ \Sigma \end{bmatrix} = [-\Sigma \quad I]$$

$$\times \begin{bmatrix} A + \Delta BR^{-1}B^T(I + \Delta A^T)^{-1}Q & -BR^{-1}B^T(I + A^T\Delta)^{-1} \\ -(I + \Delta A^T)^{-1}Q & -(I + \Delta A^T)^{-1}A^T \end{bmatrix} \begin{bmatrix} I \\ \Sigma \end{bmatrix}$$

$$= -\Sigma \left[A + \Delta BR^{-1}B^T(I + \Delta A^T)^{-1}Q \right]$$

$$+ \Sigma \left[BR^{-1}B^T(I + \Delta A^T)^{-1} \right] \Sigma$$

$$- (I + \Delta A^T)^{-1}Q - (I + \Delta A^T)^{-1}A^T\Sigma$$

$$= -\Sigma \left[\frac{(I + \Delta A)}{\Delta} + \Delta BR^{-1}B^T(I + \Delta A^T)^{-1}Q \right]$$

$$+ \Sigma BR^{-1}B^T(I + \Delta A^T)^{-1}\Sigma$$

$$- (I + \Delta A^T)^{-1}Q + \left[\frac{(I + \Delta A^T)^{-1}\Sigma}{\Delta} \right]$$

$$= (I + \Delta A^T)^{-1} \left\{ -(I + \Delta A^T)\Sigma \left(\frac{I + \Delta A}{\Delta} \right) \right.$$

$$- \Delta(I + \Delta A^T)\Sigma BR^{-1}B^T(I + \Delta A^T)^{-1}Q$$

$$+ (I + \Delta A^T)\Sigma BR^{-1}B^T(I + \Delta A^T)^{-1}\Sigma$$

$$\left. - Q + \frac{\Sigma}{\Delta} \right\}$$

$$= (I + \Delta A^T)^{-1} \left\{ -(I + \Delta A^T)\Sigma \left[\frac{I + \Delta A}{\Delta} \right. \right.$$

$$\left. - \Delta BR^{-1}B^T(I + \Delta A^T)^{-1}\left(\frac{\Sigma}{\Delta} - Q \right) \right] + \left(\frac{\Sigma}{\Delta} - Q \right) \right\}$$

$$= (I + \Delta A^T)^{-1} \left\{ -(I + \Delta A^T)\Sigma \left(\frac{I + \Delta A}{\Delta} \right) \right.$$

$$\left. + \left[\Delta(I + \Delta A^T)\Sigma BR^{-1}B^T(I + \Delta A^T)^{-1} + I \right]\left(\frac{\Sigma}{\Delta} - Q \right) \right\}$$

(11.3.27)

Hence substituting (11.3.27) into the left side of (11.3.24) gives

$$\left(\frac{\Sigma}{\Delta} - Q \right) = \left[I + \Delta(I + \Delta A^T)\Sigma BR^{-1}B^T(I + \Delta A^T)^{-1} \right]^{-1}$$

$$\times (I + \Delta A^T)\Sigma \left(\frac{I + \Delta A}{\Delta} \right) \qquad (11.3.28)$$

318   Optimal Control   Chap. 11

Applying the matrix inversion lemma to the first term on the right side of (11.3.28) gives

$$\left(\frac{\Sigma}{\Delta} - Q\right) = \left[I - \Delta(I + \Delta A^T)\Sigma B R^{-1}(I + \Delta B^T \Sigma B R^{-1})^{-1} \right.$$
$$\left. \times B^T(I + \Delta A^T)^{-1}\right]$$
$$\times \left[(I + \Delta A^T)\Sigma\left(\frac{I + \Delta A}{\Delta}\right)\right]$$
$$= (I + \Delta A^T)\Sigma\left(\frac{I + \Delta A}{\Delta}\right)$$
$$- \Delta(I + \Delta A^T)\Sigma B(R + \Delta B^T \Sigma B)^{-1} B^T \Sigma\left(\frac{I + \Delta A}{\Delta}\right) \quad (11.3.29)$$

Equation (11.3.29) immediately simplifies to the ARE (11.3.2), (11.3.3).

(iii) Let $X = [X_{11}^T \ X_{21}^T]^T (\in \mathbb{R}^{2n \times n})$ be the matrix formed of the generalized eigenvectors corresponding to the eigenvalues of M inside or on the stability boundary. Then

$$M \begin{bmatrix} X_{11} \\ X_{21} \end{bmatrix} = \begin{bmatrix} X_{11} \\ X_{21} \end{bmatrix} \Lambda \quad (11.3.30)$$

where Λ is a diagonal matrix of eigenvalues (more generally a Jordan form matrix in the case of repeated eigenvalues).

From (11.3.30) we have

$$M \begin{bmatrix} I \\ X_{21} X_{11}^{-1} \end{bmatrix} = \begin{bmatrix} X_{11} \Lambda X_{11}^{-1} \\ X_{21} \Lambda X_{11}^{-1} \end{bmatrix} \quad (11.3.31)$$

Hence, equation (11.3.24) is satisfied with $\Sigma = X_{21} X_{11}^{-1}$. Thus this is a solution of the ARE.

Now using (11.3.23), the top equation in (11.3.31) can be written

$$A + \Delta B R^{-1} B^T (I + \Delta A^T)^{-1} \left[Q - \frac{\Sigma}{\Delta}\right] = X_{11} \Lambda X_{11}^{-1} \quad (11.3.32)$$

or using (11.3.29)

$$A - BR^{-1}B^T\left[I - \Delta\Sigma B(R + \Delta B^T \Sigma B)^{-1}B^T\right]\Sigma(I + A\Delta) = X_{11} \Lambda X_{11}^{-1} \quad (11.3.33)$$

Hence rearranging this equation we obtain

$$X_{11} \Lambda X_{11}^{-1} = A - B\left\{R^{-1}\left[I - \Delta B^T \Sigma B(R + \Delta B^T \Sigma B)^{-1}\right]B^T \Sigma(I + A\Delta)\right\}$$
$$= A - BR^{-1}[R + \Delta B^T \Sigma B - \Delta B^T \Sigma B]$$
$$\times [R + \Delta B^T \Sigma B]^{-1} B^T \Sigma(I + A\Delta)$$
$$= A - B[R + \Delta B^T \Sigma B]^{-1} B^T \Sigma(I + A\Delta)$$
$$= A - BL \quad (11.3.34)$$

Thus, we see that the matrix on the right side of (11.3.33) is the "A" matrix of the closed-loop system. Hence Λ represents the Jordan form of this matrix and X_{11}

represents the corresponding matrix of (generalized) eigenvectors. Thus, if we choose Λ to contain roots inside or on the stability boundary then (11.3.25) will give the strong solution to the ARE.

(iv) Established in the proof of (iii)

(v) Immediate ▽▽▽

Next, we clarify the conditions under which the solutions of the RDE (11.2.19) converge to the strong solution of the ARE.

Theorem 11.3.2

(i) Stabilizability only.

Subject to $(\Sigma_f - \Sigma_s) \geq 0$, then $\lim_{t \to \infty} \Sigma(t) = \Sigma_s$ if and only if (A, B) is stabilizable, where $\Sigma(t)$ is the solution of the RDE with final condition Σ_f, and Σ_s is the unique *strong* solution of the ARE.

(ii) Stabilizability and no unobservable modes on stability boundary.

Subject to $\Sigma_f > 0$, then the stabilizability of (A, B) and the nonexistence of unobservable modes of (C, A) on the stability boundary are necessary and sufficient conditions for $\lim_{t \to \infty} \Sigma(t) = \Sigma_s$ (exponentially fast), where $\Sigma(t)$ is the solution of the RDE with final condition Σ_f, and Σ_s is the unique *stabilizing* solution of the ARE.

Proof: See references given at the end of the chapter. Note that this result is the dual of the result given in Theorem 10.7.2. ▽▽▽

Example 11.2.1. Continued

For this example, we have

1. (A, B) is stabilizable and therefore the strong solution ($P = 3.70$) of the ARE exists and is unique.
2. $Q = 1 = 1*1 = C^T C$. Thus (C, A) is completely observable and thus the strong solution is the only nonnegative solution of the ARE.
3. The strong solution is stabilizing and is positive definite.

11.4 FREQUENCY DOMAIN INTERPRETATIONS OF LQ OPTIMAL CONTROL

In the case of optimal state estimation, we found that the algebraic Riccati equation has an interpretation in terms of spectral factorization. In this section, we present the dual result for the LQ optimal control problem. We discuss several implications of this result.

We first note that the dual of the spectral factorization result given in Lemma 10.8.1 is given in:

Lemma 11.4.1

If L and Σ satisfy (11.3.2), (11.3.3), then

$$B^T(\gamma I - A)^{-T}Q\left(-\frac{\gamma}{1+\gamma\Delta}I - A\right)^{-1}B + R =$$

$$\left(I + B^T(\gamma I - A)^{-T}L^T\right)(R + \Delta B^T\Sigma B)\left(I + B^T\left(-\frac{\gamma}{1+\gamma\Delta}I - A\right)^{-T}L^T\right)^T$$

(11.4.1)

Proof: By duality from Lemma 10.8.1. $\triangledown\triangledown\triangledown$

To interpret the result given in Lemma 11.4.1, we consider a simplified case in which u is a scalar and

$$Q = C^TC \tag{11.4.2}$$

where C is a row vector. Note that this is equivalent to minimization of the following criterion:

$$J = \int_0^\infty (y^2 + Ru^2)\, dt \tag{11.4.3}$$

where

$$y = Cx \tag{11.4.4}$$

In this case we have the following result:

Lemma 11.4.2

The control that minimizes (11.4.3), (11.4.4) satisfies

$$N(\gamma)N\left(-\frac{\gamma}{1+\Delta\gamma}\right) + RM(\gamma)M\left(-\frac{\gamma}{1+\Delta\gamma}\right)$$

$$= (R + \Delta B^T\Sigma B)M^*(\gamma)M^*\left(-\frac{\gamma}{1+\Delta\gamma}\right) \tag{11.4.5}$$

where $N(\gamma)/M(\gamma)$ is the input–output transfer function; that is,

$$\frac{N(\gamma)}{M(\gamma)} = C(\gamma I - A)^{-1}B \tag{11.4.6}$$

and where $M^*(\gamma)$ is the closed loop characteristic polynomial:

$$M^*(\gamma) = \det(\gamma I - A + BL) \tag{11.4.7}$$

Proof: Lemma 11.4.1 together with the preceding definitions lead to

$$\frac{N(\gamma)N\left(-\frac{\gamma}{1+\gamma\Delta}\right)}{M(\gamma)M\left(-\frac{\gamma}{1+\gamma\Delta}\right)} + R = (R + \Delta B^T \Sigma B)\left(1 + B^T(\gamma I - A)^{-T}L\right)$$

$$\times \left(1 + B^T\left(\left(-\frac{\gamma}{1+\Delta\gamma}I - A\right)^{-T}L\right)^T\right) \quad (11.4.8)$$

Now

$$M^*(\gamma) = \det(\gamma I - A + BL)$$
$$= \det(\gamma I - A)\det\left(I + (\gamma I - A)^{-1}BL\right)$$
$$= M(\gamma)\det\left(1 + B^T(\gamma I - A)^{-T}L^T\right) \quad (11.4.9)$$

The result follows directly from (11.4.8) and (11.4.9). ▽▽▽

This result can be interpreted in the context of pole assignment by state variable feedback, since it quantifies the relationship between the closed loop poles, the open loop poles and zeros, and the weighting in the cost function. Indeed, since $R + \Delta B^T \Sigma B$ is a scalar, Equation (11.4.5) shows that the closed loop poles can be obtained by spectral factorization without first finding Σ.

The result can be used to study the limiting causes as $R \to 0$ and $R \to \infty$. For the case $R \to 0$, we see that the closed loop poles converge to the plant zeros (if stable) or to the reflection of the plant zeros through the stability boundary (if unstable) plus $(n - m)$ poles at $-1/\Delta$, where n, m are the degree of denominator and numerator, respectively. For the case $R \to \infty$, we see that the closed loop poles converge to the plant poles (if stable) or to the reflection of the plant poles through the stability boundary (if unstable).

Lemma 11.4.2 also gives rise to the following corollary.

Corollary 11.4.1. Gain and Phase Margins of the LQ Optimal Controller

For the case of complete state measurements, the controller that minimizes (11.4.3) has the following properties:

1. For all ω,

$$\left|\sigma\left(\frac{e^{j\omega\Delta} - 1}{\Delta}\right)\right|^2 \geq \alpha \quad (11.4.10)$$

where

$$\alpha \triangleq \frac{R}{R + \Delta B^T \Sigma B} \quad (11.4.11)$$

and $\sigma(\gamma)$ is the return difference:

$$\sigma(\gamma) \triangleq 1 + L(\gamma I - A)^{-1}B = \frac{M^*(\gamma)}{M(\gamma)} \tag{11.4.12}$$

2. The gain margin is at least

$$\frac{1}{1 - \sqrt{\alpha}}$$

3. The phase margin is at least

$$\cos^{-1}(1 - \tfrac{1}{2}\alpha)$$

▽▽▽

Proof:

1. From Lemma 11.4.2 it follows that

$$\frac{N(\gamma)}{M(\gamma)} \frac{N\left(-\frac{\gamma}{1 + \Delta\gamma}\right)}{M\left(-\frac{\gamma}{1 + \Delta\gamma}\right)} + R$$

$$= (R + \Delta B^T \Sigma B) \frac{M^*(\gamma)}{M(\gamma)} \frac{M^*\left(-\frac{\gamma}{1 - \Delta\gamma}\right)}{M\left(-\frac{\gamma}{1 + \Delta\gamma}\right)}$$

$$= (R + \Delta B^T \Sigma B)\sigma(\gamma)\sigma\left(-\frac{\gamma}{1 + \Delta\gamma}\right) \tag{11.4.13}$$

For

$$\gamma = \frac{e^{j\omega\Delta} - 1}{\Delta}, \qquad -\frac{\gamma}{1 + \Delta\gamma} = \gamma^*$$

(that is, the conjugate of γ), and the result then follows.

2/3. From part 1, it follows that the Nyquist diagram for the open loop LQ optimal controller does not enter the region illustrated in Figure 11.4.1. The results then follow by considering the geometry of Figure 11.4.1. ▽▽▽

Note, in the continuous time case ($\Delta = 0$), that $\alpha = 1$, and thus we are assured of infinite gain margin and 60° phase margin. In discrete time (unless the relative degree is zero!), it is impossible to have infinite gain margin. However, by sampling rapidly it is clear that we can make α arbitrarily close to 1, and so the gain margin may be made arbitrarily large.

It should be remembered that the results of this section *depend on the availability of the states* for feedback. In practice, when some form of observer is used, it is rarely true that we obtain the gain and phase margins indicated by Corollary 11.4.1. See also Section 10.12.

We illustrate the preceding results in Example 11.4.1.

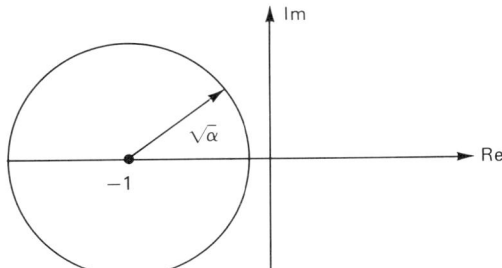

Figure 11.4.1 Forbidden region on the Nyquist diagram for an LQ optimal controller.

Example 11.4.1

Consider a plant having the following state space model:

$$\delta x = \begin{bmatrix} -1 & 0 \\ 1 & 0 \end{bmatrix} x + \begin{bmatrix} 1 \\ 0 \end{bmatrix} u \tag{11.4.14}$$

$$y = \begin{bmatrix} -1 & 1 \end{bmatrix} x$$

where the sampling period is $\Delta = 0.1$. This system has transfer function

$$\frac{N(\gamma)}{M(\gamma)} = \frac{(-\gamma + 1)}{\gamma(\gamma + 1)} \tag{11.4.15}$$

We wish to find the control $u = -Lx$ that minimizes the cost function

$$J = \underset{0}{\overset{\infty}{S}} \left[u^2(t) + y^2(t) \right] dt \tag{11.4.16}$$

Using Lemma 11.4.2, we have that

$$(1 + \Delta B^T \Sigma B) M^*(\gamma) M^*\left(-\frac{\gamma}{1 + \Delta \gamma}\right)$$

$$= N(\gamma) N\left(-\frac{\gamma}{1 + \Delta \gamma}\right) + M(\gamma) M\left(-\frac{\gamma}{1 + \Delta \gamma}\right)$$

$$= \frac{(0.9\gamma^4 - 0.1\gamma^3 - \gamma^2) + (-0.11\gamma^3 - 1.09\gamma^2 + 0.2\gamma + 1)}{(1 + \Delta \gamma)^2}$$

$$= \frac{0.9(\gamma + 1.2103)(\gamma - 1.3770)(\gamma + 0.7838)(\gamma - 0.8505)}{(1 + \Delta \gamma)^2}$$

$$= 1.1102(\gamma + 1.2103)(\gamma + 0.7838)$$

$$\times \left(-\frac{\gamma}{1 + \Delta \gamma} + 1.2103\right)\left(-\frac{\gamma}{1 + \Delta \gamma} + 0.7838\right) \tag{11.4.17}$$

Thus

$$M^*(\gamma) = \gamma^2 + 1.9941\gamma + 0.9486 \tag{11.4.18}$$

and

$$(R + \Delta B^T \Sigma B) = 1.1102 \tag{11.4.19}$$

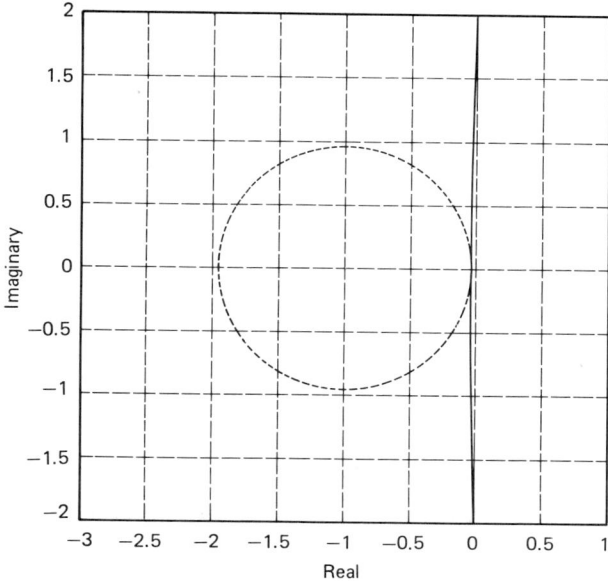

Figure 11.4.2 Nyquist plot for Example 11.4.1.

From the closed loop poles given in (11.4.18), it is easy to show that the optimal feedback gain is

$$L = \begin{bmatrix} 0.9941 & 0.9486 \end{bmatrix} \qquad (11.4.20)$$

Also, from (11.4.19) and Corollary 11.4.1, we have that $\alpha = 0.9$, and thus we expect to have a gain margin greater than 19.5 and a phase margin greater than 56.6° (compared with ∞ and 60°, respectively, for the continuous time case). The open loop Nyquist diagram is shown in Figure 11.4.2. It can be shown that this control loop has a gain margin of 20.07 and a phase margin of about 87°. We leave it as an exercise for the reader to verify, using the algebraic Riccati equation, that (11.4.20) is the optimal gain, and the optimal cost matrix is

$$\Sigma = \begin{bmatrix} 1.1102 & 1.0540 \\ 1.0540 & 3.1020 \end{bmatrix} \qquad (11.4.21)$$

∇∇∇

11.5 INTERNAL MODEL PRINCIPLE IN LQ OPTIMAL CONTROL

The discussion so far has emphasized regulation to the origin in the state space. However, it is easy to extend the ideas to cover the usual tracking considerations involved in practical control system design. For example, the internal model principle can be included in the design by suitably augmenting the plant as shown in Figure 11.5.1.

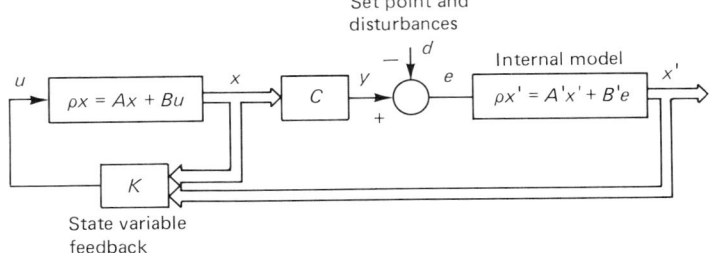

Figure 11.5.1 Augmented state model to include internal model principle.

The composite model is

$$\rho \begin{pmatrix} x \\ x' \end{pmatrix} = \begin{bmatrix} A & 0 \\ B'C & A' \end{bmatrix} \begin{pmatrix} x \\ x' \end{pmatrix} + \begin{pmatrix} B \\ 0 \end{pmatrix} u \quad (11.5.1)$$

We then have the following result, which supports the design of LQ controllers for the composite plant.

Lemma 11.5.1

A sufficient condition for the composite plant (11.5.1) to be completely reachable (from u) is that (i) the original plant is completely reachable (from u), and (ii) if λ is an eigenvalue of A' with corresponding left eigenvector q_2, then $q_2 B' T(\lambda) \neq 0$, where $T(\gamma)$ is the original plant transfer function.

Proof: Suppose

$$\begin{pmatrix} A & 0 \\ B'C & A' \end{pmatrix}, \quad \begin{pmatrix} B \\ 0 \end{pmatrix}$$

is *not* reachable; using Theorem 8.3.3, there exists $[q_1 \quad q_2] \neq 0$ and λ such that

(i) $\quad [q_1 \quad q_2] \begin{bmatrix} B \\ 0 \end{bmatrix} = 0$; that is $q_1 B = 0 \quad (11.5.2)$

and

(ii) $\quad [q_1 \quad q_2] \begin{bmatrix} A & 0 \\ B'C & A' \end{bmatrix} = \lambda [q_1 \quad q_2] \quad (11.5.3)$

That is,

$$q_1 A + q_2 B'C = \lambda q_1 \quad (11.5.4)$$

and

$$q_2 A' = \lambda q_2 \quad (11.5.5)$$

That is, λ is an eigenvalue of the internal model (A') or $q_2 = 0$.

Now if q_2 is zero, then (11.5.4), (11.5.2) imply that the original plant is unreachable. Alternatively, if $q_2 \neq 0$, λ is an eigenvalue of the internal model and

326   Optimal Control   Chap. 11

q_2 is the corresponding left eigenvector. Also, from (11.5.4) we have

$$q_1(\lambda I - A) = q_2 B'C \tag{11.5.6}$$

Multiplying (11.5.6) on the right by $\text{adj}(\lambda I - A)B$ gives

$$\det(\lambda I - A) q_1 B = q_2 B'C \, \text{adj}(\lambda I - A) B \tag{11.5.7}$$

Using (11.5.2) in (11.5.7) gives

$$q_2 B'C \, \text{adj}(\lambda I - A) B = q_2 B' T(\lambda) = 0 \tag{11.5.8}$$

▽▽▽

The necessity of the conditions given in Lemma 11.5.1 is explored in Problem 11.3. The result has simple interpretations in special cases. For example, we immediately have the following corollary.

Corollary 11.5.1

In the case of a scalar output plant and a single integrator internal model, sufficient conditions for complete reachability of the composite plant is that $B' \neq 0$ and the original plant have nonzero dc gain.

Proof: Immediate from Lemma 11.5.1 since $q_2 \neq 0$ and B' are scalars.   ▽▽▽

These results show that, under reasonable conditions, augmentation of the plant by an internal model will preserve reachability. Then the results of Section 11.3 show that it is possible to design an exponentially stable controller via optimization of a linear quadratic criterion. Finally, provided it is constructed properly, the internal model will ensure that the tracking error goes to zero.

11.6 LINEAR QUADRATIC GAUSSIAN (LQG) STOCHASTIC OPTIMAL CONTROL

So far in this chapter we have considered only cases where the complete state vector can be directly measured. In cases where it is not possible to measure the state vector, it seems reasonable that we should replace the true states by state estimates when generating the feedback control signal. For a linear system, Theorem 9.4.1 holds, which ensures that the poles of the resulting closed loop system are simply the poles of the state observer plus those that would have been obtained by feeding back the true states. Indeed, if we hypothesize that the system is described by a linear model with Gaussian noise sources, then it is possible to show that the optimal observer to use is the Kalman filter. To explain this a little further, consider the following stochastic state space model:

$$\rho x = Ax + Bu + v_1 \tag{11.6.1}$$

$$y = Cx + v_2 \tag{11.6.2}$$

where v_1 and v_2 are mutually uncorrelated white noise processes having spectral densities Γ and Ω, respectively, and where $x(t_0)$ is a zero mean Gaussian random variable, independent of v_1, v_2 and having covariance P_0.

Let us assume that the class of admissible feedback signals is such that u can be expressed as a function of past values of y. We wish to determine the input from the admissible class of functions to minimize a quadratic criterion of the form

$$J = E\left\{ x(t_f)^T \Sigma_f x(t_f) + \int_{t_0}^{t_f} \left[x(t)^T Q x(t) + u(t)^T R u(t) \right] dt \right\} \quad (11.6.3)$$

where E denotes expected value. This problem is known as the LQG (linear quadratic Gaussian) optimal control problem. For this problem we have the following result.

Theorem 11.6.1. Stochastic Separation Theorem

The solution to the LQG optimal control problem is

$$u(t) = -L(t+\Delta)\hat{x}(t) \quad (11.6.4)$$

where $L(t)$ satisfies (11.2.18) to (11.2.21); that is,

$$L = [R + \Delta B^T \Sigma B]^{-1} B^T \Sigma [I + A\Delta] \quad (11.6.5)$$

$$\bar{\rho}\Sigma = Q + A^T \Sigma + \Sigma A + \Delta A^T \Sigma A - L^T [R + \Delta B^T \Sigma B] L \quad (11.6.6)$$

$$\Sigma(t_f) = \Sigma_f \quad (11.6.7)$$

and $\bar{\rho}$ is the reverse time generalized derivative.

In Equation (11.6.4), $\hat{x}(t)$ is the optimal state estimate given by

$$\rho\hat{x} = A\hat{x} + H[y - C\hat{x}] \quad (12.6.8)$$

$$H = (\Delta A + I)PC^T [\Delta CPC^T + \Gamma]^{-1} \quad (11.6.9)$$

and

$$\rho P = \Omega + PA^T + AP + \Delta APA^T - H[\Gamma + \Delta CPC^T]H^T \quad (11.6.10)$$

$$P(t_0) = P_0 \quad (11.6.11)$$

Proof: See the references given at the end of the chapter. ▽▽▽

The key point about the preceding result is that it shows that the solution to the LQG optimal control problem is to use the optimal state estimator together with the same feedback as would have been applied had the states been measured. This is an entirely reasonable result from an intuitive point of view. Of course, it does rely on the assumptions of linearity, Gaussian noise, and quadratic optimization criterion.

11.7 OTHER OPTIMIZATION CRITERIA

In all of the preceding discussion we have used a quadratic optimization criterion. Lest this gives the impression that this is the only criterion that can be used, we conclude this chapter by observing that the difficult part in LQ design is the choice of the weighting matrices Q and R. Indeed, a wide variety of responses can be obtained by adjusting these weighting matrices. Hence, the design problem has simply been shifted to another level: what values of Q and R give a good design? More will be said about the design problem in Chapter 13 (although this will be primarily focused on the influence of different closed loop pole locations). Other optimization criteria can also be contemplated. A very useful observation in this context is the fact established in Chapter 9 that there exists a characterization of all stabilizing controllers in the form

$$C_s = \tilde{U}_s^{-1}\tilde{V}_s = V_s U_s^{-1} \tag{11.7.1}$$

where

$$\tilde{U}_s = \tilde{R} + \Omega\tilde{N}; \qquad \tilde{V}_s = \tilde{S} - \Omega\tilde{M} \tag{11.7.2}$$

$$U_s = R + N\Omega; \qquad V_s = S - M\Omega \tag{11.7.3}$$

and where Ω is any stable proper transfer function,

$$\begin{bmatrix} \tilde{R} & \tilde{S} \\ -\tilde{N} & \tilde{M} \end{bmatrix} \begin{bmatrix} M & -S \\ N & R \end{bmatrix} = I \tag{11.7.4}$$

with the plant transfer function expressed in fractional form as

$$G = NM^{-1} = \tilde{M}^{-1}\tilde{N} \tag{11.7.5}$$

Also, we recall from Theorem 9.6.2 that the closed loop system is characterized by transfer functions that are linear in Ω; that is,

$$e = G_1 \omega_y = -G_1 y^* \tag{11.7.6}$$

$$e = G_2 \omega_u \tag{11.7.7}$$

$$u = G_3 \omega_y = -G_3 y^* \tag{11.7.8}$$

$$u = G_4 \omega_u \tag{11.7.9}$$

where

$$G_1 = -M[\tilde{R} - \Omega\tilde{N}] \tag{11.7.10}$$

$$G_2 = -N[\tilde{R} + \Omega\tilde{N}] \tag{11.7.11}$$

$$G_3 = -M[\tilde{S} - \Omega\tilde{M}] \tag{11.7.12}$$

$$G_4 = -N[\tilde{S} - \Omega\tilde{M}] \tag{11.7.13}$$

These transfer functions all have the form

$$G_i = T_1^i - T_2^i \Omega T_3^i, \qquad i = 1, \ldots, 4 \tag{11.7.14}$$

where T_1, T_2, T_3 are known and Ω is any stable proper transfer function.

The affine nature of (11.7.14) can be exploited to specify the control system problem in terms of the minimization of various functions of G_i. For example, there has been considerable interest recently in H^∞ designs, where we choose Ω to minimize

$$J = \|T_1^i - T_2^i \Omega T_3^i\|_{H_\infty} \tag{11.7.15}$$

where the H infinity norm has been defined in Section 10.10.

The motivation for the H_∞ criterion is as follows: Say we have an error transfer function of the form given in (11.7.14) with input v and output r, that is

$$R(\gamma) = G_i(\Omega, \gamma) V(\gamma) \tag{11.7.16}$$

Then, if we allow $v(t)$ to be any signal having constrained energy, it follows by the same arguments as in Section 10.10 that the value of Ω minimizing the maximum $\mathcal{L}_2$ norm of $r(t)$ is given by

$$\Omega^* = \underset{(\Omega \text{ Stable})}{\arg\min} \|G_i(\Omega, \cdot)\|_\infty \tag{11.7.17}$$

where

$$\|G_i(\Omega, \cdot)\|_\infty = \sup_\Omega \sigma_{\max}\left\{G_i\left(\Omega, \frac{e^{j\omega\Delta} - 1}{\Delta}\right)\right\} \tag{11.7.18}$$

where

$$\sigma_{\max}\{G\} = \text{maximum singular value of } G$$
$$= (\text{maximum eigenvalue of } GG^T)^{1/2} \tag{11.7.19}$$

As a preliminary to describing H_∞ optimization, we first describe a state space method for computing the H_∞ norm. Let $C(\gamma I - A)^{-1} B + D$ be a state space realization of $G(\gamma)$ and for simplicity assume $(I + \Delta A)$ is nonsingular. We denote the state space realization by

$$\begin{bmatrix} A & B \\ C & D \end{bmatrix}.$$

Let α be a positive real number such that α^2 is not an eigenvalue of $(D^T - \Delta B^T (I + \Delta A^T)^{-1} C^T) D$. We then define the following generalized Hamiltonian matrix

$$M_\alpha \triangleq \begin{bmatrix} A & 0 \\ 0 & -A^T(I + \Delta A^T)^{-1} \end{bmatrix}$$
$$+ \begin{bmatrix} B & 0 \\ 0 & -(I + \Delta A^T)^{-1} C^T \end{bmatrix} \begin{bmatrix} -D & \alpha I \\ \alpha I & -(D^T - \Delta B^T(I + \Delta A^T)^{-1} C^T) \end{bmatrix}^{-1}$$
$$\times \begin{bmatrix} C & 0 \\ 0 & B^T(I + \Delta A^T)^{-1} \end{bmatrix}$$
$$\triangleq \begin{bmatrix} W & X \\ Y & Z \end{bmatrix} \tag{11.7.20}$$

where

$$W \triangleq A - BR^{-1}\left(D^T - \Delta B^T(I + \Delta A^T)^{-1}C^T\right)C \qquad (11.7.21)$$

$$X \triangleq -\alpha BR^{-1}B^T(I + \Delta A^T)^{-1} \qquad (11.7.22)$$

$$Y \triangleq \alpha(I + \Delta A^T)^{-1}C^T S^{-1} C \qquad (11.7.23)$$

$$Z \triangleq -A^T(I + \Delta A^T)^{-1} + (I + \Delta A^T)^{-1} C^T D R^{-1} B^T (I + \Delta A^T)^{-1} \qquad (11.7.24)$$

and

$$R \triangleq \left(D^T - \Delta B^T(I + \Delta A^T)^{-1} C^T\right) D - \alpha^2 I \qquad (11.7.25)$$

$$S \triangleq D\left(D^T - \Delta B^T(I + \Delta A^T)^{-1} C^T\right) - \alpha^2 I \qquad (11.7.26)$$

Note that M_α satisfies the requirements of a generalized Hamiltonian matrix as defined in Equation (11.3.22).

We also have the following result:

Lemma 11.7.1

Assume that A does not have an eigenvalue on the stability boundary and that $(I + \Delta A)$ is nonsingular. Let α be some positive number and assume that R and S in (11.7.25), (11.7.26) are nonsingular.

(i) If $\gamma_0 = (e^{j\omega_0 \Delta} - 1)/\Delta$ is some complex number on the stability boundary, then α is a singular value of $G(\gamma_0)$ if and only if $(M_\alpha - \gamma_0 I)$ is singular.

(ii) A state space realization for $[\alpha^2 I - G(\gamma)^* G(\gamma)]$ is

$$\begin{bmatrix} M_\alpha & & -BR^{-1} \\ & & (1/\alpha)(I + \Delta A^T)^{-1} C^T D R^{-1} \\ -R^{-1}NC & -\alpha R^{-1} B^T (I + \Delta A^T)^{-1} & -R^{-1} \end{bmatrix}$$

where $N = D^T - \Delta B^T (I + \Delta A^T)^{-1} C^T$ and R, S are as in (11.7.25), (11.7.26).

Proof: (i) (Only if) Let α be a singular value of $G(\gamma_0)$. Then there exists non-zero vectors u, v such that (11.7.27), (11.7.28) hold with $w = 0$

$$G(\gamma_0) u = \alpha v \qquad (11.7.27)$$

$$G(\gamma_0)^* v = \alpha u - \tfrac{1}{\alpha} w \qquad (11.7.28)$$

where

$$G(\gamma_0) = C(\gamma_0 I - A)^{-1} B + D$$

$$G(\gamma_0)^* = \left\{ C\left[\frac{-\gamma_0}{1 + \gamma_0 \Delta} I - A\right]^{-1} B + D \right\}^T$$

Sec. 11.7   Other Optimization Criteria

So

$$C(\gamma_0 I - A)^{-1} Bu + Du = \alpha v \qquad (11.7.29)$$

$$B^T\left(\frac{-\gamma_0}{1+\gamma_0\Delta}I - A^T\right)^{-1} C^T v + D^T v = \alpha u - \tfrac{1}{\alpha}w \qquad (11.7.30)$$

We note from (11.7.27), (11.7.28) that

$$[\alpha^2 I - G(\gamma_0)^* G(\gamma_0)] u = w \qquad (11.7.31)$$

We next define state variables r and s by

$$r = (\gamma_0 I - A)^{-1} Bu \qquad (11.7.32a)$$

$$s = -\left(\gamma_0 I + (I + \Delta A^T)^{-1} A^T\right)^{-1} (I + \Delta A^T)^{-1} C^T v \qquad (11.7.32b)$$

Now Equation (11.7.30) can be written as

$$-B^T(1+\gamma_0\Delta)\left[\gamma_0(I+\Delta A^T)+A^T\right]^{-1} C^T v + D^T v = \alpha u - \tfrac{1}{\alpha}w \qquad (11.7.33)$$

$$-B^T(1+\gamma_0\Delta)\left[\gamma_0 I + (I+\Delta A^T)^{-1} A^T\right]^{-1}[I+\Delta A^T]^{-1} C^T v + D^T v$$
$$= \alpha u - \tfrac{1}{\alpha}w \qquad (11.7.34)$$

$$-\Delta B^T\left(\tfrac{1}{\Delta} + \gamma_0\right)\left[\gamma_0 I + (I+\Delta A^T)^{-1} A^T\right]^{-1}[I+\Delta A^T]^{-1} C^T v + D^T v$$
$$= \alpha u - \tfrac{1}{\alpha}w \qquad (11.7.35)$$

$$-\Delta B^T\left\{\gamma_0 I + (I+\Delta A^T)^{-1} A^T + \left[\tfrac{1}{\Delta}I - (I+\Delta A^T)^{-1} A^T\right]\right\}$$
$$\times\left\{\gamma_0 I + (I+\Delta A^T)^{-1} A^T\right\}^{-1}[I+\Delta A^T]^{-1} C^T v + D^T v = \alpha u - \tfrac{1}{\alpha}w \qquad (11.7.36)$$

$$\left(-\Delta B^T[I+\Delta A^T]^{-1} C^T + D^T\right)^v u - B^T\left\{I - (I+\Delta A^T)^{-1}\Delta A^T\right\}$$
$$\times\left\{\gamma_0 I + (I+\Delta A^T)^{-1} A^T\right\}^{-1}[I+\Delta A^T]^{-1} C^T v = \alpha u - \tfrac{1}{\alpha}w \qquad (11.7.37)$$

$$-B^T[I+\Delta A^T]^{-1}\left[\gamma_0 I + (I+\Delta A^T)^{-1} A^T\right]^{-1}[I+\Delta A^T]^{-1} C^T v$$
$$+\left(D^T - \Delta B^T[I+\Delta A^T]^{-1} C^T\right)v = \alpha u - \tfrac{1}{\alpha}w \qquad (11.7.38)$$

Combining (11.7.32b) and (11.7.38) gives

$$B^T[I+\Delta A^T]^{-1} s + \left[D^T - \Delta B^T(I+\Delta A^T)^{-1} C^T\right] v = \alpha u - \tfrac{1}{\alpha}w \qquad (11.7.39)$$

Hence, from (11.7.29), (11.7.32a) and (11.7.39) we have

$$\begin{bmatrix} u \\ v \end{bmatrix} = \begin{bmatrix} -D & \alpha I \\ \alpha I & -(D^T - \Delta B^T(I + \Delta A^T)^{-1} C^T) \end{bmatrix}^{-1}$$

$$\times \left\{ \begin{bmatrix} C & 0 \\ 0 & B^T(I + \Delta A^T)^{-1} \end{bmatrix} \begin{bmatrix} r \\ s \end{bmatrix} + \begin{bmatrix} 0 \\ +\frac{1}{\alpha} \end{bmatrix} w \right\} \quad (11.7.40)$$

Thus for $w = 0$, $[r^T \ s^T] \neq 0$. Also, from (11.7.31), (11.7.32) with $w = 0$

$$(\gamma_0 I - A)r = Bu \quad (11.7.41)$$

$$\left(\gamma_0 I + (I + \Delta A^T)^{-1} A^T\right) s = -(I + \Delta A^T)^{-1} C^T v \quad (11.7.42)$$

Hence combining (11.7.40) with (11.7.41), (11.7.42) we have for $w = 0$,

$$\left\{ \begin{bmatrix} A & 0 \\ 0 & -A^T(I + \Delta A^T)^{-1} \end{bmatrix} \right.$$

$$+ \begin{bmatrix} B & 0 \\ 0 & -(I + \Delta A^T)^{-1} C^T \end{bmatrix} \begin{bmatrix} -D & \alpha I \\ \alpha I & -(D^T - \Delta B^T(I + \Delta A^T)^{-1} C^T) \end{bmatrix}^{-1}$$

$$\left. \times \begin{bmatrix} C & 0 \\ 0 & B^T(I + \Delta A^T)^{-1} \end{bmatrix} \right\} \begin{bmatrix} r \\ s \end{bmatrix}$$

$$= \gamma_0 \begin{bmatrix} r \\ s \end{bmatrix} \quad (11.7.43)$$

Thus, noting (11.7.20), we have

$$M_\alpha \begin{bmatrix} r \\ s \end{bmatrix} = \gamma_0 \begin{bmatrix} r \\ s \end{bmatrix} \quad (11.7.44)$$

(If) Let M_α have an eigenvalue γ_0 on the stability boundary. Then (11.7.44) holds for some $[r^T \ s^T] \neq 0$. Define u, v by (11.7.40) with $w = 0$. Then $[u^T \ v^T] \neq 0$. From (11.7.40), (11.7.43) we have (11.7.29), (11.7.30) with $w = 0$. This establishes that α is a singular value of $G(\gamma_0)$.

(ii) From (11.7.40) we have

$$\begin{bmatrix} u \\ v \end{bmatrix} = \begin{bmatrix} -D & \alpha I \\ \alpha I & -N \end{bmatrix}^{-1} \left\{ \begin{bmatrix} C & 0 \\ 0 & B^T(I + \Delta A^T)^{-1} \end{bmatrix} \begin{bmatrix} r \\ s \end{bmatrix} + \begin{bmatrix} 0 \\ -\frac{1}{\alpha} \end{bmatrix} w \right\}$$

$$= \begin{bmatrix} -R^{-1}N & -\alpha R^{-1} \\ -\alpha S^{-1} & -DR^{-1} \end{bmatrix} \left\{ \begin{bmatrix} C & 0 \\ 0 & B^T(I + \Delta A^T)^{-1} \end{bmatrix} \begin{bmatrix} r \\ s \end{bmatrix} + \begin{bmatrix} 0 \\ -\frac{1}{\alpha} \end{bmatrix} w \right\} \quad (11.7.45)$$

where R, S are as in (11.7.25) and (11.7.26) and

$$N = D^T - \Delta B^T(I + \Delta A^T)^{-1} C^T \quad (11.7.46)$$

Sec. 11.7   Other Optimization Criteria   333

Simplifying (11.7.45) gives

$$\begin{bmatrix} u \\ v \end{bmatrix} = \begin{bmatrix} -R^{-1}NC & -\alpha R^{-1}B^T(I + \Delta A^T)^{-1} \\ -\alpha S^{-1}C & -DR^{-1}B^T(I + \Delta A^T)^{-1} \end{bmatrix} \begin{bmatrix} r \\ s \end{bmatrix} + \begin{bmatrix} R^{-1} \\ \frac{1}{\alpha}DR^{-1} \end{bmatrix} w \quad (11.7.47)$$

Substituting (11.7.47) into (11.7.32a), (11.7.32b) gives

$$\gamma_0 \begin{bmatrix} r \\ s \end{bmatrix} = M_\alpha \begin{bmatrix} r \\ s \end{bmatrix} + \begin{bmatrix} -BR^{-1} \\ +\frac{1}{\alpha}(I + \Delta A^T)^{-1}C^T DR^{-1} \end{bmatrix} w \quad (11.7.48)$$

$$u = \begin{bmatrix} -R^{-1}NC & -\alpha R^{-1}B^T(I + \Delta A^T)^{-1} \end{bmatrix} \begin{bmatrix} r \\ s \end{bmatrix} - R^{-1}w \quad (11.7.49)$$

The result follows on comparing (11.7.48), (11.7.49) with (11.7.31). ▽▽▽

Lemma 11.7.1 immediately implies the following result:

Lemma 11.7.2

Under the hypothesis that

(i) A is stable and $(I + \Delta A)$ nonsingular
(ii) $\alpha > \min\{\sigma_{\max}(D - CA^{-1}B), \sigma_{\max}(D - \Delta C(2I + \Delta A)^{-1}B)\}$
(iii) α^2 is not an eigenvalue of $(D^T - \Delta B^T(I + \Delta A^T)^{-1}C^T)D$.

Then the H_∞ norm of G is less than α if and only if M_α has no eigenvalues on the stability boundary.

Proof: (If) Since (ii) holds then either

$$\lim_{w \to 0} \sigma_{\max}\left[G\left(\frac{e^{j\omega\Delta} - 1}{\Delta}\right)\right] < \alpha \quad (11.7.50)$$

or

$$\lim_{w \to \pi/\Delta} \sigma_{\max}\left[G\left(\frac{e^{j\omega\Delta} - 1}{\Delta}\right)\right] < \alpha \quad (11.7.51)$$

We assume the contrary, that is $\|G\|_\infty > \alpha$. Since $\sigma_{\max} G(\frac{e^{j\omega\Delta} - 1}{\Delta})$ is a continuous function of $\omega \in [0, \pi/\Delta]$, and since by assumption $\|G\|_\infty > \alpha$, then it follows from (11.7.50), (11.7.51) that there exists some γ_0 on the stability boundary such that

$$\sigma_{\max} G(\gamma_0) = \alpha \quad (11.7.52)$$

The result then follows from Lemma 11.7.1.
 (Only if) Immediate. ▽▽▽

Lemma 11.7.2 suggests a simple iterative procedure for computing the H_∞ norm by sequential adjustment of α; checking the eigenvalues of M_α at each step.

We are now in a position to discuss the H_∞ design problem. For simplicity, we will consider only the case when the complete state vector is measured. (The extension to the case of output feedback uses a second Riccati equation to design a state estimator. This and other related issues are discussed in the references given at the end of the chapter.) Our principal purpose here is to show that continuous and discrete H_∞ design can be simultaneously treated using a delta formulation.

Consider a state space description of the system in the form:

$$\rho x = Ax + B_1 w + B_2 u; \qquad x(0) = x_0 \qquad (11.7.53)$$

$$z = Cx + Du \qquad (11.7.54)$$

where w represents a disturbance and u a controllable input. To further simplify the presentation, we assume that $[A, B_2]$ is stabilizable, $[C, A]$ is detectable and $(I + \Delta A)^{-1}$ exists.

In the above system, the transfer function from the disturbance input w to the output z is a function of the choice of the input u. Motivated by implementation requirements, we will restrict the input to be a linear function of the current state of the system, and we denote this class of inputs by U.

Our objective is to choose $u \in U$ so as to minimize the $\mathscr{L}_2$ norm of the output z for the worst possible $\mathscr{L}_2$ bounded disturbance input w. Thus we seek

$$\hat{u}^* = \arg\min_{u \in U} \sup_{\|w\|_2 \le 1} \|z\|_2 \qquad (11.7.55)$$

We will approach this problem by choosing a scalar α and then by seeking an input $\hat{u} \in U$ which achieves the following result:

$$\|z\|_2 \le \alpha \|w\|_2 \text{ for all } w; \text{ and } \|z\|_2 = \alpha \|\hat{w}\|_2 \text{ for some } \hat{w} \qquad (11.7.56)$$

Then α is the H_∞ norm of the transfer function linking w and z. An iterative solution to (11.7.55) is obtained by successively reducing α until a solution for $\hat{u}$ can no longer be found such that (11.7.56) is satisfied.

We will present a solution to the above problem via an indirect route. In particular, we will show that the solution to the problem, as described above, is related to the following "game" problem:

Let $J(w, u)$ be defined as

$$J(w, u) = \frac{1}{2} S_0^T (z^T z - \alpha^2 w^T w) \, dt \qquad (11.7.57)$$

where, as usual, the symbol S denotes generalized integral—see Section 4.4. We then seek a saddlepoint solution w^*, u^* such that the w player maximizes the cost while the u player minimizes it; that is,

$$J(w^*, u^*) = \min_u \max_w J(w, u) \qquad (11.7.58)$$

We will express the solution (w^*, u^*) to (11.7.58) in the form of state feedback control laws. It will turn out that for $x(0) = 0$, $J(w^*, u^*) = 0$. We also define $\hat{u}$ as the value of this state feedback law when w takes any value (that is, w is not necessarily w^*). $\hat{w}$ is defined similarly. We then show that when $u = \hat{u}$ but w is not necessarily $\hat{w}$, the cost decreases; that is, for $x(0) = 0$, we have

$$J(w, \hat{u}) \leq J(w^*, u^*) = 0 \tag{11.7.59}$$

Finally, substituting (11.7.59) into (11.7.57) we see that, with $u = \hat{u}$, (11.7.56) is satisfied. To develop this program, we first give the necessary conditions for the saddlepoint solution.

Lemma 11.7.3

Necessary conditions for the saddlepoint w^*, u^*, x^*, z^* are:
For $t = 0, 1, \ldots, T - \Delta$

$$D^T z^*(t) + B_2^T \lambda^*(t + \Delta) = 0 \tag{11.7.60}$$

$$-\alpha^2 w^*(t) + B_1^T \lambda^*(t + \Delta) = 0 \tag{11.7.61}$$

where λ^* satisfies the following "costate" equation for $t = 0, 1, \ldots, T - \Delta$

$$\rho \lambda^*(t) = -A^T \lambda^*(t + \Delta) - C^T z^*(t) \tag{11.7.62}$$

with

$$\lambda^*(T) = 0 \tag{11.7.63}$$

Proof: The optimization in (11.7.58) needs to be carried out subject to the constraint (11.7.53). We therefore add this constraint to the cost function using a Lagrange multiplier $\lambda(t + \Delta)$. This gives

$$J(w, u)' = J(w, u) + \int_0^T \lambda(t + \Delta)^T [Ax(t) + B_1 w(t) + B_2 u(t) - \rho x(t)] \, dt \tag{11.7.64}$$

Integrating the last term in (11.7.64) by parts (see Lemma 4.4.1, part viii) we obtain

$$J(w, u)' = J(w, u) + \int_0^T \lambda(t + \Delta)^T [Ax(t) + B_1 w(t) + B_2 u(t)]$$

$$- \lambda(T)^T x(T) + \lambda(0)^T x(0)$$

$$+ \int_0^T [\rho \lambda(t)]^T x(t) \, dt$$

$$= J(w, u) + \int_0^T \{ \lambda(t + \Delta)^T [Ax(t) + B_1 w(t) + B_2 u(t)]$$

$$+ [\rho \lambda^*(t)]^T x(t) \} \, dt$$

$$- \lambda(T)^T x(T) + \lambda(0)^T x(0) \tag{11.7.65}$$

Then, setting $\partial J'/\partial u$, $\partial J'/\partial w$, $\partial J'\partial x$ equal to zero for $t = 0, 1, \ldots, T - \Delta$, gives (11.7.60) to (11.7.62) respectively. Also, setting $\partial J'/\partial x(t)$ to zero gives (11.7.63). ▽▽▽

Further insight into the necessary conditions given in Lemma 11.7.3 are presented in Lemma 11.7.4.

Lemma 11.7.4

(a) For all w and u the cost function (11.7.54) can be expressed as

$$J(w, u) = \tfrac{1}{2} x(0)^T \lambda^*(0) + \tfrac{1}{2} \|z - z^*\|_2^2$$

$$- \frac{\alpha^2}{2} \|w - w^*\|_2^2 \qquad (11.7.66)$$

where z^*, w^* are the particular time functions satisfying Lemma 11.7.3 and $\|z\|_2$ denotes the L_2 norm of z, i.e.

$$\|z\|_2^2 \triangleq \underset{0}{\overset{T}{S}} z(t)^T z(t) \, dt \qquad (11.7.67)$$

(b) $\quad J(w^*, u^*) = \tfrac{1}{2} x(0)^T \lambda^*(0) \leq J(w^*, u) \qquad (11.7.68)$

and hence $J(w^*, u^*) = 0$ for $x(0) = 0$.

Proof:

(a) Consider the saddlepoint of the game. Equations (11.7.53), (11.7.54) become:

$$\rho x^*(t) = Ax^*(t) + B_1 w^*(t) + B_2 u^*(t) \qquad (11.7.69)$$
$$z^*(t) = Cx^*(t) + Du^*(t) \qquad (11.7.70)$$

These equations, together with (11.7.60) to (11.7.63) yield

$$z^*(t)^T z^*(t) - \gamma^2 w^*(t)^T w^*(t)$$
$$= \left(x^*(t)^T C^T + u^*(t)^T D^T\right) z^*(t) - \gamma^2 w^*(t)^T w^*(t)$$
$$= x^*(t)^T \left[-\rho \lambda^*(t) - A^T \lambda^*(t + \Delta)\right]$$
$$\quad - u^*(t)^T B_2^T \lambda^*(t + \Delta)$$
$$\quad - w^*(t)^T B_1^T \lambda^*(t + \Delta)$$
$$= -x^*(t)^T [\rho \lambda^*(t)] - [Ax^*(t) + B_1 w^*(t) + B_2 u^*(t)]^T \lambda^*(t + \Delta)$$
$$= -x^*(t)^T [\rho \lambda^*(t)] - [\rho x^*(t)]^T \lambda^*(t + \Delta)$$
$$= -\rho \left[x^*(t)^T \lambda^*(t)\right] \qquad (11.7.71)$$

Hence

$$\int_0^T \left[z^*(t)^T z^*(t) - \gamma^2 w^*(t)^T w^*(t)\right] dt$$
$$= -\left[x^*(t)^T \lambda^*(t)\right]_0^T$$
$$= x(0)^T \lambda^*(0) \qquad (11.7.72)$$

Similarly, we have

$$\int_0^T \left[z(t)^T z^*(t) - \gamma^2 w(t)^T w^*(t)\right] dt = x(0)^T \lambda^*(0) \qquad (11.7.73)$$

Finally,

$$\tfrac{1}{2}\left\{\|z - z^*\|_2^2 - \alpha^2 \|w - w^*\|_2^2\right\}$$
$$= \frac{1}{2} \int_0^T \left\{\left(z(t)^T z(t) - \alpha^2 w(t)^T w(t)\right)\right.$$
$$+ \left(z^*(t)^T z^*(t) - \alpha^2 w^*(t)^T w^*(t)\right)$$
$$\left. - 2\left(z(t)^T z^*(t) - \alpha^2 w(t)^T w^*(t)\right)\right\} dt$$
$$= J(w, u) + \tfrac{1}{2} x(0)^T \lambda^*(0) - x(0)^T \lambda^*(0) \qquad (11.7.74)$$

This immediately implies (11.7.66).
(b) With $u = u^*$, $w = w^*$, then $z = z^*$. Hence (11.7.66) implies

$$\tfrac{1}{2} x(0)^T \lambda^*(0) = J(w^*, u^*) \leq J(w^*, u) = \tfrac{1}{2} x(0)^T \lambda^*(0) + \tfrac{1}{2}\|z - z^*\|_2^2 \qquad (11.7.75)$$

▽▽▽

A difficulty with the necessary conditions given in Lemma 11.7.3 is that the boundary conditions for $x(t)$ are defined at time $t = 0$, whereas the boundary conditions for $\lambda(t)$ are defined at time $t = T$. This kind of problem is generally termed a "Two Point Boundary Value Problem." These problems are, in general, very difficult to solve. However, since the equations in Lemma 11.7.3 are linear, we can obtain an explicit solution as outlined below. To simplify the presentation, we assume that D has full column rank. We also carry out some preliminary transformations to convert the problem into a simplified form.

Lemma 11.7.5

Consider the system description of (11.7.53) (11.7.54) where D has full column rank. We can then transform the system into the following form via a state feedback transformation:

$$\rho x = \overline{A} x + B_1 w + B_2 \overline{u} \qquad (11.7.76)$$
$$\overline{z} = \overline{C} x + \overline{D} \overline{u} \qquad (11.7.77)$$

where
$$\bar{z} \triangleq Ez \quad (11.7.78)$$
with E a nonsingular transformation such that
$$ED \triangleq \begin{bmatrix} I \\ 0 \end{bmatrix} \quad (11.7.79)$$
Also, in (11.7.76), (11.7.77) we have
$$\bar{A} \triangleq A - B_2 D^T E^T EC \quad (11.7.80)$$
$$\bar{C} \triangleq EC - EDD^T E^T EC \quad (11.7.81)$$
$$\bar{D} \triangleq ED \quad (11.7.82)$$
$$\bar{u} \triangleq u + D^T E^T ECx \quad (11.7.83)$$
For the above transformed system we have
$$\bar{D}^T [\bar{C} \quad \bar{D}] = [0 \quad I] \quad (11.7.84)$$

Proof: We define E by (11.7.79). Then, premultiplying (11.7.54) by E gives
$$\bar{z} = ECx + \bar{D}u \quad (11.7.85)$$
Now define $\bar{u}$ by (11.7.83). Substituting (11.7.83) into (11.7.85) and (11.7.53) gives the result. Equation (11.7.84) follows from (11.7.81), (11.7.82), (11.7.79).

Note that state feedback of the form (11.7.83) does not change the detectability or stabilizability of the system. ▽▽▽

In view of Lemma (11.7.5), we can assume, without loss of generality, that the following condition holds for the model (11.7.53), (11.7.54) provided D has full column rank
$$D^T [C \quad D] = [0 \quad I] \quad (11.7.86)$$
We next express the two point boundary value problem in a slightly different form. Using (11.7.86), we see that $D^T z = u$ and $C^T z = C^T Cx$. Then, using the fact that $\lambda(t + \Delta) = (1 + \Delta\rho)\lambda(t)$, Equation (11.7.62) can be written as (where we have dropped the superscript * for convenience):
$$(I + A^T \Delta)\rho\lambda(t) = -A^T \lambda(t) - C^T Cx(t) \quad (11.7.87)$$
Similarly, substituting (11.7.60), (11.7.61) into (11.7.53) gives
$$\rho x(t) = Ax(t) + \left(\frac{1}{\alpha^2} B_1 B_1^T - B_2 B_2^T\right)(1 + \Delta\rho)\lambda(t) \quad (11.7.88)$$
Equation (11.7.87) and (11.7.88) can be written together as
$$\begin{bmatrix} I & -\Delta\left(\frac{1}{\alpha^2} B_1 B_1^T - B_2 B_2^T\right) \\ 0 & (I + A^T \Delta) \end{bmatrix} \begin{bmatrix} \rho x(t) \\ \rho \lambda(t) \end{bmatrix}$$
$$= \begin{bmatrix} A & \frac{1}{\alpha^2} B_1 B_1^T - B_2 B_2^T \\ -C^T C & -A^T \end{bmatrix} \begin{bmatrix} x(t) \\ \lambda(t) \end{bmatrix} \quad (11.7.89)$$

Sec. 11.7 Other Optimization Criteria

or

$$\rho \begin{bmatrix} x(t) \\ \lambda(t) \end{bmatrix} = M_\infty \begin{bmatrix} x(t) \\ \lambda(t) \end{bmatrix}; \qquad \begin{bmatrix} x(0) \\ \lambda(t) \end{bmatrix} = \begin{bmatrix} x_0 \\ 0 \end{bmatrix} \qquad (11.7.90)$$

where M_∞ is the following generalized Hamiltonian matrix

$$M_\infty = \begin{bmatrix} I & -\Delta\left(\frac{1}{\alpha^2}B_1B_1^T - B_2B_2^T\right) \\ 0 & (I + A^T\Delta) \end{bmatrix}^{-1} \begin{bmatrix} A & \frac{1}{\alpha^2}B_1B_1^T - B_2B_2^T \\ -C^TC & -A^T \end{bmatrix}$$

$$= \begin{bmatrix} A - \Delta\left(\frac{1}{\alpha^2}B_1B_1^T - B_2B_2^T\right)(I + A^T\Delta)^{-1}C^TC & \left(\frac{1}{\alpha^2}B_1B_1^T - B_2B_2^T\right)(I + A^T\Delta)^{-1} \\ -(I + A^T\Delta)^{-1}C^TC & -(I + A^T\Delta)^{-1}A^T \end{bmatrix}$$
$$(11.7.91)$$

From (11.7.90) we can write

$$\begin{bmatrix} x(t) \\ \lambda(t) \end{bmatrix} = \begin{bmatrix} \phi_{11}(t,T) & \phi_{12}(t,T) \\ \phi_{21}(t,T) & \phi_{22}(t,T) \end{bmatrix} \begin{bmatrix} x(T) \\ \lambda(T) \end{bmatrix} \triangleq \Phi(t,T) \begin{bmatrix} x(T) \\ \lambda(T) \end{bmatrix} \qquad (11.7.92)$$

where $\Phi(t,T)$ is the state transition matrix associated with M_∞. That is

$$\rho\Phi(t,T) = M_\infty \Phi(t,T); \qquad \Phi(T,T) = I \qquad (11.7.93)$$

We note that, as α increases, M_∞ converges to the generalized Hamiltonian matrix associated with the linear quadratic optimal control problem given in (11.3.23). Moreover, it is well known, that $\phi_{11}(t,T)^{-1}$ exists for the latter problem under observability and stabilizability assumptions. It therefore seems reasonable to conclude that if we choose α large enough, then $\phi_{11}(t,T)^{-1}$ will exist for the game problem. Under this assumption, from (11.7.92) we have

$$\lambda(t) = \Sigma(t)x(t) \qquad (11.7.94)$$

where

$$\Sigma(t) = \phi_{21}(t,T)\phi_{11}(t,T)^{-1}; \qquad \Sigma(T) = 0 \qquad (11.7.95)$$

It is readily shown by "differentiating" (11.7.95) and using (11.7.93) that $\Sigma(t)$ satisfies the following matrix Riccati equation:

$$-\rho\Sigma(t - \Delta) = C^TC + A^T\Sigma(t) + \Sigma(t)A + \Delta A^T\Sigma(t)A$$
$$+ (I + A\Delta)^T\Sigma(t)\left[I - \Delta\left(\frac{1}{\alpha^2}B_1B_1^T - B_2B_2^T\right)\Sigma(t)\right]^{-1}$$
$$\times \left[\frac{1}{\alpha^2}B_1B_1^T - B_2B_2^T\right]\Sigma(t)(I + A\Delta) \qquad (11.7.96)$$

Also, from (11.7.94), (11.7.60), (11.7.64)

$$\lambda(t + \Delta) = \Sigma(t + \Delta)\left[(I + \Delta A)x(t) + \Delta B_1 w(t) + \Delta B_2 u(t)\right]$$
$$= \Sigma(t + \Delta)\left[(I + \Delta A)x(t) + \frac{\Delta}{\alpha^2}B_1B_1^T\lambda(t + \Delta)\right.$$
$$\left. - \Delta B_2 B_2^T \lambda(t + \Delta)\right] \qquad (11.7.97)$$

Hence

$$\left(I - \Delta\Sigma(t+\Delta)\left[\frac{B_1 B_1^T}{\alpha^2} - B_2 B_2^T\right]\right)\lambda(t+\Delta) = \Sigma(t+\Delta)[I + A\Delta]x(t) \tag{11.7.98}$$

Substituting (11.7.98) into (11.7.60), (11.7.67) gives

$$u^*(t) = -L_\infty(t)x^*(t) \tag{11.7.99}$$

where

$$L_\infty(t) = B_2^T\left[I - \Delta\Sigma(t+\Delta)\left(\frac{B_1 B_1^T}{\alpha^2} - B_2 B_2^T\right)\right]^{-1}\Sigma(t+\Delta)(I + \Delta A) \tag{11.7.100}$$

and

$$w^*(t) = Q_\infty(t)x^*(t) \tag{11.7.101}$$

where

$$Q_\infty(t) = \frac{1}{\alpha^2}B_1^T\left[I - \Delta\Sigma(t+\Delta)\left(\frac{B_1 B_1^T}{\alpha^2} - B_2 B_2^T\right)\right]^{-1}\Sigma(t+\Delta)(I + \Delta A) \tag{11.7.102}$$

Equations (11.7.99) and (11.7.101) define the joint saddle point conditions for the input and disturbance. However, we are interested in the case when the disturbance takes any value in L_2. We therefore define $\hat{u}(t)$ as in (11.7.99) but where $x^*(t)$ is replaced by the state vector $x(t)$ corresponding to any disturbance input, that is

$$\hat{u}(t) = -L_\infty(t)x(t) \tag{11.7.103}$$

where $L_\infty(t)$ is given by (11.7.100)
Similarly, we define

$$\hat{\omega}(t) = Q_\infty(t)x(t) \tag{11.7.104}$$

where $Q_\infty(t)$ is given by (11.7.102)
We then have the following results.

Lemma 11.7.6

The cost function of the finite time game problem is given by

$$J(w, u) = \tfrac{1}{2}x(0)^T\Sigma(0)x(0) + \|u - \hat{u}\|_2^2 - \alpha^2\|w - \hat{w}\|_2^2$$
$$+ \frac{\Delta}{2}\int_0^T [B_1(w(t) - \hat{w}(t)) + B_2(u(t) - \hat{u}(t))]^T$$
$$\times \Sigma(t+\Delta)[B_1(w(t) - \hat{w}(t)) + B_2(u(t) - \hat{u}(t))]\, dt \tag{11.7.105}$$

where $\hat{u}(t)$ and $\hat{w}(t)$ are given by (11.7.103), (11.7.104).

Proof: By some algebraic manipulation using (11.7.103), (11.7.104) and (11.7.57).

Lemma 11.7.7

The cost function for the finite time game problem satisfies the following condition for all $w, u \in L^2[0, T]$:
$$J(w, \hat{u}) \leq J(\hat{w}, \hat{u}) \qquad (11.7.106)$$
if and only if
$$\alpha^2 I - \Delta B_1^T \Sigma(t + \Delta) B_1 \geq 0 \quad \text{for all} \quad T \in [0, T - \Delta] \qquad (11.7.107)$$

Proof: It follows from (11.7.105) that
$$J(w, \hat{u}) = -\frac{1}{2} \int_0^T (w(t) - \hat{w}(t))^T (\alpha^2 I - \Delta B_1^T \Sigma(t + \Delta) B_1)(w(t) - \hat{w}(t)) \, dt$$
$$+ \tfrac{1}{2} x(0)^T \Sigma(0) x(0) \qquad (11.7.108)$$
and
$$J(\hat{w}, \hat{u}) = \tfrac{1}{2} x(0)^T \Sigma(0) x(0) \qquad (11.7.109)$$

The "if" part of the result then follows immediately. Conversely, if (11.7.107) is not true then there exists a time t' and a vector v such that $v^T(\alpha^2 I - \Delta B_1^T \Sigma(t' + \Delta) B_1) v < 0$ and hence by putting $w(t) = \hat{w}(t) + i(t') v$ we can make the first term on the right hand side of (11.7.108) positive. [Here $i(\cdot)$ denotes the unit pulse function.]

We note that $\hat{u}(t)$ is a linear function of $x(t)$ and thus belongs to U (defined after (11.7.54)). Then, with $x(0) = 0$, we see that (11.7.106) implies (11.7.56). We have thus achieved our objective of finding an input $\hat{u} \in U$ which achieves
$$\|z\|_2 \leq \alpha \|w\|_2 \quad \text{for all } w; \text{ and } \|z\|_2 = \alpha \|\hat{w}\|_2 \quad \text{for some } \hat{w} \qquad (11.7.109)$$
Then, for $u = \hat{u}$, the norm of the transfer function from w to z is $\leq \alpha$. We can then successively reduce the norm by reducing α until (11.7.107) is no longer satisfied; thus approaching the solution of (11.7.55).

We next consider the limiting case as $T \to \infty$. By analogy with the linear quadratic optimal control problem, the solution $\Sigma(t)$ to (11.7.96) will converge (under suitable conditions) to a finite limit, Σ_s, which is a solution to the following Algebraic Riccati Equation:
$$C^T C + A^T \Sigma + \Sigma A + \Delta A^T \Sigma A$$
$$+ (I + A\Delta)^T \Sigma \left[I - \Delta \left(\frac{1}{\alpha^2} B_1 B_1^T - B_2 B_2^T \right) \Sigma \right]^{-1}$$
$$\times \left(\frac{1}{\alpha^2} B_1 B_1^T - B_2 B_2^T \right) \Sigma (I + A\Delta) = 0 \qquad (11.7.110)$$

Also, by analogy with the linear quadratic problem, the solutions of equation (11.7.110) are related to the eigenstructure of the generalized Hamiltonian matrix

M_∞ defined in (11.7.91). In particular if $\begin{bmatrix} X_{11} \\ X_{21} \end{bmatrix} \in \mathbb{R}^{2n \times n}$ spans the nth order stable invariant subspace of M_∞, and X_{11} is nonsingular, then the "stabilizing" solution of (11.7.110) is given by

$$\Sigma_s = X_{21} X_{11}^{-1} \tag{11.7.111}$$

It can also be shown, using an identical argument to Lemma 11.3.1 part IV, that A' is asymptoticlly stable where

$$A' = A - \left(\frac{B_1 B_1^T}{\alpha^2} - B_2 B_2^T \right) \left[I - \Delta \Sigma_s \left(\frac{B_1 B_1^T}{\alpha^2} - B_2 B_2^T \right) \right]^{-1} \Sigma_s (I + \Delta A) \tag{11.7.112}$$

Also, for future reference, we define $\hat{u}(t)$ as the feedback policy given in (11.7.99) when $w(t)$ is not necessarily $w^*(t)$ and $T \to \infty$. That is

$$\hat{u}(t) \triangleq -L_\infty x(t) \tag{11.7.113}$$

where

$$L_\infty = B_2^T \left[I - \Delta \Sigma_s \left(\frac{B_1 B_1^T}{\alpha^2} - B_2 B_2^T \right) \right]^{-1} \Sigma_s (I + \Delta A) \tag{11.7.114}$$

Likewise, we define

$$\hat{w}(t) \triangleq Q_\infty x(t) \tag{11.7.115}$$

where

$$Q_\infty = \frac{B_1^T}{\alpha^2} \left[I - \Delta \Sigma_s \left(\frac{B_1 B_1^T}{\alpha^2} - B_2 B_2^T \right) \right]^{-1} \Sigma_s (I + \Delta A) \tag{11.7.116}$$

We then have the following result:

Theorem 11.17.1

Provided

i) there exists a stabilizing solution $\Sigma_s = X_{21} X_{11}^{-1}$, and
ii) $\Sigma_s \geq 0$ with $\alpha \geq \bar{\sigma}(\sqrt{\Delta} \Sigma_s^{1/2} B_1)$
where $\bar{\sigma}(\cdot)$ denotes the maximum singular value.
then with $u = \hat{u}$.

a) The closed loop system is stable, and
b) $\|G_{zw}\|_\infty \leq \alpha$
where G_{zw} is the transfer function from w to z.

Proof: a) Let $\bar{A} = A - B_2 L_\infty$.

It follows that

$$M_\infty \begin{bmatrix} I \\ \Sigma_s \end{bmatrix} = \begin{bmatrix} I \\ \Sigma_s \end{bmatrix} \left(\bar{A} + \frac{B_1 B_1^T}{\alpha^2} \Sigma_s \left(I - \Delta \frac{B_1 B_1^T}{\alpha^2} \Sigma_s \right)^{-1} (I + \Delta \bar{A}) \right)$$

with $\begin{bmatrix} I \\ \Sigma_s \end{bmatrix}$ spans the stable invariant subspace of M_∞. Hence

$$\bar{A} + \frac{B_1}{\alpha} \left(I - \Delta \frac{B_1^T}{\alpha} \Sigma_s \frac{B_1}{\alpha} \right)^{-1} \frac{B_1^T}{\alpha} \Sigma_s (I + \Delta \bar{A})$$

is asymptotically stable and

$$\left[\left(I - \Delta \frac{B_1^T}{\alpha} \Sigma_s \frac{B_1}{\alpha} \right)^{-1/2} \frac{B_1^T}{\alpha} \Sigma_s (I + \Delta \bar{A}), \bar{A} \right]$$

is detectable. Rearranging (11.7.110), we have

$$\bar{A}^T \Sigma_s + \Sigma_s \bar{A} + \Delta \bar{A}^T \Sigma_s \bar{A} + \bar{C}^T \bar{C}$$

$$+ (I + \Delta \bar{A}^T) \Sigma_s \frac{B_1}{\alpha} \left[I - \Delta \frac{B_1^T}{\alpha} \Sigma_s \frac{B_1}{\alpha} \right]^{-1} \frac{B_1^T}{\alpha} \Sigma_s (I + \Delta \bar{A}) = 0$$

where $\bar{C} = C - DL_\infty$. This is a Lyapunov equation with $\Sigma_s \geq 0$, and

$$\left[\left(I - \Delta \frac{B_1^T}{\alpha} \Sigma_s \frac{B_1}{\alpha} \right)^{-1/2} \frac{B_1^T}{\alpha} \Sigma_s (I + \Delta \bar{A}), \bar{A} \right]$$

detectable. Hence the closed loop system matrix $\bar{A}$ is asymptotically stable, by the strengthened Lyapunov stability theorem.

b) Similar to the proof of Lemma (11.7.7). ▽▽▽

Remark 11.7: An alternative motivation for the result in Theorem 11.7.1 can be obtained as follows:

As before, we define $\bar{A} \triangleq A - B_2 L_\infty$ and $\bar{C} = C - DL_\infty$ and note that, provided (11.7.113) is satisfied, the transfer function from w to z has state space realization $\begin{bmatrix} \bar{A} & B_1 \\ \bar{C} & 0 \end{bmatrix}$. Now the end conditions of Lemma 11.7.2 (ii) are satisfied provided

$$\alpha \geq \min\left\{ \bar{\sigma}(\bar{C}\bar{A}^{-1}B_1), \bar{\sigma}(\Delta \bar{C}(2I + \Delta \bar{A})^{-1} B_1) \right\} \tag{11.7.115}$$

Subject to this condition, then by Lemma 11.7.2, the transfer function from w to z has H_∞ norm less than α if and only if the following Hamiltonian matrix has no eigenvalues on the stability boundary:

$$M'_\alpha = \begin{bmatrix} \bar{A} - \frac{\Delta}{\alpha^2} B_1 B_1^T (I + \Delta \bar{A}^T)^{-1} \bar{C}^T \bar{C} & \frac{1}{\alpha} B_1 B_1^T (I + \Delta \bar{A}^T)^{-1} \\ -\frac{1}{\alpha} (I + \Delta \bar{A}^T)^{-1} \bar{C}^T \bar{C} & -\bar{A}^T (I + \Delta \bar{A}^T)^{-1} \end{bmatrix} \quad (11.7.116)$$

We now note that

$$\begin{bmatrix} I & 0 \\ -\Sigma_s & \alpha I \end{bmatrix} M'_\alpha \begin{bmatrix} I & 0 \\ -\Sigma_s & \alpha I \end{bmatrix}^{-1} = \begin{bmatrix} I & 0 \\ -\Sigma_s & \alpha I \end{bmatrix} M'_\alpha \begin{bmatrix} I & 0 \\ \frac{1}{\alpha}\Sigma_s & \frac{1}{\alpha}I \end{bmatrix}$$

$$= \begin{bmatrix} X_1 & Y_1 \\ 0 & Z_1 \end{bmatrix} \quad (11.7.117)$$

and

$$\begin{bmatrix} I & 0 \\ -\Sigma_s & I \end{bmatrix} M_\infty \begin{bmatrix} I & 0 \\ \Sigma_s & I \end{bmatrix} = \begin{bmatrix} X_2 & Y_2 \\ 0 & Z_2 \end{bmatrix} \quad (11.7.118)$$

$$X_1 = \bar{A} + \frac{1}{\alpha^2} B_1 B_1^T \left(I - \frac{\Delta}{\alpha^2} \Sigma_s B_1 B_1^T \right)^{-1} \Sigma_s (I + \Delta \bar{A}) \quad (11.7.119)$$

$$Y_1 = \frac{1}{\alpha^2} B_1 B_1^T (I + \Delta \bar{A}^T)^{-1} \quad (11.7.120)$$

$$Z_1 = \left[-\bar{A}^T - \frac{1}{\alpha^2} \Sigma_s B_1 B_1^T \right] (I + \Delta \bar{A}^T)^{-1} \quad (11.7.121)$$

$$X_2 = A + \left(\frac{1}{\alpha^2} B_1 B_1^T - B_2 B_2^T \right) \left[I - \Delta \Sigma_s \left(\frac{1}{\alpha^2} B_1 B_1^T - B_2 B_2^T \right) \right]^{-1} \Sigma_s (I + A\Delta) \quad (11.7.122)$$

$$Y_2 = \left(\frac{1}{\alpha^2} B_1 B_1^T - B_2 B_2^T \right) (I + \Delta A^T)^{-1} \quad (11.7.123)$$

$$Z_2 = \left[-A^T - \Sigma_s \left(\frac{1}{\alpha^2} B_1 B_1^T - B_2 B_2^T \right) \right] (I + \Delta A^T)^{-1} \quad (11.7.124)$$

and indeed it can be readily shown that

$$X_1 = X_2 \quad (11.7.125)$$

$$Z_1 = Z_2 \quad (11.7.126)$$

We thus see that M'_α and M_∞ have the same eigenalues.

Thus using Lemma 11.7.2, provided M_∞ has no eigenvalues on the stability boundary and provided (11.7.115) is satisifed, then with $u = \hat{u}$ the transfer function from w to z has H_∞ norm less than or equal to α. ▽▽▽

As in the finite time case, the result in Theorem 11.7.1 can be used to choose the input so as to reduce the H_∞ norm by successively reducing α until the conditions given in the Theorem can no longer be satisfied.

11.8 SUMMARY

The main points covered in this chapter were:

- Introduction of the quadratic design criterion

$$J = x(t_f)^T \Sigma_f x(t_f) + \int_{t_0}^{t_f} \left(x(t)^T Q x(t) + u(t)^T R u(t) \right) dt \qquad (11.2.2)$$

- The control minimizing the quadratic design criterion is

$$u(t) = -L(t + \Delta) x(t) \qquad (11.2.17)$$

$$L(t) = \left(R + \Delta B^T \Sigma(t) B \right)^{-1} B \Sigma(t) (I + A\Delta) \qquad (11.2.18)$$

where $\Sigma(t)$ satisfies a reverse time Riccati equation

$$\bar{\rho} \Sigma(t) = Q + A^T \Sigma(t) + \Sigma(t) A + \Delta A^T \Sigma(t) A$$
$$- L(t)^T \left[R + \Delta B^T \Sigma(t) B \right] L(t) \qquad (11.2.19)$$

$$\Sigma(t_f) = \Sigma_f \qquad (11.2.20)$$

- The LQ optimal control problem is the dual of the (minimum variance) optimal filtering problem.
- As the optimization horizon tends to ∞, the solutions of the Riccati equation of optimal control tend (under reasonable conditions) to well-defined solutions of the algebraic Riccati equation (Theorem 11.3.2).
- In the case of complete state measurements, LQ optimal control guarantees a minimum gain and phase margin, which tend to ∞ and $60°$, respectively, as $\Delta \to 0$ (Corollary 11.4.1).
- Tracking performance can be achieved by adjoining states to the original system so as to implement the internal model principle.
- In the case of a linear system having Gaussian noise, the linear quadratic Gaussian (LQG) optimal control is given by

$$u(t) = -L(t + \Delta) \hat{x}(t)$$

where $L(t)$ is the same as when the states are measured and where $\hat{x}(t)$ is the state estimate provided by the Kalman filter.
- By exploiting the affine nature of the characterization of all stabilizing controllers, other design criteria can be used, for example, H^∞.
- Optimal control design using an H_∞ criterion has been shown to require the solution of a Riccati equation together with an iterative search on scalar variable.

11.9 REFERENCES

Further results on LQ design may be found in:

ANDERSON, B. D. O., and J. B. MOORE (1971) *Linear Optimal Control.* Prentice-Hall, Englewood Cliffs, N.J.

ÅSTROM, K. J. (1970) *Introduction to Stochastic Control Theory*. Academic Press, New York.

KWAKERNAAK, H., and R. SIVAN (1972) *Linear Optimal Control Systems*. Wiley-Interscience, New York.

STENGEL, R. F. (1986) *Stochastic Optimal Control*. Wiley, New York.

The stochastic separation theorem was first discussed in:

SIMON, H. A. (1956) "Dynamic programming under uncertainty with a quadratic criterion function." *Econometrica*, 24, p. 74.

H^∞ design is discussed at length in:

BOYD, S., V. BALAKRISHNAN, and P. KABAMBA (1988) "On computing the H_∞-norm of a transfer matrix" to appear. *Math. Control, Signals and Systems*.

DOYLE, J., K. GLOVER, P. KHARGONEAR, and B. FRANCIS (1989) "State space solutions to standard H_2 and H_∞ control problems." *IEEE Transactions on Automatic Control*, vol. 34, no. 8, pp. 831–847.

FRANCIS, B. (1987) *A Course in H^∞ Control Theory*, Lecture Notes in Control and Information Sciences, Vol. 88. Springer-Verlag, New York.

GLOVER, K. (1984) "All optimal Hankel norm approximations of linear multivariable systems and their L_∞ error bounds." *Int. Journal of Control*, Vol. 39, pp. 1115–1193.

LEE, L. (1989) "A Unified Approach to Discrete and Continuous H^∞ Design Problems" M.E. Thesis, Electrical Engineering, University of Newcastle.

LIMEBEER, D. J. N., B. D. O. ANDERSON, P. P. KHARGONEKAR and M. GREEN. (1989) "A Game Theoretic Approach to H^∞ Control for Time Varying Systems" Tech. Report, Department of Electrical Engineering, Imperial College, London.

11.10 PROBLEMS

1. Consider the problem of stabilizing an unstable system. Let the system be

    $$\rho x = Ax + Bu$$

    and assume that A has all its eigenvalues outside the stability boundary.

    (a) Consider the following optimal control criterion:

    $$J = x(t_f)^T \Sigma_f x(t_f) + \int_{t_0}^{t_f} u^*(t)^T R u^*(t)\, dt$$

    Show that if Σ_f is positive definite then the infinite horizon solution to this problem gives a stable feedback law.

    (b) Show that the optimal control obtained in part (a) achieves

    $$\lim_{t \to \infty} x(t) = 0$$

    and minimizes

    $$\lim_{t_0 \to -\infty} \int_{t_0}^{t_f} u(t)^T R u(t)$$

2. Consider the following simple control problem:

    $$\rho x = u, \qquad x(t_0) = 1$$

Suppose we want to minimize $\lim_{t_0 \to -\infty} J(x_0, t_f)$, where

$$J(x_0, t_f) = x^2(t_f) + \int_{t_0}^{t_f} u^2(t)$$

(a) Show that the optimal cost converges to zero as $t_0 \to -\infty$.

(b) Explain the result in part (a) by considering the finite horizon case and then letting $t_0 \to -\infty$.

3. Consider the LQ optimal control problem with internal model as in Section 11.5. Show that:

   (a) Reachability of the original plant is a necessary condition for reachability of the composite plant.

   (b) For the case where A and A' have no common eigenvalues, show that condition (ii) of Lemma 11.5.1 is also necessary. *Hint:* (1) If the original system is unreachable, then there exists a $q_1 \neq 0$ and λ such that $q_1(\lambda I - A) = 0$ and $q_1 B = 0$; then consider

   $$(q_1 \;\; 0) M(\lambda), \qquad \text{where } M(\lambda) = \begin{bmatrix} B & (\lambda I - A) & 0 \\ 0 & -B'C & (\lambda I - A') \end{bmatrix}$$

   (2) If condition (ii) is not satisfied, then there exists $q_2 \neq 0$, λ such that $q_2(\lambda I - A') = 0$ and $q_2 B'C \operatorname{adj}(\lambda I - A)B = 0$. Then consider $[q_2 B'C \operatorname{adj}(\lambda I - A), q_2 \det(\lambda I - A)] M(\lambda)$. Finally, note that $q_2 \det(\lambda I - A) \neq 0$.

4. Consider the following continuous time system:

   $$\frac{d}{dt} x = x + u$$

   (a) Discretize the system with sampling period Δ.

   (b) Consider the following cost function:

   $$J = \int_{t_0}^{t_f} S x^2 + R u^2 \, dt$$

   (c) Find the optimal control in terms of Δ and R.

   (d) Investigate the nature of the optimal control and closed loop systems as Δ and R take different values.

*5. Consider the finite horizon LQ optimal controller (11.2.17) to (11.2.19). Show that if

   $$V(t) = x(t)^T \Sigma(t) x(t)$$

   then

   $$\rho V(t) = -x(t)^T \big(Q + L(t + \Delta)^T R L(t + \Delta) \big) x(t)$$

   Hint:

   (i) $\rho V(t) = (\rho x(t))^T \Sigma(t + \Delta) x(t) + x(t)^T \Sigma(t + \Delta)(\rho x(t))$
   $+ \Delta (\rho x(t))^T \Sigma(t + \Delta)(\rho x(t))$
   $+ x(t)^T (\rho \Sigma(t)) x(t)$

   (ii) $\rho \Sigma(t) = -\bar{\rho} \Sigma(t + \Delta)$

   (iii) Use the fact that $\rho x(t) = [A - BL(t + \Delta)] x(t)$. See (11.2.17) and $\rho x = Ax + Bu$.

*6. Consider the infinite horizon LQ optimal controller (11.3.2), (11.3.3). Show that if

   $$V(t) = x(t)^T \Sigma x(t)$$

then
$$\rho V(t) = -x(t)^T(Q + L^TRL)x(t)$$
Hint: Use the fact from (11.3.3) that $B^T\Sigma(I + A\Delta) = (R + \Delta B^T\Sigma B)L$.

7. Consider the continuous time system
$$\frac{d}{dt}x = \begin{bmatrix} 0 & 1 \\ -1 & 0 \end{bmatrix} x + \begin{bmatrix} 0 \\ 1 \end{bmatrix} u$$
$$y = \begin{bmatrix} 1 & 0 \end{bmatrix} x$$

Find the controller that minimizes the cost function
$$J = \int_0^\infty (y^2(t) + Ru^2(t)) \, dt$$

where:

(a) $R = 0.1$
(b) $R = 1$
(c) $R = 10$

8. Find the gain and phase margins for your answers to Problem 7(a), (b), and (c).

9. Consider the discrete time system with transfer function
$$H(\gamma) = \frac{1}{\gamma - 1}, \qquad \Delta = 0.1$$

Show that the controller that minimizes the cost function
$$J = \underset{0}{\overset{\infty}{S}}(y^2(t) + Ru^2(t)) \, dt$$

has a closed loop pole at
$$\gamma = \frac{1 + R - \sqrt{(1 + R)(1 + 441R)}}{22R}$$

What are the limiting positions of the closed loop pole as $R \to \infty$ and $R \to 0$?

10. If $H(\gamma) = (-\gamma + 1)/(\gamma^2 + \gamma)$ and $\Delta = 0.2$, find the closed loop poles for the optimal controller with cost function as in Problem 9 and $R = 0, 0.1, 1, 10,$ and ∞.

11. Consider the discrete time multivariable system
$$\rho x = \begin{bmatrix} -1 & 0 \\ 0 & +1 \end{bmatrix} x + \begin{bmatrix} 2 & 1 \\ 1 & 1 \end{bmatrix} u, \qquad \Delta = 0.05$$
$$y = \begin{bmatrix} 3 & 3 \\ -1 & 1 \end{bmatrix} x$$

Find the controller that minimizes the cost function
$$J = \underset{0}{\overset{\infty}{S}}\left(y^T(t)y(t) + u(t)^T \begin{bmatrix} 1 & 0 \\ 0 & 10 \end{bmatrix} u(t) \right) dt.$$

Express your controller in the form $u = -Ly$.

12. Suppose a certain plant has a discrete time transfer function
$$\frac{Y(\gamma)}{U(\gamma)} = H(\gamma) = \frac{1}{1 + \gamma}, \qquad \Delta = 0.1$$

subject to a constant additive output disturbance, d. Augment the plant with an integrator as suggested in Section 11.5.

(a) Find the controller that minimizes

$$J = \int_0^\infty \left(u^2(t) + e^2(t) \right) dt$$

when $d = 0$.

(b) Explain why this cost function is ill defined if $d \neq 0$.

(c) Show that the controller given by part (a) gives $e(t) \to 0$ in the case where $d \neq 0$.

13. Consider the system described by

$$\rho x = \begin{bmatrix} -0.1 & 0.2 \\ -0.2 & 0.3 \end{bmatrix} x + \begin{bmatrix} 0.1 \\ 0.2 \end{bmatrix} u, \qquad \Delta = 1$$

$$y = \begin{bmatrix} 1 & 2 \end{bmatrix} x$$

Find the controller gains that minimize the cost function

$$J = 10 y^2(10) + \int_0^{10} \left(u^2(t) + 0.1 y^2(t) \right) dt$$

14. For the continuous time plant

$$\frac{d}{dt} x = u$$

$$y = x$$

show that the control starting at time $-T$ that minimizes

$$J = 10 y^2(0) + \int_{-T}^0 \left(u^2(t) + y^2(t) \right) dt$$

is

$$u(t) = -L(t) y(t), \qquad t \leq 0$$

where

$$L(t) = \frac{11 e^{-2t} + 9}{11 e^{-2t} - 9}$$

Hint: Show that $\Sigma(t) = L(t)$ and

$$\frac{d}{dt} \Sigma(t) = \Sigma^2(t) - 1$$

$$\Sigma(0) = 10$$

12

Parameter Estimation

12.1 INTRODUCTION

In this chapter we will consider the problem of estimating the various constants that appear in a dynamic model for a system. We will do this by processing the input–output data from the plant. The essential idea is to compare the observed system response $y(t)$ with the output of a parameterized model $\hat{y}(\theta, t)$ over some interval of time. We then adjust the parameters θ in the model until the model output $\hat{y}(\theta, t)$ is a good approximation to the observed system output. This process is shown diagrammatically in Figure 12.1.1.

The user has to make several choices in this estimation procedure. Examples of the choices are:

- What form should the model $\hat{y}(\theta, t)$ take?
- What criterion should be used to compare $\hat{y}(\theta, t)$ with $y(t)$?
- What experimental conditions (sampling rate, input signals, and the like) should be used?

Typical choices might be:

Model: A linear differential equation.
Criterion: The integral square error; that is,

$$\hat{\theta} = \text{Arg min} \int_0^T [y(t) - \hat{y}(t, \theta)]^2 \, dt$$

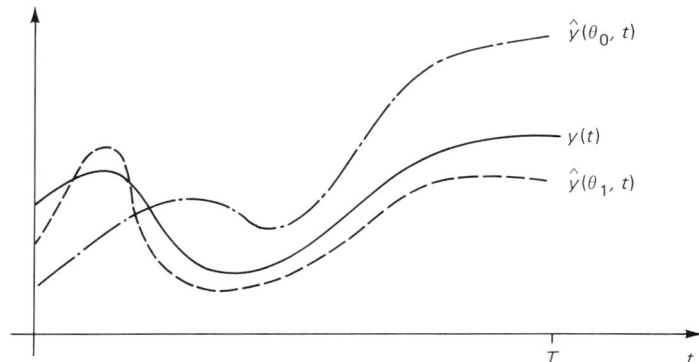

Figure 12.1.1 Comparison of $y(t)$ with $\hat{y}(\theta, t)$.

An important consideration in the user choices is the resultant set of properties that the estimator has. To study these properties, it is usually necessary to be more precise about the nature of the data generating mechanism. A typical assumption is that the data can be exactly modeled as $y(t) = \hat{y}(\theta_*, t) + \eta(t) + \omega(t)$, where θ_* denotes a particular parameter value and $\eta(t)$ and $\omega(t)$ represenrt error terms. These errors are usually described in terms of their characteristics. Typical characteristics might be:

1. $\omega(t)$ is white noise.
2. $|\omega(t)|$ is bounded by a constant d.
3. $|\eta(t)|$ is bounded by a function of the past inputs to the system.

The quality of a particular estimate for θ is then judged in terms of its properties under these assumptions about the modeling errors. Thus, if the modeling error is assumed to be a white noise process, we might ask whether or not the expected value of $\hat{\theta}$ is equal to θ_*. We call this property *unbiasedness*. Alternatively, if the errors are simply bounded, then we might ask that $|\hat{\theta} - \theta_*|$ be small if $|\omega(t)|$ is small. Finally, in some cases we may not be interested in the parameter errors but only in whether or not a $\hat{\theta}$ can be found such that the predicted output $\hat{y}(\hat{\theta}, t)$ is close to $y(t)$.

In this chapter, we will describe various parameter estimation algorithms, and we will investigate their properties under a variety of assumptions about the nature of the modeling error $\eta(t)$ or $\omega(t)$. We will also investigate what happens when on-line parameter estimation is combined with on-line control system synthesis. We will call these algorithms *adaptive controllers*.

12.2 LINEAR REGRESSIONS FOR DYNAMICAL SYSTEMS

The estimation methods that we will study in this chapter basically depend on our ability to rearrange the model so that the predicted output is describable as a linear function of a parameter vector θ; that is, there exists some vector of measured

variables, $\phi(t)$, such that the model output $\hat{y}(t, \theta) \triangleq \bar{y}(t)$ can be expressed as
$$\bar{y}(t) = \phi(t)^T \theta \tag{12.2.1}$$
In some cases, it is straightforward to express the model in this form. For example, say we wish to model the data by a sine wave of known frequency ω_0, but of unknown amplitude and phase; that is,
$$\bar{y}(t) = A \sin(\omega_0 t + \alpha) \tag{12.2.2}$$
Equation (12.2.2) is linear in A but is nonlinear in the unknown phase α. Thus it does not immediately fit into the form (12.2.1). However, a linear representation can be achieved by reexpressing the model (12.2.2) in the following form:
$$\bar{y}(t) = \theta_1 \sin \omega_0 t + \theta_2 \cos \omega_0 t \tag{12.2.3}$$
Equation (12.2.3) is now of the form (12.2.1), where
$$\phi(t)^T = [\sin \omega_0 t, \cos \omega_0 t] \tag{12.2.4}$$

It also turns out that the representation (12.2.1) applies to a large class of dynamical system models, as we will now show.

12.2.1 Discrete Time Shift Operator Linear Models

We saw in Chapter 3 that discrete time shift operator models can be expressed in the form
$$A(q)y(t) = B(q)u(t) \tag{12.2.5}$$
where
$$A(q) = q^n + a_{n-1}q^{n-1} + \cdots + a_0$$
$$B(q) = b_m q^m + \cdots + b_0$$
Equation (12.2.5) can be explicitly expressed as
$$y(t) = -a_{n-1}y(t-1) \ldots - a_0 y(t-n) + b_m u(t-n+m)$$
$$+ \cdots + b_0 u(t-n) \tag{12.2.6}$$
which is of the form (12.2.1), where $\bar{y}(t) = y(t)$ and
$$\phi(t)^T = [y(t-1), \ldots, y(t-n), u(t-n+m), \ldots, u(t-n)] \tag{12.2.7}$$
and
$$\theta^T = [-a_{n-1}, \ldots, -a_0, b_m, \ldots, b_0] \tag{12.2.8}$$

12.2.2 Unified Linear Models

For unified linear systems, the model corresponding to (12.2.5) is
$$A(\rho)y(t) = B(\rho)u(t) \tag{12.2.9}$$
where $\rho \triangleq d/dt$ (continuous time) or δ (discrete time):
$$A(\rho) = \rho^n + a_{n-1}\rho^{n-1} + \cdots + a_0$$
$$B(\rho) = b_m \rho^m + \cdots + b_0$$

When the generalized derivatives of $y(t)$ and $u(t)$ are available, then the model (12.2.9) is immediately in the required form, where $\bar{y}(t) = \rho^n y(t)$ and

$$\phi(t)^T = [\rho^{n-1} y(t), \ldots, y(t), \rho^m u(t), \ldots, u(t)] \tag{12.2.10}$$

$$\theta^T = [-a_{n-1}, \ldots, -a_0, b_m, \ldots, b_0] \tag{12.2.11}$$

In cases when the derivatives of $y(t), u(t)$ are not available, we can use filtered derivatives in place of the raw derivatives. To see how this is compatible with (12.2.9), let us operate on this equation by a stable nth-order filter $1/E(\rho)$ to give

$$A(\rho)\left[\frac{y(t)}{E(\rho)}\right] = B(\rho)\left[\frac{u(t)}{E(\rho)}\right] \tag{12.2.12}$$

This equation is of the form (12.2.1), where $\bar{y}(t) = \rho^n y_f(t)$, θ is as in (12.2.11), and

$$\phi(t)^T = [\rho^{n-1} y_f(t), \ldots, y_f(t), \rho^m u_f(t), \ldots, u_f(t)] \tag{12.2.13}$$

$$y_f(t) = \frac{1}{E(\rho)} y(t) \tag{12.2.14}$$

$$u_f(t) = \frac{1}{E(\rho)} u(t) \tag{12.2.15}$$

Actually, there are many other equivalent ways of describing the model (12.2.12) in the form of a linear regression. For example, if we add

$$\frac{E(\rho) - A(\rho)}{E(\rho)} y(t)$$

to both sides of (12.2.12), we obtain

$$y(t) = \frac{E(\rho)}{E(\rho)} y(t) = \frac{E(\rho) - A(\rho)}{E(\rho)} y(t) + \frac{B(\rho)}{E(\rho)} u(t) \tag{12.2.16}$$

This is of the form of (12.2.1), where $\bar{y}(t) = y(t)$, $\phi(t)$ is as in (12.2.13), and

$$\theta^T = [e_{n-1} - a_{n-1}, \ldots, e_0 - a_0, b_m, \ldots, b_0] \tag{12.2.17}$$

It is interesting to note that the regression vector ϕ for the preceding two models can be readily generated via the following nth-order state space models:

$$\rho \phi_y = \bar{E} \phi_y + \beta y \tag{12.2.18}$$

$$\rho \phi_u = \bar{E} \phi_u + \beta u \tag{12.2.19}$$

$$\phi_y^T = [\rho^{n-1} y_f \cdots y_f] \tag{12.2.20}$$

$$\phi_u^T = [\rho^{n-1} u_f \cdots u_f] \tag{12.2.21}$$

$$\bar{E} = \begin{bmatrix} -e_{n-1} & & -e_0 \\ 1 & & \\ & \ddots & \\ & & 1 & 0 \end{bmatrix}, \quad \beta = \begin{bmatrix} 1 \\ 0 \\ \vdots \\ 0 \end{bmatrix} \tag{12.2.22}$$

Finally, we note that the shift operator model of Section 12.2.1 is actually a special case of this unified model in which we take $\rho = q$ and $E(q) = q^n$.

12.2.3 Nonlinear Models

In the preceding discussion we have referred to linear dynamic systems. However, the key point about the linear regression (12.2.1) is linearity in the parameters and not necessarily linearity of the dynamics. Hence, the model (12.2.1) can be used for certain classes of nonlinear systems.

As a simple example, consider the following bilinear system

$$\rho y = -ay + byu \qquad (12.2.23)$$

This model is of the form (12.2.1), where $\bar{y} = \rho y$ and

$$\phi^T = [y, yu] \qquad (12.2.24)$$

$$\theta^T = [-a, b] \qquad (12.2.25)$$

The same idea also applies to many other nonlinear systems. For example, consider the one degree of freedom robot shown in Figure 12.2.1. Suppose the center of mass of the link plus load is at position (x_{cm}, y_{cm}) when $\alpha = 0$, the total mass is M, and the effective inertia about the origin is J_{eff}. Then the dynamic equation describing the motion of the robot (including gravity in the negative y direction) is

$$J_{eff}\ddot{\alpha} = \tau + gMy_{cm}\sin\alpha - gMx_{cm}\cos\alpha \qquad (12.2.26)$$

This equation may be rewritten in the linear regression form, (12.2.1), where

$$\bar{y} = \tau$$

$$\phi(t)^T = [\ddot{\alpha} \quad -g\sin\alpha \quad g\cos\alpha]$$

and

$$\theta^T = [J_{eff} \quad My_{cm} \quad Mx_{cm}]$$

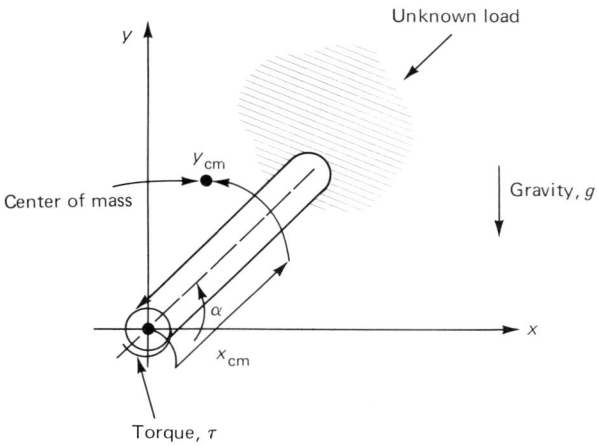

Figure 12.2.1 One degree of freedom robot.

This idea can be extended to more general robot systems; see the references given at the end of the chapter.

12.2.4 Models Which Are Nonlinear in the Parameters

We next briefly consider the more general problem of estimating parameters in models when the parameters appear nonlinearly. Suppose that

$$y = f(\psi, \theta) \tag{12.2.27}$$

where ψ and y are known functions of time and θ is a vector of unknown parameters. The determination of θ such that some cost function is minimized is, in general, a difficult nonlinear optimization problem. We propose an approximate solution based on a Taylor series expansion of (12.2.27) about some prior estimate $\bar{\theta}$; that is,

$$y \approx f(\psi, \bar{\theta}) + \left(\frac{\partial f}{\partial \theta}\bigg|_{\psi, \bar{\theta}}\right)(\theta - \bar{\theta}) \tag{12.2.28}$$

If we let $\bar{y} = f(\psi, \bar{\theta})$ and

$$\phi^T = \frac{\partial f}{\partial \theta}\bigg|_{\psi, \bar{\theta}} \tag{12.2.29}$$

then (12.2.28) may be written as

$$y \approx \bar{y} + \phi^T(\theta - \bar{\theta}) = (\bar{y} - \phi^T\bar{\theta}) + \phi^T\theta \tag{12.2.30}$$

We then define a "pseudo" observation $\bar{y}'$ by $\bar{y}' = y - \bar{y} + \phi^T\bar{\theta}$ and note from (12.2.30) that

$$\bar{y}' \simeq \phi^T\theta \tag{12.2.31}$$

This equation is of the form of the linear regression given in (12.2.1). Obviously, the closer $\bar{\theta}$ is to θ then the better this approximation will be. Hence, it is desirable to make the linearization about the best currently available estimate of the parameters.

We next show how the parameters in the linear regression model (12.2.1) can be estimated via least squares.

12.3 LEAST SQUARES ESTIMATION

To get the basic idea, let us first consider a trivial problem in which y is represented as

$$y(t) = \theta + \nu(t) \tag{12.3.1}$$

where we assume $\nu(t)$ is a zero mean noise signal. Our estimation problem is then, given a set of measurements of $y(\tau)$ for $0 \leq \tau < t$, to find a good estimate, $\hat{\theta}$, of θ. One common measure of goodness of an estimate is the least squares cost function:

$$C(\hat{\theta}, t) = \frac{1}{2}\int_0^t (y(\tau) - \hat{\theta})^2 d\tau \tag{12.3.2}$$

Minimization of this cost function can be achieved by setting the partial derivative to zero:

$$\frac{\partial C}{\partial \hat{\theta}} = -\int_0^t \left(y(\tau) - \hat{\theta} \right) d\tau$$
$$= 0 \tag{12.3.3}$$

or, since $\hat{\theta}$ is not a function of τ,

$$\hat{\theta} = \frac{1}{t} \int_0^t y(\tau) \, d\tau \tag{12.3.4}$$

Equation (12.3.4) describes the intuitively reasonable result that we should average the data when measuring a constant in noise. Note also that, under the assumption that $\nu(t)$ has zero sample mean, we have

$$\lim_{t \to \infty} \hat{\theta}(t) = \lim_{t \to \infty} \frac{1}{t} \int_0^t y(\tau) \, d\tau$$
$$= \lim_{t \to \infty} \frac{1}{t} \int_0^t (\theta + \nu(\tau)) \, d\tau$$
$$= \theta \tag{12.3.5}$$

Now let us extend the problem slightly. Suppose that instead of (12.3.1) we have [compare with Equation (12.2.1)]

$$y(t) = \phi(t)^T \theta + \nu(t) \tag{12.3.6}$$

where $y(t), \phi(t)$ are known functions of time t, and θ is a vector of unknown parameters. By analogy with the previous case (12.3.2), we let

$$C(\hat{\theta}, t) = \frac{1}{2} \int_0^t \left(y(\tau) - \phi(\tau)^T \hat{\theta} \right)^2 d\tau \tag{12.3.7}$$

In this case, if we set $\partial C / \partial \hat{\theta} = 0$, we obtain

$$\frac{\partial C}{\partial \hat{\theta}} = -\int_0^t \phi(\tau) \left(y(\tau) - \phi(\tau)^T \hat{\theta} \right) d\tau$$
$$= 0 \tag{12.3.8}$$

or

$$\left[\int_0^t \phi(\tau) \phi(\tau)^T d\tau \right] \hat{\theta} = \left[\int_0^t \phi(\tau) y(\tau) \, d\tau \right] \tag{12.3.9}$$

Provided $\int_0^t \phi(\tau)\phi(\tau)^T d\tau$ is nonsingular, a unique solution to (12.3.9) exists. However, in cases where the preceding condition does not hold, it is possible for a family of solutions to (12.3.9) to exist. For example, suppose

$$v^T \phi(\tau) = 0 \quad \forall \tau, \qquad 0 \leq \tau < t \tag{12.3.10}$$

and let $\hat{\theta}_p$ be a solution to (12.3.9); then

$$\hat{\theta} = \hat{\theta}_p + \alpha v \tag{12.3.11}$$

is also a solution for any scalar α. In this case, we usually need to add some extra information to the problem formulation. For example, we might know a value $\hat{\theta}_0$ such that the true parameters are "close" to this value. This case can be handled by a slight generalization of the cost function (12.3.7) to measure both the error between y and $\hat{y}$ and between $\hat{\theta}$ and $\hat{\theta}_0$. A suitable quadratic function is

$$C'(\hat{\theta}, t) = \frac{1}{2}\left\{\int_0^t \frac{1}{\Gamma}\left(y(\tau) - \phi(\tau)^T\hat{\theta}\right)^2 d\tau + \left(\hat{\theta} - \hat{\theta}_0\right)^T P_0^{-1}\left(\hat{\theta} - \hat{\theta}_0\right)\right\} \quad (12.3.12)$$

where Γ is a positive scalar and where P_0^{-1} is a positive definite, symmetric matrix. (The reason for using the inverse and the significance of the subscript zero will be clarified in the next section.) Minimization of this cost function yields the following unique solution:

$$\hat{\theta} = P(t)\left\{P_0^{-1}\hat{\theta}_0 + \int_0^t \frac{1}{\Gamma}\phi(\tau)y(\tau)\,d\tau\right\} \quad (12.3.13)$$

where

$$P(t) = \left\{P_0^{-1} + \int_0^t \frac{1}{\Gamma}\phi(\tau)\phi(\tau)^T\,d\tau\right\}^{-1} \quad (12.3.14)$$

12.4 RECURSIVE LEAST SQUARES

Equations (12.3.13) and (12.3.14) essentially solve the least squares estimation problem. However, it is often the case that additional data arrives either continuously or sequentially in time. If we use Equation (12.3.13), then the complete calculation must be redone whenever we want a new set of estimates. However, it seems plausible that, if only one extra piece of data has arrived, that it may be possible to obtain an updated parameter estimate without having to redo the entire calculation. To see that this is indeed the case, we first consider the continuous time situation where $\rho = d/dt$. Then differentiating (12.2.13) gives the following differential equation, which describes the evolution of $\hat{\theta}$:

$$\frac{d}{dt}\hat{\theta} = \frac{d}{dt}(P)\left\{P_0^{-1}\hat{\theta}_0 + \int_0^t \frac{1}{\Gamma}\phi(\tau)y(\tau)\,d\tau\right\} + \frac{1}{\Gamma}P(t)\phi(t)y(t) \quad (12.4.1)$$

Also, from (12.3.14) we have

$$\frac{d}{dt}(P^{-1}) = \frac{1}{\Gamma}\phi\phi^T \quad (12.4.2)$$

Then, since

$$\frac{d}{dt}(PP^{-1}) = 0 = \left(\frac{d}{dt}P\right)P^{-1} + P\frac{d}{dt}(P^{-1})$$

we have

$$\frac{d}{dt}P = -P\frac{d}{dt}(P^{-1})P = -\frac{1}{\Gamma}P\phi\phi^T P \quad (12.4.3)$$

Substituting (12.4.3) into (12.4.1) gives

$$\frac{d}{dt}\hat{\theta} = -\frac{1}{\Gamma}P(t)\phi(t)\phi(t)^T P(t)\left\{P_0^{-1}\hat{\theta}_0 + \int_0^t \frac{1}{\Gamma}\phi(\tau)y(\tau)\,d\tau\right\}$$
$$+ \frac{1}{\Gamma}P(t)\phi(t)y(t) \tag{12.4.4}$$

Finally, using Equation (12.3.13), we have

$$\frac{d}{dt}\hat{\theta} = \frac{1}{\Gamma}P\phi e \tag{12.4.5}$$

$$\frac{d}{dt}P = -\frac{1}{\Gamma}P\phi\phi^T P \tag{12.4.6}$$

where

$$e = y - \phi^T\hat{\theta} \tag{12.4.7}$$

Equations (12.4.5) and (12.4.6) are a set of coupled differential equations that describe the evolution of $\hat{\theta}$ and P. The initial conditions for these equations are obtained from (12.3.14) and (12.3.13) as

$$P(0) = P_0 \tag{12.4.8}$$

$$\hat{\theta}(0) = \hat{\theta}_0 \tag{12.4.9}$$

The preceding derivation can be readily extended to the discrete and unified cases by replacing d/dt by ρ and using generalized derivatives. Applying a generalized derivative to both sides of (12.3.13), we obtain

$$\rho\hat{\theta}(t) = \rho P(t)\left\{P_0^{-1}\hat{\theta} + \underset{0}{\overset{t}{S}}\frac{1}{\Gamma}\phi(\tau)y(\tau)\,d\tau\right\}$$
$$+ \frac{1}{\Gamma}P(t)\phi(t)y(t) + \frac{\Delta}{\Gamma}(\rho P(t))\phi(t)y(t) \tag{12.4.10}$$

Using the matrix inversion lemma (see Problem 1) on (12.4.10), we also have

$$\rho P(t) = \frac{-P(t)\phi(t)\phi(t)^T P(t)}{\Gamma + \Delta\phi(t)^T P(t)\phi(t)} \tag{12.4.11}$$

Then, substituting (12.4.11) in (12.4.10) and rearranging, we have

$$\rho\hat{\theta}(t) = \frac{P(t)\phi(t)\left(y(t) - \phi(t)^T\hat{\theta}(t)\right)}{\Gamma + \Delta\phi(t)^T P(t)\phi(t)} \tag{12.4.12}$$

Also, from (12.3.13) and (12.3.14) it is clear that

$$\hat{\theta}(0) = \hat{\theta}_0 \tag{12.4.13}$$

and

$$P(0) = \left(P_0^{-1}\right)^{-1} = P_0 \tag{12.4.14}$$

Note that Equations (12.4.11) and (12.4.12), together with initial conditions (12.4.13) and (12.4.14), give a recursive form for the least squares algorithm. This recursive least squares algorithm no longer involves, explicitly, the solution of a linear system of equations, but needs only simple matrix multiplication and scalar division.

We see that there are two update laws involved; one generates (recursively) the parameter estimates (12.4.12), and the other updates the matrix P, where P contains directional information relating to future "search" directions for $\hat{\theta}$.

12.5 PROPERTIES OF THE RECURSIVE LEAST SQUARES ALGORITHM

In this section we will establish certain general properties of the recursive least squares algorithm. In particular, we will study two types of properties.

First, we will establish properties of the prediction error $e(t)$ between the observed data $y(t)$ and the model output $\hat{y}(t, \hat{\theta})$. These properties require no particular assumption regarding the nature of the regression vector $\phi(t)$.

Second, we will establish properties of the parameter error $\tilde{\theta}(t)$ between the true parameter vector θ_0 and the estimated parameter vector $\hat{\theta}(t)$. In particular, we will show that, provided the regression vector satisfies certain properties, $\tilde{\theta}(t)$ converges to zero. The sufficient conditions on $\phi(t)$ that ensure this parameter convergence will in turn be linked to properties of the system input.

We recall that the data are generated by a model of the form (12.2.1); that is,

$$y(t) = \phi(t)^T \theta \tag{12.5.1}$$

We will basically analyze the properties of the estimation algorithm (12.4.11), (12.4.12). However, to make the results more widely applicable, we slightly generalize this algorithm as follows:

$$\rho \hat{\theta} = \frac{\alpha P \phi (y - \phi^T \hat{\theta})}{\Gamma + \Delta \phi^T P \phi} \tag{12.5.2}$$

and

$$\rho P = \frac{-\alpha P \phi \phi^T P}{\Gamma + \Delta \phi^T P \phi} + \Omega \tag{12.5.3}$$

where

$\alpha(t) = $ a (time-varying) gain, $\alpha(t) \in [0, 1]$

$\Gamma(t) = $ a (time-varying) normalization term, $\Gamma(t) > 0$

and where $\Omega(t)$ represents a modification to the "covariance" update (12.4.11), with

$$\Omega(t) = \Omega^T(t) \geq 0 \tag{12.5.4}$$

[The term covariance update is used since $P(t)$ can be shown to correspond to a covariance matrix under suitable conditions; see Section 12.7.2.] Note that this general algorithm reduces to the algorithm introduced earlier by making the choices $\alpha = 1$, $\Gamma(t) = \Gamma$ (fixed), and $\Omega = 0$.

12.5.1 Prediction Error Properties

The algorithm (12.5.2), (12.5.3) has been motivated as a generalization of the recursive least squares algorithm. However, up to this stage we have not discussed the properties of this general class of algorithms. It turns out these algorithms have two key properties which hold under very general conditions. Loosely speaking, these properties are: first, that the estimated parameters get closer to the true parameters at each step, and second, the predicted output converges to the true output. These results are precisely stated in Theorem 12.5.1.

Theorem 12.5.1

Consider the general algorithm (12.5.2), (12.5.3), where $y(t)$ is generated by the ideal model (12.5.1). Under these circumstances the algorithms have the following properties:

(i) *Parameter error reduction*: $\tilde{\theta}(t)^T P(t)^{-1} \tilde{\theta}(t)$ is a nonincreasing function and thus

$$\|\tilde{\theta}(t)\|^2 \leq \frac{\lambda_{\max}(P(t))}{\lambda_{\min}(P(0))} \|\tilde{\theta}(0)\|^2 \qquad (12.5.5)$$

where

$$\tilde{\theta}(t) \triangleq \hat{\theta}(t) - \theta$$

(ii) *Prediction error convergence*: The normalized prediction error $\tilde{e}$ belongs to $\mathscr{L}_2^+$, where

$$\tilde{e} \triangleq \left[\frac{\alpha}{\Gamma + \Delta\phi^T P \phi}\right]^{1/2} (y - \phi^T \hat{\theta}) \qquad (12.5.6)$$

(iii) *Parameter change convergence*: Provided $(\alpha\phi^T P^2 \phi)/(\Gamma + \Delta\phi^T P \phi)$ is bounded, then $\rho\hat{\theta}$ belongs to $\mathscr{L}_2^+$.
$\mathscr{L}_2^+$ denotes the set of vector functions $l(\cdot)$ such that $S_0^\infty l(\tau)^T l(\tau)\,d\tau < \infty$.

Proof: Consider the scalar function

$$V(t) = \tilde{\theta}(t)^T P(t)^{-1} \tilde{\theta}(t) \qquad (12.5.7)$$

We first note that the algorithm ensures that $P(t)$ is positive semidefinite for all time. Then, since $e = y - \phi^T \hat{\theta} = -\phi^T \tilde{\theta}$ and since $\Omega \geq 0$, it follows by some lengthy algebra (see Problem 21) that

$$\rho V \leq \frac{-\alpha e^2}{\Gamma + \Delta\phi^T P \phi} = -\tilde{e}^2 \qquad (12.5.8)$$

with equality if $\Omega = 0$.

Part (i) follows since (12.5.8) shows that V is nonincreasing.

Integrating both sides of (12.5.8) shows that $\tilde{e}$ belongs to $\mathscr{L}_2^+$, which establishes part (ii).

Also, from (12.5.2), we have

$$(\rho\hat{\theta})^T(\rho\hat{\theta}) = \frac{\alpha^2 e^2 \phi^T P^2 \phi}{(\Gamma + \Delta\phi^T P \phi)^2} = \left(\frac{\alpha \phi^T P^2 \phi}{\Gamma + \Delta\phi^T P \phi}\right)\tilde{e}^2 \qquad (12.5.9)$$

Part (iii) then follows from part (ii). ▽▽▽

Theorem 12.5.1 has the following interpretations: the first property implies that the parameter error $\tilde{\theta}$ is nonincreasing in "size" (where size is measured by $\|\cdot\|_{P^{-1}}$). For example, in the special case when $P_0 = (1/\varepsilon)I$ and $\Omega = 0$, it follows that $\|\tilde{\theta}(t)\| \leq \|\tilde{\theta}(0)\|$. (See Problem 7.) The second result establishes that the normalized prediction error tends to zero in an $\mathscr{L}_2$ sense. Thus we note that, for bounded ϕ, P, Γ, and α^{-1}, the prediction error $y - \hat{y}$ decays to zero in an $\mathscr{L}_2$ sense, regardless of any further properties on ϕ. The third result demonstrates that the fluctuations in the parameter estimates also tend to zero in the $\mathscr{L}_2$ sense.

12.5.2 Parameter Convergence

Note that the preceding results do not necessarily imply that the estimated parameters converge to the true values. This difficulty can be traced to the fact that the regression vector, ϕ, may not span the whole space. To take a simple example, if we are performing a linear regression on sets of x–y data, we cannot obtain the correct parameters if all the data points have the same x coordinate. This idea applies also to dynamic system identification where we are attempting to estimate the transfer function in a parametric manner. In this case, we cannot completely identify the transfer function unless we have information on the frequency response at a sufficient number of data points. This is clarified in the following results. The first lemma establishes conditions on the regression vector ϕ under which parameter convergence can be assured. Lemma 12.5.2 gives sufficient conditions on the plant input u to ensure that ϕ has the required properties.

Lemma 12.5.1. Persistent Spanning

Suppose that P, ϕ, Γ, and α^{-1} are bounded in the estimator (12.5.2), (12.5.3), where the data are generated by (12.5.1). Then a sufficient condition for parameter convergence to the true values is that the regression vector ϕ satisfies the following persistent spanning condition: there exists t_1, $\varepsilon > 0$, such that for any t

$$\int_t^{t+t_1} \phi(\tau)\phi(\tau)^T \, d\tau \geq \varepsilon I \qquad (12.5.10)$$

Proof: Since ϕ, Γ, α^{-1}, and P are bounded, Theorem 12.5.1 establishes that $e(\cdot) \triangleq y - \phi^T\hat{\theta}$ and $\rho\hat{\theta}(\cdot)$ belong to $\mathscr{L}_2^+$. This in turn implies that for any t_1

$$\lim_{t \to \infty} \left\{ \int_t^{t+t_1} e^2(\tau) \, d\tau \right\} = 0 \qquad (12.5.11)$$

and

$$\lim_{t \to \infty} \left\{ \int_t^{t+t_1} (\rho \hat{\theta}(\tau))^T (\rho \hat{\theta}(\tau)) \, d\tau \right\} = 0 \qquad (12.5.12)$$

Equation (12.5.11) may be rewritten as

$$\int_t^{t+t_1} \tilde{\theta}^T(\tau) \phi(\tau) \phi(\tau)^T \tilde{\theta}(\tau) \, d\tau \to 0 \qquad (12.5.13)$$

or

$$\tilde{\theta}(t)^T \left(\int_t^{t+t_1} \phi(\tau) \phi(\tau)^T \, d\tau \right) \tilde{\theta}(t)$$

$$+ \left(\int_t^{t+t_1} 2(\tilde{\theta}(\tau) - \tilde{\theta}(t))^T \phi(\tau) \phi(\tau)^T \, d\tau \right) \tilde{\theta}(t)$$

$$+ \int_t^{t+t_1} (\tilde{\theta}(\tau) - \tilde{\theta}(t))^T \phi(\tau) \phi(\tau)^T ((\tilde{\theta}(\tau) - \tilde{\theta}(t))) \, d\tau \to 0 \qquad (12.5.14)$$

Now

$$\|\tilde{\theta}(\tau) - \tilde{\theta}(t)\| = \left\| \int_t^\tau \rho \hat{\theta}(t_0) \, dt_0 \right\|$$

$$\leq \left\{ (\tau - t) \int_t^\tau (\rho \hat{\theta}(t_0))^T (\rho \hat{\theta}(t_0)) \, dt_0 \right\}^{1/2} \qquad (12.5.15)$$

using Schwarz's inequality.

From (12.5.12) and the definition of $\tilde{\theta}$, it follows that $[\tilde{\theta}(\tau) - \tilde{\theta}(t)] \to 0$ (for $\tau < t + t_1$) as $t \to \infty$. This implies that the last two terms on the left side of (12.5.14) tend to zero; that is,

$$\tilde{\theta}(t)^T \left(\int_t^{t+t_1} \phi(\tau) \phi(\tau)^T \, d\tau \right) \tilde{\theta}(t) \to 0 \qquad (12.5.16)$$

Then, provided (12.5.10) is satisfied, it is clear that $\tilde{\theta}(t) \to 0$. ∇∇∇

Lemma 12.5.1 gives some insight into the parameter convergence problem. However, the condition (12.5.10) is expressed in terms of the regression vector ϕ. To further clarify the result, we will translate this condition into a condition on the input signal, u. Thus consider the unified linear model discussed in Section 12.2.2:

$$A(\rho) y(t) = B(\rho) u(t) \qquad (12.5.17)$$

where

$$A(\rho) = \rho^n + a_{n-1}^0 \rho^{n-1} + \cdots + a_0^0 \qquad (12.5.18)$$

$$B(\rho) = b_{n-1}^0 \rho^{n-1} + \cdots + b_0^0 \qquad (12.5.19)$$

We further assume that $A(\rho)$ is stable.

We first note that from (12.2.18) to (12.2.22) we can write

$$\phi(t)^T = \frac{1}{E(\rho)}\left[\rho^{n-1}y(t),\ldots,y(t),\rho^{n-1}u(t),\ldots,u(t)\right]$$

$$= \frac{1}{A(\rho)E(\rho)}\left[\rho^{n-1}A(\rho)y(t),\ldots,A(\rho)y(t),\right.$$

$$\left.\rho^{n-1}A(\rho)u(t),\ldots,A(\rho)u(t)\right]$$

$$= \frac{1}{A(\rho)E(\rho)}\left[\rho^{n-1}B(\rho)u(t),\ldots,B(\rho)u(t),\right.$$

$$\left.\rho^{n-1}A(\rho)u(t),\ldots,A(\rho)u(t)\right] \qquad (12.5.20)$$

Using (12.2.9), we have

$$\phi = \frac{1}{AE}\begin{bmatrix} 0 & b^0_{n-1} & \cdots & b^0_0 & \cdots & \cdots \\ \vdots & & \ddots & & \ddots & \vdots \\ \vdots & & & b^0_{n-1} & & b^0_0 \\ 1 & a^0_{n-1} & & a^0_0 & & \vdots \\ \vdots & \ddots & \ddots & & \ddots & \\ \cdots & \cdots & 1 & a^0_{n-1} & \cdots & a^0_0 \end{bmatrix}\begin{bmatrix} \rho^{2n-1} \\ \rho^{2n-2} \\ \vdots \\ \vdots \\ \rho \\ 1 \end{bmatrix} u \qquad (12.5.21)$$

The $2n \times 2n$ matrix on the right side of (12.5.21) will be recognized as the transpose of the eliminant matrix S associated with the polynomials $A(\rho)$, $B(\rho)$ (modulo the ordering of the columns). If these polynomials are coprime, then S will be nonsingular.

Next we consider the matrix given in (12.5.10):

$$M = \int_t^{t+t_1} \phi(\tau)\phi(\tau)^T d\tau \qquad (12.5.22)$$

For simplicity, we assume that the input $u(t)$ has period T and has the following Fourier series expansion:

$$u(t) = \sum_{k=1}^N G_k \cos(\omega_k t + \alpha_k) \qquad (12.5.23)$$

where $\omega_k = n_k(2\pi/T)$ for n_k distinct integers in the range $0, (T/2\Delta)$.

We then have the following result, which shows that, provided the input contains a sufficient number of sinusoidal components, parameter convergence is guaranteed.

Lemma 12.5.2. Persistent Excitation

A sufficient condition for an input to be persistently exciting [that is, for $u(t)$ to be such that the regression vector $\phi(t)$ is persistently spanning] is that it be periodic with not less than n distinct sinusoids and that $A(\rho)$, $B(\rho)$ be relatively prime.

Proof: The matrix M is singular if and only if there exists a $2n$ vector $v \neq 0$ such that $v^T M v = 0$. Substituting (12.5.23) into (12.5.21) and using the orthogonality of the sine waves, where we select t_1 as the period, T gives

$$\phi(\tau) = S \sum_{k=1}^{N} \text{Re} \left\{ \frac{G_k}{A(z_k)E(z_k)} \begin{bmatrix} z_k^{2n-1} \\ z_k^{2n-2} \\ \vdots \\ 1 \end{bmatrix} e^{j(\omega_k t + \alpha_k)} \right\} \quad (12.5.24)$$

where

$$z_k = \frac{e^{j\omega_k \Delta} - 1}{\Delta} \quad (12.5.25)$$

Now consider

$$s = v^T M v \quad (12.5.26)$$

Using (12.5.22), (12.5.24) in (12.5.26), we have

$$s = \frac{t_1}{2} \sum_{k=1}^{N} \beta_k v^T S \begin{bmatrix} z_k^{2n-1} \\ z_k^{2n-2} \\ \vdots \\ 1 \end{bmatrix} \left[(z_k^*)^{2n-1} (z_k^*)^{2n-2} \cdots 1 \right] S^T v \quad (12.5.27)$$

where

$$\beta_k = \frac{G_k^2}{|A(z_k)E(z_k)|^2} \quad (12.5.28)$$

When $A(\rho)$, $B(\rho)$ are relatively prime, then S is nonsingular. Hence $v \neq 0$ implies $W = S^T v \neq 0$. Thus, if we write W in terms of its entries as

$$W^T \triangleq [w_{2n-1}, w_{2n-2}, \ldots, w_0] \quad (12.5.29)$$

then s in (12.5.27) can be expressed as

$$s = \frac{t_1}{2} \sum_{k=1}^{N} \beta_k |W(z_k)|^2 \quad (12.5.30)$$

where

$$\overline{W}(\rho) = w_{2n-1}\rho^{2n-1} + w_{2n-2}\rho^{2n-2} + \cdots + w_0 \quad (12.5.31)$$

Since $\overline{W}(\rho)$ is a polynomial of degree (at most) $2n - 1$ with real coefficients, it has at most $2n - 1$ zeros. However, from (12.5.30) we see that $s = 0$ implies, since $\beta_k > 0$, that $\overline{W}(z_k) = 0$ for $k = 1, \ldots, n$. This would then imply that $\overline{W}(\rho)$ has $2n$ distinct zeros: $z_1, \ldots, z_n, z_1^*, \ldots, z_n^*$. This contradiction establishes that $s \neq 0$ for any $v \neq 0$. ▽▽▽

The preceding result is intuitively reasonable since each sinusoid yields two pieces of information: the gain and phase of the transfer function at the appropriate frequency. Thus n sinusoids are sufficient to uniquely determine the $2n$ parameters

($a_{n-1}, \ldots, a_0, b_{n-1}, \ldots, b_0$). The result in Lemma 12.5.2 is actually true even when the signals are nonperiodic; see Problem 20.

Our analysis has been carried out under idealized conditions. We next turn to various practical modifications to the algorithm to enhance its performance. In particular, in the next section we consider the situation where certain prior knowledge is available about the parameters. Subsequent sections show how nonideal system characteristics such as noise, undermodeling, and parameter time variations may be handled.

12.6 INCORPORATION OF PRIOR KNOWLEDGE

In the basic algorithm presented so far, we have assumed that all parameters are unknown and need to be estimated. Consider now the case where some of the parameters in the model are known. As before we will assume the output is described by

$$y = \phi^T \theta \qquad (12.6.1)$$

Suppose we know $\theta_1, \theta_2, \ldots, \theta_m$, and $\theta_{m+1}, \ldots, \theta_n$ are unknown. This case can be simply handled as follows. We define

$$\bar{y} = y - \sum_{i=1}^{m} \phi_i \theta_i \qquad (12.6.2)$$

$$\bar{\phi} = [\phi_{m+1}, \ldots, \phi_n]^T \qquad (12.6.3)$$

and

$$\bar{\theta} = [\theta_{m+1}, \ldots, \theta_n]^T \qquad (12.6.4)$$

Equation (12.6.1) can then be rewritten as

$$\bar{y} = \bar{\phi}^T \bar{\theta} \qquad (12.6.5)$$

Estimation can now be performed based on (12.6.5), where the dimensionality of the problem has been reduced by m. One situation where this frequently occurs is when it is known that a plant contains one or more integrators. In this case, $a_0 = 0$ and this information may be included in the model.

A slight generalization of this idea is to the case when the parameters θ satisfy m linear relationships; that is,

$$M\theta = K \qquad (12.6.6)$$

where M is a known ($m \times n$) matrix and K is a known m vector. Then, assuming that the rows in M are likely independent, we form a matrix $\mathscr{B}$ as follows

$$\mathscr{B} = \begin{bmatrix} M \\ \cdots \\ N \end{bmatrix} \qquad (12.6.7)$$

where N is an arbitrary (($n - m) \times n$) matrix except that $\mathscr{B}$ should be nonsingular.

We then define

$$\bar{\theta} = \mathcal{B}\theta \tag{12.6.8}$$

and

$$\bar{\phi} = \mathcal{B}^{-T}\phi \tag{12.6.9}$$

in which case

$$y = \bar{\phi}^T\bar{\theta} \tag{12.6.10}$$

with the constraint $\bar{\theta}_1 = K_1, \bar{\theta}_2 = K_2, \ldots, \bar{\theta}_m = K_m$. Thus we have transformed the problem into one of the partial knowledge of parameters.

An example where this type of prior knowledge may occur is when we have knowledge of the location of poles and/or zeros of a dynamic system. For example, a zero at $\rho = z$ is a constraint of the form

$$[z^{n-1}, z^{n-2}, \ldots, 1]\begin{bmatrix} b^{n-1} \\ b^{n-2} \\ \vdots \\ b_0 \end{bmatrix} = 0 \tag{12.6.11}$$

Another situation where prior knowledge can be used is when the true parameters are known to lie inside a given region. In this case it is relatively easy to modify the estimator so that $\hat{\theta}$ remains inside the given region.

Given any estimate $\hat{\theta}$ and a convex region ($\mathscr{C}$ say) in $\mathbb{R}^n$, we can always project $\hat{\theta}$ into $\mathscr{C}$ if it lies outside $\mathscr{C}$. The difficulty is that this projection should be performed in such a way that the key estimator properties (see Theorem 12.5.1) are retained. This is achieved by the following scheme.

If $\hat{\theta}$ is outside $\mathscr{C}$, we select $\hat{\theta}'$ as

$$\hat{\theta}' = \arg\min_{\hat{\theta}' \in \mathscr{C}} \left\{ (\hat{\theta} - \hat{\theta}')^T P^{-1} (\hat{\theta} - \hat{\theta}') \right\} \tag{12.6.12}$$

Special cases of this general scheme together with a recursive algorithm are given in Problems 15 and 16.

The reason for incorporating prior knowledge as outlined here is twofold. First, by incorporating prior knowledge, we expect superior rates of convergence for the prediction errors and parameters. Second, it is important in some applications to ensure that bad data, for example, low excitation in ϕ or a noise spike, do not take the parameter estimates into forbidden regions. For example, very large values of parameters or very small estimates of gain may lead to problems in applications such as adaptive control.

12.7 DEALING WITH NOISE IN PARAMETER ESTIMATION

We next want to consider the effect of noise on the parameter estimation procedure. We thus modify the model of (12.5.1) to include noise as follows:

$$y(t) = \phi(t)^T\theta + \eta \tag{12.7.1}$$

where η denotes additive errors in the measurements. We will consider next various types of noise and show how various algorithm modifications can be included to deal with them.

12.7.1 Deterministic Disturbances

If η is a deterministic disturbance component, then we saw in Chapter 3 that η can be modeled as

$$S(\rho)\eta = 0 \tag{12.7.2}$$

If $S(\rho)$ is known, then prefiltering can be performed to eliminate this component. This is achieved as follows. We choose $Q(\rho)$ as a stable polynomial operator, having degree not smaller than the degree of S, and define

$$Q(\rho)y_f \triangleq S(\rho)y \tag{12.7.3}$$

and

$$Q(\rho)\phi_f \triangleq S(\rho)\phi \tag{12.7.4}$$

Operating on both sides of (12.7.1) by $S(\rho)/Q(\rho)$ and using (12.7.2) gives

$$y_f = \phi_f^T \theta \tag{12.7.5}$$

Thus, we see that, in this case, the disturbance has been removed from the filtered model and hence the previous algorithms and analysis are now applicable. It is very common to encounter constant load disturbances and hence it is almost always a good idea to, at least, consider $S(\rho) = \rho$ in (12.7.2). In this case the prefilter given in (12.7.3), (12.7.4) becomes a simple high-pass filter.

For the case where $S(\rho)$ is unknown, it is possible to estimate those additional parameters by augmenting the system model. For example, consider a linear model of the form

$$A(\rho)y(t) = B(\rho)u(t) + \eta'(t) \tag{12.7.6}$$

where $\eta'(t)$ is a disturbance having (unknown) nulling polynomial of degree equal to r. Multiplying (12.7.6) by the appropriate polynomial $S(\rho)$ gives

$$\overline{A}(\rho)y(t) = \overline{B}(\rho)u(t) \tag{12.7.7}$$

where

$$\overline{A}(\rho) = S(\rho)A(\rho), \qquad \overline{B}(\rho) = S(\rho)B(\rho)$$

Parameter estimation can then be carried out on the model (12.7.7) in the normal way. It is interesting to note that Theorem 12.5.1 and Lemma 12.5.1 apply without modification to the model (12.7.7). On the other hand, Lemma (12.5.2) does not directly apply since $\overline{A}(\rho)$ and $\overline{B}(\rho)$ are not relatively prime. However, it can be shown that relative primeness is not necessary, and indeed, provided the disturbance $\eta'(t)$ is present and the original polynomials $A(\rho)$, $B(\rho)$ are relatively prime, parameter convergence for the polynomials $\overline{A}(\rho)$, $\overline{B}(\rho)$ can be guaranteed.

12.7.2 White Noise

In the case where $\eta(t)$ is a white noise process, we can gain insight into the least squares procedure by associating it with an equivalent Kalman filtering problem, as in Example 10.3.2. Thus the model can be rewritten as

$$\rho x(t) = Ax(t) + v_1(t) \tag{12.7.8}$$

$$y(t) = C(t)x(t) + v_2(t) \tag{12.7.9}$$

where $x(t) \equiv \theta(t)$, $v_1(t) = 0$, $C(t) \equiv \phi(t)$, $A = 0$, and $v_2(t) \equiv \eta(t)$. Substituting into Lemma 10.3.1 for the optimal filter gives

$$\rho\hat{\theta} = \frac{P\phi(y - \phi^T\hat{\theta})}{\Gamma + \Delta\phi^T P\phi} \tag{12.7.10}$$

$$\rho P = \frac{P\phi\phi^T P}{\Gamma + \Delta\phi^T P\phi} \tag{12.7.11}$$

where Γ is the spectral density for $v_2(t) = \eta(t)$.

This estimator can be seen to be identical to the least squares estimator (12.4.11), (12.4.12). Hence this latter estimator inherits the optimality properties of the Kalman filter; that is, $\hat{\theta}$ is the best linear unbiased estimate of θ, and when η is gaussian, $\hat{\theta}$ is the minimum variance estimate. Also, again by analogy with the Kalman filter, we see that $P(t)$ (equivalent to $\Sigma(t)$ in the Kalman filter) is the covariance of the parameter estimates $\hat{\theta}$. This explains the terminology "covariance matrix" for $P(t)$.

12.7.3 Colored Noise with Known Spectra

When the noise $\eta(t)$ is colored but has known spectral properties, then prefiltering can be used to reduce the problem to the white noise case. If this is not done, then in general the estimate $\hat{\theta}$ generated by least squares will be nonoptimal. In particular, if ϕ and η are correlated, then $\hat{\theta}$ will be biased.

Colored noise can generally be described as filtered white noise. Thus η might be described by a model of the form

$$\eta(t) = \frac{C(\rho)}{D(\rho)} \nu(t) \tag{12.7.12}$$

where $C(\rho)$, $D(\rho)$ are stable polynomials in ρ and $\nu(t)$ represents white noise. In cases where reasonable prior estimates of $C(\rho)$ and $D(\rho)$ are available, we can simply prefilter the data to form

$$\phi_f = \frac{D(\rho)}{C(\rho)} \phi \tag{12.7.13}$$

$$y_f = \frac{D(\rho)}{C(\rho)} y \tag{12.7.14}$$

Then, operating on (12.7.1) by $D(\rho)/C(\rho)$ and using (12.7.12), we have

$$y_f = \phi_f^T \theta + \frac{D(\rho)}{C(\rho)} \eta = \phi_f^T \theta + \nu \tag{12.7.15}$$

Thus, if we work with the filtered signals, the white noise optimality properties are regained.

12.7.4 Colored Noise with Unknown Spectra

To extend the preceding ideas to the case when the spectral properties of the noise are unknown, the basic strategy is to estimate the parameters in the noise model as well as in the input–output transfer function. To see how this can be done, consider the following linear stochastic system described in state space form:

$$\rho x = Fx + Gu + v_1 \tag{12.7.16}$$

$$y = Cx + v_2 \tag{12.7.17}$$

where v_1 and v_2 are white noise processes of spectral density Ω and Γ, respectively.

The model (12.7.16), (12.7.17) is difficult to work with for parameter estimation purposes. Thus, we first convert the model to a more convenient form by finding the steady-state Kalman filter corresponding to (12.7.16), (12.7.17). This filter has the following innovations form:

$$\rho \hat{x} = F\hat{x} + Gu + K(y - C\hat{x}) \tag{12.7.18}$$

If we let v denote the innovations signal $y - C\hat{x}$, then (12.7.18), can be expressed as

$$\rho \hat{x} = F\hat{x} + Gu + Kv \tag{12.7.19}$$

$$y = C\hat{x} + v \tag{12.7.20}$$

Now, if we parameterize the model in observer canonical form, then we can eliminate $\hat{x}$ in (12.7.19), (12.7.20) to yield the following stochastic input–output model (see Section 8.8.3):

$$A(\rho)y = B(\rho)u + C(\rho)v \tag{12.7.21}$$

where

$$A(\rho) = \rho^n + a_{n-1}\rho^{n-1} + \cdots + a_0$$

$$B(\rho) = b_{n-1}\rho^{n-1} + \cdots + b_0$$

$$C(\rho) = \rho^n + c_{n-1}\rho^{n-1} + \cdots + c_0$$

and where v represents the innovations process, which by the properties of the Kalman filter is a white noise signal.

Now, if we prefilter by $1/E(\rho)$ as in (12.2.12), we obtain

$$A(\rho)\left[\frac{y}{E(\rho)}\right] = B(\rho)\left[\frac{u}{E(\rho)}\right] + C(\rho)\left[\frac{v}{E(\rho)}\right] \tag{12.7.22}$$

Clearly, if a reasonable prior estimate of $C(\rho)$ is known, then the best choice for $E(\rho)$ is $C(\rho)$ since this gives a model in which the errors are nearly white. Given the model (12.7.22), we can then proceed to estimate the unknown polynomials $A(\rho)$, $B(\rho)$, and $C(\rho)$. Two possible ways of doing this are outlined next.

(i) Nonlinear regression

The model (12.7.22) can be rearranged into the form

$$y = \frac{C(\rho) - A(\rho)}{C(\rho)} y + \frac{B(\rho)}{C(\rho)} u + v$$

$$= \frac{D(\rho)}{C(\rho)} y + \frac{B(\rho)}{C(\rho)} u + v \quad (12.7.23)$$

This model has white noise error but is nonlinear in the unknown parameters $d_0, \ldots, d_{n-1}, b_0, \ldots, b_{n-1}$, and $c_0, \ldots, c_{n-1}$. Hence the nonlinear regression algorithm of Section 12.2.4 can be used. Note that the elements of ϕ corresponding to (12.2.29) satisfy the following equations:

$$\phi \triangleq \begin{pmatrix} \phi_d \\ \phi_b \\ \phi_c \end{pmatrix} \quad (12.7.24)$$

where

$$(\phi_d)_i = \frac{\rho^i}{\hat{C}(\rho)} y \quad (12.7.25)$$

$$(\phi_b)_i = \frac{\rho^i}{\hat{C}(\rho)} u \quad (12.7.26)$$

$$(\phi_c)_i = \frac{-\rho^i}{\hat{C}(\rho)^2} [\hat{D}(\rho) y + \hat{B}(\rho) u] = \frac{-\rho^i}{\hat{C}(\rho)} \hat{y} \quad (12.7.27)$$

where $\hat{y}$ is the predicted output.

These filters are implemented using the on-line estimates of $\hat{\theta}$ as they evolve. Note that it is generally necessary to project $\hat{\theta}$ into a region in which $\hat{C}(\rho)$ is Hurwitz so that the filters (12.7.25) to (12.7.27) are stable. The preceding algorithm is generally known as a *prediction error method* in the estimation literature. It is known to be (locally) convergent to the optimal (maximum likelihood) estimate of θ under suitable conditions; see the references at the end of the chapter.

(ii) Extended linear regression

There are two problems with the nonlinear regression algorithm just described. First, we must project to keep $\hat{C}(\rho)$ stable and this may be difficult, especially for high-order polynomials. Second, due to the nonlinear dependence of y on θ, local minima can occur. These two difficulties can be overcome by expanding $1/C(\rho)$ in terms of a known polynomial stable $E(\rho)$. This possibility is explored in the following result.

Lemma 12.7.1

Consider the stochastic operator model (12.7.22). Provided we know an $e \in (0, 1/\Delta]$ such that the zeros ξ_i of the polynomial $C(\rho)$ satisfy

$$|\xi_i + e| < e, \qquad \text{for } i = 1, \ldots, n \tag{12.7.28}$$

then, given any $\varepsilon > 0$, there exists a polynomial $E'(\rho)$ [independent of $A(\rho), B(\rho), C(\rho)$] such that the model can be expressed as

$$A'(\rho)\left[\frac{y}{E(\rho)}\right] = B'(\rho)\left[\frac{u}{E(\rho)}\right] + v + v' \tag{12.7.29}$$

In (12.7.29) $B'(\rho)/A'(\rho)$ is a not necessarily minimal model of the input–output transfer function, with $A'(\rho)$ monic. Also, v is a white noise process, and v' has variance less than or equal to ε.

Proof: Choose any Hurwitz polynomial $E(\rho)$ of degree n. Then, by a partial fraction expansion, we can write (where for simplicity we assume nonrepeated zeros of C)

$$\frac{E(\rho)}{C(\rho)} = 1 + \sum_{i=1}^{n} \frac{\alpha_i}{\rho + \xi_i}$$

$$= 1 + \sum_{i=1}^{n} \frac{\alpha_i}{(\rho + e) + (\xi_i - e)}$$

$$= 1 + \sum_{i=1}^{n} \frac{\alpha_i}{\rho + e} \left\{ \frac{1}{1 + \left(\frac{\xi_i - e}{\rho + e}\right)} \right\} \tag{12.7.30}$$

Expanding the term in parentheses by a power series, we have

$$\frac{E(\rho)}{C(\rho)} = 1 + \sum_{i=1}^{n} \frac{\alpha_i}{\rho + e} \left\{ 1 - \left(\frac{\xi_i - e}{\rho + e}\right) + \left(\frac{\xi_i - e}{\rho + e}\right)^2 \cdots \right\} \tag{12.7.31}$$

This power series is absolutely convergent for all ρ on the stability boundary if and only if (12.7.28) holds.

From (12.7.30), we can write

$$\frac{1}{C(\rho)} = \frac{1}{E(\rho)}\left[1 + \frac{F(\rho)}{(\rho + e)^m} + R(\rho)\right] \tag{12.7.32}$$

where $F(\rho)/[(\rho + e)^m]$ represents the first m terms in the expansion (12.7.30) and the proper transfer function $R(\rho)$ represents the remainder.

We now define

$$\frac{F'(\rho)}{E'(\rho)} \triangleq \frac{1}{E(\rho)}\left[1 + \frac{F(\rho)}{(\rho+e)^m}\right] \qquad (12.7.33)$$

Note that $E'(\rho) = E(\rho)(\rho + e)^m$ and $F'(\rho)$ is a monic polynomial of degree m. Operating on (12.7.21) by $F'(\rho)/E'(\rho)$ gives

$$A'(\rho)\left[\frac{y}{E'(\rho)}\right] = B'(\rho)\left[\frac{u}{E'(\rho)}\right] + \frac{C(\rho)F'(\rho)}{E'(\rho)}v \qquad (12.7.34)$$

where

$$A'(\rho) \triangleq A(\rho)F'(\rho), \qquad B'(\rho) \triangleq B(\rho)F'(\rho)$$

Using (12.7.32) in (12.7.33), we obtain

$$A'(\rho)\left[\frac{y}{E'(\rho)}\right] = B'(\rho)\left[\frac{u}{E'(\rho)}\right] + \left[1 - \frac{C(\rho)}{E(\rho)}R(\rho)\right]v \qquad (12.7.35)$$

The result then follows by defining v' as

$$v' \triangleq -\frac{C(\rho)}{E(\rho)}R(\rho)v \qquad (12.7.36)$$

Note that the variance of v' can be made as small as desired by suitable choice of m in (12.7.32). ∇∇∇

The key implication of the preceding result is that the model (12.7.29) is arbitrarily close to a model with white noise. Thus it is possible to use ordinary linear least squares on the extended model (12.7.29).

Note that, although the model is in general nonminimal, there is no difficulty finding the transfer function of the system. It is clearly desirable to choose $E(\rho)$ as close as possible to $C(\rho)$ since then the residues α_i will be small. Thus the overparameterization m needed will, in turn, be small. It is also desirable to choose e as close as possible to the zeros ξ_i of $C(\rho)$ since this guarantees rapid convergence of the power series (12.7.31). This also suggests that, if the zeros of $C(\rho)$ are known to be widely spaced in approximately known regions, then it is desirable to define $E'(\rho)$ as

$$E'(\rho) = E(\rho)\left[(\rho + e_1)^{m_1}(\rho + e_2)^{m_2} \cdots\right] \qquad (12.7.37)$$

where $e_1, e_2, \ldots$ are chosen such that for each i there exists an e_j that is "close" to ξ_i.

The shift operator form of this algorithm has a long history in which E is chosen as q^n; that is, the value of e is chosen as 0 in the z-plane (that is, $1/\Delta$ in the δ-plane). This choice works for any $C(q)$ since (12.7.28) is automatically satisfied by stable zeros in the z-plane. However, the choice $E = q^n$ is usually a very poor choice since α_i will not be small and the power series (12.7.31) will converge slowly, as discussed previously. These difficulties should be contrasted with the more powerful result revealed in the unified continuous/discrete version of the algorithm discussed here, which includes a more appropriate choice of $E(\rho)$.

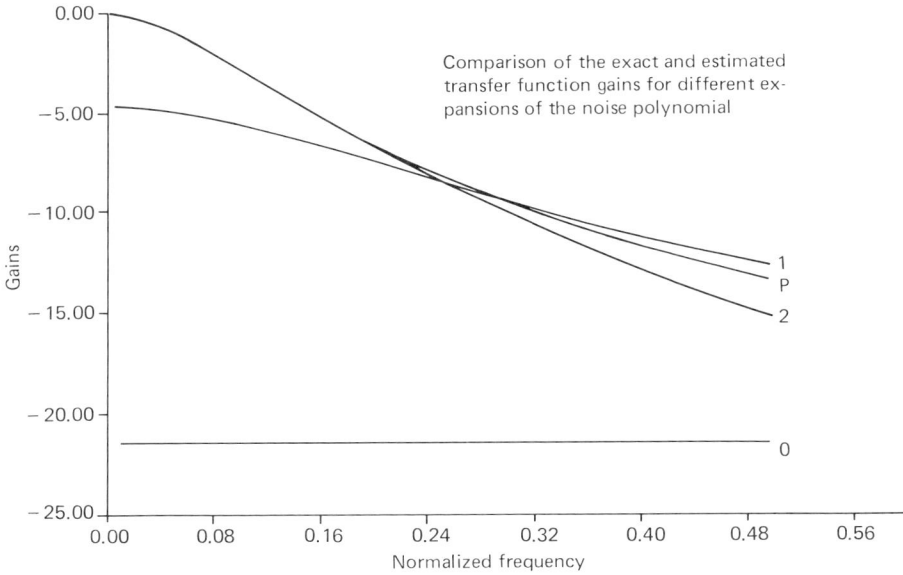

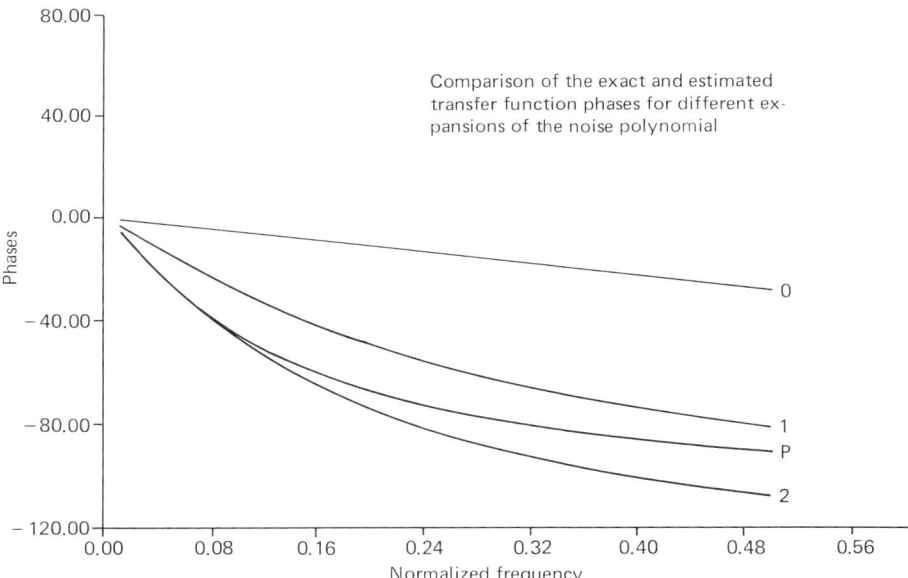

Figure 12.7.1 Comparison of estimated gain and phases of frequency response.

Example 12.7.1

To illustrate the preceding algorithm, consider the following discrete system having white measurement noise and $\Delta = 1$:

$$\delta x = -0.1x + 0.1u$$
$$y = x + \omega$$

The input–output or operator form of this system is

$$A(\delta)y = B(\delta)u + C(\delta)\omega$$

where

$$A(\delta) = \delta + a_0 = \delta + 0.1$$
$$B(\delta) = b_0 = 0.1$$
$$C(\delta) = \delta + c_0 = \delta + 0.1$$

This system was simulated for 2000 samples with u and ω two independent white sequences uniformly distributed between ± 0.5. Three estimation algorithms were tested to find the plant transfer function $B(\delta)/A(\delta)$:

[0] Least squares with $E(\delta) = \delta + 1$. Note that this is equivalent to using $E(q) = q$, which is a common choice for shift operator models.

[1] Least squares with $E(\delta) = \delta + 0.2$. Note that this underestimates the time constant of the noise filter by a factor of 2.

[2] Least squares with $E(\delta) = (\delta + 0.2)^2$. Note that we also increase the degrees of $A(\delta)$ and $B(\delta)$ accordingly to give

$$A'(\delta) = \delta^2 + a_1\delta + a_0$$
$$B'(\delta) = b_1\delta + b_0$$

The estimated frequency responses are shown in Figure 12.7.1 together with the true system frequency response. Table 12.7.1 gives the estimated transfer functions. The following observations are evident in the table and figure.

[0] This result is unacceptable by any standard. Note that $E(q) = q^n$ is sometimes suggested in the literature but we strongly counsel against this choice in this book.

[1] Here the estimated model is fair, although we do note that the dc gain has been underestimated by a factor of almost 2 and the time constant has been underestimated by a factor of 2.

[2] We see that by increasing the order of A and B by 1 and using $E(\delta) = (\delta + 0.2)^2$ we get quite an acceptable model. Indeed, if we approximate the estimated model about $\gamma = 0$, we find

$$\frac{B(\delta)}{A(\delta)} \simeq \frac{0.1}{\delta + 0.11}$$

TABLE 12.7.1 ESTIMATED TRANSFER FUNCTIONS

Case	$E(\delta)$	Number of Parameters in Model	Transfer Function
True system P	—	3	$\dfrac{0.1}{\gamma + 0.1}$
0	$\delta + 1$	2	$\dfrac{0.0838}{\gamma + 0.9856}$
1	$\delta + 0.2$	2	$\dfrac{0.1099}{\gamma + 0.1865}$
2	$(\delta + 0.2)^2$	4	$\dfrac{0.0668\gamma + 0.0359}{\gamma^2 + 0.4370\gamma + 0.0382}$

Further improvements, if desired, would be achieved by increasing the order of E, A, and B further. ▽▽▽

12.8 DEALING WITH UNDERMODELING IN PARAMETER ESTIMATION

Another common source of errors in modeling arises due to the fact that the chosen model does not give a complete description of the system. This difficulty is generally referred to as undermodeling. This problem is very widespread. Indeed it can be argued that, in practice, it is unrealistic to expect any model to give a complete description of a real process.

In this section we will discuss three aspects of the undermodeling problem. First, we suggest modifications to the basic parameter estimation procedure that avoid the catastrophic behavior that can occur in the presence of undermodeling. We then show that these modifications preserve the key properties of the estimation algorithm. This gives credibility to the use of the modified algorithms in practice. Finally, we show how the effect of undermodeling on the results of parameter estimation can be quantified. This last step is particularly important since it is always desirable to associate with any estimated model some measure of the fidelity or accuracy of that model in relation to the true system.

12.8.1 Algorithm Modifications

The key difficulty with errors arising from undermodeling is that they lack structure. Indeed, if this were not the case, they would hardly qualify as undermodeling errors.

Now consider the general recursive least algorithm given in (12.5.2), i.e.,

$$\rho\hat{\theta} = \frac{\alpha P\phi(y - \phi^T\hat{\theta})}{\Gamma + \Delta\phi^T P\phi} \qquad (12.8.1)$$

together with the model in (12.7.1), where η corresponds to modeling errors.

If we substitute (12.7.1) into (12.8.1), then with $\tilde{\theta} = \hat{\theta} - \theta$ we have

$$\rho\tilde{\theta} = -\frac{\alpha P\phi\phi^T\tilde{\theta}}{\Gamma + \Delta\phi^T P\phi} + \frac{\alpha P\phi\eta}{\Gamma + \Delta\phi^T P\phi} \tag{12.8.2}$$

This is a linear differential or difference equation forced by the last term. When ϕ and η are correlated (as will almost always be the case with undermodeling) then it can be seen from (12.8.2) that the term $\phi\eta$ on the right-hand side can cause $\tilde{\theta}$ to drift. This problem is particularly significant when ϕ is not persistently exciting, since the first term on the right-hand side of (12.8.2) is zero in certain directions and hence (12.8.2) acts like a pure integrator.

When parameter estimators are used with on-line control design to produce an adaptive controller (see Section 12.10), then this drift phenomenon can lead to "bursting" as follows. When the parameter estimates reach sensible values, the control law becomes satisfactory and the system response settles down. This causes lack of excitation in ϕ and the parameters (slowly) drift as explained above. After a little while, the parameter estimates become poor and the control law becomes unsatisfactory. Then there is considerable (and unwanted) excitation in ϕ and the estimates return to sensible values thus repeating the cycle.

Probably the best currently known way of overcoming these problems is to stop the drifting from occurring by turning the parameter estimator off whenever the prediction errors are small. The resulting algorithm is generally said to have a *dead zone*. To construct such an algorithm, we need to have some idea of the "size" of the unmodeled error so as to set the switching threshold in the dead zone.

To establish bounds on the unmodeled response requires knowledge of an overbounding function, $d(t)$. One simple case is when the term $\eta(t)$ has a known maximum amplitude. Another case of interest is where we allow for undermodeling in the plant transfer function. To illustrate, say we have a linear system having nominal (modeled) transfer function and multiplicative unmodeled dynamics $\overline{H}$ and additive unmodeled dynamics $\overline{\overline{H}}$; that is, we express y as

$$y = \left[H_0(1 + \overline{H}) + \overline{\overline{H}}\right]u = \frac{B_a}{A_a}u \tag{12.8.3}$$

where

$$H_0 = \frac{B}{A} \tag{12.8.4}$$

The model (12.8.3) can be rewritten as

$$Ay = Bu + \eta \tag{12.8.5}$$

where u, y denote the plant input and output, respectively, and η denotes an unmodeled component, which is given by

$$\eta = \left[B\overline{H} + A\overline{\overline{H}}\right]u \tag{12.8.6}$$

We now introduce an appropriate filter as in Section 12.2.2. We thus define the following filtered variables $\bar{y}_f, \bar{u}_f$ corresponding to y and u, respectively. Applying this filter to (12.8.5) yields

$$A\bar{y}_f = B\bar{u}_f + \bar{\eta}_f \tag{12.8.7}$$

where $\bar{\eta}_f$ is a filtered version of η, and the filter has relative degree at least n. A bound on the size of this residual component can now be obtained, as in the following result:

Lemma 12.8.1

Provided the unmodeled components are stable (that is, provided $\overline{H}$ and $\overline{\overline{H}}$ are stable strictly proper transfer functions) and provided the prefilter is stable, then $\eta_f(t)$ is bounded by a function $d(t)$ as follows. There exist $\sigma > 0$, $\varepsilon > 0$ such that

$$|\eta_f(t)| \leq \varepsilon \int_0^t E(-\sigma, t-\tau)|u(\tau)|\,d\tau \triangleq d(t) \tag{12.8.8}$$

where $E(-\alpha, s)$ is a stable exponential.

Proof: Immediate on using the stability of $\overline{H}$ and $\overline{\overline{H}}$ and the prefilter. (See Problem 18.) ∇∇∇

Thus we see that $\eta_f(t)$ is bounded by the output of a linear system driven by the modulus of the plant inputs. In the sequel, we assume that we have available an overbounding function, $d(t)$, for the modeling errors, where

$$|\eta_f(t)| \leq d(t) \tag{12.8.9}$$

where

$$y(t) = \phi(t)^T \theta + \eta_f(t) \tag{12.8.10}$$

and $\eta_f(t)$ denotes the filtered modeling errors. Then one way of choosing the dead zone rule is as follows. We pick $a \in (0, 1)$, let $\bar{p} = \sup_{t \in [0, \infty)}\{\|P(t)\|_2\}$ and select

$$k > \left(\frac{1 + \Delta \bar{p}}{1 + \Delta(1-a)\bar{p}}\right)^{1/2} \tag{12.8.11}$$

Define (for $r \geq 0$ and any e)

$$f(r, e) = \begin{cases} e - r & e > r \\ 0 & |e| \leq r \\ e + r & e < -r \end{cases} \tag{12.8.12}$$

and

$$g(r, e) = \begin{cases} \dfrac{f(r, e)}{e} & |e| > r \\ 0 & |e| < r \end{cases} \tag{12.8.13}$$

The functions $f(\cdot, \cdot)$ and $g(\cdot, \cdot)$ are illustrated in Figure 12.8.1

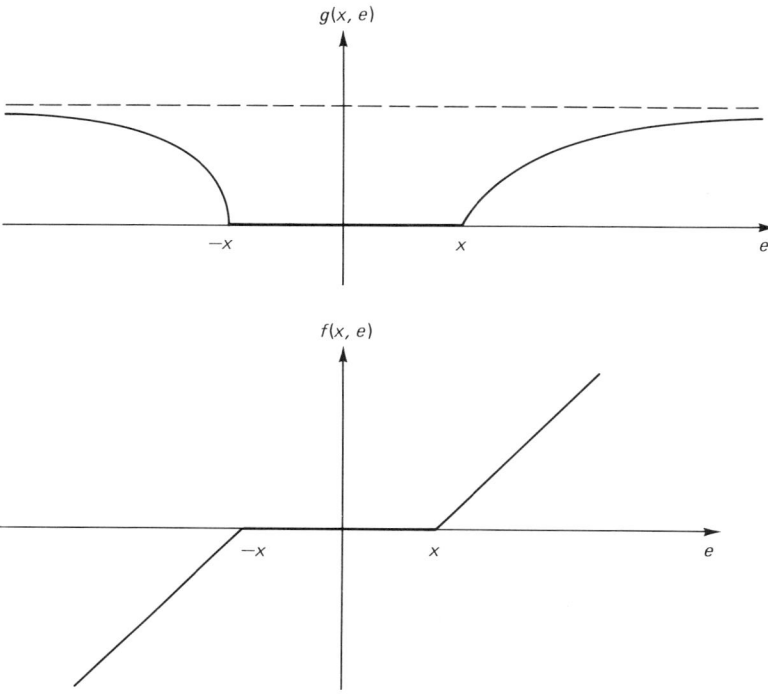

Figure 12.8.1 Dead zone functions in Equations (12.8.13) and (12.8.14).

We now define a least squares estimator with dead zone as in (12.8.1) with $\Gamma = 1 + \phi^T\phi$ and where $\alpha(t)$ implements a variable gain as follows:

$$\alpha(t) = ag(kd(t), e(t)) \qquad (12.8.14)$$

where $e(t)$ is the prediction error $y(t) - \phi(t)^T\hat{\theta}(t)$. This algorithm avoids parameter drift and generally leads to an estimator having desirable practical properties. Some of these properties will be derived in the next section.

12.8.2 Properties of Estimator with Dead Zone

In Section 12.5, results were derived detailing the properties of the estimation algorithms under ideal conditions. In this section we will derive the corresponding results for the nonideal case when modeling errors are present. In particular, we extend Theorem 12.5.1 to cover the dead zone algorithm of Equations (12.5.2), (12.5.3), and (12.8.14). Loosely speaking, the dead zone algorithm has two key properties: first, as in the ideal case, the parameter error decreases at each step, and second, the prediction error decreases until it reaches the size of the dead zone. These results are made precise in Lemma 12.8.2.

Lemma 12.8.2

The estimator (12.5.2), (12.5.3) with $\Gamma = 1 + \phi^T\phi$, α defined as in (12.8.14), and Ω chosen such that P is bounded, has the following properties, assuming (12.8.9) holds:

(i) *Parameter error reduction*: $V = \tilde{\theta}^T P^{-1} \tilde{\theta}$ is a nonincreasing function.

(ii) *Prediction error convergence*:

$$\tilde{f}(t) \triangleq \frac{f(kd(t), e(t))}{\left[1 + \phi(t)^T \phi(t)\right]^{1/2}} \in \mathscr{L}_2^+ \tag{12.8.15}$$

(iii) *Parameter change convergence*:

$$\rho\hat{\theta} \in \mathscr{L}_2^+ \tag{12.8.16}$$

Proof: In this case we have

$$\rho V \leq \frac{ag}{\Gamma + \Delta\phi^T P\phi}\left(\eta_f^2 \frac{\Gamma + \Delta\phi^T\phi}{\Gamma + \Delta(1 - ag)\phi^T P\phi} - e^2\right) \tag{12.8.17}$$

(with equality if Ω is zero). Since $g(\cdot, \cdot) \leq 1$, $a < 1$, and $\|P\|_2 \leq \bar{p}$, we then have

$$\rho V \leq \frac{ag}{\Gamma + \Delta\phi^T P\phi}\left(\frac{\eta_f^2(1 + \Delta\bar{p})}{1 + \Delta(1 - a)\bar{p}} - e^2\right) \tag{12.8.18}$$

Now $|\eta_f| \leq d$, and $g = 0$ if $|e| < kd$, so $g\eta_f^2 \leq g(e^2/k^2)$; that is,

$$\rho V \leq -\frac{ag}{\Gamma + \Delta\phi^T P\phi}\left(1 - \frac{1 + \Delta\bar{p}}{1 + k^2\Delta(1 - a)\bar{p}}\right)e^2 \tag{12.8.19}$$

In view of (12.8.11), we then have that there exists $a' > 0$ such that

$$\rho V \leq -\frac{a'ge^2}{\Gamma + \Delta\phi^T P\phi} \tag{12.8.20}$$

$$\leq -\frac{a'f^2}{1 + \phi^T\phi + \Delta\phi^T P\phi} \tag{12.8.21}$$

since $ge = f$ and $fe \geq f^2$. The results then follow essentially as that in the proof of Theorem 12.5.1. See also Problem 22. ▽▽▽

A simple case of this result is explored in Problem 13.

Note that the dead zone estimator analyzed in Lemma 12.8.2 has very similar properties in the nonideal case, as does the original estimator in the ideal case. In particular, property (i) essentially guarantees that parameter drift away from the true parameters is avoided.

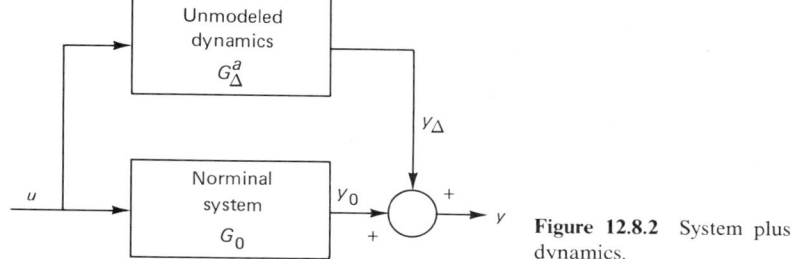

Figure 12.8.2 System plus unmodeled dynamics.

12.8.3 Quantification of Undermodeling on Estimation

The dead zone algorithm avoids catastrophic behavior in the presence of undermodeling and poor excitation. However, it is obvious that, even when these steps are taken, the estimated model can never capture the true system model. Hence there will be some error between the estimated model (here denoted $\hat{G}$) and the true system model (here denoted G_T). It is important in estimation to have some measure of the size of this error, since this will affect the fidelity of the conclusions drawn from using the estimated model. Our purpose in this section is to show how these errors can be quantified based on some *a priori* knowledge of the inherent ability of the given model to describe the given system.

We consider the basic system structure given earlier in Equations (12.8.3) and (12.8.4). For simplicity, we lump all the model uncertainty into the additive component. Thus we will describe the true system G_T as a composition of a nominal system plus additive unmodeled dynamics, as shown in Figure 12.8.2. The nominal system is assumed to be described by a rational function of the complex variable γ belonging to a class $\mathcal{G}$. Members of this class are assumed to have the following structure:

$$G(\gamma, \theta) \triangleq \frac{B(\theta, \gamma)}{A(\theta, \gamma)} \tag{12.8.22}$$

where

$$A(\theta, \gamma) \triangleq \gamma^n + a_{n-1}\gamma^{n-1} + \cdots + a_0 \tag{12.8.23}$$

$$B(\theta, \gamma) \triangleq b_m\gamma^m + b_{m-1}\gamma^{m-1} + \cdots + b_0 \tag{12.8.24}$$

$$\theta \triangleq (a_{n-1}, a_{n-2}, \ldots, a_0, b_m, b_{m-1}, \ldots, b_1)^T \tag{12.8.25}$$

Thus the nominal model $G_0(\gamma)$ is simply $G(\gamma, \theta_0)$. We will denote the estimated model by $\hat{G}(\gamma) = G(\gamma, \hat{\theta})$. From Figure 12.8.2, we have

$$y = G_0 u + G_\Delta^a u$$

$$= \frac{B(\theta_0, \rho)}{A(\theta_0, \rho)} u + G_\Delta^a u \tag{12.8.26}$$

Multiplying (12.8.26) by $A(\theta_0, \rho)/E(\rho)$ gives

$$\frac{A(\theta_0, \rho)}{E(\rho)} y = \frac{B(\theta_0, \rho)}{E(\rho)} u + \frac{A(\theta_0, \rho)}{E(\rho)} G_\Delta^a u \qquad (12.8.27)$$

where $E(\rho)$ is a suitable observer polynomial of the same degree as A.

Equation (12.8.27) can be rearranged as

$$y = \frac{E(\rho) - A(\theta_0, \rho)}{E(\rho)} y + \frac{B(\theta_0, \rho)}{E(\rho)} u + \frac{A(\theta_0, \rho)}{E(\rho)} G_\Delta^a u$$

$$= \phi(t)^T \theta_0' + \eta(t) \qquad (12.8.28)$$

where

$$\phi(t)^T = \left[\frac{\rho^{n-1} y(t)}{E(\rho)}, \ldots, \frac{y(t)}{E(\rho)}; \frac{\rho^m u(t)}{E(\rho)}, \ldots, \frac{u(t)}{E(\rho)} \right] \qquad (12.8.29)$$

$$\theta_0' = \left[e_{n-1} - a_{n-1}^0, \ldots, e_0 - a_0^0; b_m^0, \ldots, b_0^0 \right]^T \qquad (12.8.30)$$

$$\eta(t) = A(\theta_0, \rho) \frac{G_\Delta^a(\rho)}{E(\rho)} u(t) \qquad (12.8.31)$$

We will assume that some form of on-line least squares estimation is used to generate $\hat{\theta}$. For example, we may use the dead zone algorithm of Equations (12.5.2), (12.5.3), and (12.8.14). All algorithms of this type can be expressed in the general form

$$\rho \hat{\theta}_t = K_t \varphi_t e_t \qquad (12.8.32)$$

where K_t is an algorithm gain (which will be zero inside the dead zone) and e_t is the prediction error given by

$$e_t = y_t - \varphi_t^T \hat{\theta}_t \qquad (12.8.33)$$

Substituting (12.8.28) into (12.8.33) gives

$$e_t = -\varphi_t^T \tilde{\theta}_t + \eta_t \qquad (12.8.34)$$

where

$$\tilde{\theta}_t = \hat{\theta}_t - \theta_0 \qquad (12.8.35)$$

Substituting (12.8.35) into (12.8.32) gives

$$\rho \tilde{\theta}_t = -K_t \varphi_t \varphi_t^T \tilde{\theta}_t + K_t \varphi_t \eta_t \qquad (12.8.36)$$

We notice that Equation (12.8.36) represents a linear time varying system. The solution of this equation can be obtained in closed form as

$$\tilde{\theta}(t) = \Phi(t, 0) \tilde{\theta}(0) + \int_0^t \Phi(t, \tau + \Delta) K(\tau) \phi(\tau) \eta(\tau) \, d\tau \qquad (12.8.37)$$

where $\Phi(t, \tau)$ denotes the state transition matrix corresponding to (12.8.36). As an example, it can be shown (see Problem 19) that the state transition matrix for the exponential weighted least squares algorithm given in Section 12.9 is

$$\Phi(t, \tau) = P(t) P(\tau)^{-1} E(-\sigma, t - \tau) \qquad (12.8.38)$$

Also, in view of (12.8.31), $\eta(t)$ is related to the system input $u(t)$ by a linear time invariant system. If we let $h(\tau)$ denote the impulse response of the unmodeled dynamics G_Δ^a, then $\eta(t)$ can be expressed as

$$\eta(t) = \int_0^t h(t - \tau - \Delta)x(\tau)\,d\tau \tag{12.8.39}$$

where $x(t)$ is related to the input by

$$x(t) = \frac{A(\theta_0, \rho)}{E(\rho)} u(t) \tag{12.8.40}$$

To simplify the subsequent presentation, we will assume sampled data, in which case equation (12.8.39) can be written compactly in matrix notation as

$$S \triangleq UH\Delta \tag{12.8.41}$$

where S denotes a vector of $\eta(0), \ldots, \eta([N-1]\Delta)$, H is the vector of the sampled impulse response h of G_Δ^a, and U is a matrix formed from the filtered inputs; specifically,

$$H \triangleq [h(0), \ldots, h[N-1]\Delta]^T \tag{12.8.42}$$

$$U = \begin{bmatrix} x(0) & x(-\Delta) & \cdots & x[1-N]\Delta \\ x(\Delta) & x(0) & \cdots & x[2-N]\Delta \\ \vdots & & & \\ x[N-1]\Delta & x[N-2]\Delta & \cdots & x(0) \end{bmatrix} \tag{12.8.43}$$

and $x(k)$ is as given in (12.8.40). In practice, we will need to replace the unknown polynomial $A_0(\rho)$ by an estimate, for example, $A(\hat{\theta}, \rho)$.

For simplicity, we assume that the data not eliminated by the dead zone are rich, in which case the first term on the right side of (12.8.37) can be neglected. Thus we can write the solution of (12.8.37) in the form

$$\tilde{\theta}(N\Delta) = LUH\Delta \tag{12.8.44}$$

where L is a matrix formed from the impulse response from η to $\tilde{\theta}$:

$$L \triangleq [\Phi(N\Delta, \Delta)K(0)\phi(0), \ldots, \Phi(N\Delta, N\Delta)K([N-1]\Delta)\phi([N-1]\Delta)] \tag{12.8.45}$$

Equation (12.8.44) quantifies the error in $\tilde{\theta}$ as a function of the input. However, we are frequently interested in the error in the estimated frequency response. Thus we will be interested in computing G_Δ^l, which represents the error between the true transfer function $G_T(\gamma)$ and the estimated transfer function $G(\gamma, \hat{\theta})$; that is,

$$G_\Delta^l(\gamma) \triangleq G(\gamma, \hat{\theta}) - G_T(\gamma) \tag{12.8.46}$$

where, from Figure 12.8.2,

$$G_T(\gamma) = G_0(\gamma, \theta_0) + G_\Delta^a(\gamma) \tag{12.8.47}$$

Substituting (12.8.47) into (12.8.46), we obtain
$$G'_\Delta(\gamma) = [G(\gamma, \hat{\theta}) - G_0(\gamma, \theta_0)] - G^a_\Delta(\gamma) \tag{12.8.48}$$
Note that if $\hat{\theta} \neq \theta_0$ then it is likely that $G(\gamma, \hat{\theta}) \neq G(\gamma, \theta_0)$. However, it still may be true that $G'_\Delta(\gamma)$ is small (or even zero for certain inputs) due to the term $G^a_\Delta(\gamma)$ in (12.8.48).

We will evaluate the first term in (12.8.48) via a Taylor's series expansion as follows:
$$\hat{G}(\gamma) - G_0(\gamma) \simeq V(\gamma)^T \tilde{\theta} \tag{12.8.49}$$
where $V(\gamma)$ is a column vector such that
$$V(\gamma)^T \triangleq \left.\frac{\partial G(\gamma, \theta)}{\partial \theta}\right|_{\theta=\hat{\theta}} \tag{12.8.50}$$

To obtain the corresponding frequency response, we simply replace γ by $(e^{j\omega\Delta} - 1)/\Delta$. We then have the following result, which quantifies the a posteriori errors resulting from estimation:

Lemma 12.8.3

The magnitude of the a posteriori uncertainty G'_Δ between the estimated transfer function $\hat{G}$ and the true transfer function G_T is given by
$$|G'_\Delta|^2 = |\hat{G} - G_T|^2 \simeq (V^T LU - F^T) HH^T (V^T LU - F^T)^\# \Delta^2 \tag{12.8.51}$$
where $(\ )^\#$ denotes conjugate transpose and
$$F \triangleq \begin{bmatrix} 1 & e^{-j\omega\Delta} & \cdots & e^{-(N-1)j\omega\Delta} \end{bmatrix}^T \tag{12.8.52}$$

Proof: We first note that, using (12.8.48),
$$|\hat{G} - G_T|^2 = |\hat{G} - G_0 + G_0 - G_T|^2 = |\hat{G} - G_0 - G^a_\Delta|^2$$
$$= |\hat{G} - G_0|^2 + |G^a_\Delta|^2 - (\hat{G} - G_0)(G^a_\Delta)^* - (\hat{G} - G_0)^* G^a_\Delta \tag{12.8.53}$$

Also, from (12.8.44), (12.8.49), we have
$$\hat{G} - G_0 = V^T LUH\Delta \tag{12.8.54}$$
$$G^a_\Delta = F^T H\Delta \tag{12.8.55}$$
since premultiplication of an impulse response vector by F^T yields the Fourier transform. Substituting (12.8.54), (12.8.55) into (12.8.53) and rearranging gives (12.8.51). ▽▽▽

The problem of evaluating $|G'_\Delta|^2$ using (12.8.51) now depends on the a priori assumptions regarding G^a_Δ. We will assume that we have available a bound on the corresponding impulse response $h(\tau)$. Thus we assume we know a vector $\bar{H} \triangleq [\bar{h}(0), \ldots, \bar{h}(N-1)]^T$ such that
$$\bar{h}(i) \geq |h(i)|, \quad \text{for all } i \tag{12.8.56}$$

Using the triangle inequality, we then have from (12.8.51) the following bound for G_Δ^l:

$$|G_\Delta^l| \leq \left\{ \left[\Sigma |\beta_{1i} \bar{h}_i| \right]^2 + \left[\Sigma |\beta_{2i} \bar{h}_i| \right]^2 \right\}^{1/2} \tag{12.8.57}$$

where β_{1i} and β_{2i} denote the real and imaginary parts, respectively, of the ith component of the vector $(V^T LU - F^T)$.

We illustrate the use of this result by some simple examples.

Example 12.8.1

Consider a true (*but unknown*) system having continuous time transfer function

$$G_T(s) = \frac{15}{(s+1)(s+15)} \tag{12.8.58}$$

A sampling period of $\Delta = 0.01$ second was chosen. Also, the nominal model for the

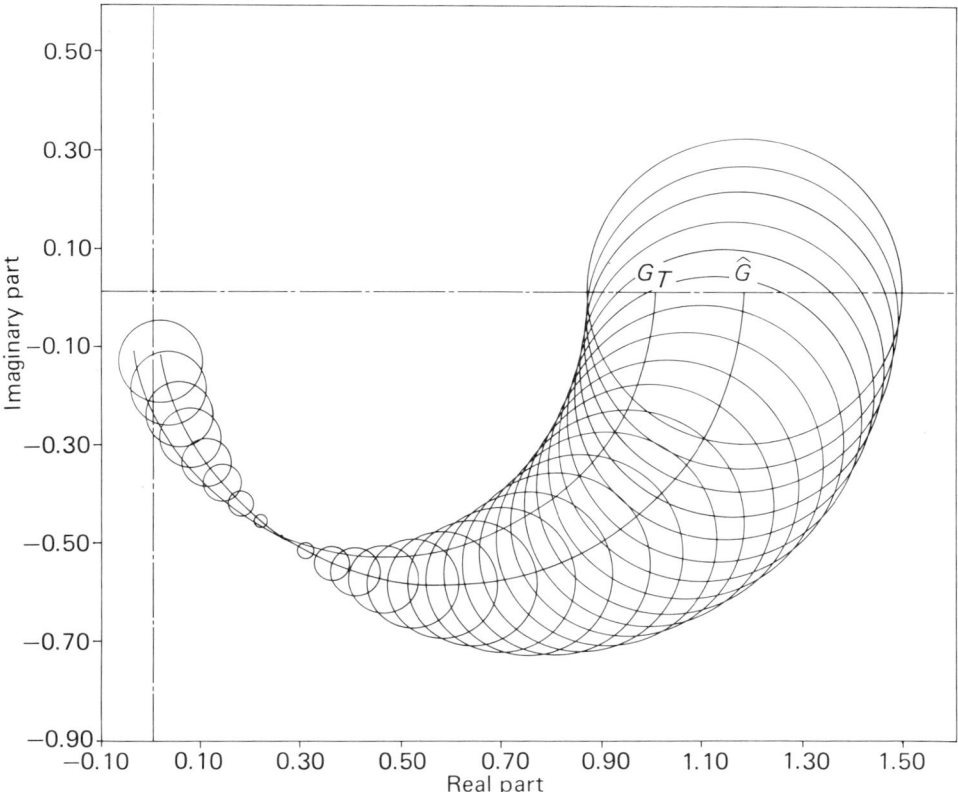

Figure 12.8.3 Estimation errors shown on a Nyquist diagram.

system will be taken as

$$G_0(\gamma) = \frac{b_0}{\gamma + a_0} \tag{12.8.59}$$

For simplicity, the estimation procedure will be taken to be block least squares as in Section 12.3. We choose the filter polynomial $E(\gamma)$ to be

$$E(\gamma) = \gamma + 2 \tag{12.8.60}$$

To estimate the effect of the modeling error we use (12.8.57). We will describe the undermodeling by assuming about 10% frequency domain modeling error and a "bandwidth" of about 10 rad sec^{-1} for the undermodeled dynamics. This is consistent with a choice for $\overline{H}$ such that $\overline{h}(t) = e^{-10t}$.

The input was chosen as a sine wave of frequency 1.5 rad sec^{-1}. The results are shown in Figure 12.8.3. We see from the figure that the bound (12.8.57) accurately captures the difference between the estimated transfer function $\hat{G}$ and the true but *unknown* transfer function G_T. We note that the least squares algorithm finds the true system at the test frequency. This is in accord with intuition, since G_Δ^a is invariant and is therefore captured by any model at a single frequency. ▽▽▽

Example 12.8.2

Consider the true (*but unknown*) system having continuous time transfer function

$$G_T(s) = \frac{e^{-s}}{s + 1} \tag{12.8.61}$$

A sampling period of $\Delta = 0.01$ second was chosen. Also, the nominal model was assumed to be a third-order *rational* model as follows:

$$G_0(\gamma) = \frac{b_2\gamma^2 + b_1\gamma + b_0}{\gamma^3 + a_2\gamma^2 + a_1\gamma + a_0} \tag{12.8.62}$$

Again, we use standard least squares with filter polynomial

$$E(\gamma) = (\gamma + 3)^3 \tag{12.8.63}$$

The input was a random signal in the frequency band $(1.2, 3)$ rad sec^{-1}. The regression vector was prefiltered using a bandpass filter in the same band.

We assume about 10% frequency domain error and a "bandwidth" of 0.5 rad sec^{-1}, since we can forecast that the errors will be dominated by low-frequency terms due to the presence of the bandpass filter. Thus we take $\overline{h}(t) = 0.05e^{-0.5t}$, which is consistent with the preceding specifications.

The result are shown in Figure 12.8.4. Again we see from the figure that the bound (12.8.57) accurately captures the difference between the estimated transfer function $\hat{G}$ and the true but *unknown* function G_T.

For this example, the result shows that the true system is not found at any frequency due to the complex nature of the input. However, the uncertainty bands indicate how the a posteriori errors are distributed in frequency as a result of the estimation procedure. ▽▽▽

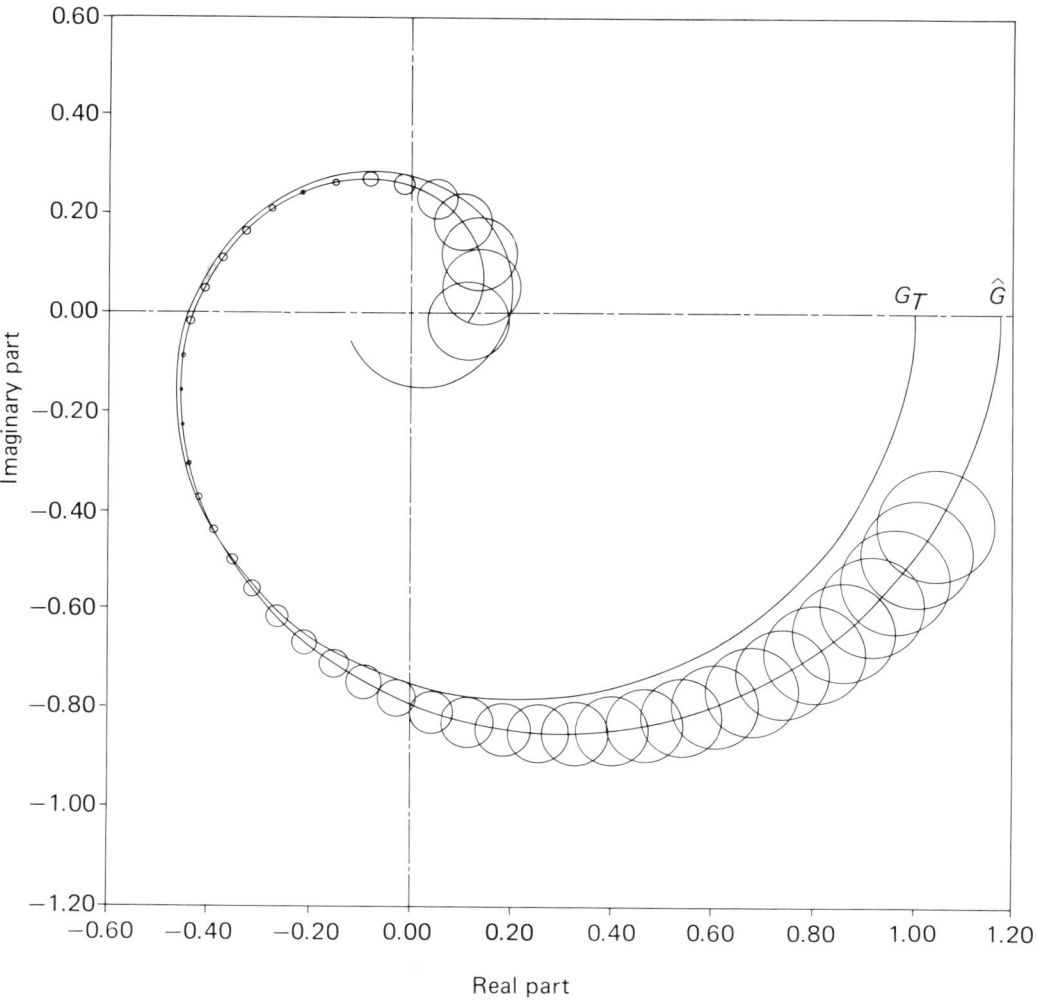

Figure 12.8.4 Estimation errors shown on a Nyquist diagram.

12.9 DEALING WITH PARAMETER TIME VARIATIONS

So far in this chapter we have implicitly assumed that the underlying system is time invariant. However, often the real motivation for on-line parameter estimation is that the system parameters are time varying. The raw least squares algorithm of Section 12.3 is inappropriate in this case because the estimated parameters are simply some kind of average value applicable to the complete observation period. It seems that a more appropriate estimate in the time varying case would be one that emphasized more recent data and somehow discounted very old data.

We will show next that the general algorithm of (12.5.2), (12.5.3) has the feature of emphasizing more recent data provided Ω is chosen appropriately. There are several ways that this algorithm can be motivated. We will give several possible motivations beginning from different perspectives.

12.9.1 Kalman Filter Formulation

The Kalman filter interpretation of the least squares algorithm suggests ways in which the estimator might be modified to handle time varying parameters. Say we model the parameter time variations by adding process noise to the model given in Section 12.7.2. This leads to

$$\rho \theta(t) = \omega(t) \tag{12.9.1}$$

and

$$y(t) = \phi(t)^T \theta(t) + \nu(t) \tag{12.9.2}$$

where $\omega(t), \nu(t)$ are zero mean, white noise processes, and

$$E\left[\omega(t)\omega(t)^T\right] \triangleq Q = \frac{1}{\Delta}\Omega \tag{12.9.3}$$

$$E\left[\nu(t)^2\right] \triangleq \sigma^2 = \frac{1}{\Delta}\Gamma \tag{12.9.4}$$

As in Section 12.7.2, this model is exactly as described in Chapter 10 for the Kalman filter. Thus, the optimal linear estimator for $\theta(t)$ is given as in Lemma 10.3.1, where $A = 0$, $S = 0$, $C = \phi(t)^T$. Substituting these quantities into Lemma 10.3.1 gives

$$\rho\hat{\theta} = \frac{P\phi(y - \phi^T\hat{\theta})}{\Gamma + \Delta\phi^T P\phi} \tag{12.9.5}$$

$$\rho P = \frac{-P\phi\phi^T P}{\Gamma + \Delta\phi^T P\phi} + \Omega \tag{12.9.6}$$

Comparing this with the result in (12.7.10), (12.7.11) for the time invariant case show that the only change necessary is to add Ω to the P-update equation. This prevents P from going to zero and allows the algorithm to track time varying parameters as desired.

12.9.2 Exponential Data Weighting

Another approach used to account for time variations is to modify the cost function upon which we base the estimator. The usual way in which this is done is to introduce weighting in the cost function, which will cause recent data to have a greater influence on the estimates.

For simplicity, we normalize the cost function by putting $\Gamma = 1$. Then define

$$C''(\hat{\theta}, t) = \frac{1}{2} \int_0^t w(t,\tau)\left(y(\tau) - \phi(\tau)^T \hat{\theta}\right)^2 d\tau$$
$$+ \frac{1}{2} w(t,0)(\hat{\theta} - \hat{\theta}_0)^T P_0^{-1}(\hat{\theta} - \hat{\theta}_0) \quad (12.9.7)$$

where $w(t,\tau)$ denotes a data weighting function. Minimization of this cost function leads to the following estimator:

$$\hat{\theta}(t) = \left\{ w(t,0) P_0^{-1} + \int_0^t w(t,\tau)\phi(\tau)\phi(\tau)^T d\tau \right\}^{-1}$$
$$\times \left\{ w(t,0) P_0^{-1} \hat{\theta}_0 + \int_0^t w(t,\tau)\phi(\tau)y(\tau) d\tau \right\} \quad (12.9.8)$$

One common form of data weighting is exponential data weighting, where we use

$$w(t,\tau) = E(-\sigma, t - \tau) \quad (12.9.9)$$

where σ is a constant and $0 < \sigma < 1/\Delta$. In this case we may rewrite (12.5.8) in recursive form as (see Problem 3)

$$\rho \hat{\theta} = \frac{P\phi(y - \phi^T \hat{\theta})}{1 + \Delta \phi^T P \phi} \quad (12.9.10)$$

and

$$\rho P = \frac{1}{1 - \Delta \sigma}\left(\sigma P - \frac{P\phi\phi^T P}{1 + \Delta \phi^T P \phi} \right) \quad (12.9.11)$$

Note that this algorithm reduces to the one given in (12.4.11), (12.4.12) when $\sigma = 0$ (no forgetting) and $\Gamma = 1$ (noise variance normalized to 1). A minor disadvantage of (12.9.10), (12.9.11) for the case of $\sigma \neq 0$ is that the "size" of P is roughly proportional to σ. This can be overcome by scaling P to give $\bar{P} = (1/\sigma)P$. The algorithm can then be expressed in terms of $\bar{P}$ as

$$\rho \hat{\theta} = \frac{\sigma \bar{P} \phi (y - \phi^T \hat{\theta})}{1 + \Delta \sigma \phi^T \bar{P} \phi} \quad (12.9.12)$$

and

$$\rho \bar{P} = \frac{\sigma}{1 - \Delta \sigma}\left[\bar{P} - \frac{\bar{P}\phi\phi^T \bar{P}}{1 + \Delta \sigma \phi^T \bar{P} \phi} \right] \quad (12.9.13)$$

In this form, $\bar{P}$ is roughly equal to the inverse of the average value of $(\phi\phi^T)$ over an interval depending on σ. This tends to be a constant matrix, which can be checked against the signal levels in ϕ. The constant σ appearing in the equation for $\hat{\theta}$ essentially determines the bandwidth of the estimator, as we will show in Section 12.9.5.

One remaining problem with the algorithm (12.9.12), (12.9.13) is that, under low excitation conditions, $\bar{P}$ can grow without bound. For this reason it is desirable, in practice, to monitor the size of $\overline{P}$ and to ensure that excessively large values are avoided by an appropriate algorithm modification, for example, by keeping trace $\overline{P}$ constant. [This is readily done in discrete time by simply putting $\overline{P}' = (C/\text{trace } \overline{P})\overline{P}$ after the $\overline{P}$ update.] A method that works also for the continuous time case will be given in Section 12.9.6, part (vi).

12.9.3 Gradient Algorithm

The key point about the algorithms given in Sections 12.9.1 and 12.9.2 is that they prevent the P matrix from going to zero. Thus the algorithm gain is kept "alive" to track time varying parameters. A very simple way of achieving the same end result is to simply fix P at some positive definite matrix $\overline{P}$. This gives the following gradient algorithm (sometimes known as normalized LMS algorithm):

$$\rho\hat{\theta} = \frac{\overline{P}\phi(y - \phi^T\hat{\theta})}{\Gamma + \Delta\phi^T\overline{P}\phi} \tag{12.9.14}$$

A special choice for $\overline{P}$ is σI, which leads to

$$\rho\hat{\theta} = \frac{\sigma\phi(y - \phi^T\hat{\theta})}{\Gamma + \Delta\sigma\phi^T\phi} \tag{12.9.15}$$

The name gradient arises from the fact that the right side of (12.9.8) is a scalar multiple of the negative gradient of the instantaneous cost function $(y - \phi^T\hat{\theta})^2$.

This algorithm is very simple, having a small amount of computation per update step. However, convergence is generally slower than for least squares types of algorithms, especially if $\overline{P}$ is chosen poorly.

12.9.4 Resetting Algorithm

Another way of stopping the matrix P from going to zero is to periodically reset it to some nonzero value. Indeed, a good time to reset the algorithm is when the parameters undergo a large change, since this causes the algorithm to have rapid initial convergence. Thus, it is desirable to include some form of fault detection algorithm along with this scheme to guide when the covariance matrix P is reset.

12.9.5 Exponential Forgetting and Resetting Algorithm

In all the preceding algorithms, P was prevented from going to zero, thus keeping the algorithm "alive" to track time varying parameters. One difficulty, however, is that with poor excitation these modifications can actually cause P to grow very large. This is equally undesirable, since the algorithm then becomes excessively sensitive to noise. Thus some form of restriction on the maximum value of P is also required. This can be achieved in a number of ways, for example, by holding the trace of P constant or by resetting P whenever it exceeds some threshold. Another alternative is to allow the algorithm to smoothly reset P to some sensible value whenever the excitation is poor. This leads to another algorithm, as outlined next.

Consider the basic algorithm of (12.2.13), (12.2.14). We see from those equations that in the case of pure least squares P is essentially the inverse of the integral of $\phi\phi^T$. To ensure that this integral does not blow up, we replace the integral by an average over some interval. This average is conveniently computed via a low-pass filter as follows:

$$\rho \overline{M} = -\sigma \overline{M} + \sigma \phi \phi^T \qquad (12.9.16)$$

A problem with this equation is that it is not possible to ensure that $\overline{M}$ will be nonsingular without having restrictive conditions on ϕ. To guarantee nonsingularity for all ϕ, we add a small offset to the right side of (12.9.16), giving

$$\rho \overline{M} = -\sigma \overline{M} + \sigma \phi \phi^T + \sigma \varepsilon I \qquad (12.9.17)$$

Arguing as in Section 12.4, we see that with $\overline{P} = \overline{M}^{-1}$ we now have

$$\rho \overline{P} = \sigma \overline{P} - \frac{\sigma \overline{P} \phi \phi^T \overline{P}}{\Gamma + \Delta \sigma \phi^T \overline{P} \phi} - \sigma \varepsilon \overline{P}^2 \qquad (12.9.18)$$

A desirable feature of the algorithm given in (12.9.18) is that $\overline{P}$ is prevented from going to ∞ by the last term. Indeed, when excitation is poor, $\overline{P}$ automatically resets itself to $(1/\varepsilon)I$. One remaining point is that if ϕ becomes very large then it is possible for $\overline{P}$ to become very small. This can be avoided by normalizing the quadratic term in ϕ on the right side of (12.9.18) to give

$$\rho \overline{P} = \sigma \overline{P} - \frac{\sigma \overline{P} \phi \phi^T \overline{P}}{1 + \beta \phi^T \phi + \Delta \sigma \phi^T \overline{P} \phi} - \varepsilon \sigma \overline{P}^2 \qquad (12.9.19)$$

The corresponding update equation for $\hat{\theta}$ is exactly as in (12.9.12):

$$\rho \hat{\theta} = \frac{\sigma \overline{P} \phi e}{1 + \beta \phi^T \phi + \Delta \sigma \phi^T \overline{P} \phi} \qquad (12.9.20)$$

where

$$e = y - \phi^T \hat{\theta} \qquad (12.9.21)$$

We will call the algorithm (12.9.19) to (12.9.21) the *exponential forgetting and resetting algorithm* (EFRA). This algorithm has many desirable features when applied to time varying systems, including the fact that $\overline{P}$ is automatically bounded from above and below and that, in the absence of excitation in ϕ, $\overline{P}$ resets itself to $(1/\varepsilon)I$. This and other features are explored in Problem 12.6.

12.9.6 A General Class of Algorithms

All the algorithms described previously can be conveniently summarized in the following generic algorithm:

$$\rho \hat{\theta} = \frac{\alpha P \phi (y - \phi^T \hat{\theta})}{\Gamma + \Delta \phi^T P \phi} \qquad (12.9.22)$$

$$\rho P = \frac{-\alpha P \phi \phi^T P}{\Gamma + \Delta \phi^T P \phi} + \Omega \qquad (12.9.23)$$

where $0 \leq \alpha(t) \leq 1$ and where $\Omega(t)$ represents a modification to the covariance update with $\Omega(t) = \Omega(t)^T \geq 0$.

In practice, we need to "tune" the various free parameters in (12.9.22), (12.9.23) so as to give the required trade-off between rapidly tracking parameter time variations and giving noise smoothing. (A little more will be said on this topic in the next section.) By making suitable choices of α, β, and Ω, the algorithm (12.9.22), (12.9.23) reduces to the various algorithms presented in Sections 12.9.1 to 12.9.5. Specifically, we have:

(i) **Gradient algorithm**

$$\Omega = \frac{\alpha P \phi \phi^T P}{\Gamma + \Delta \phi^T P \phi}, \qquad \alpha \text{ constant}$$

(ii) **Recursive least squares**

$$\Omega = 0, \qquad \alpha = 1$$

(iii) **Least squares with forgetting**

$$\alpha = 1, \qquad \Omega = \left(\frac{1}{1 - \Delta\lambda}\right)\left(\lambda P - \frac{\Delta P \phi \phi^T P}{\Gamma + \Delta \phi^T P \phi}\right), \qquad \lambda \leq 1$$

(iv) **Kalman filter modification**

$$\alpha = 1, \qquad \Omega > 0$$

(v) **Covariance resetting**

$$\alpha = 1, \qquad \Omega(t) = (P_0 - P(t + \Delta)) \sum_{j=1}^{\infty} i_\delta(t - t_j)$$

(vi) **Constant trace**

$$\Omega = \frac{\alpha}{C_1} \frac{(\phi^T P^2 \phi) P}{\Gamma + \Delta \phi^T P \phi}, \qquad C_1 = \text{Trace}(P_0)$$

(vii) **Exponential forgetting and resetting algorithm (EFRA)**

$$\Omega = \sigma P - \varepsilon \sigma P^2$$
$$\alpha = \sigma, \qquad \Gamma = 1 + \beta \phi^T \phi$$

12.9.7 Response Time Considerations

In this section we will give additional insights into the algorithms described for the time varying case by making a rough connection with linear systems analysis. The algorithms are nonlinear and thus, to make the argument simple, we will need to make some approximations. Also, for simplicity we consider only the continuous time case, but similar results hold for the discrete case in view of the close connection that always exists between these cases.

By way of illustration, we consider the basic P update equation given in (12.9.18). The exact solution of this equation is

$$\bar{P}(t) = \left[e^{-\sigma t}P_0^{-1} + \varepsilon I + \overline{\phi\phi^T}\right]^{-1} \quad (12.9.24)$$

where

$$\overline{\phi\phi}^T \triangleq \int_0^t \sigma e^{-\sigma(t-\tau)}\phi(\tau)\phi(\tau)^T \, d\tau \quad (12.9.25)$$

Asymptotically, and provided $\overline{\phi\phi}^T$ is not too small, we have

$$\bar{P} \simeq \left[\overline{\phi\phi}^T\right]^{-1} \quad (12.9.26)$$

Now consider Equation (12.9.20), with $\Delta = 0$; that is,

$$\rho\dot{\hat{\theta}} = \sigma\bar{P}\phi e \quad (12.9.27)$$

where

$$e = y - \phi^T\hat{\theta} \quad (12.9.28)$$

Now let us assume that $y = \phi^T\theta_0$ for some fixed value θ_0. Then substituting (12.9.28) into (12.9.27) gives

$$\rho\dot{\tilde{\theta}} = -\sigma\bar{P}\phi\phi^T\tilde{\theta} \quad (12.9.29)$$

where

$$\tilde{\theta} = \hat{\theta} - \theta_0 \quad (12.9.30)$$

If in equation (12.9.29) we assume that $\tilde{\theta}$ and $\bar{P}$ vary slowly relative to ϕ, then equation (12.9.29) is equivalent to

$$\rho\dot{\bar{\tilde{\theta}}} \simeq -\sigma\bar{P}\overline{\phi\phi}^T\bar{\tilde{\theta}} \quad (12.9.31)$$

where $\bar{\tilde{\theta}}$ denotes the average value of $\tilde{\theta}$. Assuming that the average value of $\phi\phi^T$ is essentially constant, then we have from (12.9.26), (12.9.31) that

$$\rho\dot{\bar{\tilde{\theta}}} \simeq -\sigma\bar{\tilde{\theta}} \quad (12.9.32)$$

Thus we see that $1/\sigma$ roughly determines the time constant of the parameter estimator. A small value of σ will give good noise rejection but slow response to parameter time variations. Conversely, a large value of σ gives fast response to parameter time variations but poor noise rejection.

From Equation (12.9.32), if the true parameters are time varying with rate of variation small compared with σ, then good tracking will occur. On the other hand, if jump changes in parameters can occur, then the initial transient response can be improved by resetting $\bar{P}$ to a larger value immediately following the detection of a change.

12.10 ADAPTIVE CONTROL

One use that on-line parameter estimation can be put to is in adaptive control. The basic idea of these algorithms is to combine an on-line estimation method with on-line control law synthesis procedure. This kind of algorithm is very attractive since it offers the capability of automatic initial tuning of a control system or of retuning of the system should the plant parameters subsequently change. Adaptive control algorithms have been the subject of intensive study over the past 30 years. However, it is only recently that successful industrial applications have been reported. This recent success can be attributed to a maturation in the associated theory and to advances in microcomputer technology, which have facilitated the implementation of the algorithms.

In this section we will present a typical adaptive control algorithm and establish some of its properties. We will also give several illustrative examples of adaptive control.

12.10.1 Certainty Equivalence

The basic architecture of an adaptive controller is shown in Figure 12.10.1. Note that the parameter estimator provides an on-line estimate of the plant parameters as well as a measure of the uncertainty associated with those estimates. This uncertainty is then combined with any a priori uncertainty to give an on-line model estimate together with a measure of its fidelity. These are then used in a robust

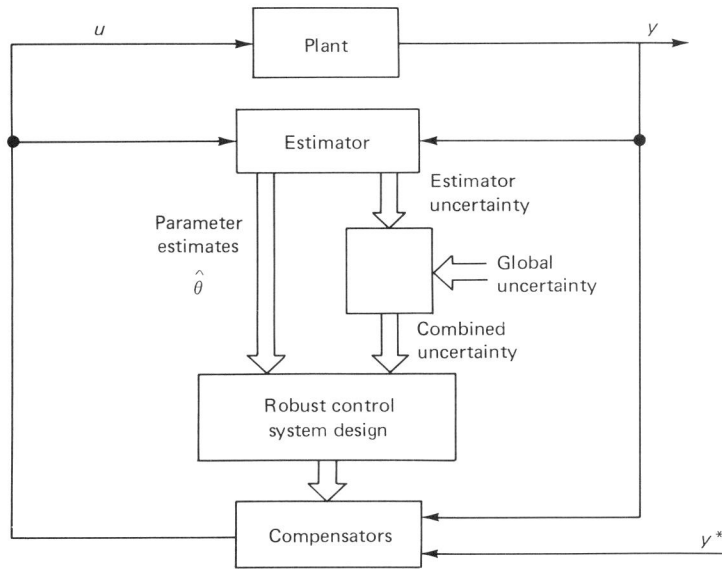

Figure 12.10.1 Block diagram for an adaptive robust controller.

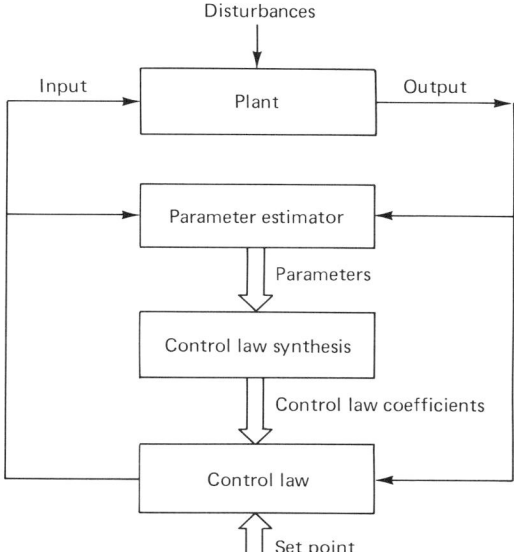

Figure 12.10.2 Basic architecture of a certainty equivalence adaptive controller.

control system design procedure to produce the feedback and feedfoward compensator.

In some cases, the general structure given in Figure 12.10.1 is simplified by ignoring the model uncertainty. In this case, the on-line control synthesis block treats the estimated model as if it were the true plant. This strategy is generally known as *certainty equivalence* adaptive control and is illustrated in Figure 12.10.2.

12.10.2 An Illustrative Algorithm

There are obviously many choices that could be made for the parameter estimation and control law synthesis modules. Here we will make a relatively simple selection of algorithms: least squares with dead zone for the parameter estimation module and pole assignment with internal model for the control law synthesis module. The system model that we will consider here is

$$Ay = Bu + \eta + d \qquad (12.10.1)$$

where A and B are polynomials in the operator ρ, η represents unmodeled errors, and d represents a deterministic disturbance.

The deterministic component d is assumed to satisfy a homogeneous model of the type discussed in Chapter 3; that is,

$$Sd = 0 \qquad (12.10.2)$$

where S is a known monic polynomial of degree $\partial(S)$ having nonrepeated zeros on the stability boundary. We suggest that the data be first prefiltered by a bandpass filter. To be specific, we will assume this filter has the form $GS/(EJQ)$. Applying this filter to (12.10.1) gives

$$\bar{y} = \phi^T \theta + \eta_f \tag{12.10.3}$$

where

$$\phi^T = \left[y_f \cdots \rho^{n-1} y_f, u_f \cdots \rho^m u_f \right] \tag{12.10.4}$$

$$\bar{y} = \frac{GS}{JQ} y, \qquad \bar{u} = \frac{GS}{JQ} u \tag{12.10.5}$$

$$y_f = \frac{1}{E} \bar{y}, \qquad u_f = \frac{1}{E} \bar{u} \tag{12.10.6}$$

and θ a vector of nominal system parameters.

Given this model, we can apply the dead zone algorithm of Section 12.8.1. This algorithm generates on-line estimates of the polynomials $\hat{A}$ and $\hat{B}$. These estimates are then used in a certainty equivalence form of the pole assignment design. Thus, let $\hat{A}$ be a desired closed loop polynomial (which can depend on $\hat{A}$ and $\hat{B}$). We then solve the following on-line pole assignment equation.

$$\hat{A}\hat{L}GS + \hat{B}\hat{P} = \hat{A}^* \tag{12.10.7}$$

for $\hat{L}$ and $\hat{P}$. Then implement the control law as

$$\hat{L} u_f = -\hat{P} z_f' \tag{12.10.8}$$

where

$$z_f' = \frac{1}{EJQ}(y - y^*)$$

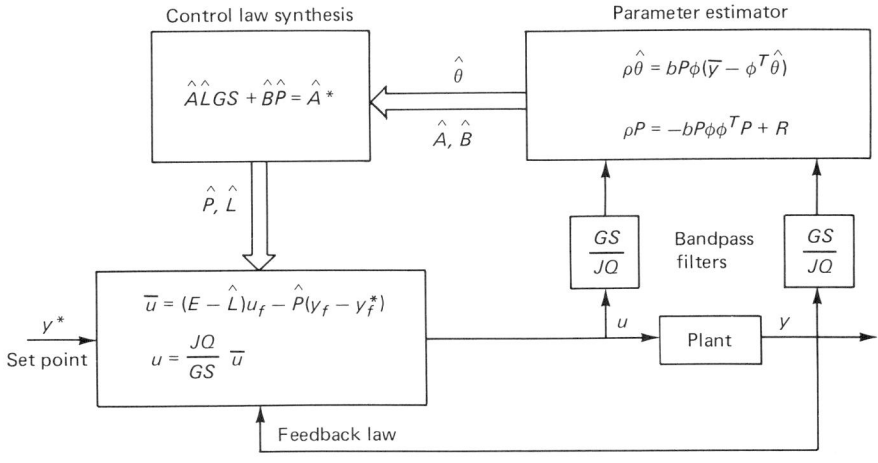

Figure 12.10.3 Parametric adaptive controller.

Equation (12.8.8) may be rewritten as

$$\bar{u} = (E - \hat{L})u_f - \hat{P}z'_f \qquad (12.10.9)$$

$$u = \frac{JQ}{GS}\bar{u} \qquad (12.10.10)$$

The final adaptive control law is as illustrated in Figure 12.10.3.

This algorithm can be shown to have certain desirable features. For completeness, a full analysis is given in Appendix F. The key conclusions are that, provided the unmodeled error is sufficiently small, all signals remain bounded and the output converges to the desired output y^* provided it satisfies the homogeneous model

$$S(\rho)y^* = 0 \qquad (12.10.11)$$

We present next two examples that illustrate some practical features of adaptive control.

12.10.3 Electromechanical Servo Example

Here we consider the adaptive control of an electromechanical servo kit, a Feedback Ltd. type ES1B servo. The servo kit was arranged as shown in Figure 12.10.4.

u	-10 to $+10$ V input voltage
V_T	Tacho generator output voltage
V_a	Armature voltage
θ_m	Motor angle
θ	Shaft angle after reduction gearbox
y	-10 to $+10$ V voltage representing θ

Note that the servo kit arrangement includes several nonideal features, including drifting dc offsets in the servo amplifier summing junction (the servo amplifier is several decades old, and uses valve technology), gearbox backlash, and nonlinear friction in the motor. The relevant equations for the dc motor are

$$\tau_m = k_m I_f i_a \qquad (12.10.12)$$

and

$$V_a = L_a \frac{di_a}{dt} + R_a i_a + k_m I_f \omega_a \qquad (12.10.13)$$

where

τ_m = torque acting on the motor armature and shaft
k_m = machine constant
I_f = dc field current
i_a = armature current
L_a = armature inductance
ω_a = armature angular velocity

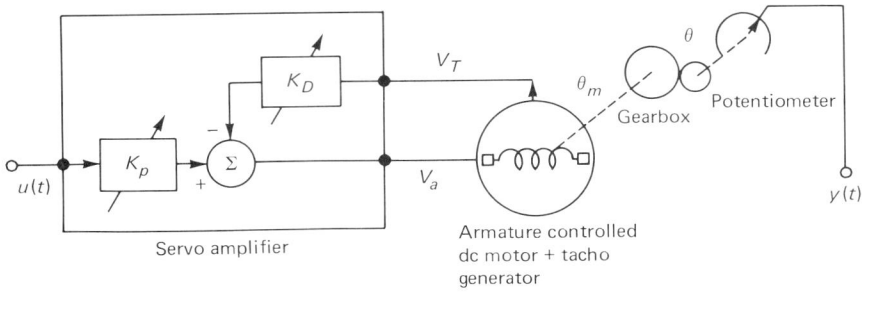

u	—10 to +10 V input voltage
V_T	Tacho generator output voltage
V_a	Armature voltage
θ_m	Motor angle
θ	Shaft angle after reduction gearbox
y	—10 to +10 V voltage representing θ

Figure 12.10.4 Servo kit arrangement used for the adaptive controller trials.

The tacho generator voltage is given by

$$V_T = k_T \omega_a \qquad (12.10.14)$$

If we let J_m be the total mass moment of inertia, referred to the motor shaft, then Newton's laws of motion give that

$$J_m \frac{d\omega_a}{dt} = \tau_m \qquad (12.10.15)$$

Combining (12.10.12), (12.10.13), and (12.10.15), we can show that ω_a is related to V_a by the following differential equation:

$$L_a \frac{d^2\omega_a}{dt^2} + R_a \frac{d\omega_a}{dt} + \frac{k_m^2 I_f^2}{J_m} \omega_a = \frac{k_m I_f}{J_m} V_a \qquad (12.10.16)$$

In most dc machines, the armature inductance L_a is kept to a minimum. Usually, the "electrical" time constant, L_a/R_a is on the order of milliseconds. Since the other time constants in this example are on the order of hundreds of milliseconds, we ignore the electrical time constant and take

$$L_a \approx 0 \qquad (12.10.17)$$

In this case we get the following reduced order model:

$$\frac{d\omega_a}{dt} + \frac{k_m^2 I_f^2}{R_a J} \omega_a = \frac{k_m I_f}{R_a J} V_a \qquad (12.10.18)$$

or

$$(1 + T_a D)\omega_a = K_a V_a \qquad (12.10.19)$$

where

$$T_a = \frac{R_a J}{k_m^2 I_f^2}$$

$$D = \frac{d}{dt}$$

and

$$K_a = \frac{1}{k_m I_f}$$

Because of the arrangement shown in Figure 12.10.4, and using (12.10.14), the relationship between ω_a and $u(t)$ is

$$(1 + T_a' D)\omega_a = K_a' u \qquad (12.10.20)$$

where

$$T_a' = \frac{T_a}{1 + K_a K_T K_D}$$

and

$$K_a' = \frac{K_p K_a}{1 + k_a K_T K_D}$$

Thus we see that by varying K_p and/or K_D we can vary the effective gain and time constant of the dc motor model (12.10.20). Thus we have a "hardware simulation" of a motor with parameters that vary.

The relationship between the output y and the input u can thus be seen to be

$$\frac{d}{dt}\left(1 + T_a' \frac{d}{dt}\right) y = \frac{K_a'}{N} u = Ku \qquad (12.10.21)$$

where N is the gear ratio.

For the servo kit used, T_a' is on the order of a few hundred milliseconds. For this reason and also because of the hardware constraints, we used a 20-msec sampling period. In our discrete time model, we neglect the sampling zeros, since it is clear that these are in the far-left half-plane. The discrete time delta model used was thus

$$\delta(\delta + \hat{a}) y = \hat{b} u \qquad (12.10.22)$$

The structure of the control law used is shown in Figure 12.10.5. The controller is strictly proper and uses a slow fixed parallel integrator. A simple form of antiintegral windup, to be discussed in Chapter 14, was implemented by using a code section of the form:

```
:sum := sum + zeta * delta * error;  {error is e(t)}
IF  sum > sum_maximum;  {if integrator output is too large}
    THEN sum := sum_maximum;  {saturate it}
IF  sum < -(sum_maximum) {similarly if output is too small}
    THEN sum := -(sum_maximum);
```

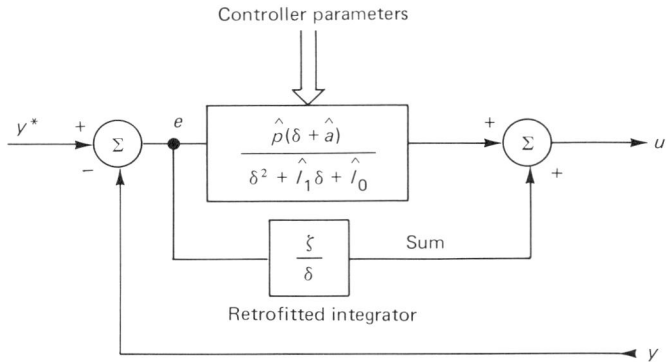

Figure 12.10.5 Controller structure for the adaptive controller.

The controller parameters $\hat{a}$, $\hat{p}$, $\hat{l}_1$, and $\hat{l}_0$ were calculated from the estimated parameters $\hat{a}$ and $\hat{b}$ by the certainty equivalence principle based on the following pole assignment equation:

$$\delta(\delta + \hat{a})(\delta^2 + \hat{l}_1\delta + \hat{l}_0) + \hat{b}\hat{p}(\delta + \hat{a}) = (\delta + \hat{a})A^*(\delta) \qquad (12.10.23)$$

Equation (12.10.23) can be simply solved as follows:

$$\hat{p} = \frac{a_0^*}{\hat{b}} \qquad (12.10.24)$$

$$\hat{l}_0 = a_1^* \qquad (12.10.25)$$

and

$$\hat{l}_1 = a_2^* \qquad (12.10.26)$$

The structure of the filters for the parameter estimation is shown in Figure 12.10.6. This figure should be compared with Figure 12.10.3. The least squares estimator was implemented in code of the following form, where we have included matrix and vector arithmetic, without explicitly showing the loops involved.

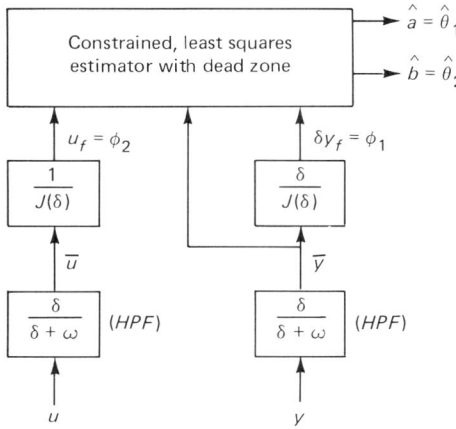

Figure 12.10.6 Filters for the parameter estimator.

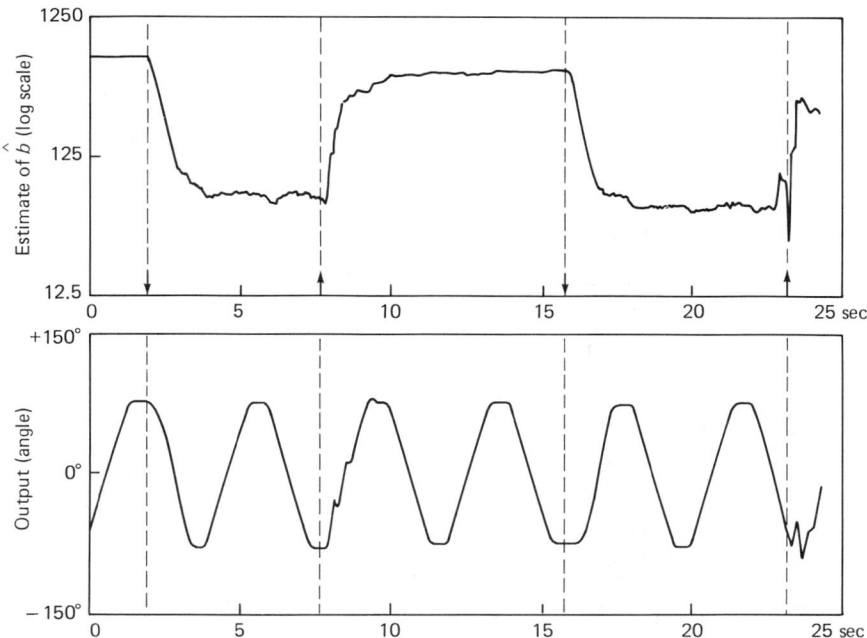

Figure 12.10.7 Adaptive controller response.

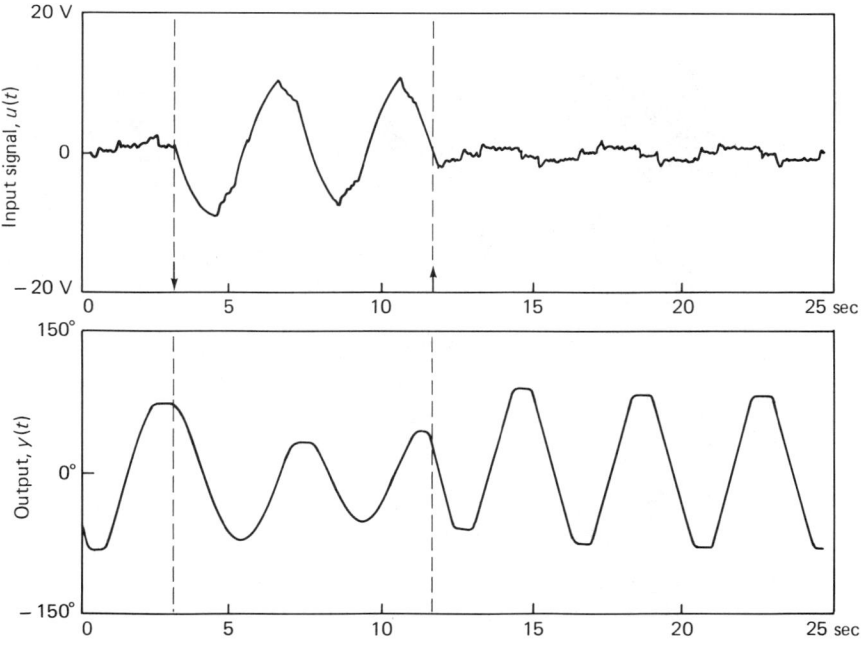

Figure 12.10.8 Fixed three-term controller response.

Sec. 12.10 Adaptive Control

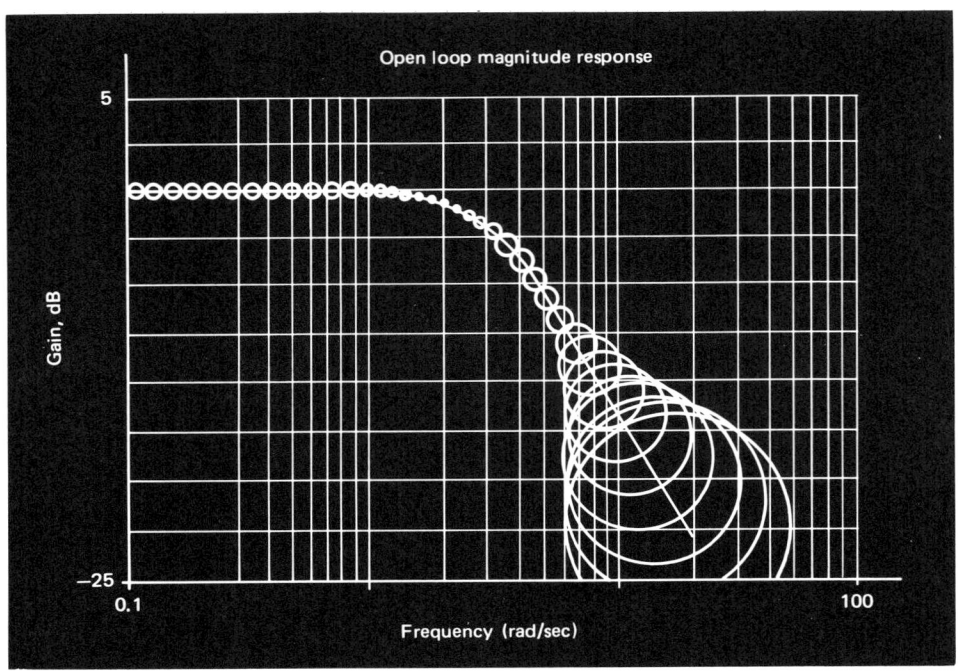

Figure 12.10.9 Magnitude of estimated frequency response.

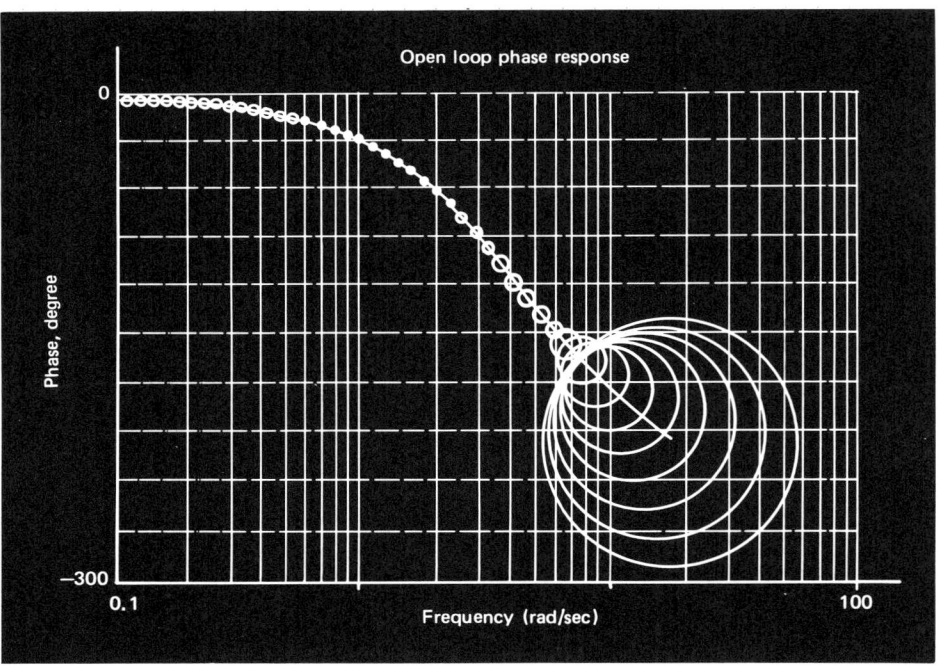

Figure 12.10.10 Phase of estimated frequency response.

```
     ⋮
     prediction_error := y_bar − j[1] d * y_f − j[0] * y_f − phi' * theta_hat;
        { j[0] is the constant coefficient in J, etc.}
     p_phi := p * phi;
        { calculating p_phi reduces the computations}
     past_p := p;
        { store current value of the covariance matrix}
     estimator_gain := (abs(prediction_error) − epsilon) * alpha /
        (abs(prediction_error) * (1.0 + phi' * p − phi));
        { calculate the estimator_gain, b, including a deadzone}
     IF  estimator_gain > 0 THEN BEGIN
        {we're in the deadband if estimator_gain < 0}
     theta_hat := theta_hat + delta * estimator_gain * prediction_error * p − phi;
        {update theta_hat}
           ip := p − delta * estimator_gain * p_phi * p_phi';
        {update p with R = 0}
     IF  theta_hat > max_theta THEN
           project (theta_hat, max_theta, p);
     IF  theta_hat < min_theta THEN
           project (theta_hat, min_theta, p)
     END;  { IF estimator_gain > 0}
     p := p + delta * (beta1 * I + beta2 * past_p − beta3 * past_p * past_p);
        { add exponential forgetting and resetting}
     ⋮
```

Note that the adaptive controller requires the specification of several constants. Most of these are noncritical, and simple rules of thumb may be developed for their selection. For example, we choose

$$J(\delta) = \delta^2 + 50\delta + 625 = (\delta + 25)^2 \qquad (12.10.27)$$

and

$$A^*(\delta) = \delta^3 + 40\delta^2 + 475\delta + 2500 \qquad (12.10.28)$$

From (12.10.28) we see that the dominant closed loop poles have been placed with a bandwidth of roughly 10 rad/sec, with a damping factor of 0.75. The polynomial $J(\delta)$ is then selected to have a roll-off at roughly 20 rad/sec.

The performance of the algorithm is illustrated in the chart recording reproduced in Figure 12.10.7. The arrows ↓ and ↑ in the figure represent order of magnitude, step decreases and increases (respectively) in the gain on the servo kit. The set point y^* was a truncated triangular wave. Note that the adaptive controller maintains a consistent, high-level performance, excepting a short transient that occurs when the parameters change.

By comparison, a robust PID control law, tuned to give acceptable results under the worst case (that is, higher gain), has the performance illustrated in Figure

12.10.8. Clearly, when the gain is reduced, the performance is lost, even though stability (in this case) is retained.

12.10.4 Heat Exchanger Example

This system has a significant time delay that can vary. The time delay was modeled via a rational approximation. Hence there will be explicit undermodeling under all conditions. For this reason it was decided to use the more general form of adaptive control shown in Figure 12.10.1. A sampling period of 20 msec was used and all models were expressed in delta form.

A parametric model of the following form was fitted to the on-line data to the system:

$$G(\gamma) = \frac{b_1\gamma + b_0}{\gamma^2 + a_1\gamma + a_0} \qquad (12.10.29)$$

The estimated parameters were then used to obtain an on-line estimate of the

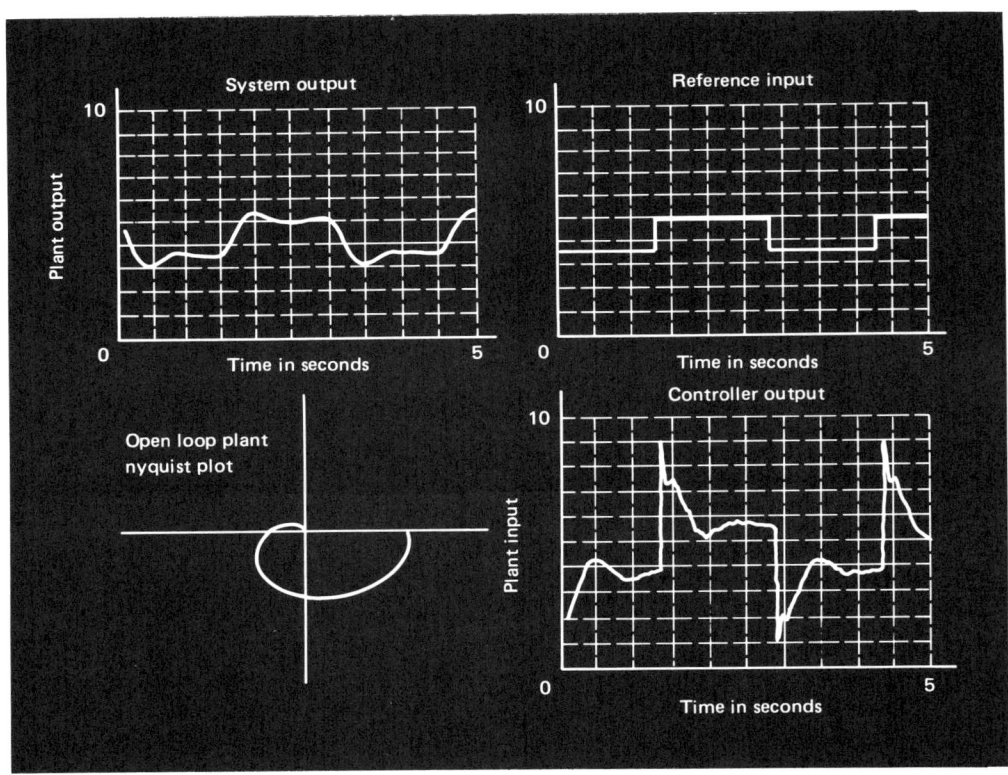

Figure 12.10.11 Adaptive control performance using certainty equivalence.

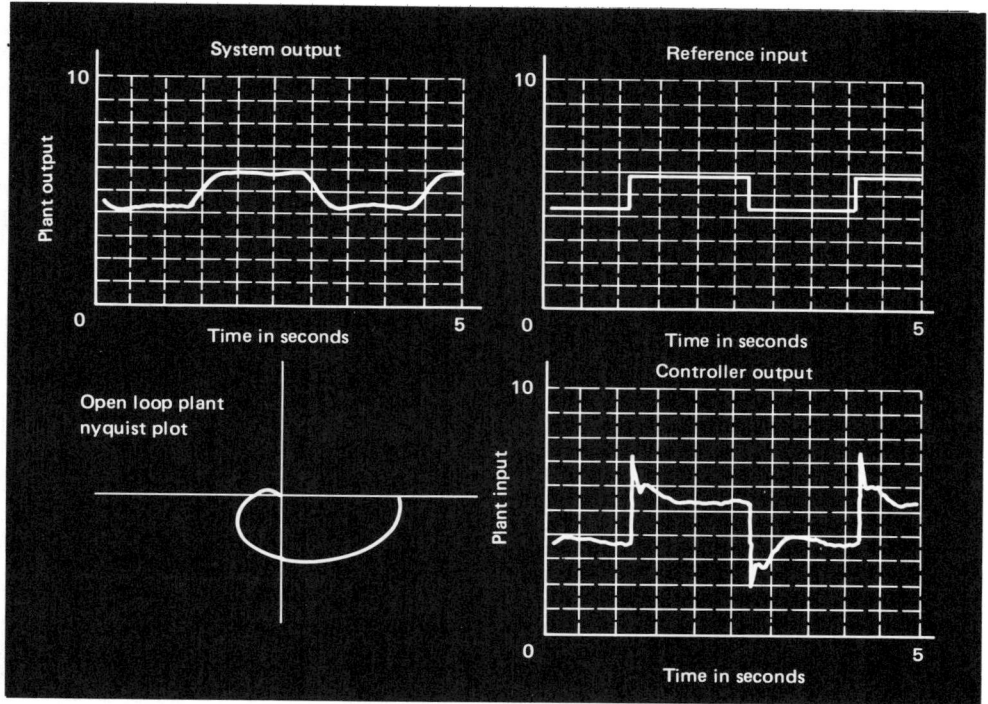

Figure 12.10.12 Adaptive control performance using a posteriori uncertainty.

system's frequency response. This frequency response is refreshed at the 20-msec rate.

Figures 12.10.9 and 12.10.10 show the frequency response of the estimated model together with the a posteriori uncertainty bounds calculated using Equation (12.8.57).

The on-line control system design was based on the estimated frequency response by solving the pole assignment equation via a least squares fit in the frequency domain. The design objective was to achieve a rapid transient response with little or no overshoot.

Figure 12.10.11 shows the performance of the resultant controller based on a certainty equivalence design. The figure shows four plots that are updated at the sampling rate. These plots are top left, system output response; top right, set point; bottom right, system input; bottom left, Nyquist diagram. It can be seen from this figure that the certainty equivalence controller gives about 25% overshoot.

Figure 12.10.12 shows the performance of the controller when the design accounts for the estimated model uncertainty. Note that the overshoot has now been eliminated.

Due to the fact that the system frequency response is estimated on line, the responses given in Figures 12.10.11 and 12.10.12 are maintained in the face of substantial variation in the plant gain, time constant, and pure time delay.

12.11 SUMMARY

The key points covered in this chapter were:

- Estimation is relatively easy for models that are linear in the parameters such that the output can be expressed as

$$\bar{y}(t) = \phi(t)^T \theta \tag{12.2.1}$$

- We showed how the response of linear dynamic systems could be expressed in the above form (Section 12.2).
- In Section 12.3 we introduced the concept of least squares estimation, which minimizes the cost function

$$C(\hat{\theta}, t) = \frac{1}{2} \underset{0}{\overset{t}{S}} \left(y(\tau) - \phi(\tau)^T \hat{\theta} \right)^2 d\tau \tag{12.3.7}$$

- We showed how the estimate $\hat{\theta}$ minimizing (12.3.7) could be obtained recursively leading to the unified estimator:

$$\rho \hat{\theta} = \frac{P\phi(y - \phi^T \hat{\theta})}{\Gamma + \Delta \phi^T P \phi} \tag{12.4.12}$$

$$\rho \hat{P} = \frac{-P\phi \phi^T P}{\Gamma + \Delta \phi^T P \phi} \tag{12.4.11}$$

- Various modifications to allow for time varying parameters were introduced in Section 12.9, leading to the following general estimator structure:

$$\rho \hat{\theta} = \frac{\alpha P\phi(y - \phi^T \hat{\theta})}{\Gamma + \Delta \phi^T P \phi} \tag{12.5.2}$$

$$\rho P = \frac{-\alpha P\phi \phi^T P}{\Gamma + \Delta \phi^T P \phi} + \Omega \tag{12.5.3}$$

- Deterministic properties of the above estimator were derived under the ideal modeling assumption $y = \phi^T \theta$ (Theorem 12.5.1). In particular, it was shown that
  (i) $\tilde{\theta}^T P^{-1} \tilde{\theta}$ is nonincreasing
  (ii), (iii) $\tilde{e}$ and $\rho \hat{\theta}$ are both $\mathscr{L}_2^+$ functions, where $\tilde{e}$ is the normalized error

$$\tilde{e} = \left(\frac{\alpha}{\Gamma + \Delta \phi^T P \phi} \right)^{1/2} (y - \phi^T \hat{\theta}) \tag{12.5.6}$$

- It was shown that, in the ideal case, parameter convergence to the true value could be guaranteed provided the input contained sufficient sinusoidal components (Lemma 12.5.2).
- Various algorithm enhancements were described, including utilization of prior knowledge (Section 12.6).
- The estimator was shown to inherit the optimality properties of the Kalman filter when the noise is white (Section 12.7.2).

- Properties of the parameter estimator were described for the nonideal case (Section 12.8.2). These properties are very close to those obtained in the ideal case (Theorem 12.5.1).
- A method for quantifying the effect of model uncertainty on the performance of the parameter estimator was developed (subsection 12.8.3). In particular, the following bound was obtained on the difference between the estimated transfer function $\hat{G}(\gamma)$ and the true transfer function $G_T(\gamma)$:

$$|\hat{G}(\gamma) - G_T(\gamma)| \leq \left\{ \left[\Sigma|\beta_{ii}\bar{h}_i|\right]^2 + \left[\Sigma|\beta_{2i}\bar{h}_i|\right]^2 \right\}^{1/2} \qquad (12.8.57)$$

where $\bar{h}$ is an overbound on the impulse response of the unmodeled dynamics.
- It is shown how parameter estimation can be combined with on-line control law synthesis to yield a parameter adaptive controller.
- Two practical examples of adaptive control were described.

12.12 REFERENCES

Additional material on parameter estimation is contained in:

GOODWIN, G. C., and R. L. PAYNE (1977) *Dynamic System Identification: Experimental Design and Data Analysis.* Academic Press, New York.

GOODWIN, G. C. and K. S. SIN (1984) *Adaptive Filtering Prediction and Control.* Prentice-Hall, Englewood Cliffs, N.J.

LJUNG, L. (1987) *System Identification: Theory for the User.* Prentice-Hall, Englewood Cliffs, N.J.

LJUNG, L., and T. SÖDERSTRÖM (1983) *Theory and Practice of Recursive Identification.* MIT Press, Cambridge, Mass.

NORTON, J. P. (1987) *System Identification.* Academic Press, New York.

Specific algorithms for the case of time varying parameters are contained in:

FOGEL, E., and Y. F. HUANG (1982) "On the value of information in system identification—Bounded noise case." *Automatica*, Vol. 18, pp. 229–238.

FORTESCUE, T. R., L. S. KERSHENBAUM, and B. E. YDSTIE (1981) "Implementation of self-tuning regulators with variable forgetting factors." *Automatica*, Vol. 17, pp. 831–835.

SALGADO, M. E., G. C. GOODWIN, and R. H. MIDDLETON (1988) "A modified least squares algorithm incorporating exponential resetting and forgetting." *Int. J. Control*, Vol. 47, no. 2, pp. 477–491.

Techniques for quantifying the effect of undermodeling on parameter estimation are contained in:

GOODWIN, G. C., and M. E. SALGADO (1989) "Quantification of uncertainty in estimation using an embedding principle." 1989 American Control Conference, Pittsburgh.

GOODWIN, G. C., D. Q. MAYNE, and M. E. SALGADO (1989) "Uncertainty, Information and Estimation," Plenary address, IFAC Symposium on Adaptive Systems in Control and Signal Processing, Glasgow, U.K.

Background to adaptive control may be found in:

ÅSTRÖM, K. J. (1983) "Theory and applications of adaptive control—A survey." *Automatica*, Vol. 19, pp. 471–486.

ÅSTRÖM, K. J. and T. HAGGLAND (1984) "Automatic tuning of simple regulators with specifications on phase and amplitude margins." *Automatica*, Vol. 20,. No. 5, pp. 645–653.

ÅSTRÖM, K. J., and B. WITTENMARK (1989) *Adaptive Control*, Addison-Wesley, Reading, Mass.

ÅSTRÖM, K. J., and B. WITTENMARK (1973) "On self tuning regulators." *Automatica*, Vol. 9, pp. 187–199.

CARUTHERS, F. P., and H. LEVENSTEIN (eds.) (1963) *Adaptive Control Systems*. Pergamon Press, Elmsford, N.Y.

CHALAN, V. V. (1987) *Adaptive Control Systems*. Marcel Dekker, New York.

GOODWIN, G. C., and K. S. SIN (1984) *Adaptive Filtering Prediction and Control*. Prentice-Hall, Englewood Cliffs, N.J.

HARRIS, C. J., and S. A. BILLINGS (ed.) (1981) *Self Tuning and Adaptive Control*. Peter Peregrinus, London.

LANDAU, I. D. (1974) "A survey of model reference adaptive techniques in theory and applications." *Automatica*, Vol. 10, pp. 353–379.

NARENDRA, K. S., and R. V. MONOPOLI (eds.) (1980) *Applications of Adaptive Control*. Academic Press, New York.

SEBORG, D. E., S. L. SHAH, and T. F. EDGAR (1983) "Adaptive strategies for process control —A survey." *Proc. Annual AIChE Meeting*, Washington, D.C.

TZYPKIN, Ya. Z. (1968) *Adaptation and Learning in Automatic Systems*. Nanka, Moscow.

UNBERHAUEN, H. (ed.) (1980) *Methods and Applications in Adaptive Control*. Springer-Verlag, New York.

Recent publications dealing with the theory of adaptive control for time varying systems are:

DE LARMINAT, P., and H. F. RAYNAUD (1986) "A robust solution to the stabilizability problem in indirect passive adaptive control." *Proc. 25th IEEE Conference on Decision and Control*, pp. 462–467.

KREISSELMEIER, G. (1986) "Adaptive control of a class of slowly time varying plants," *Systems Control Letters*, Vol. 8, No. 2, pp. 97–103.

MIDDLETON, R. H., and G. C. GOODWIN (1988) "Adaptive control of time varying linear systems." *IEEE Trans. Auto. Control*, Vol. 33, No. 2, pp. 150–155.

TSAKALIS, K., and P. A. IOANNOU (1986) "Adaptive control of linear time-varying plants." *Proc. IFAC Workshop on Adaptive Systems in Control and Signal Processing*, Lund, Sweden.

Further background in the adaptive control of the servo system is contained in:

GOODWIN, G. C., and R. H. MIDDLETON (1987) "Continuous and Discrete Adaptive Control." Chapter in *Control and Dynamic Systems*, Vol. 25, edited by C. T. Leondes, Academic Press, New York.

GOODWIN, G. C., D. Q. MAYNE, and M. E. SALGADO (1989) "Uncertainty, Information and Estimation," Plenary address, IFAC Symposium on Adaptive Systems in Control and Signal Processing, Glasgow.

Linear in the parameters models for nonlinear robotic systems are described in:

AN CHAE, H., C. G. ATKESON, and J. M. HOLLERBACK (1984), "Estimation of inertial parameters of rigid body links of manipulators," *Proc. 24th Conference on Decision and Control*.

KHASLA, P. K., and T. KANADE (1984) "Parameter identification for robot arms," *Proc. 24th Conference on Decision and Control*.

12.13 PROBLEMS

1. Verify (12.4.11). *Hint:*

$$P(t) = \left\{ P_0^{-1} + \int_0^t \frac{1}{\Gamma} \phi(\tau) \phi(\tau)^T d\tau \right\}^{-1}$$

   Hence $\rho P^{-1} = (1/\Gamma) \phi \phi^T$. Then show that with (12.4.11) we have $\rho[PP^{-1}] = 0$.

2. Consider a deterministic system, $y = \phi^T \theta$. Show that the following properties hold:
   (a) $P^{-1} \tilde{\theta}$ is an invariant of the recursive least squares estimator.
   (b) $P^{-1}(t) \hat{\theta}(t) = P_0^{-1} \hat{\theta}_0 + P_{LS}^{-1}(t) \hat{\theta}_{LS}(t)$, where $P_{LS}(t)$, $\hat{\theta}_{LS}(t)$ denote the least squares estimates (that is, the estimates when $P_0^{-1} = 0$).

3. Verify that the least squares estimator with exponential data weighting (12.9.8), (12.9.9) can be written in recursive from as (12.9.10), (12.9.11). *Hint:* Let $P_t = \{w(t,0) P_0^{-1} + \int_0^t w(t,\tau) \phi(\tau) \phi(\tau)^T d\tau\}^{-1}$.

4. Suppose $\lambda_{\min} P$ is zero; that is, there exists a $v \neq 0$ such that $Pv = 0$. Show that the general least squares estimator of (12.9.22), (12.9.23) is then insensitive to parameter changes in the direction v, in the following sense: $\rho(v^T \hat{\theta}) = 0$.

5. Show that algorithms (ii), (iii), and (vi) of Section 12.9.6 do *not* give a bound on P^{-1}; that is, $\lambda_{\min}(P)$ may be arbitrarily close to zero. *Hint:* Let $\phi(t) = v$ (constant), with $P_0 = kI$. Show that v is always an eigenvector of $P(t)$. Then evaluate $\rho(v^T P v)$.

*6. Show that the EFRA algorithm, given in (12.9.19), where the constants satisfy

$$\Delta \sigma < \frac{\epsilon \beta}{1 + \epsilon \beta}$$

   has the following properties if $\bar{\sigma} I \leq P_0 \leq \bar{\nu} I$, where

$$\bar{\sigma} \triangleq \frac{1}{\epsilon + (1/\beta)}$$

   and

$$\bar{\nu} \triangleq \frac{1}{\epsilon}$$

   (i) $\bar{\sigma} I \leq P(t) \leq \bar{\nu} I$ for all t.
   (ii) If $\phi(t) = 0$ for all t, then $P(t) \to \bar{\nu} I$.
   (iii) Ω, as defined for EFRA, satisfies $\Omega \geq 0$.
   Hints: (i) $\rho P \geq \sigma[P - (P^2/\bar{\sigma})]$; $\rho P \geq \sigma[P - (P^2/\bar{\nu})]$. (ii) if $\phi(t) = 0$, then the eigenvectors of P are invariant. Thus let $v_i = \lambda_i - \bar{\nu}$ and show $\rho v_i = -\sigma \epsilon \lambda_i v_i$.

7. Show that recursive least squares, in the noise free case and with $P_0 = (1/\epsilon) I$, has the property $\|\tilde{\theta}(t)\|_2 \leq \|\tilde{\theta}_0\|_2$.

8. The standard normal distribution $P(z)$ is defined to be

$$P(z) = \frac{1}{\sqrt{2\pi}} \int_z^\infty e^{-x^2/2}\, dx$$

This integral cannot be explicitly evaluated for general z; however, standard statistical tables give the following numerical values for $P(z)$:

z	0	0.5	1.0	1.5	2.0	2.5	3.0
$P(z)$	0.5	0.3085	0.1587	0.0668	0.0228	0.0062	0.0013

(a) Find a least squares approximation, $\hat{P}(z)$, to $P(z)$, where $\hat{P}(z)$ has the form
$$\hat{P}(z) = \theta_0 + \theta_1(\tan^{-1} z) + \theta_2(\tan^{-1} z^3)$$

(b) Repeat part (a), except constrain $\theta_0, \theta_1, \theta_2$ so that $\hat{P}(z)$ has the property
$$\hat{P}(-z) = 1 - \hat{P}(z)$$

(c) Compare your approximations in parts (a) and (b) to tables of the normal distribution.

9. Let $\phi^T = [1\ t]$, $\theta^T = [y_0\ y_1]$, and $y = \phi^T\theta$ + noise. Show that the gradient estimator
$$\rho\hat{\theta} = \frac{\alpha\phi e}{\phi^T\phi}, \qquad e = y - \phi^T\hat{\theta}$$

is equivalent to

$$\rho\hat{y} = \hat{y}_1 + \alpha\left(\frac{1 + \Delta t + t^2}{1 + t^2}\right)(y - \hat{y})$$

and

$$\rho\hat{y}_1 = \alpha\left(\frac{t}{1 + t^2}\right)(y - \hat{y})$$

where $\hat{y} = \phi^T\hat{\theta}$ and $\hat{y}_1 = [0\ 1]\hat{\theta}$.

10. Which of the following ϕ vectors have the following properties:
   (i) They are persistently exciting.
   (ii) They are sufficiently exciting to allow convergence of the least squares estimator (12.3.9). *Hint:* This least squares estimator converges if $\int_0^t \phi\phi^T\, d\tau$ is nonsingular for any t.

   (a) $\phi^T(t) = \left[1, \dfrac{1}{1 + t}\right]$
   (b) $\phi^T(t) = \left[\dfrac{1}{1 + t}, \dfrac{-2}{3 + t}\right]$
   (c) $\phi^T(t) = [1, t]$
   (d) $\phi^T(t) = [\sin\omega t, \cos\omega t]$, $\quad 0 < \omega < \pi/\Delta$
   (e) $\phi^T(t) = [e^{-t}, e^{-2t}]$

11. Show that the recursive least squares estimator (12.4.11), (12.4.12) can be rewritten as
$$\rho\hat{\theta} = [(1 + \Delta\rho)P]\phi e$$
$$\rho[P^{-1}] = \phi\phi^T/\Gamma$$

12. Consider the problem of estimating a signal $y(t)$, where
$$y(t) = y_0 + y_1 t + \nu$$

Given data at samples of period Δ, show that the least squares estimator for $y(t)$ (with $P_0^{-1} = 0$) for this system can be written as

$$\rho \hat{y} = \hat{y}_1 + k_0(t)(y - \hat{y})$$
$$\rho \hat{y}_1 = k_1(t)(y - \hat{y})$$

where

$$k_0(t) = \frac{4}{t + \Delta} \quad \text{and} \quad k_1(t) = \frac{6}{(t + \Delta)(t + 2\Delta)}$$

Hint: Formulate the problem as one of estimating y_0 and y_1 and write down the recursive least squares estimator for this system. Since $\phi(t)$ is known in advance, $P(t)$ may be precalculated. Then take $\hat{y} = \phi^T \hat{\theta}$ and use the result of Problem 11.

13. (a) Suppose we have a system $y = \phi^T \theta + v$, where the only information available regarding $v(t)$ is that there exists a known $\varepsilon > 0$, such that $|v(t)| \le \varepsilon$ for all t. Given $\hat{\theta}(t)$, $y(t)$, and $\phi(t)$, we now wish to find $\hat{\theta}(t + \Delta)$ such that $\|\hat{\theta}(t + \Delta) - \hat{\theta}(t)\|_2^2$ is a minimum, subject to the constraint that $|y(t) - \phi(t)^T \hat{\theta}(t + \Delta)| \le \varepsilon$. Show that this can be achieved by the following estimator:

$$\rho \hat{\theta} = \frac{\phi f(e)}{\Delta \phi^T \phi}, \qquad e = y - \phi^T \hat{\theta}$$

where

$$f(e) = \begin{cases} e - \varepsilon, & \text{for } e > \varepsilon \\ 0, & |e| \le \varepsilon \\ e + \varepsilon, & \text{for } e < \varepsilon \end{cases}$$

(b) Show that the estimator and system in part (a) give the following properties:
   (i) $\|\tilde{\theta}\|_2$ is a nonincreasing function.
   (ii) $\rho \hat{\theta}$ and $\tilde{f} \triangleq f(e)/\|\phi\|$ belong to $\mathscr{L}_2$.

14. (a) Give an example of a function $\theta(t)$ that has the properties:
   (i) $\hat{\theta}(t)$ is bounded; and
   (ii) $\rho \hat{\theta}(t)$ belongs to $\mathscr{L}_2$; yet which does *not* have a limit (that is, $\lim_{t \to \infty} \hat{\theta}(t)$ does not exist). *Hint:* A scalar function suffices for this part.

   (b) Repeat part (a) for the case where:
   (i) $\|\hat{\theta}(t)\|_2$ is a nonincreasing function; and
   (ii) $\rho(\hat{\theta}(t))$ belongs to $\mathscr{L}_2$. *Hint:* A scalar function is *not* sufficient in this case.)

15. Show that the discrete time, convex region projection given in Equation (12.6.12) preserves the parameter estimator properties of Theorem 12.5.1. *Hint:* Show that (12.6.12) causes V to be more decrescent.

16. Consider a single linear inequality constraint $w^T \theta \le k$ in a discrete time least squares based estimator. Show that the algorithm

$$\hat{\theta}(t + \Delta) = \hat{\theta}(t) + \frac{\Delta P \phi e}{1 + \Delta \phi^T P \phi}$$

$$P'(t) = P(t + \Delta) = P(t) - \frac{\Delta P(t) \phi(t) \phi(t)^T P(t)}{1 + \Delta \phi(t)^T P(t) \phi(t)}$$

$$\hat{\theta}'(t + \Delta) = \hat{\theta}(t + \Delta) + \frac{P'w}{w^T P'w} \min\{0, (k - w^T \hat{\theta}')\}$$

gives the projection required by (12.6.12).

17. Repeat Problem 16 for the following continuous time least squares algorithm with projection

$$\dot{\theta} = P\phi e + h(k - w^T\hat{\theta})\frac{Pw}{w^TPw}\min\{0, -w^TP\phi e\}$$

$$\dot{P} = -P\phi\phi P$$

where

$$h(a) = \begin{cases} 0, & a > 0 \\ 1, & a \leq 0 \end{cases}$$

Hint: If $w^T\hat{\theta} = k$, show that $(d/dt)(w^T\hat{\theta}) \leq 0$.

18. Prove Lemma 12.8.1. *Hint:* Let $h(t)$ denote the impulse of this operator, and evaluate $\eta_f(t)$ using convolution.

19. Consider the exponential weighted least squares algorithm given in (12.9.10), (12.9.11). Show that the state transition matrix for this algorithm is given by

$$\Phi(t,\tau) = P(t)P(\tau)^{-1}E(-\sigma, t-\tau)$$

Hint: The state transition matrix for the time varying linear system $\rho x = A(t)x$ satisfies

$$\rho\Phi(t,\tau) = A(t)\Phi(t,\tau), \qquad \Phi(\tau,\tau) = I$$

Then use the product rule to show that the $\Phi(t,\tau)$ given above satisfies

$$\rho\Phi(t,\tau) = \frac{-P(t)\phi(t)\phi(t)^T}{1 + \Delta\phi(t)^T P(t)\phi(t)}\Phi(t,\tau)$$

as required.

20. Consider the matrix M given in Equation (12.5.22), where the input signal is a sum of sine waves; that is,

$$u(t) = \sum_{k=1}^{N} G_k \cos(\omega_k t + \alpha_k)$$

(a) Use the orthogonality of sine waves of different frequencies to show that M can be written as

$$M = S\left[\sum_{k=1}^{N} \beta_k M_k\right] S^T$$

where M_k has ilth element

$$[M_k]_{il} = \begin{cases} \omega_k^{4n-i-l}\left[\operatorname{sinc}\left(\frac{\omega_k\Delta}{2}\right)\right]^{4n-i-l}(-1)^{\frac{i-l}{2}}\cos((i-l)\sigma_k), \\ \qquad\qquad\qquad\qquad\qquad\qquad\qquad\qquad \text{for } (i-l) \text{ even} \\ \omega_k^{4n-i-l}\left[\operatorname{sinc}\left(\frac{\omega_k\Delta}{2}\right)\right]^{4n-i-l}(-1)^{\frac{i-l+1}{2}}\sin((i-l)\sigma_k), \\ \qquad\qquad\qquad\qquad\qquad\qquad\qquad\qquad \text{for } (i-l) \text{ odd} \end{cases}$$

$$\operatorname{sinc}(x) \triangleq \frac{\sin x}{x}$$

$$\sigma_k \triangleq \tan^{-1}\left(2\sin\left(\frac{\omega_k\Delta}{2}\right)\right)$$

and

$$\beta_k = \frac{TG_k^2}{2\left|A\left(\frac{e^{j\omega_k\Delta}-1}{\Delta}\right)E\left(\frac{e^{j\omega_k\Delta}-1}{\Delta}\right)\right|^2}$$

(b) Show that as $\Delta \to 0$ the above result converges to the continuous time result in which

$$M = \begin{bmatrix} \omega_k^{4n-2} & 0 & -\omega_k^{4n-4} & 0 & \omega_k^{4n-6} & \cdots \\ 0 & \omega_k^{4n-4} & 0 & -\omega_k^{4n-6} & & \\ -\omega_k^{4n-4} & 0 & \omega^{4n-6} & & & \\ 0 & -\omega_k^{4n-6} & & & & \\ \omega_k^{4n-6} & & & & & -\omega_k^2 \\ \vdots & & & & \omega_k^2 & 0 \\ \vdots & & & -\omega_k^2 & 0 & 1 \end{bmatrix}$$

21. Extend Lemma 12.5.2 to the case of nonperiodic signals composed of sine waves. *Hint:* The key difference in this case is that the signals are not exactly orthogonal. Thus we need to modify (12.5.29). Note that the cross-coupling terms have bounded influence as $t_1 \to \infty$, while the other terms have unbounded influence.

22. Consider the general estimator (12.5.2), (12.5.3) in the ideal case where (12.5.1) holds. Prove (12.5.8). *Hints:*
    (1) Show that if

    $$\rho(A) = BC^T$$

    then

    $$\rho(A^{-1}) = \frac{-A^{-1}BC^TA^{-1}}{1 + \Delta C^T A^{-1}B}$$

    This is frequently called the matrix inversion lemma; see Problem 12.1.

    (2) Thus show that

    $$\rho(P^{-1}) \le \frac{\alpha\phi\phi^T}{\Gamma + \Delta(1-\alpha)\phi^T P\phi}$$

    with equality if $\Omega = 0$.

    (3) Take $\Omega = 0$ and show that

    $$\rho(P^{-1}\tilde{\theta}) = 0$$

    (4) Thus establish (12.5.8) on noting from part (2) that

    $$\rho V|_{\Omega \ge 0} \le \rho V|_{\Omega = 0}$$

23. Consider the dead zone estimator discussed in Lemma 12.8.2. Prove that (12.8.17) holds. *Hint:* Take $\Omega = 0$ and show that

    $$\rho(P^{-1}\tilde{\theta}) = \frac{\alpha\phi\eta_f}{\Gamma + \Delta(1-\alpha)\phi^T P\phi}$$

24. Consider the model with errors given in (12.8.10); that is, $y = \phi^T\theta + \eta_f$. An algorithm based on normalization for dealing with $\eta_f(t)$ is the following:

$$\rho\hat{\theta} = \frac{\alpha P\phi(\bar{y} - \phi^T\hat{\theta})}{\Gamma + \phi^T P\phi}$$

$$\rho P = \frac{-\alpha P\phi\phi^T P}{\Gamma + \phi^T P\phi} + \Omega$$

The key idea in normalization is to choose Γ in this algorithm so as to reduce the algorithm gain when $\eta_f(t)$ is large. This can be achieved by choosing Γ as some scaled function of $d(t)$, where $d(t)$ overbounds $|\eta_f(t)|$. This ensures that

$$\frac{\eta_f(t)^2}{\Gamma} \leq \varepsilon^2$$

Under these conditions, establish the following set of properties for the estimator (suitably modified to ensure that P^{-1}, P, and $\hat{\theta}$ are bounded). There exists a $k_1 < \infty$ such that for any $t_0, t > 0$:

(i) $\displaystyle\int_{t_0}^{t_0+t} \frac{e^2(\tau)}{\Gamma(\tau) + \phi^T(\tau)P(\tau)\phi(\tau)}\, d\tau \leq k_1 + \varepsilon^2 t$

(ii) $\displaystyle\int_{t_0}^{t_0+t} (\rho\hat{\theta}(\tau))^T(\rho\hat{\theta}(\tau))\, d\tau \leq \alpha^2(\lambda_{\max} P)(k_1 + \varepsilon^2 t)$

Hint: Define V as in (12.6.4). Then show that

$$\rho V \leq \frac{\alpha\eta_f^2}{\Gamma(\tau) + (1 - \alpha\Delta)\phi^T P\phi} - \frac{\alpha e^2}{\Gamma(\tau) + \phi^T P\phi}$$

(with equality if $\Omega = 0$). Then show that

$$\rho V \leq \alpha\left(\frac{-e^2}{\Gamma(\tau) + \phi^T P\phi}\right) + \alpha\varepsilon^2$$

Integrate both sides of this equation from $t_0 + t$ to give

$$\int_{t_0}^{t_0+t} \frac{\alpha e(\tau)^2}{\Gamma(\tau) + \phi^T(\tau)P(\tau)\phi(\tau)}\, d\tau \leq V(t_0) - V(t_0 + t) + \alpha\varepsilon^2 t$$

Show that V is bounded and then let

$$k_1 = \sup_{t \in [0, \infty)} \{V(t)\}$$

25. Complete the proof of Lemma F.1.1 of Appendix F. *Hint:* If

$$H_u = \begin{bmatrix} \dfrac{BH + A\overline{\overline{H}}}{V} \end{bmatrix}$$

is proper and exponentially stable, then the impulse response can be written as $K_u i_\Delta(\tau) + h_u(\tau)$, where $i_\Delta(\tau)$ is the impulse function, and $h_u(\tau)$ is bounded by an exponential.

26. Extend Lemma F.1.1 to the discrete time case.

27. Verify that $\Omega(t)$ as in Lemma F.2.2:
    (a) Is uniformly positive definite; that is, there exists a constant c, d such that $dI \geq \Omega(t) \geq cI$ for all time.

(b) Satisfies Equation (F.2.11). *Hint:*

(a) $\Omega(t) = \int_0^\infty E(A^T(t), \tau) \Gamma E(A(t), \tau) \, d\tau$

In view of conditions (i) and (iii) in Lemma F.2.2,

$\|E(A^T(t), \tau)\| \leq c_0 E(-\sigma_0, t)$

(b) Equation (F.2.10) can be rewritten as a system of linear equations.

*28. Extend Theorem F.2.1(i) to the normalized algorithm of Problem 12.23.

29. Verify F.2.26 in Theorem F.2.1.

30. Show that (F.2.39) implies (F.2.40). *Hint:* If $f \in \mathscr{L}_2^+$, then

$$\int_0^t E(-\sigma_0, t-\tau) f^2(\tau) \, d\tau = \int_0^{t/2} E(-\sigma_0, t-\tau) f^2(\tau) \, d\tau$$

$$+ \int_{t/2}^t E(-\sigma_0, t-\tau) f^2(\tau) \, d\tau \to 0$$

Then use Schwarz's inequality.

31. Show that if $v(t) \in \mathscr{L}_2^+$ and $\rho v(t)$ is bounded, then $v(t) \to 0$. (This result is used in the proof of Theorem F.2.1.)

13

Design Considerations

13.1 INTRODUCTION

So far in the book we have described a number of analysis tools for control and estimation. We have also seen how it is possible to obtain specific control and estimation algorithms by techniques such as optimization and pole placement. However, these techniques fall short of being a complete solution to the design problems since they do not expose the trade-offs and fundamental limitations inherent in control systems design.

In this chapter, we will outline some of the trade-offs and limitations associated with the design problem. In particular, we will develop guidelines that show how the tools and techniques developed earlier can be used in practice. As a simple illustration, we give some indication of suitable locations for closed loop poles, as opposed to the problem of designing the feedback to achieve these locations. For simplicity, we limit our considerations to single-input, single-output problems. However, similar conclusions apply to the multivariable case.

A key factor in control system design is the role played by uncertainty. We will show in this chapter how feedback, feedforward, and parameter estimation can be used to achieve performance objectives in the face of uncertainty. As in previous chapters, we will treat both the continuous and discrete cases simultaneously using our unified notation. We will also specifically point to the design penalties arising from the sampling process in discrete time systems.

13.2 THE CASE FOR HIGH-GAIN FEEDBACK

Consider the simple feedback control system shown in Figure 13.2.1, where the symbols have the following meaning:

- G plant transfer function
- C controller transfer function
- y^* desired output; that is, set point for y
- e error signal
- u system input
- d output disturbance
- y plant output
- v output measurement noise
- y_m measured plant output

For the moment, we will take $v = 0$. In this case, it is trivial to see that the output y can be expressed as

$$y = G_{CL} y^* + Sd \tag{13.2.1}$$

where

$$G_{CL} \triangleq \frac{GC}{1 + GC}, \quad S \triangleq \frac{1}{1 + GC} \tag{13.2.2}$$

From this expression it is clear that y will be close to y^* over that frequency range in which GC is large. We will loosely define the bandwidth b_ω of the closed loop system as that set of frequencies for which $|GC| \geq 1$. Thus we see that for good tracking and disturbance rejection it is desirable to have a large bandwidth.

Next we consider the sensitivity of the closed loop transfer function to variations in the plant. We measure this sensitivity via the following partial derivative:

$$\frac{\partial G_{CL}}{\partial G} = \frac{C}{(1 + GC)^2} \tag{13.2.3}$$

We see that the sensitivity of the closed loop to variations in the plant is reduced if $|GC|$ is large. This again suggests that it is desirable to have as wide a bandwidth as possible. This very well known fact is illustrated in Example 13.2.1.

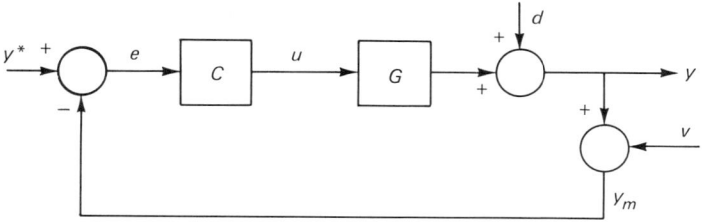

Figure 13.2.1 Elementary feedback system.

Example 13.2.1

Consider the design of a simple power amplifier having a desired gain of unity and a bandwidth of ω_b rad sec^{-1}. Say we have an amplifier available, having a nominal bandwidth of $\omega_b/10$ rad sec^{-1}. The open loop solution to this problem would be to precede the power amplifier by a low-power preamplifier having transfer function G_a, where

$$G_a(s) = \frac{(10s/\omega_b) + 1}{(s/\omega_b) + 1} \tag{13.2.4}$$

This gives the open loop design shown in Figure 13.2.2.

It is instructive to compute the various signals u and y if y^* is a unit step input applied at time $t = 0$. These are shown in Figure 13.2.3 for two cases: (a) when the power amplifier has exactly the nominal transfer function, that is, $G(s) = 1/[(10s/\omega_b) + 1]$, and (b) when the power amplifier has a slightly different bandwidth due to parameter variations, that is, $G(s) = 1/[(5s/\omega_b) + 1]$. It can be seen from the figure that the open loop design leads to 60% overshoot when the plant parameters change. The feedback solution to this problem is shown in Figure 13.2.4 where the preamplifier is simply a gain.

Note that to increase the closed loop bandwidth by 10:1 we have to increase the loop gain by approximately 10:1 to achieve the desired final gain (in electronic amplifier design this is usually called the gain/bandwidth product rule). Thus, the gain of the preamplifier is approximately 10.

The responses are again shown in Figure 13.2.3 for the case (a) when the nominal plant is $1/[(10s/\omega_b) + 1]$ and (b) when the nominal plant is $1/[(5s/\omega_b) + 1]$ as before. The dramatic reduction in sensitivity to parameter variations resulting from the use of feedback is evident in the figure. ▽▽▽

Finally, we consider the question of control of unstable plants. Clearly, feedback is necessary to stabilize these systems. If in addition to stability we consider the transient response of the closed loop system, then it turns out that it is necessary to have a lower limit on the bandwidth. To show why this is the case, we consider the transform of the error signal e:

$$E(\gamma) = \frac{1}{1 + G(\gamma)C(\gamma)} Y^*(\gamma) \tag{13.2.5}$$

Now suppose y^* is a unit step, and $G(\gamma)$ [or $C(\gamma)$] has a real, unstable pole at $\gamma = \alpha > 0$. Then, provided the closed loop is stable, it follows that

$$E(\alpha) = 0 \tag{13.2.6}$$

Again, using the stability of the closed loop and with $Y^*(\gamma) = 1/\gamma$, we note that $\gamma = \alpha$ is within the region of convergence for the transform of e. Thus, from

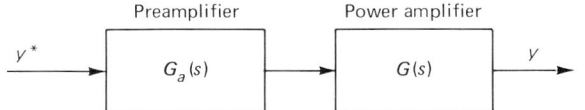

Figure 13.2.2 Open loop compensator.

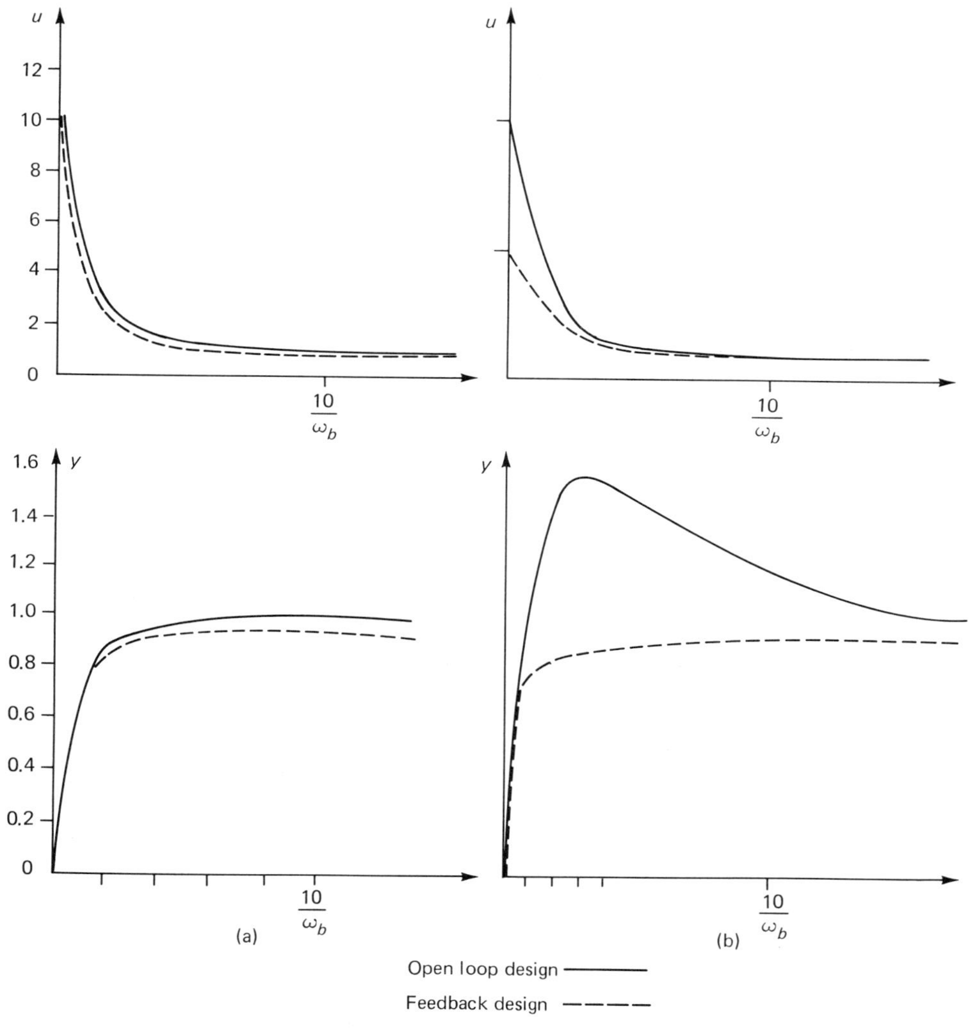

Figure 13.2.3 Sensitivity of open and closed loop designs.

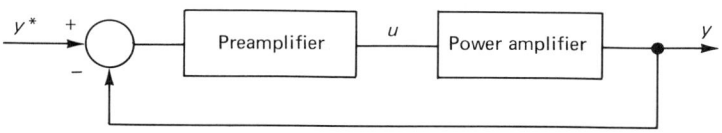

Figure 13.2.4 Closed loop compensator.

Sec. 13.2   The Case for High-Gain Feedback

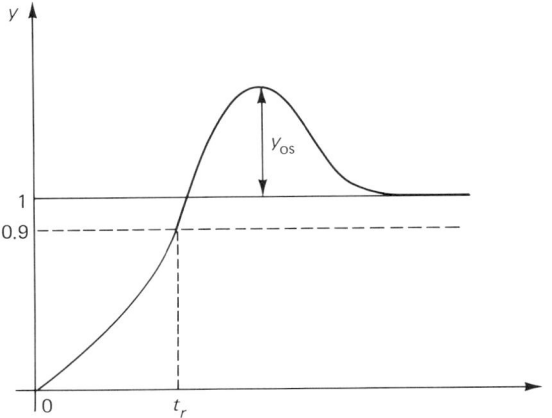

Figure 13.2.5 Rise time and overshoot for closed loop.

(13.2.6) and the definition of the transform, we obtain the integral equality:

$$\int_0^\infty E(\alpha, -t) e(t)\, dt = 0 \tag{13.2.7}$$

Clearly, from (13.2.7), $e(t)$ cannot have a single sign; that is, unless $e(t)$ is zero for all time, $e(t)$ must be negative at some times and positive at other times. We thus note that a unity feedback system (one of the form shown in Figure 13.2.1) that has a real, right half-plane open loop pole must have overshoot in its step response.

The amount of overshoot is related to the rise time and the location of the right half-plane pole, as we now show. We loosely define the rise time t_r as the time taken for the output response to a step change in the reference to reach and stay above 90% of its desired value. This is illustrated in Figure 13.2.5.

We approximate a bound on the step response over the interval 0 to t_r by

$$y(t) \leq \frac{0.9t}{t_r}, \qquad t \in [0, t_r] \tag{13.2.8}$$

or

$$e(t) \geq \left(1 - \frac{0.9t}{t_r}\right), \qquad t \in [0, t_r] \tag{13.2.9}$$

Using (13.2.9) in the integral equality (13.2.7), we obtain

$$-\int_{t_r}^\infty E(\alpha, -t) e(t)\, dt \geq \int_0^{t_r} E(\alpha, -t)\left(1 - \frac{0.9t}{t_r}\right) dt \tag{13.2.10}$$

It follows from this expression that long rise times [relative to the location of the unstable pole and hence relative to the exponential decay in (13.2.10)] inevitably lead to large overshoot. This is made more precise next. From (13.2.10) and the definition of the overshoot y_{os} in Figure 13.2.5, it follows that

$$y_{os} \int_{t_r}^\infty E(\alpha, -t)\, dt \geq \int_0^{t_r} E(\alpha, -t)\left(1 - \frac{0.9t}{t_r}\right) dt \tag{13.2.11}$$

which leads to (see Problem 1)

$$y_{os} \geq \alpha t_r \quad (13.2.12)$$

Using the standard rules of thumb, the rise time t_r can be related to the bandwidth b_w (in radians per second) by

$$t_r b_w \approx 2.3 \quad (13.2.13)$$

Thus, to keep the overshoot y_{os} below about 40%, we require

$$b_w \geq 5\alpha \quad (13.2.14)$$

Thus, unstable poles place a lower limit on the bandwidth that should be used.

To summarize these findings, we have seen that wide bandwidth is desirable to achieve (1) good tracking, (2) good disturbance rejection, (3) low sensitivity to plant variations, and (4) good transient performance in the presence of unstable open loop poles. Thus, all these factors suggest that the closed loop bandwidth should be made as large as possible (theoretically infinite). In practice, there is an upper limit for the bandwidth imposed by other considerations, as we next show.

13.3 FACTORS LIMITING FEEDBACK BANDWIDTH

Consider again the elementary feedback system shown in Figure 13.2.1. If we consider the measurement noise v, then the component of y due to v is given by

$$y_v = -\frac{GC}{1 + GC} v \quad (13.3.1)$$

Thus the measurement noise is transferred to the output whenever $|GC| > 1$. Hence excessive bandwidth will lead to large output errors due to measurement noise.

Next we consider the input response due to the disturbance term d. If the frequency content of d lies inside the bandwidth, then it follows from (13.2.1) that it will be (approximately) eliminated from the closed loop response. This can only happen if there is a component in the input that is given by

$$u_d = -G^{-1} d$$

Thus, if d is a sinusoid of amplitude M and frequency ω, then the required input amplitude is

$$|U_{max}| = \left| \frac{M}{g_\omega} \right| \quad (13.3.2)$$

where g_ω is the plant gain at frequency ω. Similarly, the corresponding input slew rate (that is, rate of change) is

$$|\dot{U}_{max}| = \left| \frac{\omega M}{g_\omega} \right| \quad (13.3.3)$$

Typically, g_ω decreases with increasing frequency and hence a greater bandwidth will almost invariably lead to greater input amplitude and greater slew rate. Hence, if the input amplitude and/or slew rate is limited, this will place a limit on the achievable bandwidth.

Next we consider the stability of the closed loop system of Figure 13.2.1 in the face of plant uncertainties. Let G denote the true plant transfer function (as before) and let G_0 denote the nominal plant transfer function used for design. Then the stable nominal closed loop transfer function is

$$G_{CL}^0 = \frac{CG_0}{1 + CG_0} \tag{13.3.4}$$

Hence, if G_{CL}^0 is specified, then the required controller is obtained by inverting (13.3.4), which yields the following controller:

$$C_0 = \frac{G_{CL}^0}{G_0(1 - G_{CL}^0)} \tag{13.3.5}$$

When this controller is applied to the true plant, $G = G_0 + G_\Delta$, then the resulting closed loop transfer function is

$$\begin{aligned} G_{CL} &= \frac{GC_0}{1 + GC_0} \\ &= \frac{[1 + (G_\Delta/G_0)]G_{CL}^0}{1 + (G_\Delta/G_0)G_{CL}^0} \end{aligned} \tag{13.3.6}$$

where G_Δ represents (additive) modeling errors; that is,

$$G = G_0 + G_\Delta \tag{13.3.7}$$

Under the assumption that both G_{CL}^0 and G_Δ are stable, it is then evident that a necessary and sufficient condition for closed loop stability is that the locus of $(G_\Delta/G_0)G_{CL}^0$ does not encircle the -1 point. This is guaranteed provided

$$\left|\frac{G_\Delta}{G_0}\right| |G_{CL}^0| < 1, \quad \text{for all } \omega \tag{13.3.8}$$

Hence, at those frequencies for which $|G_{CL}^0|$ is near 1 we require

$$\left|\frac{G_\Delta}{G_0}\right| < 1$$

This implies that the closed loop bandwidth is limited to those frequencies such that the relative plant uncertainty $|G_\Delta/G_0|$ is significantly less than 1.

Finally, we consider the transient response of the closed loop system when the plant has unstable zeros. To gain an appreciation of what happens in this case, let the plant have a single real zero at $\sigma > 0$. Let $Y(\gamma)$ denote the transform of the output $y(t)$; then

$$Y(\gamma) = \frac{C(\gamma)G(\gamma)}{1 + C(\gamma)G(\gamma)} Y^*(\gamma) \tag{13.3.9}$$

Since $G(\sigma) = 0$, and assuming closed loop stability,

$$Y(\sigma) = 0 = \int_0^\infty E(\sigma, -t) y(t)\, dt \tag{13.3.10}$$

for any bounded $y^*(t)$. Thus consider $y^*(t)$ as a unit step. Then (13.3.10) shows that $y(t)$ cannot have a single sign; that is, $y(t)$ must undershoot. We now proceed to relate the size of this undershoot to the rise time of the system. With the rise time t_r as defined in Section 13.2 (see Figure 13.2.5), we note that

$$y(t) \geq 0.9, \quad \text{for } t \in [t_r, \infty) \tag{13.3.11}$$

We define the undershoot y_{us} as

$$y_{us} = -\inf_t \{y(t)\} \tag{13.3.12}$$

Using (13.3.12) and (13.3.11) in (13.3.10), we obtain

$$y_{us} \int_0^{t_r} E(\sigma, -t) \, dt \geq \int_{t_r}^{\infty} 0.9 E(\sigma, -t) \, dt \tag{13.3.13}$$

From (13.3.13) it is clear that, for a given σ, the undershoot must become large as t_r becomes small. In particular, it follows from (13.3.13) that

$$y_{us} \geq \frac{0.9}{E(\sigma, t_r) - 1} \tag{13.3.14}$$

Using the rule of thumb relating the bandwidth b_w and the rise time t_r (13.2.13), we note that to keep the minimum undershoot given by (13.3.14) below about 10% we require

$$b_w \leq \frac{1}{\Delta} \ln(1 + \Delta\sigma) \approx \sigma \tag{13.3.15}$$

Thus we see that unstable zeros place an upper limit on the bandwidth that can be used.

In summary, we have seen that the closed loop bandwidth is limited by several factors, including (1) measurement noise, (2) input amplitude and slew rate limits, (3) instability resulting from plant uncertainty and, (4) undershoot associated with right half-plane zeros.

13.4 FREQUENCY DOMAIN SENSITIVITY FUNCTIONS

We have seen that there are fundamental limits on the bandwidth imposed by the locations of the poles and zeros of the system. These constraints imply that the performance of a control system has certain inescapable limits that hold irrespective of the design method used. It is clearly desirable to have an appreciation of these constraints during the design process.

In Sections 13.2 and 13.3 the design constraints were motivated by time domain arguments. To gain further insight into the constraints, we next view them from a frequency domain perspective. To do this, we consider the following two sensitivity functions:

$$S = \frac{1}{1 + GC} \tag{13.4.1}$$

$$T = \frac{GC}{1 + GC} \tag{13.4.2}$$

We have seen in Sections 13.2 and 13.3 that having S small reduces the closed loop sensitivity to disturbances [see (13.2.1)] and plant variations [see (13.2.3)], while having T small reduces the closed loop sensitivity to measurement errors [see

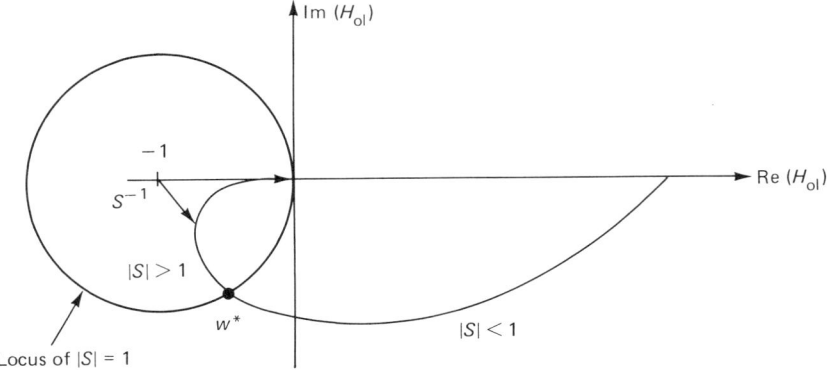

Figure 13.4.1 Typical Nyquist diagram.

(13.3.1)]. However, both sensitivity functions cannot be simultaneously small since $S + T = 1$.

We will see next that if we design the feedback to hold S small in a given frequency band then it is likely to be large in another frequency band. Heuristically, this can be seen from the Nyquist plot of the open loop transfer function $H_{ol} \triangleq GC$ as follows. Say the relative degree of H_{ol} is greater than or equal to 2; then the phase angle of H_{ol} must ultimately go below $-\pi/2$. However, S^{-1} is the vector from the -1 point to H_{ol}. Thus we have the typical Nyquist diagram as shown in Figure 13.4.1, where we see that $|S|$ is less than 1 up to ω^* but then increases above 1 beyond ω^*. As the frequency ω^* is varied, the magnitude of the peak in $|S|$ also varies. These features are made explicit in the following result for stable *closed loop* systems.

Lemma 13.4.1

(a) Continuous

Provided the open loop transfer function has relative degree ≥ 2 (or ≥ 1 plus a pure time delay), then the sensitivity function S satisfies the following integral constraint:

$$\int_0^\infty \log|S(j\omega)|\,d\omega = \pi \sum_{i \in U} p_i \qquad (13.4.3)$$

where U is the set of poles p_i in the right half plane.

(b) Discrete

Provided the open loop transfer function has relative degree ≥ 1, then the sensitivity function S satisfies the following integral constraint:

$$\int_0^{\pi/\Delta} \log\left|S\left(\frac{e^{j\omega\Delta} - 1}{\Delta}\right)\right|\,d\omega = \pi \sum_{i \in U} p_i^c \qquad (13.4.4)$$

where U is the set of unstable poles and p_i^c is the continuous counterpart of the ith unstable discrete pole p_i; that is,

$$p_i^c \triangleq \frac{1}{\Delta} \log(1 + \Delta p_i) \tag{13.4.5}$$

Proof

(a) We first note that the zeros of S are the poles of H_{ol}. For simplicity, we assume that H_{ol} has nonrepeated right half-plane poles. Now $\log S$ is analytic in the right half plane save when $S = 0$, that is, at the open loop unstable poles. Let $p_1 \ldots p_N$ be the unstable poles and consider the contour shown in Figure 13.4.2. Clearly, on the total contour $\mathscr{C} \triangleq \mathscr{C}_0 + \mathscr{C}_1 + \cdots + \mathscr{C}_{N+1}$, we have

$$\oint_{\mathscr{C}} \log S(s)\, ds = 0 \tag{13.4.6}$$

The limit as $\varepsilon \to 0$ of the portion due to $\mathscr{C}_0$ is evaluated as follows:

$$\lim_{\varepsilon \to 0} \int_{\mathscr{C}_0} \log S(s)\, ds = j \int_{-\infty}^{\infty} \log[S(j\omega)]\, d\omega$$

$$= j \int_0^{\infty} \{\log[S(j\omega)] + \log[S(-j\omega)]\}\, d\omega$$

$$= 2j \int_0^{\infty} \log|S(j\omega)|\, d\omega \tag{13.4.7}$$

Due to the relative degree assumption, we have

$$\lim_{\varepsilon \to 0} \int_{\mathscr{C}_{N+1}} \log S(s)\, ds = 0 \tag{13.4.8}$$

Finally, for $\mathscr{C}_1 \ldots \mathscr{C}_N$, we have

$$\lim_{\varepsilon \to 0} \int_{\mathscr{C}_i} \log S(s)\, ds = \int_{\mathscr{C}_i} \log(s - p_i)\, ds$$

$$= [(s - p_i)\log(s - p_i) - (s - p_i)]_{j(y_i - \varepsilon)}^{j(y_i + \varepsilon)} \tag{13.4.9}$$

where $p_i = x_i + jy_i$, $i = 1, \ldots, N$. Hence

$$\lim_{\varepsilon \to 0} \int_{\mathscr{C}_i} \log S(s)\, ds = -j2\pi x_i \tag{13.4.10}$$

The result follows from (13.4.6) on noting that any complex poles must appear in conjugate pairs.

(b) Consider the evaluation of the integral of $[\log S(\gamma)/(1 + \Delta\gamma)]$ on the contour shown in Figure 13.4.3. Clearly, on the total contour $\mathscr{C} = \mathscr{C}_0 + \mathscr{C}_1 + \cdots + \mathscr{C}_{N+2}$,

$$\oint_{\mathscr{C}} \frac{\log S(\gamma)}{1 + \Delta\gamma}\, d\gamma = 0 \tag{13.4.11}$$

The limit of the integrals on $\mathscr{C}_{N+1}$ and $\mathscr{C}_{N+2}$ are zero due to the relative degree

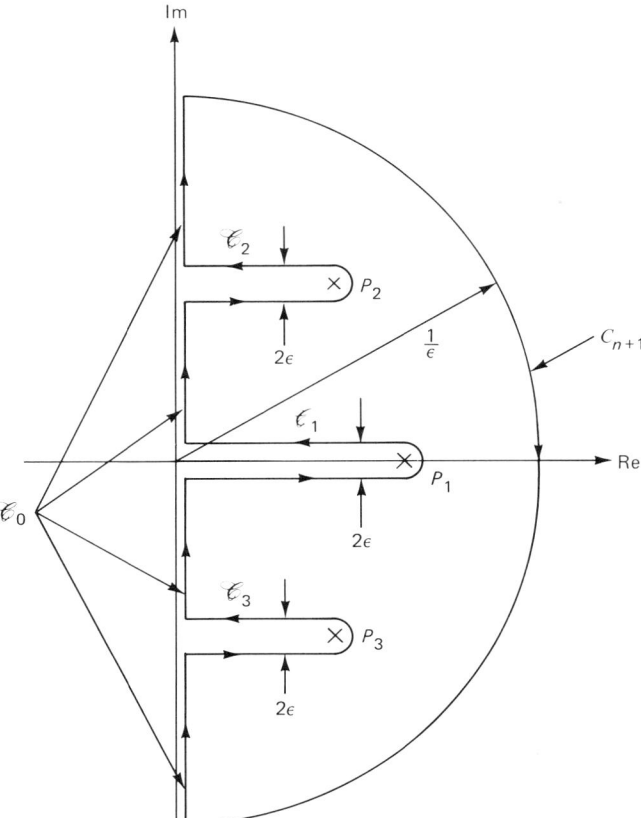

Figure 13.4.2 Integration contour.

assumption and due to the factor $1/(1 + \Delta\gamma)$ on the denominator in (13.4.11). For $\mathscr{C}_0$, we have $\gamma = (e^{j\omega\Delta} - 1)/\Delta$ with $d\gamma = je^{j\omega\Delta} d\omega = j(1 + \Delta\gamma) d\omega$. Hence

$$\lim_{\varepsilon \to 0} \int_{\mathscr{C}_0} \frac{\log S(\gamma)}{1 + \Delta\gamma} d\gamma = 2j \int_0^{\pi/\Delta} \log\left|S\left(\frac{e^{j\omega\Delta} - 1}{\Delta}\right)\right| d\omega \qquad (13.4.12)$$

Finally, for the $\mathscr{C}_i$ contours we have

$$\lim_{\varepsilon \to 0} \int_{\mathscr{C}_i} \frac{\log S(\gamma)}{1 + \Delta\gamma} d\gamma = \lim_{\varepsilon \to 0} \int_{\mathscr{C}_i} \frac{\log(\gamma - p_i)}{1 + \Delta\gamma} d\gamma$$

Then integrating by parts

$$\lim_{\varepsilon \to 0} \int_{\mathscr{C}_i} \frac{\log S(\gamma)}{1 + \Delta\gamma} d\gamma = \lim_{\varepsilon \to 0} \left[\frac{1}{\Delta} \log(1 + \Delta\gamma) \log(\gamma - p_i)\right]_{-(1/\Delta)+(1/\Delta)e^{j(\theta_i - \varepsilon)}}^{-(1/\Delta)+(1/\Delta)e^{j(\theta_i + \varepsilon)}}$$

$$- \oint_{\mathscr{C}_i} \frac{1}{\Delta} \frac{\log(1 + \Delta\gamma)}{\gamma - p_i} d\gamma \qquad (13.4.13)$$

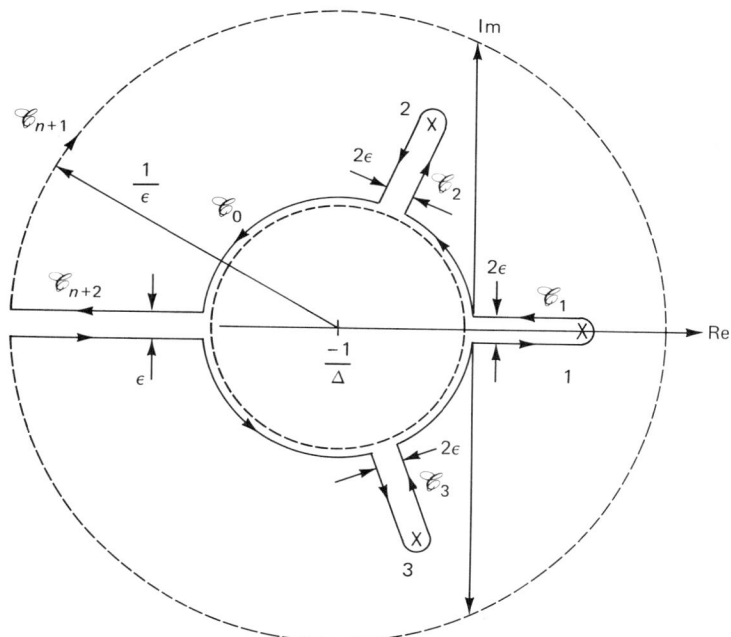

Figure 13.4.3 Discrete time integration contour.

where

$$p_i = -\frac{1}{\Delta} + r_i e^{j\theta_i}$$

Evaluating (13.4.13) using Cauchy's formula for the last term gives

$$\lim_{\varepsilon \to 0} \int_{\mathscr{C}_i} \frac{\log S(\gamma)}{1 + \Delta \gamma} d\gamma = j\frac{2\pi}{\Delta} \log e^{j\theta_i} + \frac{j2\pi}{\Delta} \log(1 + \Delta p_i)$$

$$= -\frac{2\pi}{\Delta}\theta_i + j2\pi p_i^c \qquad (13.4.14)$$

The result then follows from (13.4.11) since the real term in (13.4.14) cancels because the poles occur in conjugate pairs. ∇∇∇

The continuous and discrete time results given here are consistent for $\Delta \to 0$. In particular, if the discrete time unstable poles are due to sampled continuous time unstable poles, then the right sides of (13.4.3) and (13.4.4) are identical. Hence the penalty arising from sampling is simply the upper limit in the integral on the left side of (13.4.4).

We next translate the integral constraints (13.4.3), (13.4.4) into design constraints. To do this, we first give approximate bounds on the sensitivity function S. From experience, the sensitivity function has the general form shown in Figure 13.4.4, where $|S[(e^{j\omega\Delta} - 1)/\Delta]| \triangleq \bar{S}(\omega)$.

Sec. 13.4   Frequency Domain Sensitivity Functions

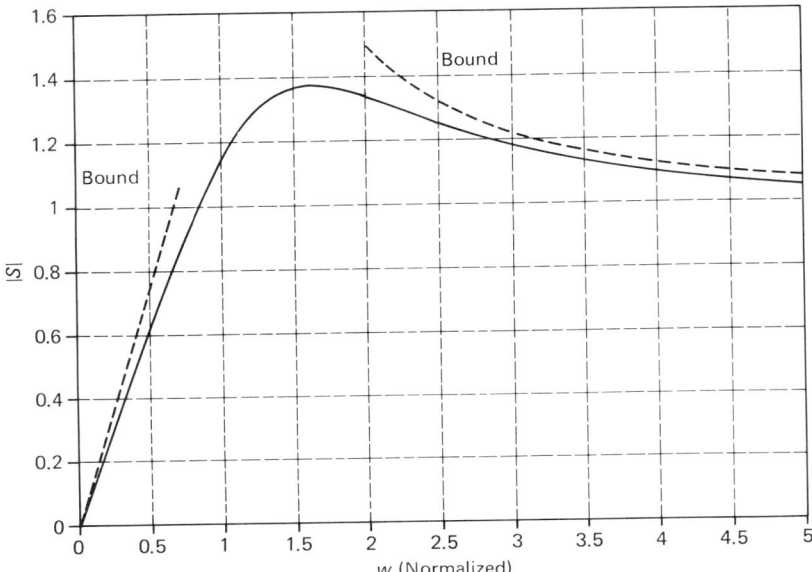

Figure 13.4.4 Approximate shape of $\bar{S}(\omega)$.

Upper bounds on $\bar{S}(\omega)$ are typically

$$\bar{S}(\omega) \leq \frac{1.5\omega}{\omega_B}, \qquad \text{for } \frac{\omega}{\omega_B} < 0.75 \tag{13.4.15}$$

$$\bar{S}(\omega) \leq 1 + \frac{2\omega_B^2}{\omega^2}, \qquad \text{for } \frac{\omega}{\omega_B} > 2 \tag{13.4.16}$$

Using (13.4.3) or (13.4.4), we have

$$\pi \sum_{i \in U} p_i^c = \int_0^{\pi/\Delta} \log \bar{S}(\omega) \, d\omega$$

$$= \int_0^{0.75\omega_B} \log \bar{S}(\omega) \, d\omega + \int_{0.75\omega_B}^{2.75\omega_B} \log \bar{S}(\omega) \, d\omega$$

$$+ \int_{2.75\omega_B}^{\pi/\Delta} \log \bar{S}(\omega) \, d\omega \tag{13.4.17}$$

Using the bounds given in (13.4.15), (13.4.16), we have

$$\int_{0.75\omega_B}^{2.75\omega_B} \log \bar{S}(\omega) \, d\omega \geq \pi \sum_{i \in U} p_i^c + 0.64\omega_B^2 \Delta \tag{13.4.18}$$

From (13.4.18), we have

$$\sup \log \bar{S}(\omega) \geq \frac{\pi}{2\omega_B} \sum_{i \in u} p_i^c + 0.32\omega_B \Delta \tag{13.4.19}$$

Thus, if we want the lower limit on the peak in $\bar{S}(\omega)$ to be not more than $\sqrt{2}$, we approximately require

$$\omega_B(1 - \omega_B\Delta) \geq 5 \sum_{i \in U} p_i^c \qquad (13.4.20)$$

Note that this design rule corresponds precisely to the rule obtained from time domain arguments [see (13.2.14)]. We also note from (13.4.20) that good sensitivity requires $\omega_B\Delta$ to be small compared with 1; that is, fast sampling is generally desirable.

Next we will investigate the corresponding constraint for zeros. To do this we will use the complementary sensitivity function $T(\gamma)$, where, as in (13.4.2),

$$T(\gamma) \triangleq \frac{GC}{1 + GC} = 1 - S(\gamma) \qquad (13.4.21)$$

As shown in (13.3.1), the function $T(\gamma)$ gives a measure of the sensitivity to measurement noise. We then have the following corollary to Lemma 13.4.1.

Corollary 13.4.1

(a) Continuous

Provided the open loop transfer function $H_{ol} \triangleq GC$ has at least one integrator and provided the closed loop is stable, the complementary sensitivity function T satisfies the following integral constraint:

$$\int_0^\infty \log\left|T\left(\frac{1}{-jv}\right)\right| dv = -\frac{\pi}{2}H_0^{-1} + \pi \sum_{i \in U'} \frac{1}{\xi_i} + \frac{\pi}{2}\tau \qquad (13.4.22)$$

where

U' = set of zeros, ξ_i, in the right half plane

τ = system time delay

H_0 = system velocity constant; that is

$$H_0 = \lim_{s \to 0} sH_{ol}(s) \qquad (13.4.23)$$

(b) Discrete

Provided the open loop transfer function $H_{ol} \triangleq GC$ has at least one integrator and provided the closed loop is stable, the complementary sensitivity function T satisfies the following integral constraint:

$$\int_0^\infty \log\left|T\left(\frac{1}{-(\Delta/2) - jv}\right)\right| dv = -\frac{\pi}{2}H_0^{-1} + \pi \sum_{i \in U'} \left(\frac{1}{\xi_i} + \frac{\Delta}{2}\right)$$
$$+ (n - m)\frac{\pi\Delta}{2} \qquad (13.4.24)$$

where n and m are the degree of the denominator and numerator, respectively.

Proof: We define
$$r \triangleq \gamma^{-1} \tag{13.4.25}$$
and then transform $T(\gamma)$ as defined in (13.4.21) as
$$T(r^{-1}) = \frac{1}{1 + H_{ol}(r^{-1})^{-1}} \tag{13.4.26}$$

We observe that zeros at ∞ map to zeros at the origin in the r plane. We then follow the proof of Lemma 13.4.1 using the contour shown in Figure 13.4.5 for the complex r plane. Note that cut-sets are also required for zeros at the origin of the r plane.

The integral around the total contour $\mathscr{C} = \mathscr{C}_0 + \mathscr{C}_1 + \cdots + \mathscr{C}_{N+1}$ is zero. The integral along $\mathscr{C}_0$ satisfies

$$\lim_{\varepsilon \to 0} \int_{\mathscr{C}_0} \log T(r^{-1})\, dr = 2j \int_0^\infty \log \left| T\left(\frac{1}{-(\Delta/2) - jv} \right) \right| dv \tag{13.4.27}$$

The contribution from each of the contours $\mathscr{C}_i$, $i = 1, \ldots, N$, can be evaluated as in Lemma 13.4.1 to give

$$\lim_{\varepsilon \to 0} \int_{\mathscr{C}_i} \log T(r^{-1})\, dr = -2j\pi \,\mathrm{Re}\left\{ \frac{1}{\xi_i} + \frac{\Delta}{2} \right\} \tag{13.4.28}$$

The contribution from $\mathscr{C}_{N+1}$ can be evaluated as follows:

$$\lim_{\varepsilon \to 0} \int_{\mathscr{C}_{N+1}} \log T(r^{-1})\, dr = \lim_{\varepsilon \to 0} \int_{\mathscr{C}_{N+1}} \log\left(\frac{1}{1 + (1/rH_0)} \right) dr$$

$$= -\lim_{\varepsilon \to 0} \int_{\mathscr{C}_{N+1}} \left(\frac{1}{rH_0} \right) dr$$

$$= \frac{j\pi}{H_0} \tag{13.4.29}$$

The final term needed for the continuous time case is a cut-set to allow for the singularity of $\log T$ caused by the time delay, $e^{-s\tau}$. This time delay causes a singularity at $r = 0$; that is, $s = \infty$. Thus we take a semicircle $\mathscr{C}_\tau$ of radius ε about the origin in the r plane. On this semicircle we have

$$\int_{\mathscr{C}_\tau} \log T(r^{-1})\, dr = \int_{\mathscr{C}_\tau} \log(e^{-\tau/r})\, dr$$

$$= \int_{\mathscr{C}_\tau} -\frac{\tau}{r}\, dr$$

$$= -j\pi\tau \tag{13.4.30}$$

The result then follows from (13.4.27) to (13.4.30) on noting that the zeros ξ_i in (13.4.28) occur in conjugate pairs. ∇∇∇

Again we see that the discrete and continuous time cases are consistent. Also note that the plant zeros in continuous time map into discrete time zeros as $\Delta \to 0$. In addition, the discrete time system will have sampling zeros in the far left plane

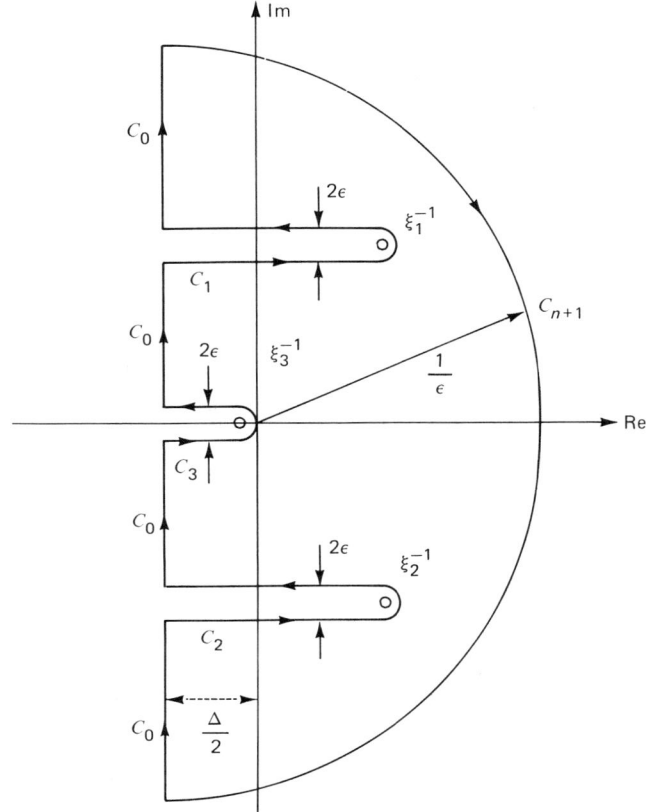

Figure 13.4.5 Contour of integration.

(see Lemma 5.5.2). These zeros map to the interval $(-\Delta/2, 0)$ and hence contribute a term of order Δ to the integral in (13.4.22). We also note that a pure time delay, τ, in continuous time maps to τ/Δ poles at $-1/\Delta$ in discrete time. Hence the relative degree increases by τ/Δ. These extra zeros at ∞ (in the γ plane) must be considered in the discrete time case. They contribute to the last term in (13.4.24) and increase its value by $(\tau/\Delta)(\pi\Delta/2)$. This is exactly equal to the last term in (13.4.22), which describes the contribution due to the time delay in continuous time.

Next we interpret the result in Corollary 13.4.1 in terms of a bandwidth constraint. To do this, we first note that the general shape of the function

$$\overline{T}(v) \triangleq \left| T\left(\frac{1}{(\Delta/2) - jv} \right) \right|$$

is as shown in Figure 13.4.6.

Approximate bounds for $\overline{T}(v)$ are (from experience) as follows:

$$\overline{T}(v) \leq 1.2 v \omega_B, \qquad \text{for } v\omega_B < 0.75 \tag{13.4.31}$$

$$\overline{T}(v) \leq \frac{v^2 \omega_B^2 + 36}{v^2 \omega_B^2 + 30}, \qquad \text{for } v\omega_B > 2 \tag{13.4.32}$$

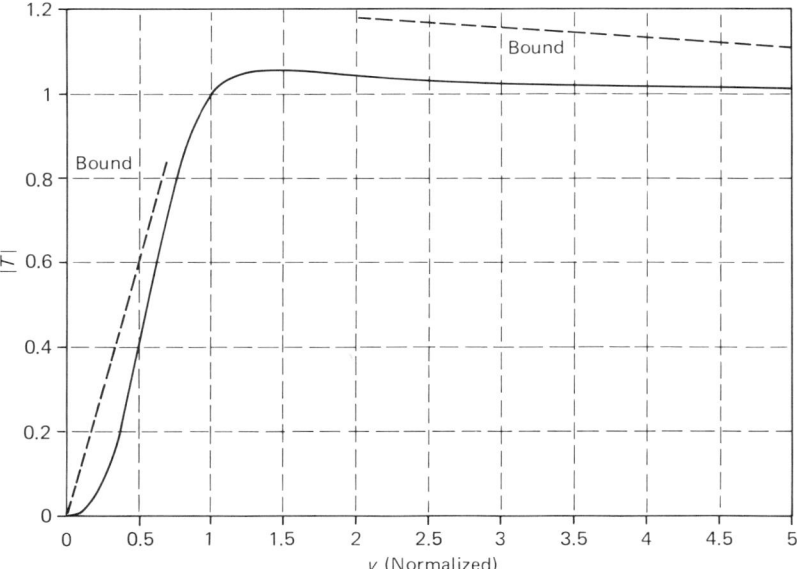

Figure 13.4.6 Approximate shape of $|T|$.

We also assume that H_{ol} falls by at least 20 dB per decade for frequencies below ω_B. Hence

$$|H_0| > \omega_B \tag{13.4.33}$$

For simplicity of exposition, we also introduce the symbol Ω to stand for the following:

$$\Omega \triangleq \sum_{i \in U'} \frac{1}{\xi_i} + \frac{\tau}{2} \qquad \text{(continuous time)} \tag{13.4.34}$$

$$\triangleq (n-m)\frac{\Delta}{2} + \sum_{i \in U'} \left(\frac{1}{\xi_i} + \frac{\Delta}{2}\right) \qquad \text{(discrete time)} \tag{13.4.35}$$

Using the bounds (13.4.31) to (13.4.33) in the integral (13.4.22) or (13.4.24) and splitting the interval of integration gives

$$\int_{1/2\omega_B}^{2/\omega_B} \log \overline{T}(v)\, dv \geq -\frac{\pi}{2\omega_B} + \pi\Omega$$

$$- \int_0^{1/2\omega_B} \log(1.2\omega_B v)\, dv$$

$$- \int_{2/\omega_B}^{\infty} \log\left(\frac{v^2\omega_B^2 + 36}{v^2\omega_B^2 + 30}\right) dv \tag{13.4.36}$$

We evaluate the integrals in the right side of (13.4.36) as follows:

$$\int_0^{1/2\omega_B} \log(1.2\omega_B v)\, dv = \frac{1}{\omega_B}[x \log 1.2x - x]_{x=0}^{x=1/2}$$

$$= -\frac{0.75}{\omega_B} \qquad (13.4.37)$$

$$\int_{2/\omega_B}^{\infty} \log\left(\frac{v^2\omega_B^2 + 36}{v^2\omega_B^2 + 30}\right) dv$$

$$= \left[v \log\left(\frac{v^2\omega_B^2 + 36}{v^2\omega_B^2 + 30}\right) + \frac{12}{\omega_B}\tan^{-1}\left(\frac{v\omega_B}{6}\right) - \frac{2\sqrt{30}}{\omega_B}\tan^{-1}\left(\frac{v\omega_B}{\sqrt{30}}\right)\right]_{2/\omega_B}^{\infty}$$

$$= \frac{1.29}{\omega_B} \qquad (13.4.38)$$

Using (13.4.38) and (13.4.37) in (13.4.36) gives

$$\int_{1/2\omega_B}^{2/\omega_B} \log \overline{T}(v)\, dv \geq -\frac{2.11}{\omega_B} + \pi\Omega \qquad (13.4.39)$$

From (13.4.39), it follows that

$$\sup(\log T(v)) \geq \frac{-(2.11/\omega_B) + \pi\Omega}{(2/\omega_B) - (1/2\omega_B)}$$

$$= -1.4 + 2\omega_B\Omega \qquad (13.4.40)$$

Thus, to keep the peak in $\overline{T}(v)$ below about $\sqrt{2}$, we require

$$\omega_B\Omega \leq 0.9 \qquad (13.4.41)$$

Note that this corresponds very closely to the bandwidth constraint given in (13.3.16) based on time domain arguments.

In the next section we will summarize all the results developed so far, together with some additional considerations, into a set of design guidelines.

13.5 FEEDBACK DESIGN GUIDELINES

The first step in design is to choose an appropriate bandwidth. This choice is subject to the following constraints.

(i) The bandwidth must be large relative to the location of unstable poles [see Section 13.2 and Equation (13.4.20)]. In particular, as discussed in Section 13.4, we suggest

$$\omega_B(1 - \omega_B\Delta) \geq 5 \sum_{i \in U} p_i^c \qquad (13.5.1)$$

where U denotes the set of unstable poles p_i^c (mapped back to continuous time).

(ii) The bandwidth must be small relative to the location of right half-lane zeros ξ_i and to the time delay τ (see Sections 13.3 and 13.4). In particular, we suggest

$$\omega_B \leq \left[\frac{\tau}{2} + \sum_{i \in U'} \left(\frac{1}{\xi_i} + \frac{\Delta}{2} \right) \right]^{-1} \qquad (13.5.2)$$

where U' denotes the set of unstable zeros ξ_i, and τ denotes the time delay in seconds for either discrete or continuous time.

(iii) The bandwidth must be small relative to the frequency, where the relative modeling error $|G_\Delta/G_0|$ approaches 1 (see Section 13.3).

(iv) The bandwidth is constrained by the input amplitude and slew rate limits (see Section 13.3).

Having chosen the bandwidth, we suggest the following additional guidelines:

(v) Any stable, well-damped zeros within the bandwidth should be canceled in the feedback controller (see Example 13.5.1).

(vi) Any stable, well-damped poles within the bandwidth may be canceled provided they are not too close to the origin compared to the bandwidth.

(vii) Let k denote the number of open loop poles in the plant plus compensator that are close to the origin relative to the bandwidth (in the sense that this distance to the origin is less than $\frac{1}{5}\omega_b$); then there should be $k-1$ closed loop poles at stable, well-damped locations whose distance to the origin is about $\frac{1}{5}\omega_b$. (This ensures that the phase shift near the gain cross-over frequency is appropriate.)

(viii) The poles not constrained by rules (v), (vi), and (vii) should be placed at stable, well-damped locations as follows:
- Not more than 2 at ω_b
- The remainder at greater than $3\omega_b$

Again this rule is motivated by phase shift considerations.

(ix) If we end up with a controller that has unstable poles or zeros, then in view of (i) and (ii) this is undesirable and suggests that alternative measurements should be used.

Example 13.5.1 motivates rule (v):

Example 13.5.1

Consider the plant

$$G(\gamma) = \frac{10(\gamma + 0.1)}{\gamma(\gamma + 2)} \qquad (13.5.3)$$

Say we want the closed loop bandwidth to be α.

(a) Zero canceling design

In this case, we cancel $B = 10\gamma + 1$, and the pole assignment equation becomes

$$\gamma(\gamma + 2)L + P = (\gamma + \alpha)^2 \qquad (13.5.4)$$

where degree $L = 0$, degree $P = 1$, and $\Delta = 0.2$ sec. The solution is

$$L = 1, \qquad P = 2(\alpha - 1)\gamma + \alpha^2$$

Thus, the controller has transfer function

$$G_c(\gamma) = \frac{2(\alpha - 1)\gamma + \alpha^2}{10(\gamma + 0.1)} \qquad (13.5.5)$$

which for $\alpha = 1$ is

$$G_c(\gamma) = \frac{1}{10(\gamma + 0.1)}$$

which is a simple lag controller.

(b) Nonzero canceling controller

In this case, the pole assignment equation becomes

$$\gamma(\gamma + 2)L + (10\gamma + 1)P = (\gamma + \alpha)^2(\gamma + 3\alpha) \qquad (13.5.6)$$

where degree $L = 1$, degree $P = 1$. The solution is

$$L = \gamma + l_0, \qquad P = p_1\gamma + p_0 \qquad (13.5.7)$$

where

$$p_0 = 3\alpha^3$$

$$p_1 = \tfrac{3}{2}\alpha^3 + \tfrac{1}{38}(\alpha - 2)^2(3\alpha - 2)$$

$$l_0 = 0.1 - \frac{(\alpha - 0.1)^2(3\alpha - 0.1)}{0.19}$$

which for $\alpha = 1$ gives

$$G_c(\gamma) = \frac{1.53\gamma + 3}{\gamma - 12.26} \qquad (13.5.8)$$

which is an unstable control law!

The following example illustrates rules (i) and (ii).

Example 13.5.2

Here we consider an inverted pendulum system of the type illustrated in Figure 13.5.1. (Note that this is a favorite problem in control textbooks but, unfortunately, suggested solutions often violate the design constraints.) Using Langrangian mechanics (see for example Kibble, 1966), a nonlinear state space model for this

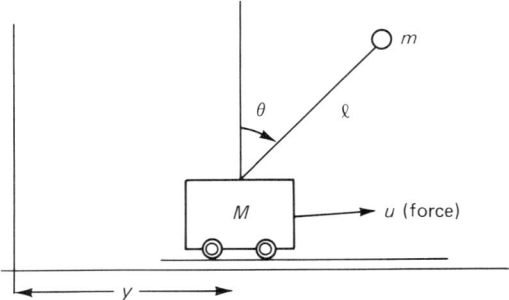

Figure 13.5.1 Inverted pendulum.

system can be shown to be

$$\ddot{y} = \frac{1}{(M/m) + \sin^2\theta}\left[\frac{u}{m} + \dot{\theta}^2 l \sin\theta - g\sin\theta\cos\theta\right] \quad (13.5.9)$$

$$\ddot{\theta} = \frac{1}{l(M/m) + \sin^2\theta}\left[-\frac{u}{m}\cos\theta - \dot{\theta}^2 l \cos\theta \sin\theta + \left(1 + \frac{M}{m}\right)g\sin\theta\right] \quad (13.5.10)$$

where g denotes the acceleration due to gravity. A linearized model for this system about the origin is readily seen to be

$$\frac{d}{dt}\begin{pmatrix}x_1\\x_2\\x_3\\x_4\end{pmatrix} = \begin{pmatrix}0 & 1 & 0 & 0\\0 & 0 & \frac{-mg}{M} & 0\\0 & 0 & 0 & 1\\0 & 0 & \frac{(M+m)g}{Ml} & 0\end{pmatrix}\begin{pmatrix}x_1\\x_2\\x_3\\x_4\end{pmatrix} + \begin{pmatrix}0\\\frac{1}{M}\\0\\-\frac{1}{Ml}\end{pmatrix}u \quad (13.5.11)$$

$$y = \begin{pmatrix}1 & 0 & 0 & 0\end{pmatrix}\begin{pmatrix}x_1\\x_2\\x_3\\x_4\end{pmatrix} \quad (13.5.12)$$

where $(x_1 \; x_2 \; x_3 \; x_4) = (y, \; \dot{y}, \; \theta, \; \dot{\theta})$.

It is relatively easy to show that this state space model is both controllable and observable (provided $M, m > 0$), and so a controller based on a state observer plus state feedback is possible. We will carry out the design in polynomial form. From (13.5.11), (13.5.12), the transfer function from u to y is

$$\frac{Y(s)}{U(s)} = \frac{K(s-b)(s+b)}{s^2(s-a)(s+a)} \quad (13.5.13)$$

where

$$K = \frac{1}{M} \quad (13.5.14)$$

$$b = +\sqrt{\frac{g}{l}} \quad (13.5.15)$$

and

$$a = \sqrt{\frac{(M+m)g}{Ml}} \quad (13.5.16)$$

Note that, in view of rules (i) and (ii), this transfer function is impractical to work with. First, it is both unstable and nonminimum phase. Second, the unstable zero occurs at a lower frequency than the unstable pole. However, the design guidelines given in (i) and (ii) suggest that the closed loop bandwidth should be above the unstable pole but below the unstable zero. This is clearly impossible in this case.

We can still design a controller for this case; however, the preceding considerations indicate that the controller will have very poor performance. To illustrate this, consider the case where $l = 1$ m, $g = 10$ ms^{-2}, and $M = m = 0.5$ kg. In this case $K = 2$, $b = \sqrt{10}$, and $a = \sqrt{20}$. Since this system has an unstable pole at $s = a = \sqrt{20}$, the closed loop bandwidth should be $\omega_B \geq 4a$. If we then use the rule of thumb $\omega_2 \approx 10\omega_B$, we suggest a sampling rate of about 25 Hz. With $\Delta = 0.04$ sec, the discrete time transfer function is

$$\frac{Y(\gamma)}{U(\gamma)} = \frac{0.0401(\gamma + 50)(\gamma + 2.9704)(\gamma - 3.3710)}{\gamma^2(\gamma + 4.0950)(\gamma - 4.8971)} \quad (13.5.17)$$

If we then assign the closed loop poles to -2.9704 (canceling the stable zero), -4.0950 (canceling the stable pole), -25 (shifting the zero at -50), -3.0 (reflecting the zero at $+3.3710$), $-0.5 \pm j0.5$ (shifting slightly the two integrators), and -15 (for the remaining closed loop pole), we obtain the following control law:

$$\frac{P(\gamma)}{L(\gamma)} = \frac{1764.7(\gamma - 0.2907)(\gamma + 0.1624)(\gamma + 4.0950)}{(\gamma - 65.4235)(\gamma + 43.6382)(\gamma + 2.9704)} \quad (13.5.18)$$

Note that this controller has unstable poles and zeros and thus violates guideline (ix). In fact a simple root locus argument shows that any controller that stabilizes the system must have this characteristic (see Problem 11). Figure 13.5.2 shows the step response when using this control on the (nonlinear) system in a unity feedback configuration, with a 0.001-m step in y^*. Note that a step in y^* of more than 0.063 m results in sufficient nonlinearity to cause the system to be unstable. Note also that if the actual mass of the cart was varied, while retaining all other parameters exact, local stability could only be retained provided

$$0.422 \text{ kg} < M < 0.561 \text{ kg} \quad (13.5.19)$$

These results are clearly unacceptable from a practical point of view. The reader may gain some appreciation of the difficulty of the control problem when only position is measured by trying to balance a broom with both eyes shut!

In view of the poor performance described, the best option is to consider this system as a single-input, two-output system, that is, to seek additional state measurements. Suppose, for example, we retain $y_1 = y$ but introduce an additional measurement y_2. Let $G_1 = N_1/d$, $G_2 = N_2/d$ denote the transfer functions from u to y_1 and y_2, respectively. We now propose a multiloop feedback structure as illustrated in Figure 13.5.3. The feedback from y_2 to u can be used to modify the

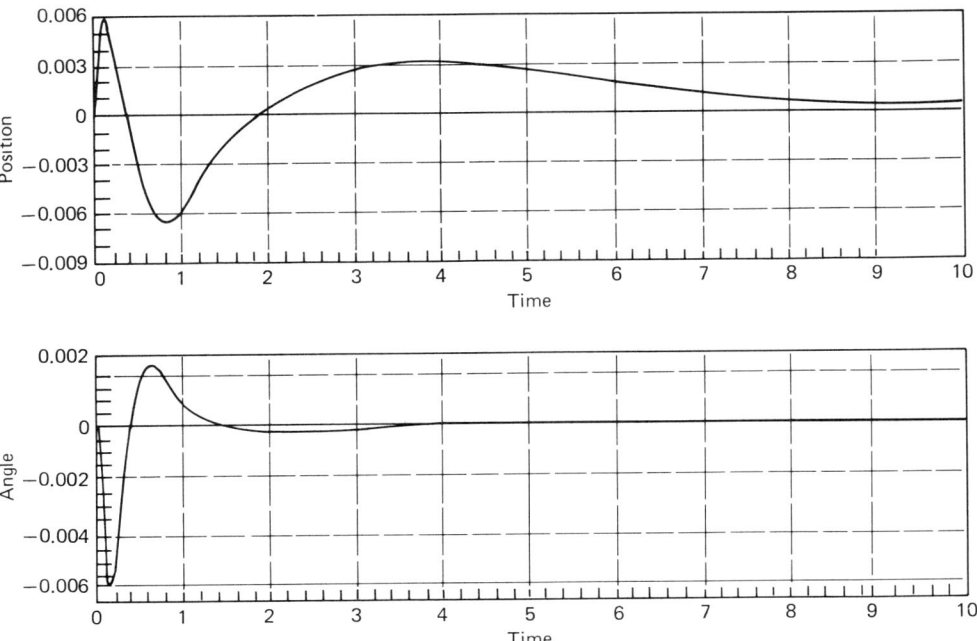

Figure 13.5.2 Step response inverted pendulum using feedback from y (position) only.

transfer function as seen from u to y_1. For example, if we use

$$u = -k_2 y_2 + v \tag{13.5.20}$$

then the closed loop transfer function from v to y_1 becomes

$$y_1 = \frac{N_1}{d + N_2 k_2} v \tag{13.5.21}$$

In the case of the inverted pendulum, a "natural" additional measurement is the angle, θ. Thus we let $y_2 = \theta$. We have already seen that in terms of the notation

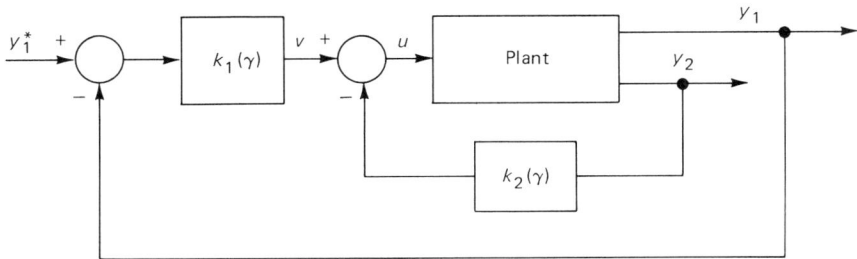

Figure 13.5.3 Cascade design for inverted pendulum.

of (13.5.21)

$$d = s^2(s-a)(s+a) \tag{13.5.22}$$
$$N_1 = K(s-b)(s+b) \tag{13.5.23}$$

We can also evaluate

$$N_2 = -\frac{1}{Ml}s^2 \tag{13.5.24}$$

The transfer function from u to y_2 is then

$$\frac{N_2}{d} = \frac{-1/Ml}{(s-a)(s+a)} = \frac{-1/Ml}{s^2 - (g/l)[1+(M/m)]} \tag{13.5.25}$$

which is the transfer function for a simple (stationary hinge) pendulum.

If we again use a 25-Hz sampling rate, we obtain the following discrete time transfer functions:

$$\frac{N_1(\gamma)}{d(\gamma)} = \frac{0.0401(\gamma+50)(\gamma+2.9704)(\gamma-3.3710)}{\gamma^2(\gamma+4.0950)(\gamma-4.8971)} \tag{13.5.26}$$

(as in 13.5.17) and

$$\frac{N_2(\gamma)}{d(\gamma)} = \frac{-0.0401(\gamma+50)\gamma^2}{\gamma^2(\gamma+4.0950)(\gamma-4.8971)} \tag{13.5.27}$$

Note that the double pole at the origin arising from the acceleration of the cart is unobservable in y_2. However, we can easily design feedback from y_2 to u so as to shift the other two poles well into the left half plane.

In our design of this inner feedback loop we shall ignore the zero due to the sampling process ($\gamma = -50$), which leaves us with an approximate transfer function,

$$\frac{\theta(\gamma)}{U(\gamma)} \approx \frac{-2.0053}{(\gamma+4.0950)(\gamma-4.8971)} \tag{13.5.28}$$

If we then assign the closed loop poles to -4.0950 (canceling the stable pole) and $-12 \pm j5$, we obtain the following control law:

$$k_2(\gamma) = \frac{-154.8457(\gamma+4.0950)}{\gamma+28.8971} \tag{13.5.29}$$

As a check, we note that this controller gives closed loop poles at $\gamma = -4.0950$, -7.4134, and -22.7968 when we use $N_2(\gamma)/d(\gamma)$ rather than the approximation in (13.5.28). The transfer function from v to y_1 then becomes (in view of 13.5.21)

$$\frac{y_1(\gamma)}{v(\gamma)} = \frac{0.0401(\gamma+50)(\gamma+28.8971)(\gamma+2.9704)(\gamma-3.3710)}{\gamma^2(\gamma+4.0950)(\gamma+7.4134)(\gamma+22.7968)} \tag{13.5.30}$$

Clearly, this gives an easier design problem. Note that the double integrator (which is unobservable from θ) could not be shifted; however, the unstable pole has been shifted by the inner feedback loop. Note also that the unstable zero is still present.

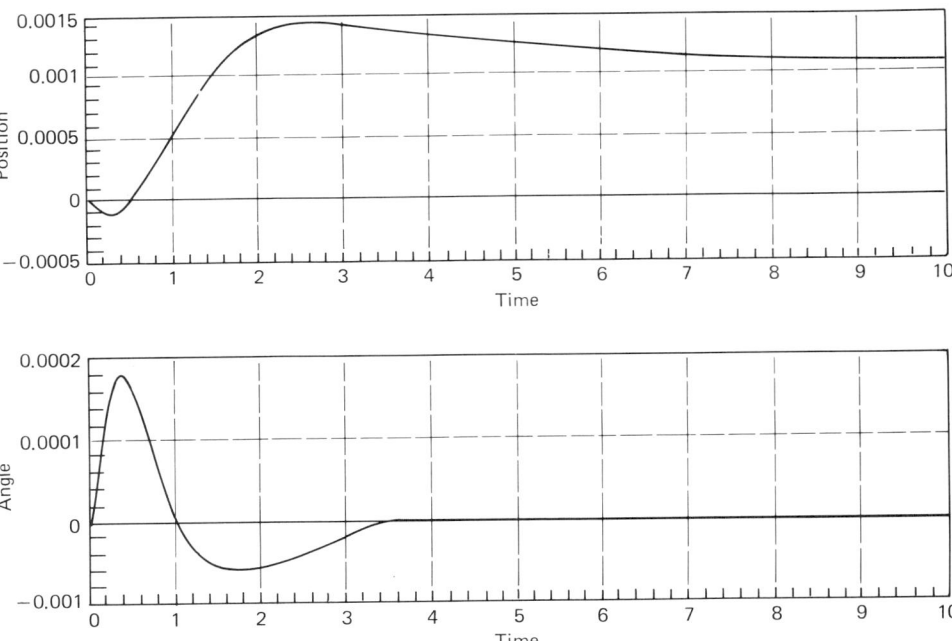

Figure 13.5.4 Step response of inverted pendulum with controller based on θ and y.

In this case, design guideline (ii) suggests that the bandwidth of the outer feedback loop should be ≤ 3 rad/sec.

This further suggests that the outer loop need only have about a 5-Hz sampling rate. If we resample the system given in (13.5.30) (where $f_s = 25$ Hz) at 5 Hz and use a discrete time zero order hold, the transfer function (at the slow rate) is (see Problem 9)

$$\frac{y_1(\gamma)}{v(\gamma)} = 0.1292 \frac{(\gamma + 7.5950)(\gamma + 5.0265)(\gamma + 2.3438)(\gamma - 4.4311)}{\gamma^2(\gamma + 5)(\gamma + 4.1386)(\gamma + 2.9559)} \quad (13.5.31)$$

$$\approx \frac{0.2371(\gamma + 2.3488)(\gamma - 4.4311)}{\gamma^2(\gamma + 2.9559)} \quad (13.5.32)$$

From (13.5.32) we suggest the following control law:

$$K_1(\gamma) = \frac{-3.8073(\gamma + 2.9559)(\gamma + 0.2)}{(\gamma + 2.3438)(\gamma + 4.9027)} \quad (13.5.33)$$

Once again, we check the exact closed loop poles using (13.5.31), which in this case are at $\gamma = -5.8657, -5.0274, -2.9559, -2.3438, -1.3160 \pm j0.3903$, and -0.2995. Figure 13.5.4 shows the step response for this multirate, multioutput digital controller. Note that, with this control law, mass values in the range

0.105 < M < 1.05 kg will give local stability, and with $M = 0.5$ kg, step changes in y^* of up to 4.35 m still give stability of the nonlinear feedback system! This is clearly orders of magnitude better than the controller designed without measurement of θ and confirms the veracity of design rules (i) and (ii).

13.6 FEEDFORWARD

So far in this chapter we have considered only the design of the feedback subsection of the control system. To complete the design, it is usually necessary to also include feedforward components. Thus we generalize the elementary feedback system of Figure 13.2.1 to the three degree of freedom controller shown in Figure 13.6.1, which includes feedforward elements from the set point and measured parts of the disturbance. In Figure 13.6.1, the symbols have the following interpretation.

y^*	desired output; that is, set-point for y
e	error signal
u	plant input
d	output disturbance (unmeasured)
d_m	disturbance (measured)
y	plant output
v	measurement errors
y_m	measured plant output
H_{FF}	set-point feedforward compensator
H_{DF}	disturbance feedforward compensator
C	error compensator
G	true plant transfer function
G_d	true disturbance transfer function

The design rules given in Section 13.5 for the error compensator C aim to achieve as wide a closed loop bandwidth as is possible within the constraints

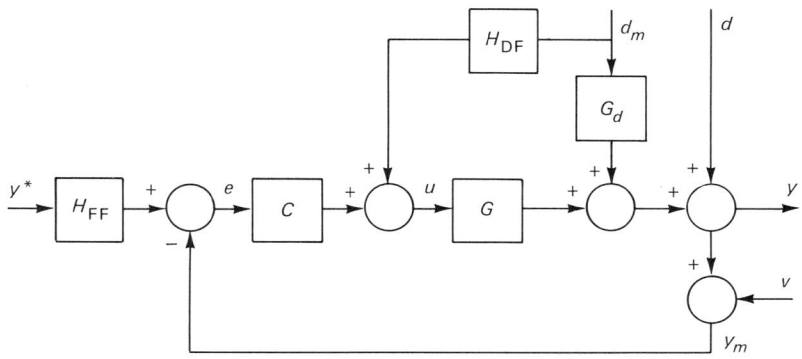

Figure 13.6.1 General control structure.

imposed by model uncertainty, unstable zeros, measurement noise, and the like. Note that this specification of closed loop bandwidth is *independent* of the actual bandwidth required between the set point y^* and the output y. The effect of having a wide bandwidth is that, *inter alia*, the effect of variations in the plant transfer function are reduced; see (13.2.3).

Having specified the achievable closed loop bandwidth ω_b, then we may use set point feedforward (H_{FF} in Figure 13.6.1) to shape the overall tracking response to the desired input–output bandwidth (ω_b^*, say). If $\omega_b^* < \omega_b$, then the feedforward compensator leads to a further reduction in the absolute effect of the uncertainty over the frequency range (ω_b^*, ω_b). On the other hand, if ω_b^* is greater than ω_b, then H_{FF} should ideally become $(1 + GC)/GC$ to give an overall unity transfer function from y^* to y. For frequencies well above b_ω, $|GC|$ will be small compared to 1, giving the desired value of H_{FF} as

$$H_{FF} \simeq \frac{1}{G_0 C}, \qquad \text{for } \omega \in (\omega_b, \omega_b^*) \tag{13.6.1}$$

where G_0 is the nominal value of the plant transfer function. If the true plant transfer function is as in (13.3.7), that is,

$$G = G_0 + G_\Delta \tag{13.6.2}$$

where G_Δ is an absolute error, then the actual transfer function from y^* to y is

$$\frac{Y}{Y^*} = \frac{H_{FF} GC}{1 + GC} \tag{13.6.3}$$

Using (13.6.1) and (13.6.3), we see that for $|GC| < 1$

$$\frac{Y}{Y^*} \simeq 1 + \frac{G_\Delta}{G_0}, \qquad \omega \in (\omega_b, \omega_b^*) \tag{13.6.4}$$

Thus the open loop uncertainty G_Δ is magnified by $1/G_0$ over the range ω_b to ω_b^*. Typically, $|G_\Delta|$ will be constant up to a very wide bandwidth, while $|G_0|$ will tend to decrease with frequency. Hence, using set point feedforward to raise the bandwidth tends to magnify the effect of plant modeling errors.

The net result of the preceding considerations is that the uncertainty in the overall input–output transfer function from y^* to y has the general characteristic shown in Figure 13.6.2.

Next we consider the effect of disturbances on output tracking. If the frequency content of the disturbances is well within the closed loop bandwidth, then it follows from (13.2.1) that they will have a small effect on the output response. On the other hand, if the frequency content of the disturbance is not small compared with the closed loop bandwidth, then disturbance feedforward may be helpful. From Figure 13.6.1 we see that the measured component of the disturbance is removed from the output if we choose the disturbance feedforward transfer function as

$$H_{DF} \simeq -\frac{G_d}{G_0} \tag{13.6.5}$$

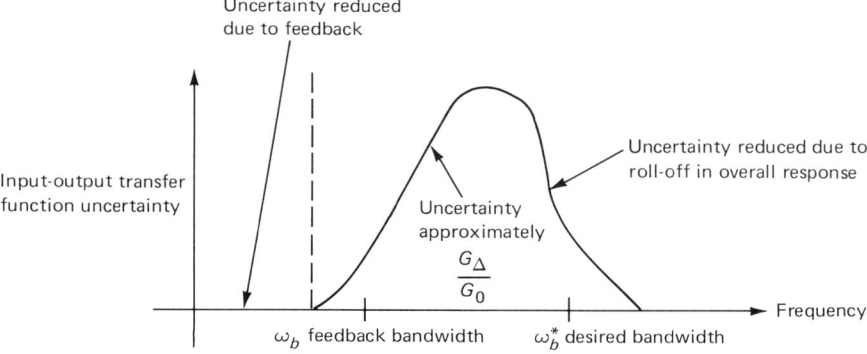

Figure 13.6.2 Overall uncertainty.

As for set-point feedforward, errors in the plant transfer function will be magnified by the operation given in (13.6.5). Indeed, we see from (13.6.1) and (13.6.5) that both set-point and disturbance feedforward essentially amount to inverting the (nominal) plant. Two observations follow: (1) the inverse must be stable and hence any unstable zeros in G_0 (which are not also unable zeros in G_d) must first be stabilized, for example, by reflection through the stability boundary, and (2) the efficacy of feedforward is critically dependent on model fidelity. Note that plant inversion and model fidelity are only required over the range of set-point (and disturbance) frequencies present. This last point will be examined further in Example 13.6.3.

Some of the preceding considerations are illustrated in the following example.

Example 13.6.1

In this example, we consider a situation where feedforward may be used to increase the tracking bandwidth ω_b^*. Suppose we have a nominal continuous time plant having transfer function

$$G(s) = \frac{Ke^{-sT}}{s+1} \tag{13.6.6}$$

where $K \in (0.5, 1)$ and $T \in (0.8, 1)$. For design purposes, we choose a nominal system with $K = 1$, $T = 1$. The system is sampled with a 200-ms period, leading to the following nominal discrete time system:

$$G(\gamma) = \frac{0.906}{(1 + \gamma\Delta)^5(\gamma + 0.906)} \tag{13.6.7}$$

Using design guideline (ii), we have that the feedback bandwidth b_ω should satisfy

$$\omega_b \leq \frac{2}{6\Delta} \simeq 1.6 \tag{13.6.8}$$

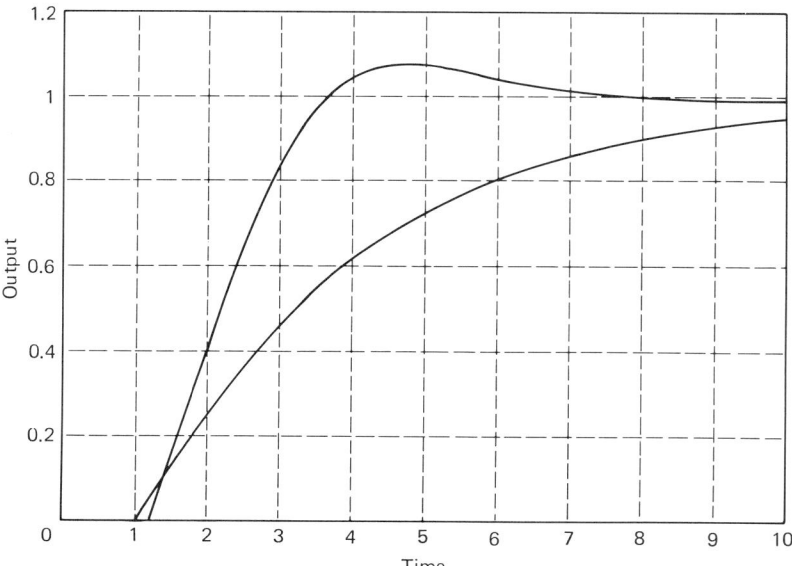

Figure 13.6.3 Step response with no FF compensation.

A feedback bandwidth of 1 rad sec^{-1} is achieved by the following error compensator:

$$C(\gamma) = \frac{0.55\gamma + 0.5}{\gamma} \qquad (13.6.9)$$

Figure 13.6.3 shows the step responses if the compensator in (13.6.9) is used on the nominal plant (upper trace) and also if $K = 0.5$, $T = 0.8$ (lower trace). Even when the plant is equal to the nominal plant, the settling time is about 6 sec. If we require a more rapid response, then feedforward may be helpful. We note that the dominant poles of the nominal closed loop system are at $-0.68 \pm j0.7$. One way of speeding up the response is to cancel these by a set-point feedforward compensator of the following form:

$$H_{FF}(\gamma) = \frac{4\gamma^2 + 5.44\gamma + 4}{\gamma^2 + 4\gamma + 4} \qquad (13.6.10)$$

The resulting overall response for the nominal system is shown in Figure 13.6.4 (upper trace). The settling time has been reduced to about 3 sec and overshoot has been eliminated. The sensitivity of the feedforward action to the model is also shown in Figure 13.6.4, where the lower trace shows the overall response when $K = 0.5$, $T = 0.8$. Note that in the latter case there is very little improvement using feedforward over the use of feedback alone. This verifies the sensitivity of feedforward design to plant variations.   ▽▽▽

444   Design Considerations   Chap. 13

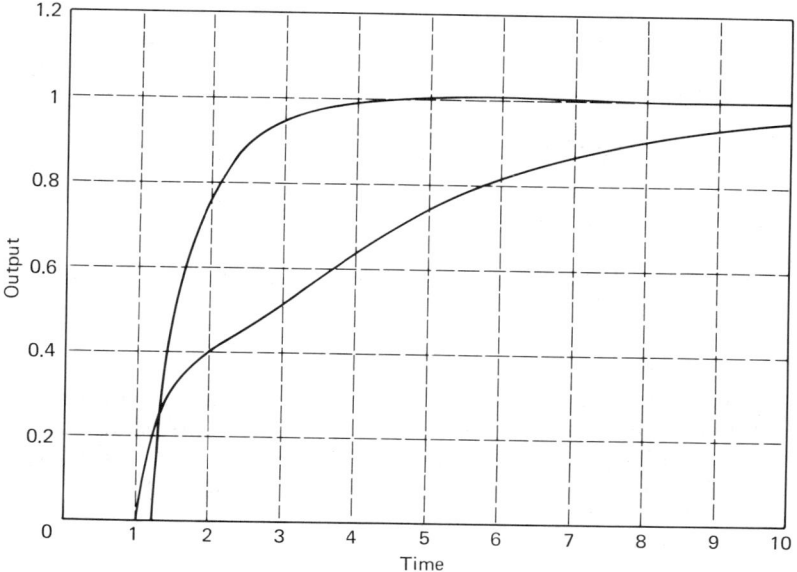

Figure 13.6.4 Step response with FF compensation.

Example 13.6.2

This example illustrates a situation where the closed loop bandwidth ω_b can be made much greater than the desired tracking bandwidth ω_b^*. The problem was suggested by Michael Masten (Texas Instruments) and Herb Cohen (U.S. Army Mathematics Steering Committee) as a test problem for advanced control ideas in a special session at the 1988 American Control Conference.

The plant transfer function is

$$G(s) = \frac{K}{s^2 + a_2 s + a_1}$$

where the parameters can take any value in the following ranges:

$K \in (0.5, 3.0)$

$a_1 \in (-2.0, 4.0)$

$a_2 \in (-0.6, 3.4)$

The set point y^* is a square wave as shown in Figure 13.6.5. Also, the system has an unmeasured input disturbance $d(t)$ of the form also shown in Figure 13.6.5.

The desired output response is specified such that the final transfer function from y^* to y should be

$$\frac{Y}{Y^*} = \frac{1}{s^2 + 1.4s + 1} \qquad (13.6.11)$$

Considering a digital control solution, a sampling period of 10 ms was chosen.

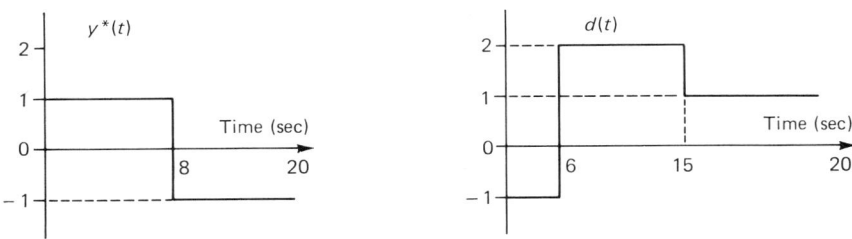

Figure 13.6.5 Set-point and input disturbance for Example 13.6.2.

In view of the fact that the relative degree of the system is 2, the sign of the high-frequency gain is known, there is no measurement noise, and there is no undermodeling, it is quite easy to design a very wide bandwidth controller. For the sake of illustration, we choose a relatively conservative value specified by the following desired closed loop characteristic polynomial:

$$A^* = (\delta^2 + 12.8\delta + 64)(\delta + 10)(\delta + 7)^2 \qquad (13.6.12)$$

The design was carried out for the nominal system where $K = 1$, $a_1 = 1$, and

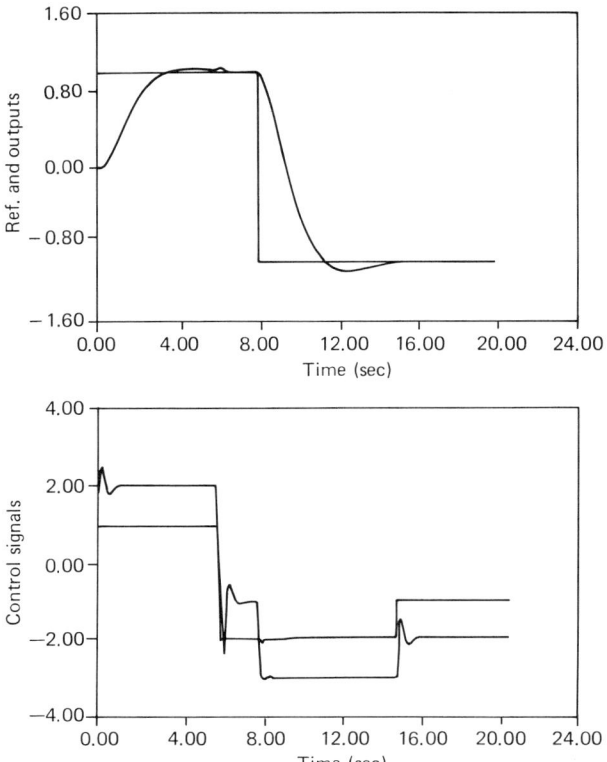

Figure 13.6.6 Input and output signals for various plants in Example 13.6.2.

$a_2 = 1.4$. This leads to the following error compensator:

$$C(\gamma) = \frac{506\gamma^3 + 4405\gamma^2 + 18364\gamma + 31360}{\gamma(\gamma^2 + 35\gamma + 3.5)} \triangleq \frac{P(\gamma)}{L(\gamma)} \qquad (13.6.13)$$

The overall response is then shaped by the following feedforward compensator, which cancels the closed loop poles and zeros and adds the desired reference model given in (13.6.11):

$$H_{FF}(\gamma) = \frac{A^*(\gamma)}{P(\gamma)(\gamma^2 + 1.4\gamma + 1)} \qquad (13.6.14)$$

Since ω_b^* is much less than ω_b, we would expect considerable insensitivity to plant variations and disturbances. This is borne out in Figure 13.6.6, which shows three curves. These are the reference, the closed loop response when the plant is the nominal system, that is, $G(s) = 1/(s^2 + 1.4s + 1)$, and the closed loop response when the plant is changed to $G'(s) = 50/(s^2 - 0.6s - 2)$. Note that the latter two curves are essentially coincident, indicating the very large insensitivity of the design to parameter variations. Figure 13.6.6 shows the corresponding input signals, which indicates that this result is achieved with a completely reasonable control effort. ▽▽▽

Example 13.6.3

In this example we consider a position servo loop, where both tacho and position signals are available. The basic setup is shown in Figure 13.6.7; it corresponds closely to a dc motor position servo with the following simplifications: we have neglected friction terms, we assume that the dc motor is current driven (this is usually achieved in practice by an inner, high-bandwidth, current loop), we have neglected the difference between a discrete time integrator and the real continuous time integration, and we have neglected gearbox compliance.

Suppose also that the output θ is required to perfectly track set points θ^*, which consist of parabolic terms, and that the bandwidth of the feedback loops is limited by unmodeled effects and the like to (say) 0.5 Hz. (Note that the problem as posed is very close to the problem of controlling a large radio telescope. In this case, structural resonances will limit the bandwidth of the feedback loop.)

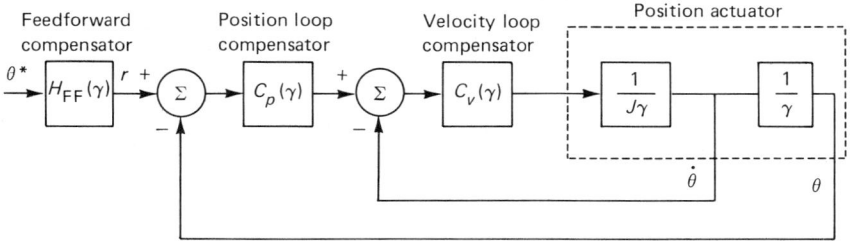

Figure 13.6.7 Position servo control loop.

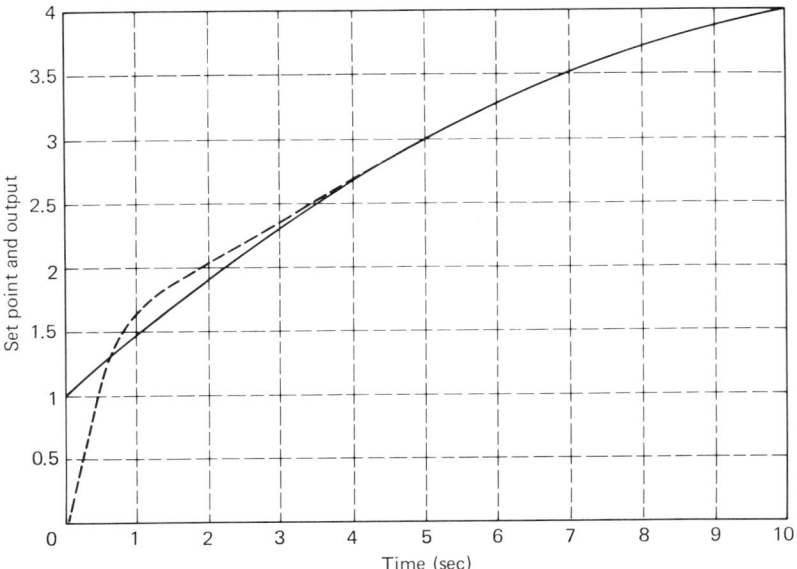

Figure 13.6.8 Feedback only tracking solution.

The pure feedback solution of this problem [that is, the solution with $H_{FF}(\gamma) = 1$] would be to force $C_p(\gamma)$ to have two integrators, so that the position loop has three integrators, which allows perfect tracking of a parabola. Given a nominal value, J_0, for J, a suitable feedback solution is

$$C_v(\gamma) = 3J_0 \tag{13.6.15}$$

and

$$C_p(\gamma) = \frac{(\gamma + 3)(\gamma + 0.5)^2}{\gamma^2(0.1\gamma + 1)} \tag{13.6.16}$$

For $J = J_0$, these controllers give a closed loop transfer function of (see Problem 13.12)

$$\frac{\theta(\gamma)}{\theta^*(\gamma)} = \frac{3\gamma^2 + 3\gamma + 0.75}{0.1\gamma^4 + \gamma^3 + 3\gamma^2 + 3\gamma + 0.75}$$

$$= \frac{1}{(0.03\gamma^2 + 0.28\gamma + 0.52)} \frac{(3\gamma^2 + 3\gamma + 0.75)}{(3\gamma^2 + 4.97\gamma + 1.43)} \tag{13.6.17}$$

Note that in this case the design of $C_p(\gamma)$ is a little difficult because of the three integrators. Figure 13.6.8 shows the tracking response when the reference input is

$$\theta^*(t) = 1 + 0.5t - 0.02t^2$$

and

$$\Delta = 0 \quad \text{(continuous time)}$$

For the feedforward solution, we use only one integrator in $C_p(\gamma)$ and design $H_{FF}(\gamma)$ to give the desired parabolic tracking. Note that if J is known we can also design $C_p(\gamma)$ with no integrators and use $H_{FF}(\gamma)$ to give the desired tracking response (see Problem 13). In this example we consider $C_p(\gamma)$ to have one integrator, for example,

$$C_p(\gamma) = \frac{(\gamma + 3)(\gamma + 0.5)}{\gamma(0.1\gamma + 1)} \tag{13.6.18}$$

which gives a closed loop transfer function from r to θ of

$$\frac{\theta(\gamma)}{R(\gamma)} = \frac{3\gamma + 1.5}{0.1\gamma^3 + \gamma^2 + 3\gamma + 1.5} \tag{13.6.19}$$

We now propose a feedforward compensator of the form

$$H_{FF}(\gamma) = \frac{n_2\gamma^2 + n_1\gamma + 1}{d_2\gamma^2 + d_1\gamma + 1} \tag{13.6.20}$$

which leads to an overall closed loop transfer function of

$$\frac{\theta(\gamma)}{\theta^*(\gamma)} = \frac{1.5 + (1.5n_1 + 3)\gamma + (1.5n_2 + 3n_1)\gamma^2 + \cdots}{1.5 + (1.5d_1 + 3)\gamma + (1.5d_2 + 3d_1 + 1)\gamma^2 + \cdots} \tag{13.6.21}$$

As discussed in Chapter 4, a closed loop system will track an nth-order polynomial set point with zero error if and only if the n lowest-order coefficients of the

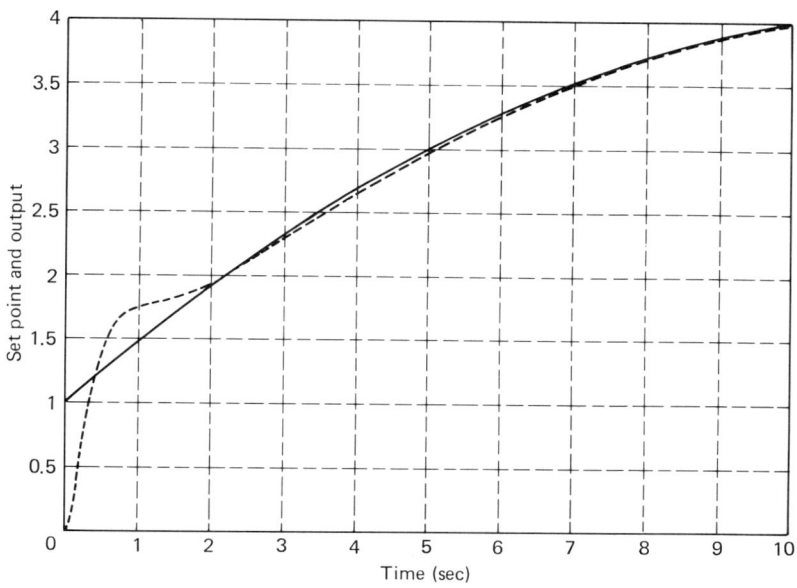

Figure 13.6.9 Tracking using feedback and feedforward.

numerator match the corresponding coefficients in the denominator of the transfer function. This rule applied to (13.6.21) gives

$$n_1 = d_1 \tag{13.6.22}$$

and

$$n_2 = d_2 + 0.6667 \tag{13.6.23}$$

Thus, for example, we could choose $n_1 = d_1 = 2$, $d_2 = 1$, and $n_2 = 1.6667$ in our feedforward compensator (13.6.20). Using this feedforward compensator with the feedback controller given in (13.6.18) gives a tracking response as shown in Figure 13.6.9.

Note from Figures 13.6.8 and 13.6.9 that the tracking performance is about the same for the pure feedback and the feedforward solutions. The feedforward solution introduces less integrators in the feedback loop, which is desirable for ease of control. The price paid for this is that the feedforward solution is more sensitive to plant uncertainty.

13.7 ON-LINE ESTIMATION

In the previous section, we argued that the overall design of a control system requires the use of 3 degrees of freedom: error compensator, set-point feedforward compensator, and disturbance feedforward compensator. Our ability to successfully design these compensators depends on the fidelity of the plant model, particularly in the case of feedforward. Large plant modeling errors significantly limit the achievable performance, for example, by reducing the achievable closed loop bandwidth (see Section 13.3) and/or by reducing the efficacy of feedforward compensation (see Section 13.6). If model uncertainty means that the achievable performance falls short of the design goal, then one option is to use on-line parameter estimation to improve model confidence.

As argued in Section 13.6, we are mainly interested in improving model fidelity in the range ω_b to ω_b^*. Hence, to focus the parameter estimator on the appropriate bandwidth, it is desirable to first prefilter all signals to this range by a bandpass filter.

If estimation is not employed, then the (robust) controller must cope with the *global uncertainty*, G_Δ^g, which includes

- Parameter errors
- Parameter time variations
- Undermodeling

If on-line estimation is used then, in principle, we can reduce the contributions to the error arising from unknown parameters or parameter time variations. Thus we are left with only two sources of error, undermodeling and the uncertainty in the estimator as discussed in Section 12.8.2. Actually, the best arrangement would be for the control system design to be based on a judicious combination of the global uncertainty G_Δ^g and the estimator uncertainty G_Δ^l, as shown in Figure 12.10.1. This kind of scheme combines prior knowledge about the plant with knowledge obtained from on-line observations.

Referring to the examples given in the previous section, Example 13.6.1 would be helped by on-line estimation since in this case the design of the feedforward compensator is sensitive to plant variations. On the other hand, Example 13.6.2 does not call for on-line estimation since excellent performance can be achieved without estimation. Furthermore, the size of the unmeasured input disturbance is such that the uncertainty of the estimator will be very large.

13.8 SUMMARY

The key points covered in this chapter were:

- The benefits of high gain feedback have been discussed and include:
  - Good tracking
  - Good disturbance rejection
  - Low sensitivity to plant variations
  - Good transient performance in the presence of unstable open loop poles.
- Factors limiting feedback bandwidth were discussed in Section 13.3 and include:
  - Measurement noise
  - Input amplitude and slew rate limits
  - Stability problems arising from plant uncertainty
  - Poor transient behavior associated with unstable zeros
- Frequency domain sensitivity functions were considered in Section 13.4, leading to the following design guidelines:

$$\omega_B(1 - \omega_B\Delta) \geq 5 \sum_{i \in U'} p_i^c \tag{13.5.1}$$

$$\omega_B \leq \left[\frac{\tau}{2} + \sum_{i \in U'}\left(\frac{1}{\xi_i} + \frac{\Delta}{2}\right)\right]^{-1} \tag{13.5.2}$$

where ω_B is the feedback bandwidth, p_i^c denote the unstable poles translated to continuous time, τ denotes the system time delay, and ξ_i denotes unstable zeros.
- Various other feedback design guidelines were discussed in Section 13.5.
- Feedforward design considerations were covered in Section 13.6.
- In cases where the feedback bandwidth is lower than that desired for set-point tracking, feedforward can be used to reduce the bandwidth. This frequently results in reduced sensitivity and a very robust design.
- In cases where the achievable performance cannot be attained through a fixed linear controller, on-line estimation may be helpful in reducing the plant uncertainty.

13.9 REFERENCES

Additional material on design may be found in many texts, including:

BODE, H. W. (1945) *Network Analysis and Feedback Amplifier Design*. Van Nostrand, Reinhold, New York.

FREUDENBERG, J., and D. P. LOOZE (1988) *Frequency Domain Properties of Scalar and Multivariable Feedback Systems*. Springer-Verlag, New York.

HOROWITZ, I. M. (1963) *Synthesis of Feedback Systems*. Academic Press, New York.

An elementary book on Lagrangian Mechanics is

KIBBLE, T. W. B. (1966) *Classical Mechanics*. McGraw-Hill, London.

13.10 PROBLEMS

1. Show that (13.2.12) follows from (13.2.11). *Hints*:
   (i) Show that
   $$\rho\left\{-\frac{(1+\alpha\Delta)}{\alpha}E(\alpha,-t)\right\} = E(\alpha,-t)$$
   and
   $$\rho\left\{-\frac{(\alpha t+1)}{\alpha}\frac{(1+\alpha\Delta)}{\alpha}E(\alpha,-t)\right\} = tE(\alpha,-t)$$
   (ii) Use (a) to evaluate the integrals in (13.2.11).
   (iii) Note that $E(\alpha, t_r) \geq 1 + \alpha t_r$.

2. Consider a discrete time plant, $\delta^2 y = (\delta - 1)u$, $\Delta = 0.1$. Design a controller of the form $(\delta + l_0)u = (p_1\delta + p_0)(y^* - y)$ to give various closed loop bandwidths. Compare the step response, gain margin, and phase margin of the various designs, and comment on the bandwidth limitation imposed by an unstable zero.

3. Consider a discrete time plant, $(\delta + 1)y = u$, $\Delta = 0.1$. Design a controller of the form $\delta u = (p_1\delta + p_0)(y^* - y)$ to give a closed loop bandwidth of 4 rad/sec. Find the gain margin, phase margin, and step response, and comment.

4. Consider a plant, $\delta^3 y = u$, $\Delta = 0.1$ with controller
   $$(\delta^2 + l_1\delta + l_0)u = (p_2\delta^2 + p_1\delta + p_0)(y^* - y)$$
   Compare two strategies for assigning the closed loop poles:
   (i) $A^* = (\delta + 5)^5$
   (ii) $A^* = (\delta + 5)^3(\delta + 0.5)^2$
   Comment on the difference in the light of design guideline (vii) of Section 13.5.

5. Suppose that the set point y^* is generated by a computer, and thus we know future values of y^*. For the case where B is stable and well damped, show how the advanced information about y^* can be used. *Hint*: If we know y^* in advance, the feedforward compensator need not be proper.

6. Consider an unstable continuous time system, with transfer function, $G(s) = (1/(s-1))$. Design several control laws using the following guidelines. Compare the designs.

(a) Continuous time controller with integral action to give a closed loop bandwidth of 5 rad/sec.

(b) As for part (a), except with a closed loop bandwidth of 1 rad/sec.

(c) Discrete time controller ($\Delta = 0.1$) with integral action. Assign the closed loop discrete time poles to $\gamma = -5$ and -0.5.

(d) Discrete time controller ($\Delta = 0.5$) with integral action, and closed loop poles assigned to -2 and -2. *Note:* $-2 = -1/\Delta$. This type of controller, where all closed loop poles are assigned to $-1/\Delta$ in the δ domain, that is, the origin in the z domain, is called deadbeat control.

7. The essential components of an overhead traveling crane are shown in Figure 13.10.P7. A trolley of mass m moves horizontally without friction on a set of rails; it is connected by a rope system to drive machinery that can apply a horizontal force $u(t)$ to the trolley. A load of mass M is supported from the trolley drum by a rope of length L. The horizontal positions of the trolley and load from some fixed vertical reference line are $v(t)$ and $z(t)$, respectively; the angle of the load rope with the vertical is $\theta(t)$. If it is assumed that L remains constant and that $\theta(t)$ is always small enough to permit the approximations $\sin\theta \simeq \theta$ and $\cos\theta \simeq 1$, the system can be modeled by the equations

$$m\ddot{v} = u(t) + Mg\theta(t)$$
$$M\ddot{z} = -Mg\theta(t)$$
$$z(t) = v(t) + L\theta(t)$$

(a) Treat the trolley force $u(t)$ as an input and define the state vector as $\underline{x}^T \triangleq [\dot{v}\ \dot{\theta}\ v\ \theta]$. Rewrite the equations in state equation form $\underline{\dot{x}} = A\underline{x}(t) + \underline{b}u(t)$. Is the system completely controllable? On physical grounds (that is, without calculations), state what natural modes you expect to find in the unforced state response of the system and why.

(b) Suppose that the state variables $\dot{v}$ and $\dot{\theta}$ cannot be measured directly. However, instruments can be installed that continuously measure the trolley position $v(t)$, the

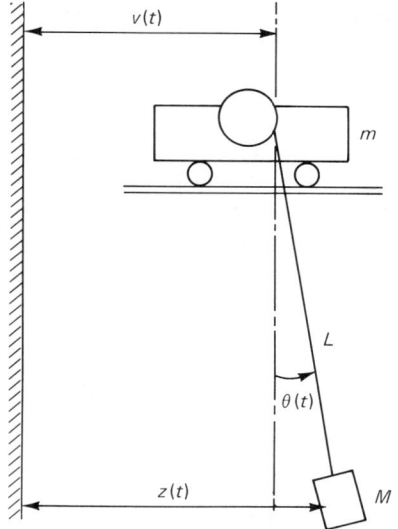

Figure 13.10.P7

trolley acceleration $\ddot{v}$, and the trolley force $u(t)$ [the latter two measurements yield a measurement of $\theta(t)$]. Thus $v(t)$ and $\theta(t)$ can be regarded as measurable system outputs. Is the system completely observable from either of the two outputs alone? If so, which one? Suppose a state observer is to be designed in order to implement a state feedback control. What practical advantages would be gained by driving the observer with both outputs instead of only one?

(c) Suppose that $m = 1$, $M = 2$, and $g = L = 10$ in consistent units.
  (i) What are the natural modes of the state response?
  (ii) Determine the equations of an observer for estimating $\dot{v}$ and $\dot{\theta}$, using v and θ and system outputs, such that the observer poles are at -2.

8. A plant has a single input $u(t)$ and a single output $v(t)$ related by

$$\ddot{v} + \alpha \dot{v} = \beta u(t)$$

where α, β are positive constants.

(a) Write the equations for a continuous time state space model of the plant in observability canonical form.
(b) For $\alpha = \beta = 1$ and a sampling interval of $\Delta = 0.1$, use your result for part (a) and derive a discrete time state equation model for the plant. Assume that $u(t)$ is produced from a discrete input sequence $u(k\Delta)$ by a zero order hold.
(c) Derive the discrete γ transfer function for the plant and hence write the operator model relating $v(k\Delta)$ and $u(k\Delta)$:

$$A(\delta)v = B(\delta)u$$

(d) Suppose the plant is to be controlled by a feedback system as shown in Figure 13.10.P8, where $y^*(k\Delta)$ is the desired output sequence and $d(k\Delta)$ a disturbance sequence. Using your answer to part (c), prove that, for any controller that results in a stable feedback system, the steady-state error $e_{SS} = \lim_{t \to \infty}\{e(t\Delta)\}$ is zero for any constant desired output and any constant disturbance; that is, integrating action in the controller is not required.
(e) Explain why, in practice, it is frequently desirable to include an integrator in the controller, even if the plant includes an integrator.

9. Consider a general system, described in state space form as

$$\rho x = A_1 x + B_1 u$$
$$y = C_1 x$$

where the sampling period is Δ_1. Suppose we resample this system at a slower rate, $\Delta_2 = N\Delta_1$, and $u(k\Delta_1)$ is generated from $u(k\Delta_2)$ using a zero-order hold [that is, $u(k\Delta_1)$ constant between slow sampling periods, Δ_2]. Show that the slowly sampled response can

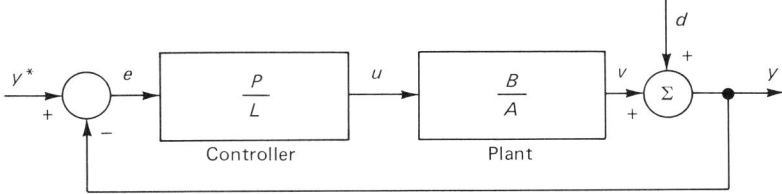

Figure 13.10.P8

be described by

$$\delta_2 x = A_2 x + B_2 u$$
$$y = Gx$$

where

$$\delta_2 = \frac{q_2 - 1}{\Delta_2}$$

$$A_2 = \frac{E_1(A_1, \Delta_2) - I}{\Delta_2}$$

and

$$B_2 = \frac{1}{\Delta_2} \int_0^{\Delta_2} E_1(A_1, \Delta_2 - \tau - \Delta_1) B_1 \, d\tau$$

Note: $E_1(\cdot, \cdot)$ is the generalized exponential at the fast sampling rate; similarly, S_1 is the generalized integral evaluated at the fast sampling rate.

10. Consider Example 13.5.2. Plot the magnitude of the sensitivity function $\bar{S}(\omega)$ for the two cases:
    (a) Where the single loop controller (13.5.18) is used
    (b) Where the cascaded feedback controller is used (see 13.5.29) [In case (b), plot the sensitivity function for the inner loop (with outer loop disconnected) and the outer loop (at the slow sampling rate) with the inner loop connected.]

11. Consider the inverted pendulum discussed in Example 13.5.2. Show that the SISO transfer function, (13.5.13), can only be stabilized by a controller with right half plane poles and zeros. *Hint*: Use a root locus argument.

12. Show that the closed loop transfer function for Example 13.6.3 with controllers as in (13.6.15), (13.6.16) is as given in (13.6.17). *Hint*: First show that the closed velocity loop has transfer function $3/(\gamma + 3)$.

13. Consider again the tracking problem of Example 13.6.3. A feedback solution (FB) was given in (13.6.16) with $H_{FF} = 1$. A feedforward solution (FF1) was also given by (13.6.18) with

$$H_{FF} = \frac{1.6667\gamma^2 + 2\gamma + 1}{\gamma^2 + 2\gamma + 1}$$

Show that either of these solutions gives zero parabolic tracking error, even if $J \neq J_0$, provided the closed loop is stable. *Note*: The integrator in the inner velocity loop is what makes FF1 have this property.

14. Another feedforward solution, FF2, to Example 13.6.3 would be to make

$$C_p(\gamma) = \frac{3(\gamma + 3)}{\gamma + 6}$$

and use $H_{FF}(\gamma)$ as in (13.6.20) to give zero parabolic tracking error.
    (a) Show that, for $J = J_0$, zero parabolic tracking error is achieved if and only if $n_1 = d_1 + \frac{2}{3}$ and $n_2 = d_2 + \frac{2}{3}d_1 + \frac{1}{9}$ in the feedforward block.
    (b) Show that the tracking error, in this case, is sensitive to changes in J.

15. Compare the responses of the three solutions FB, FF1, and FF2 (see Problems 13 and 14) to Example 13.6.3 when $J = 2J_0$. Use $\theta^*(t)$ as (13.6.18) and use $H_{FF} = (2.0539\gamma^2 + 2.0809\gamma + 1)/(\gamma^2 + 1.414\gamma + 1)$ for FF2.

14

Implementation Issues in Digital Control

14.1 INTRODUCTION

In the preceding chapters, we examined various control algorithms, emphasizing issues in the algorithm design and performance. In this chapter we wish to address some issues that arise in the implementation of control laws, particularly digital control laws.

We first consider the choice of sampling rate. We show that slow sampling has the disadvantage of information loss between samples, while rapid sampling involves numerical difficulties. Thus, in practice, the most suitable choice of sampling rate lies between these constraints.

We then describe suitable tools for the selection of various interface hardware items. In particular, we look at the choice of analog to digital converter, digital to analog converter, and antialiasing filters.

Finally, we look at several practical issues in the digital implementation of feedback systems. We study timing for proper and strictly proper control laws, bumpless transfer, and techniques for avoiding integral windup.

14.2 SELECTION OF SAMPLING RATE

In this section we examine timing issues and these lead us to argue that the sampling rate should be about ten times the closed loop bandwidth. Our reasons for suggesting this rule of thumb can be split into two parts. We will analyze the effects of slow sampling (sampling rate $f_s < 10f_B$) and show that this is deleterious from the viewpoint of control performance. Second, we show why very rapid sampling ($f_s \gg 10f_B$) is deleterious from the viewpoint of numerical accuracy.

14.2.1 Effects of Slow Sampling

We first argue that sampling at a rate significantly less than ten times the bandwidth involves, in practice, a loss of information regarding intersample behavior. Consider, for example, the case where we wish to reconstruct a signal based on the sampled data. Suppose the signal is the step response of a system with bandwidth 1 Hz, for example, the signal

$$y(t) = 1 - e^{-2\pi t} \qquad (14.2.1)$$

Figure 14.2.1 shows the sampled data versions of this signal with sampling rates $f_s = 2$, 5, and 10 Hz.

Note that reconstruction of the original signal is virtually impossible with the 2-Hz sampling rate, a little difficult with the 5-Hz sampling, but can be achieved

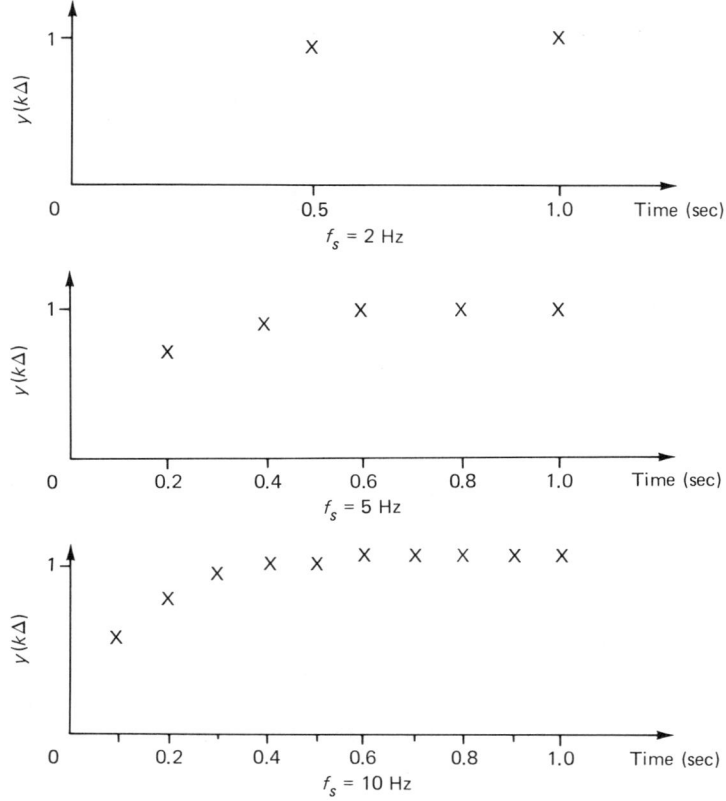

Figure 14.2.1 Sampled 1-Hz bandwidth signal.

easily with the 10-Hz sampling rate. With the 10-Hz sampling rate, a simple linear interpolation reconstruction gives an error of less than 4%.

Note that this conclusion may also be motivated by careful consideration of Shannon's reconstruction theorem. Shannon's reconstruction theorem (Theorem 3.3.1) states that a continuous time signal with a strict bandwidth limit of $\frac{1}{2}f_s$ can be completely reconstructed from sampled data, when the sampling frequency is f_s. However, the filter needed to perform the reconstruction is infinite dimensional and is not BIBO stable. Also, real signals do not have strict bandwidth limits (that is, there are still small frequency components outside the bandwidth). For these reasons we suggest that, in practice, the theoretical lower limit on the sampling rate of twice the bandwidth suggested by Shannon's reconstruction theorem should be extended to about ten times the bandwidth as in the preceding example.

As another example of the problems associated with slow sampling, consider again the antialiasing filter discussed in Section 3.2.3. There it was shown that the filter should attenuate signals at frequencies of $f_s/2$ and beyond. This means that the filter will need a cutoff frequency of less than $f_s/2$. It would then seem illogical to have a system bandwidth that is not less than the filter cutoff frequency, since otherwise the antialiasing filter becomes as significant as the system itself in determining the sampled response.

Example 14.2.1 illustrates some of the effects of slow sampling.

Example 14.2.1

Consider the case where we wish to perform digital control on an unstable plant, that is, a plant with continuous time transfer function

$$H_p(s) = \frac{0.95}{s - 1} \qquad (14.2.2)$$

Suppose also that we want a response time of about 1 sec, and the computer we use to control the system has a resolution in its D/A converter of 0.1. The response time given corresponds roughly to a bandwidth of 0.3 Hz.

With a 1-Hz sampling rate (that is, three times the desired bandwidth), the corresponding discrete time model is

$$H_{p_1}(\delta) = \frac{1.6324}{\delta - 1.7183} \qquad (14.2.3)$$

If we constrain the controller to include integral action, a suitable control law is

$$H_{c_1}(\delta) = \frac{1.7877\delta + 0.1225}{\delta} \qquad (14.2.4)$$

which gives closed loop poles at $\delta = -1$ (deadbeat) and $\delta = -0.2$ (low bandwidth integral action). The response, including D/A quantization, is shown in Figure 14.2.2. Note that the 1-Hz sampling is too slow for this example.

Alternatively, if we use a sampling rate of ten times the closed loop bandwidth (3 Hz), we obtain a discrete time model

$$H_{p_3}(\delta) = \frac{1.1275}{\delta - 1.1868}, \qquad \Delta = \frac{1}{3} \qquad (14.2.5)$$

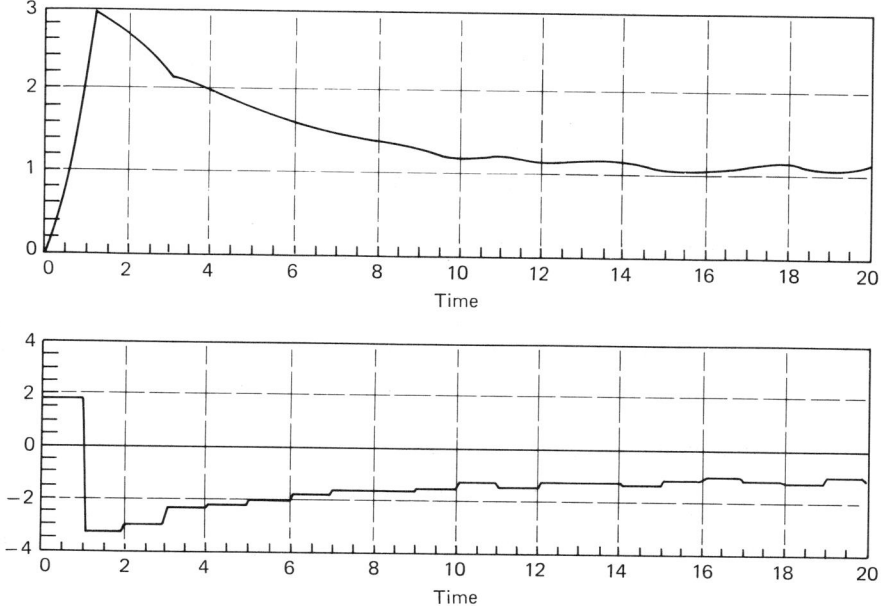

Figure 14.2.2 System response with slow sampling and input quantization.

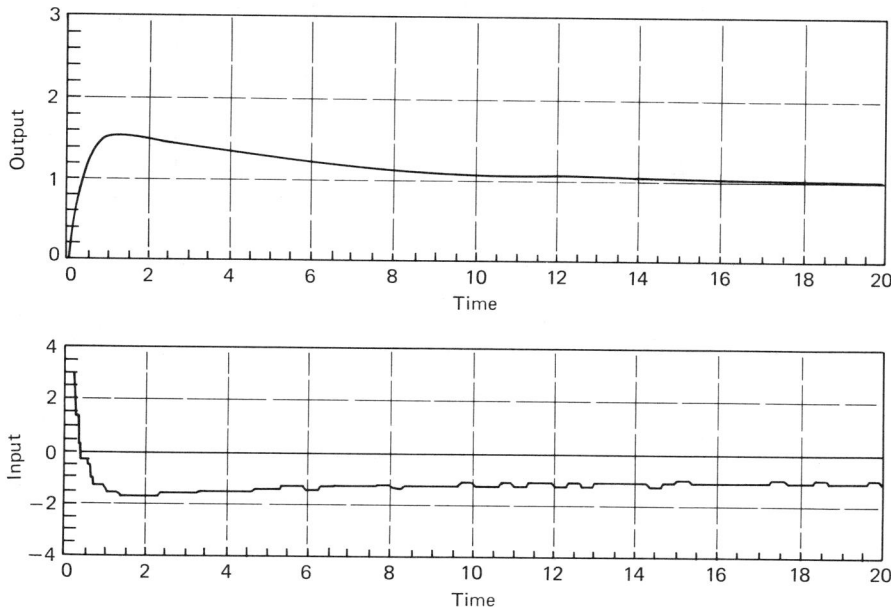

Figure 14.2.3 System response with recommended sampling and input quantization.

Sec. 14.2 Selection of Sampling Rate

In this case, a suitable control law is

$$H_{c_3}(\delta) = \frac{3.0038\delta + 0.3548}{\delta} \qquad (14.2.6)$$

which gives closed loop poles at $\delta = -2$ (1-sec response time) and $\delta = -0.2$ (same integral bandwidth as before). The response in this case is shown in Figure 14.2.3. Note that the improved performance over the 1-Hz control law is dramatic and arises mainly due to the extra information obtained using the faster sampling rate.

A further argument against slow sampling is provided by the design rules (13.5.1), (13.5.2). These show that there is a sensitivity penalty arising from the use of slow sampling.

14.2.2 Effects of Rapid Sampling

In the previous sections we argued that slow sampling involves a loss of information in a digital system, and that this usually results in poorer control performance. In this section we show that a similar loss in performance occurs if excessively fast

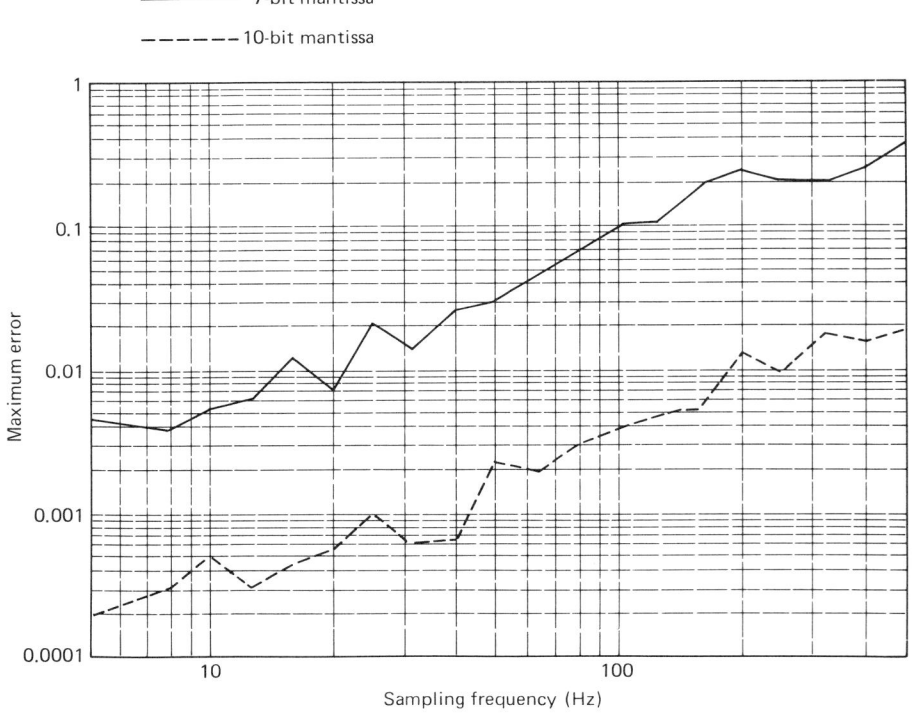

Figure 14.2.4 Maximum error between ideal and finite word length computations.

sampling is used due to numerical difficulties. Consider, for example, a discrete time system

$$\delta x = \frac{e^{-2\pi\Delta} - 1}{\Delta} x + \frac{1 - e^{-2\pi\Delta}}{\Delta} \qquad (14.2.7)$$

with zero initial conditions. Ideally, the response of this system should be

$$x(k\Delta) = 1 - e^{-2\pi k\Delta} \qquad (14.2.8)$$

The model, (14.2.7), has been programmed in a machine that simulates a computer having floating-point arithmetic of various word lengths (that is, mantissa bits). The results are shown in Figure 14.2.4 where maximum error is plotted against sampling frequency. Note that in this example the bandwidth is 1 Hz and so the rule of thumb given in the introduction suggests a sampling rate of 10 Hz. From Figure 14.2.4 we see that this leads to a maximal error of 5×10^{-4} for a 10-bit mantissa. The figure also shows that if we increase the sampling frequency by an order of magnitude the numerical errors increase by about the same factor! The reason for this can be seen as follows: Equation (14.2.7) is implemented as

$$x_{k+1} = x_k + (e^{-2\pi\Delta} - 1)x_k + (1 - e^{-2\pi\Delta}) \qquad (14.2.9)$$

Clearly, as $\Delta \to 0$, $1 - e^{-2\pi\Delta} \to 0$ and so the equations become numerically ill conditioned.

14.2.3 Recommended Sampling Rate

We have seen in the preceding sections that too slow a sampling rate leads to intersample problems, while too fast sampling leads to numerical problems. Thus, we conclude that a sensible guideline is that the sampling rate be above about 10 times the desired closed loop bandwidth. Sampling rates up to 50 times the closed loop bandwidth are often acceptable in modern, fast, high-precision computers. Indeed, fast sampling has the advantage that the interaction between the antialiasing filter and system response is insignificant. However, very rapid sampling rates above 50 times the desired loop bandwidth offer no further advantage and invariably lead to numerical difficulties.

14.3 CHOICE OF ANALOG INTERFACE HARDWARE

In this section we investigate the choice of analog interface hardware, including antialiasing filter and A/D and D/A converters.

14.3.1 Choice of Antialiasing Filter

We first discuss how we select the appropriate roll-off characteristic for an antialiasing filter. Following this, we briefly consider how we might implement this filter.

Consider the case where we have continuous time measurement noise, as indicated in Figure 14.3.1, where the noise $v(t)$ is assumed to be white with spectral

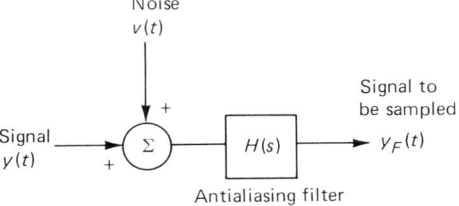

Figure 14.3.1 Diagram showing continuous time measurement noise and antialiasing filters.

density R. The signal $y(t)$ is assumed to have a flat spectrum up to the folding frequency of $f_s/2$, where f_s is the sampling frequency. The spectral density of $y(t)$ in the band $[0, f_s/2]$ will be taken to be Q. The ideal antialiasing filter is thus an ideal low-pass filter with cutoff frequency $f_s/2$. The signal to noise ratio in y_f when the ideal filter is used is denoted by S_0 and is given by

$$S_0^2 = \left(\frac{Q}{R} \right) \qquad (14.3.1)$$

In considering the performance of a practical antialiasing filter, we compute the deterioration in signal to noise ratio from the ideal given in (14.3.1) when a

TABLE 14.3.1 TABLE SHOWING PERFORMANCE OF VARIOUS ANTIALIASING FILTERS

Filter Order	Filter Type	Filter Transfer Function	Filter -3-dB Frequency	Gain at $0.9\omega_s$ (dB)	Effective Deterioration in Noise Response (dB)	Gain at $0.1\omega_s$ (dB)	Phase at $0.1\omega_s$ (degrees)
1	—	$\dfrac{0.5\omega_s}{s + 0.5\omega_s}$	$0.5\omega_s$	-6.3	3.0	-0.17	-11.3
1	—	$\dfrac{0.2\omega_s}{s + 0.2\omega_s}$	$0.2\omega_s$	-13.3	1.2	-0.97	-16.6
2	Butterworth	$\dfrac{(0.5\omega_s)^2}{s^2 + \sqrt{2}(0.5\omega_s) + (0.5\omega_s)^2}$	$0.5\omega_s$	-10.6	1.1	-0.01	-16.4
2	Butterworth	$\dfrac{(0.3\omega_s)^2}{s^2 + \sqrt{2}(0.3\omega_s) + (0.3\omega_s)^2}$	$0.3\omega_s$	-19.1	0.28	-0.05	-27.9
3	Butterworth	$\dfrac{(0.5\omega_s)^3}{s^3 + 2(0.5\omega_s)s^2 + 2(0.5\omega_s)^2 s + (0.5\omega_s)^3}$	$0.5\omega_s$	-15.4	0.64	0.00	-23.1
3	Butterworth	$\dfrac{(0.3\omega_s)^3}{s^3 + 2(0.3\omega_s)s^2 + 2(0.3\omega_s)^2 s + (0.3\omega_s)^3}$	$0.3\omega_s$	-28.6	0.08	-0.01	-39.0
∞	Averager	$\dfrac{1 - e^{-s\Delta}}{s\Delta}$	$0.317\omega_s$	-19.2	1.1	-0.14	-18

practical filter is used. The signal to noise ratio in a practical filter is given by

$$S^2 = \left(\frac{Q}{R}\right)\left[\frac{\int_{-\omega_s/2}^{\omega_s/2}|H(j\omega)|^2 d\omega}{\int_{-\infty}^{\infty}|H(j\omega)|^2 d\omega}\right] \quad (14.3.2)$$

where $H(s)$ is the filter transfer function. Thus, the deterioration in signal to noise ratio, $\mathcal{D}$, is given by

$$\mathcal{D}^2 \triangleq \frac{S_0^2}{S^2} = \frac{\int_{-\infty}^{\infty}|H(j\omega)|^2 d\omega}{\int_{-\omega_s/2}^{\omega_s/2}|H(j\omega)|^2 d\omega} \quad (14.3.3)$$

Table 14.3.1 lists a few antialiasing filter types, together with their gain at $0.9\omega_s$ and the effective decibel deterioration in the response due to noise. The frequency response at $0.9\omega_s$ is significant, in the control context, since this is the lowest frequency that gets folded to within one-tenth of the sampling frequency. We also include the gain and phase of the filter at $0.1\omega_s$ since this gives an indication of how much the antialiasing filter will affect the controller design, if we follow the advice of Section 14.2.

From the table, we suggest that, except in the most demanding applications, a second-order Butterworth filter with a cutoff frequency of about half the sampling frequency is a reasonable antialiasing filter. Note also that this filter has a negligible effect on the gain at one-tenth the sampling frequency and about 16° extra phase lag at the same frequency. This suggests that the antialiasing filter should have a small, but not negligible, effect on the controller design.

The implementation of the antialiasing filter can be performed using analog hardware for sampling periods of less than about 10 sec. For sampling periods larger than about 1 sec, large, low-leakage capacitors are required, and we suggest consideration of the scheme shown in Figure 14.3.2. The scheme shown uses

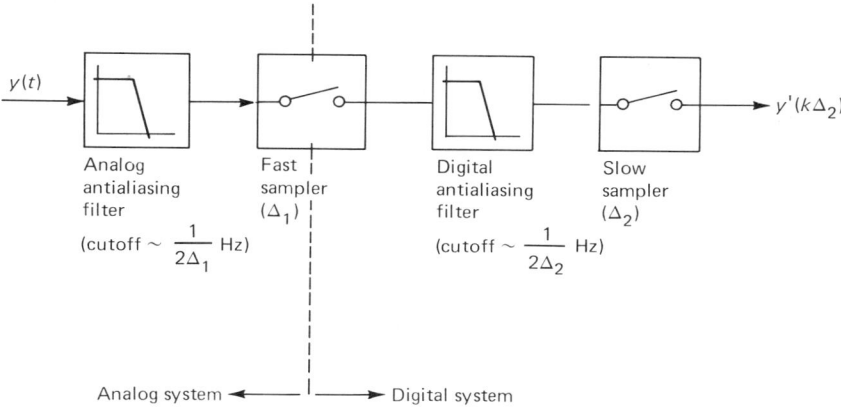

Figure 14.3.2 Possible antialiasing arrangement for systems with low bandwidth.

Sec. 14.3 Choice of Analog Interface Hardware

multirate sampling, the fast sampling rate allowing easy design of the antialiasing filter and the slow sampling rate reducing the computational burden and numerical difficulties.

14.3.2 Choice of A / D Converters

The main types of A/D converters available are discussed in Appendix A. Two main factors influence the specification: conversion time required and accuracy (that is, number of bits). The accuracy will normally be dictated by the tracking accuracy required. For example, if we want an accuracy of 0.1% of full scale, then typically we would use a converter with b bits, where

$$b = -\log_2(0.001) + 2$$
$$\approx 12 \qquad (14.3.4)$$

Note that if we are sampling a signal that is corrupted by noise, there will rarely be any advantage in extending the A/D accuracy below the noise level unless the noise is highly "colored" (or correlated). Usually, a conversion time of 10% or less of the sampling time will be desired. This will allow the majority of the sampling period to be used for computations.

14.3.3 Choice of D / A Converters

We will not consider in any detail the timing requirements for D/A converters, since in most cases conversion is essentially instantaneous. We limit our remarks to the observation that we should check that the D/A converter settling time is a small fraction of the sampling period. However, the choice of the D/A converter word length is not so simple. One way of selecting the word length is described next.

Consider the case where we have a linear discrete time feedback loop that is ideal, except for having a quantizer at the plant input (see Figure 14.3.3). The errors due to quantization can be considered as a bounded disturbance, $q(t)$, at the plant input, as shown in Figure 14.3.4. Let $Q(\gamma)$ denote the transform of $q(t)$ and $Y_q(\gamma)$ denote the transform of $y_q(t)$, which is the output response due to $q(t)$. Then

$$Y_q(\gamma) = \frac{H_p(\gamma)}{1 + H_c(\gamma)H_p(\gamma)} Q(\gamma) \qquad (14.3.5)$$

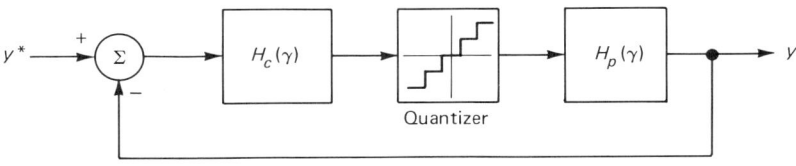

Figure 14.3.3 Ideal discrete time feedback loop, except for D/A quantization.

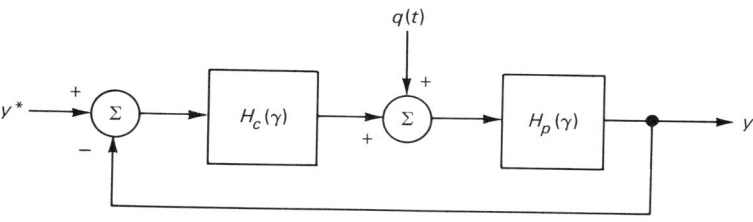

Figure 14.3.4 Equivalent realization for Figure 14.3.3 showing quantization errors as an external disturbance.

or

$$y_q(t) = \int_0^t h(t-\tau)q(\tau)\,d\tau \tag{14.3.6}$$

where

$$T\{h(t)\} = \frac{H_p(\gamma)}{1 + H_c(\gamma)H_p(\gamma)} \triangleq H(\gamma) \tag{14.3.7}$$

We will now evaluate the gain of the linear operator $H(\gamma)$ in two different ways. The first method, which is the more usual, is based on spectral analysis and relies on an assumption of white round-off noise. The second method is more conservative, requires less assumptions for its motivation, and is based on the operator norm induced by the $\mathscr{L}_\infty$ function norm.

For the first case, we make the common round-off noise assumption; that is, we assume there is sufficient activity in the feedback system for $q(t)$ to be treated as a sequence of independent identically uniformly distributed random variables. If the D/A converter has b bits (*including* sign bit) and has a range of -1 to $+1$, then

$$q(t) \in [-2^{-b}, 2^{-b}] = \left[-\frac{\varepsilon}{2}, \frac{\varepsilon}{2}\right] \tag{14.3.8}$$

where ε is the quantization level. Under these round-off assumptions, the power spectral density of $q(t)$ is

$$\Phi_q = \frac{\varepsilon^2 \Delta}{12} \tag{14.3.9}$$

Using (14.3.6), the assumption that $q(t)$ is uncorrelated, and Parseval's theorem, we can then show that the variance of y_q is

$$E[y_q^2] = \left[\int_{-1/2\Delta}^{1/2\Delta} H\left(\frac{e^{j2\pi f \Delta} - 1}{\Delta}\right) H\left(\frac{e^{-j2\pi f} - 1}{\Delta}\right) df\right] \frac{(\varepsilon)^2}{12} \Delta \tag{14.3.10}$$

The rms value of the output response may then be evaluated and the bit length chosen so that this error is within acceptable limits. For example, it should suffice that the error be less (by a few bits) than the rms accuracy requirement.

An alternative and sometimes more conservative approach is to assume only (14.3.8) and to find the maximum $y_q(t)$ possible. This may be found as follows:

$$|y_q(t)| = \left|\int_0^t h(t-\tau)q(\tau)\,d\tau\right| \qquad (14.3.11)$$

$$\leq \left(\int_0^t |h(t-\tau)|\,dt\right)\frac{\varepsilon}{2} \qquad (14.3.12)$$

$$\leq \left(\int_0^\infty |h(\tau)|\,d\tau\right)\frac{\varepsilon}{2} \qquad (14.3.13)$$

for any t. These computations are illustrated in Example 14.3.1.

Example 14.3.1. Example 14.2.1 Revisited

Suppose we are controlling the system (14.2.2) with a 3-Hz sampling rate and the controller suggested in (14.2.6). Consider the case where the maximum output range is ± 10, and we wish to have an accuracy of 0.1 in the output. We then proceed as follows. Using (14.3.7), it follows that in this case

$$H(\gamma) = \frac{1.1275\gamma}{(\gamma+2)(\gamma+0.2)} \qquad (14.3.14)$$

$$= \frac{1.2528}{\gamma+2} - \frac{0.1253}{\gamma+0.2} \qquad (14.3.15)$$

Because $\Delta = \frac{1}{3}$, we have for $t > 0$

$$h(t) = 1.2528 E(-2, t-\Delta) - 0.1253 E(-0.2, t-\Delta) \qquad (14.3.16)$$

$$= 1.2528 \cdot \left(\frac{1}{3}\right)^{(t/\Delta)-1} - 0.1253 \cdot (0.9333)^{(t/\Delta)-1} \qquad (14.3.17)$$

We then calculate numerically

$$\int_{-1/2\Delta}^{1/2\Delta} H\left(\frac{e^{j2\pi f\Delta}-1}{\Delta}\right) H\left(\frac{e^{-j2\pi f\Delta}-1}{\Delta}\right) df = \int_0^\infty h^2(t)\,dt$$

$$= 0.4772 \qquad (14.3.18)$$

and

$$\int_0^\infty |h(t)|\,dt = 0.9724 \qquad (14.3.19)$$

Then, using (14.3.10), we see that if we want an rms error of less than 0.05 (half 0.1) we need a quantization level of

$$\varepsilon \leq 0.05 \cdot \left(\frac{12}{0.4772\Delta}\right)^{1/2} \qquad (14.3.20)$$

$$= 0.4343 \qquad (14.3.21)$$

Alternatively, using (14.3.13), we will have

$$\varepsilon \leq \frac{2}{0.9724} \cdot 0.1 \tag{14.3.22}$$

$$= 0.206 \tag{14.3.23}$$

Note that the word length would be based on the input range as well as the quantization level. In this case, a suitable input range would be $+20$ to -20 units, and thus, based on (14.3.23), we suggest the use of an 8-bit D/A converter that gives a quantization level

$$\varepsilon = \frac{40}{2^8} = 0.1563 \tag{14.3.24}$$

14.4 CONTROL ALGORITHM PACKAGING

Throughout the text we have described many control algorithms. These algorithms need to be packaged with additional structures to handle practical environments. In this section we discuss some of these packaging issues, including antiintegral windup, slew rate limits, and bumpless transfer.

14.4.1 Integral Windup

Integral windup is a problem that may occur in analog or digital controllers. This problem occurs when a control law with integral action is used and the input hits an amplitude constraint (saturation limit) or a velocity constraint (slew rate limit). In this case, a large error between set point and output may persist for a long time, and this may cause the state of the integrator to wind up. This is illustrated in Example 14.4.1.

Example 14.4.1

Consider a continuous time system with transfer function

$$H(s) = \frac{1}{s(s+1)} \tag{14.4.1}$$

If we use a sampling rate of 10 Hz, then the discrete time model is

$$H_\Delta(\delta) = \frac{0.0484\delta + 0.9516}{\delta(\delta + 0.9516)} \tag{14.4.2}$$

A suitable controller for this system is

$$H_c(\delta) = \frac{17(\delta + 0.9516)(\delta + 0.5)}{(\delta + 8)(\delta)} \tag{14.4.3}$$

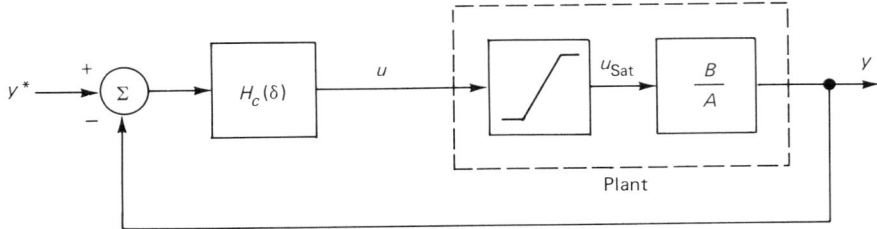

Figure 14.4.1 Control loop including plant input saturation.

Now suppose the controller output (that is, plant input) has a saturation limit of ±1 unit, as in Figure 14.4.1. Figure 14.4.2 shows the response when there is a large step change in y^*. Note that in this case the state of the integrator winds up due to the integral action in $H_c(\delta)$, and this causes the large overshoot.

One way to improve this situation is to put a limit on the rate of change of $y^*(t)$. In this example, it is reasonably clear that with the input limited to ±1 the output difference is limited as follows:

$$|\delta y(k\Delta)| \leq 1 \tag{14.4.4}$$

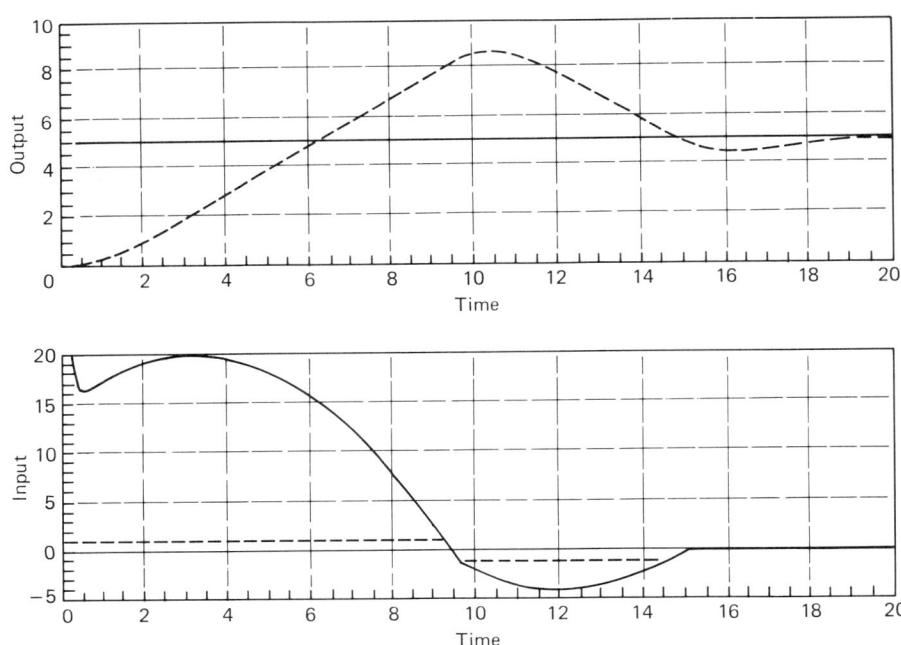

Figure 14.4.2 Closed loop step response showing the effect of integral windup.

468   Implementation Issues in Digital Control   Chap. 14

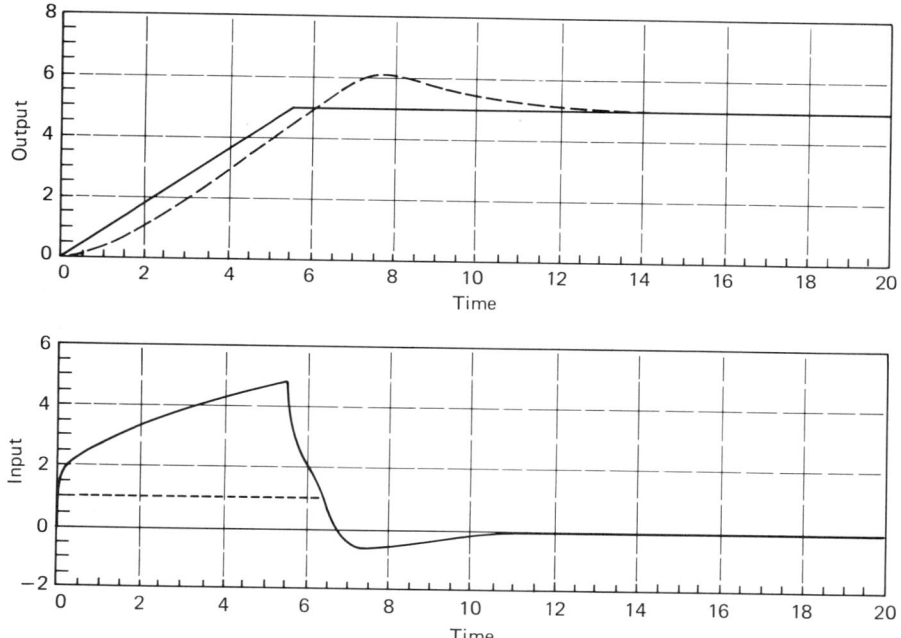

Figure 14.4.3 Closed loop step response with slew rate limit in y^*.

Thus it would seem reasonable to similarly limit changes in $y^*(k\Delta)$. This can be achieved by computer code of the following form.

```
procedure slew_rate_limit (ystar:REAL;VAR y_slew:REAL);
{ procedure to take setpoint, ystar, and past value of slew rate limited setpoint, y_slew,
and return current slew rate limited ystar in y_slew }.
CONST slew_limit = 1.0; { slew rate limit = 1 / delta }
BEGIN
    yslew := ystar;
    IF (ystar-yslew) > slew_limit THEN yslew := yslew + slew_limit;
    IF (ystar-yslew) < - slew_limit THEN yslew := yslew - slew_limit;
END;
```

Figure 14.4.3 shows the response when a rate limit of 0.9 has been imposed on y^*. Note that the integral windup has been reduced, but not eliminated, in this case.

An alternative method of overcoming integral windup is to use a nonminimal implementation of the control law. Note that the general control law equation can be implemented in a nonminimal form as shown in Figure 9.4.2. An obvious extension to this implementation, in the case where the plant has an input saturation, is the implementation shown in Figure 14.4.4.

Note from Section 9.4 that E_1 must be Hurwitz and that R is defined to be $L - E_1$. Note also that, if the software saturation levels are less or equal to the plant

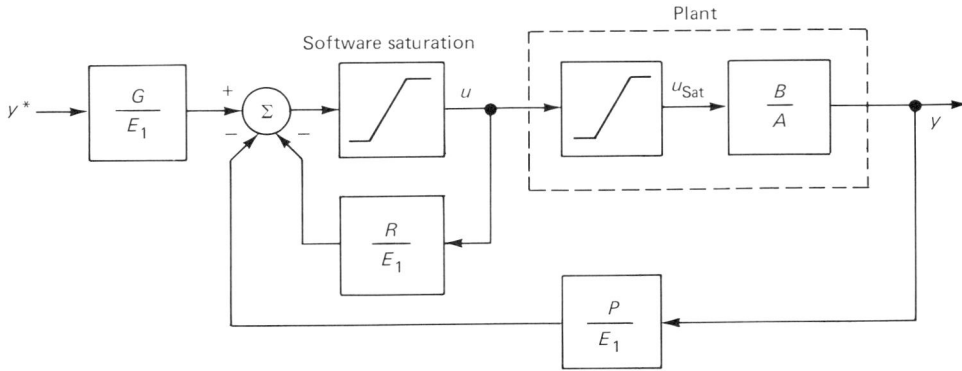

Figure 14.4.4 Nonminimal control law implementation with antiintegral windup.

saturation levels, then plant input saturation never occurs. Thus the structure shown in Figure 14.4.4 can be thought of as an observer plus state feedback, where the observer is fed current information about the input to the linear part of the plant, that is, u_{sat} in Figure 14.4.4.

Figure 14.4.5 shows the step response for the scheme shown in Figure 14.4.4 with $E = (\gamma + 2)^2$. Note that in this case, the integral windup has been eliminated.

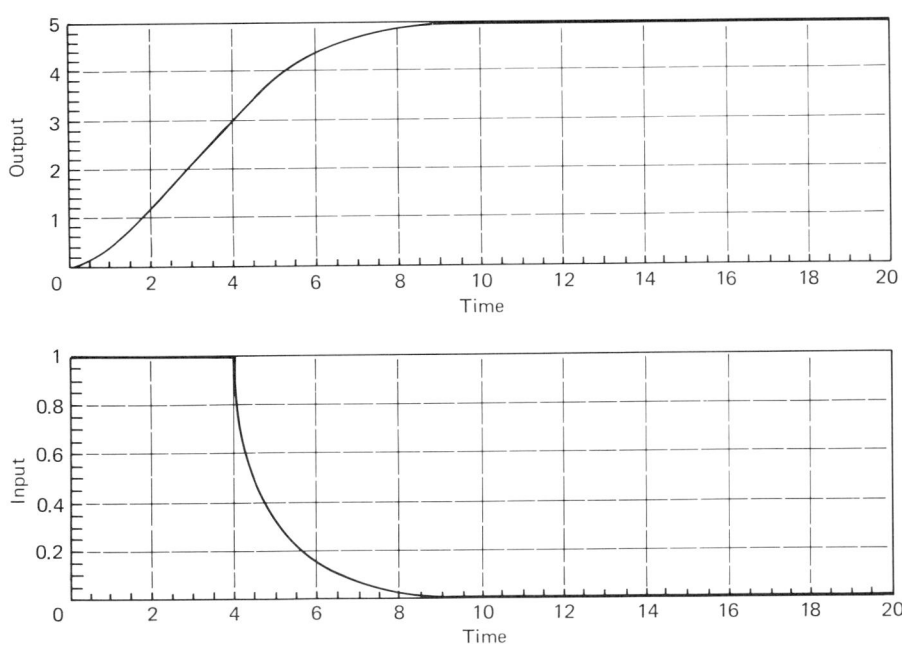

Figure 14.4.5 Response for Example 14.6.1 when using the antiintegral windup scheme of Figure 14.4.4.

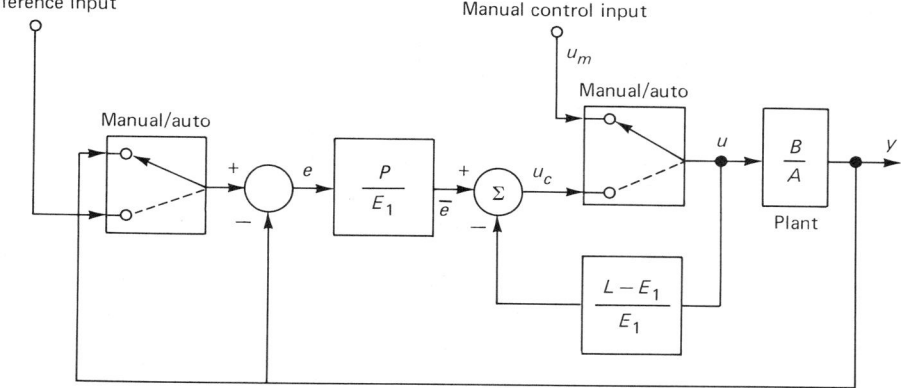

Figure 14.4.6 Block diagram showing how bumpless transfer may be achieved (switches shown in manual position).

The preceding idea can also be extended to include slew rate limits on the input. The key idea is to make the software saturation time varying as follows: let u_{max} and u_{min} be the amplitude constraints and let $(du)_{max}$ and $(du)_{min}$ be slew rate constraints on the control signal. Then at time t the upper software constraint is the minimum of $(u(t - \Delta) + \Delta(du)_{max})$ and u_{max}. Similarly, the lower constraint is the maximum of $(u(t - \Delta) + \Delta(du)_{min})$ and u_{min}. The controller structure is then implemented as in Figure 14.4.4 with the observer being fed by the true system input.

14.4.2 Bumpless Transfer

Almost invariably, practical control systems will have a manual (that is, plant operator) control setting, as well as an automatic (that is, feedback regulator) control setting. It is very important that we be able to make a smooth transition from manual control to automatic control. A smooth transition from one control mode to another is generally called *bumpless transfer*. A method suggested for achieving this is shown diagrammatically in Figure 14.4.6.

For this scheme to work, we require two main conditions. First, we require that the control law should include integral action; that is,

$$L(0) = 0 \qquad (14.4.5)$$

Second, we require that, prior to transfer, u_m should have been held at a constant value for some time. Note that when in manual mode, if u_m is a constant, then under the first condition $u_c \to u_m$ with a rate determined by E_1. Normally, E_1 should have a bandwidth that is roughly the same as the closed loop bandwidth. In this case, the convergence of u_c to u_m will have a time constant that is about the same as the closed loop time constant. Thus it will generally suffice to hold u_c constant for about five times the time constant of the closed loop system.

Under these conditions, e will be zero prior to transfer, u_c will be u_m, and so a bumpless transfer can be achieved. Note that if the reference input y^* is the same as y (prior to transfer) then there will be no transient whatsoever in the control loop. If y^* differs from the previous value of y, a step response, equivalent to shifting from $y^* = y$ (prior to transfer) to y^* (post transfer), occurs.

The scheme shown in Figure 14.4.6 essentially sets up the initial conditions of the controller to match those of the plant. In other words, the scheme shown runs a type of observer on the plant inputs and outputs at all times and adds feedback based on the observer when transfer to automatic mode occurs.

14.5 NUMERICAL ISSUES

In this section we will review some numerical issues in control law implementation. In particular, we show that, among other considerations, it is usually preferable to implement in delta rather than shift form. We also show that in high-order compensators parallel or series realizations should be used.

Numerical questions of this type have received little attention in the control literature when compared with the attention received in digital signal processing. One cause of numerical difficulties in discrete time control is the clustering of the poles and zeros for the shift operator form of the controller around the point $(1 + j0)$. This suggests that the relevant information in the controller poles and zeros is encoded in their distance from $(1 + j0)$ and that numerical properties may therefore be improved by subtracting $(1 + j0)$ from all poles and zeros in the shift operator model. As we have seen before, the delta operator is a simple way of achieving this.

14.5.1 Implementation in Delta Form

We begin by reviewing how a transfer function, or state space model, can be implemented in delta form. Suppose we wish to implement the transfer function

$$H(\delta) = \frac{n_m \delta^m + n_{m-1} \delta^{m-1} + \cdots + n_0}{\delta^n + d_{n-1} \delta^{n-1} + \cdots + d_0} = \frac{y_t}{u_t} \qquad (14.5.1)$$

where $m < n$. Then the following controller state space form will achieve the required transfer function:

$$\left. \begin{array}{l} x_t = \delta^{-1}\{A \cdot x_t + B \cdot u_t\} \\ y_t = C \cdot x_t \end{array} \right\} \qquad (14.5.2)$$

where x_t is the state vector of dimension n,

$$A = \begin{bmatrix} -d_{n-1} & -d_{n-2} & \cdots & -d_1 & -d_0 \\ 1 & 0 & \cdots & 0 & 0 \\ 0 & 1 & \cdots & 0 & 0 \\ \vdots & \vdots & & \vdots & \vdots \\ 0 & 0 & \cdots & 1 & 0 \end{bmatrix}$$

$$B^T = \begin{bmatrix} 1 & 0 & 0 & \cdots & 0 \end{bmatrix}$$

and

$$C = \begin{bmatrix} 0 & \cdots & 0 & n_m & \cdots & n_0 \end{bmatrix}$$

(14.5.3)

We thus see that the basic building block necessary to implement delta models is the function (δ^{-1}). The following Pascal subroutine shows how a transfer function with the preceding structure may be implemented.

```
TYPE vector = ARRAY [0..n-1] OF REAL;
  :
  :
PROCEDURE delta_transfer_function (
    VAR   output: REAL; input REAL;
    VAR   state: vector; numerator, denominator: vector);
{This procedure implements a discrete time delta operator transfer function. On entry,
"input" should be u(t) and "state" should be x(t). On exit, "output" will be y(t) and
"state" will be x(t + Δ). A globally defined REAL constant "delta" is used by this
procedure, and the order (i.e. the dimension of vector) "n" is also used; "numerator
[i]" and "denominator [i]" correspond to n_i and d_i in (14.5.3).}
VAR   i: INTEGER;   temp: REAL;
BEGIN
output := 0.0;
temp := input;  {temp = B*u}
FOR i := n down to 1 DO BEGIN
output := output + numerator [n - i] * state[i - 1];
    temp := temp - denominator [n - i] * state[i - 1];
    IF i > 1 THEN state[i - 1] := state [i - 1] +
        delta * state[i - 2];
    END; {end FOR loop}
{temp is now equal to the first element of (A * x + B * u)}
state[0] := state[0] + delta * temp;
END; {end of Procedure}
```

This implementation is well suited to low-order problems. For high-order systems, it is usually best to first factor the transfer function into first- and second-order blocks and then to use the procedure for each block. This will be discussed next.

14.5.2 Cascade and Parallel Realizations

For high-order systems, the response is generally sensitive to the coefficients in the standard phase variable canonical form of (14.5.3). Note that this sensitivity is a factor that applies for all operators (for example, shift, delta, and continuous time). One standard way to alleviate these sensitivity problems is to implement the transfer function as either a cascade of first- and second-order blocks or as a parallel realization of first- and second-order blocks.

The transformation to cascade form is achieved by simply factoring the numerator and denominator as follows:

$$H(\rho) = \frac{K\prod_{i=1}^{n_1}(\rho + \xi_i)\prod_{i=1}^{n_2}(\rho^2 + \alpha_i\rho + \beta_i)}{\prod_{i=1}^{n_3}(\rho + p_i)\prod_{i=1}^{n_4}(\rho^2 + \alpha'_i\rho + \beta'_i)} \quad (14.5.4)$$

The transfer function is then implemented as in Figure 14.5.1.

To obtain a parallel realization, we simply use a partial fraction expansion on the form in (14.5.4) to give (assuming distinct poles)

$$H(\rho) = \sum_{i=1}^{n_3} \frac{A_i}{\rho + p_i} + \sum_{i=1}^{n_4} \frac{B_i\rho + C_i}{\rho^2 + \alpha'_i\rho + \beta'_i} \quad (14.5.5)$$

The transfer function is then implemented as in Figure 14.5.2.

The merits of cascade and parallel realizations are illustrated in Example 14.5.1.

Example 14.5.1

In this example we consider a shift operator implementation of a fourth-order bandpass filter. Consider a standard continuous time fourth-order bandpass filter,

$$H(s) = \frac{0.0714s^2}{s^4 + 0.36s^3 + 2.0684s^2 + 0.3585s + 0.9960} \quad (14.5.6)$$

which has approximately 0-dB gain at center frequency ($\omega = 1$ rad/sec) and has -3-dB frequencies of 0.87 rad/sec and 1.14 rad/sec. Using the rule of thumb given in Section 14.2 for the selection of the sampling rate, we want $\omega_s > 10$ rad/sec. We thus use a sampling period $\Delta = 0.5$ (that is, $\omega_s \approx 12.6$ rad/sec), and obtain the

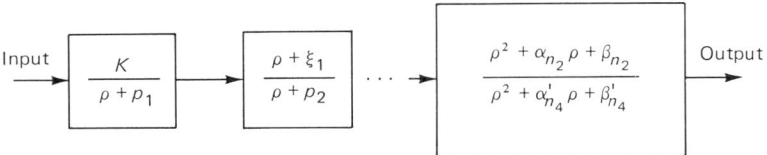

Figure 14.5.1 Cascade realization.

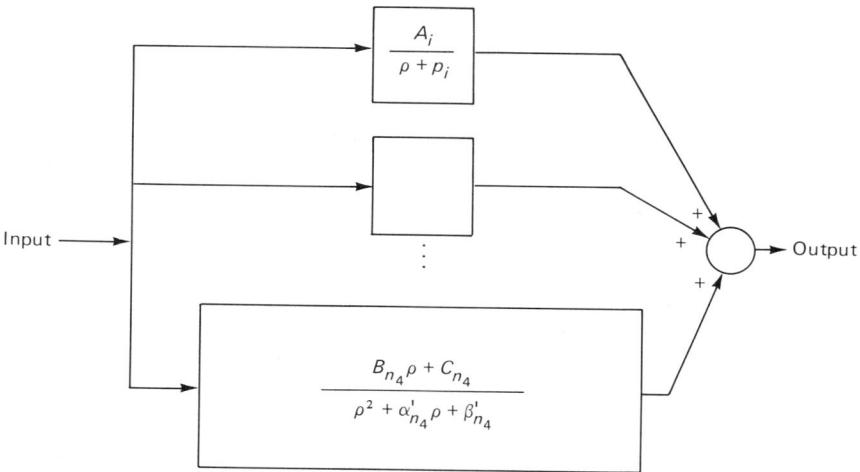

Figure 14.5.2 Parallel realization.

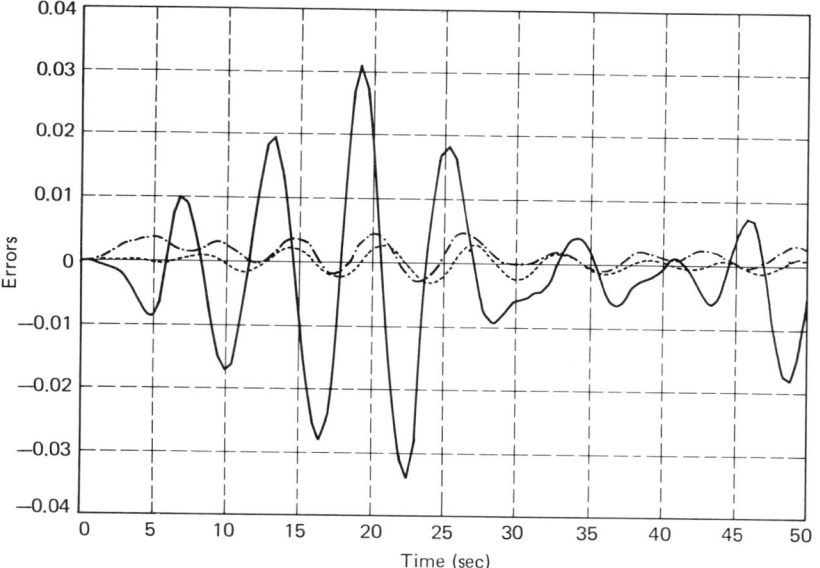

Figure 14.5.3 Numerical errors in step responses.

Sec. 14.5 Numerical Issues

following discrete time, shift operator transfer function:

$$H_Z(z) = \frac{0.0081z^3 - 0.0085z^2 - 0.0071z + 0.0076}{z^4 - 3.3522z^3 + 4.6343z^2 - 3.0628z + 0.8353} \quad (14.5.7)$$

There are several possible cascade realizations, one of which is the following:

$$H_Z(z) = \left[\frac{0.0897(z + 0.9422)}{z^2 - 1.6219z + 0.9048}\right]\left[\frac{0.0897(z^2 - 2z + 1)}{z^2 - 1.7303z + 0.9231}\right] \quad (14.5.8)$$

On the other hand, the parallel realization of $H_Z(z)$ is unique; that is,

$$H_Z(z) = \frac{0.032z + 0.0184}{z^2 - 1.6219z + 0.9048} + \frac{-0.024z - 0.0104}{z^2 - 1.7303z + 0.9231} \quad (14.5.9)$$

These three realizations of $H_Z(z)$, full (14.5.7), cascade (14.5.8), and parallel (14.5.9), have been simulated using floating-point arithmetic with a 10-bit mantissa (that is, about 0.1% relative accuracy). The differences between the ideal step response (that obtained using infinite precision arithmetic) and those responses obtained using 10-bit arithmetic with the preceding realizations are plotted in Figure 14.5.3, where the solid line represents the numerical errors obtained for the direct realization, (14.5.7), the dashed line gives the errors for the cascade realization, (14.5.8), and the dash–dot line gives the errors for the parallel realization, (14.5.9). Note the reduced numerical errors for the cascade and parallel realizations.

∇∇∇

14.5.3 Numerical Advantages of Delta Implementations

As foreshadowed in the earlier parts of the book, delta implementations are generally numerically superior to shift operator implementations. The basic reason for this is the shift in origin to the point $1 + j0$ in the Z plane. This gives better coefficient sensitivity, reduced round-off errors, and so on. The interested reader is referred to Appendix B, where some numerical issues are addressed, and to the references given at the end of the chapter.

Example 14.5.2 illustrates the advantage of delta over shift operator implementations.

Example 14.5.2

Consider a continuous time system with transfer function

$$H(s) = \frac{1}{s - 1} \quad (14.5.10)$$

With a sampling rate of 8 Hz, a suitable control law (which incorporates integral action) is

$$H'_c(\delta) = \frac{32\delta + 16}{\delta^2 + 8\delta} \quad (14.5.11)$$

or

$$H''_c(z) = \frac{4z - 3.75}{z^2 - z} \quad (14.5.12)$$

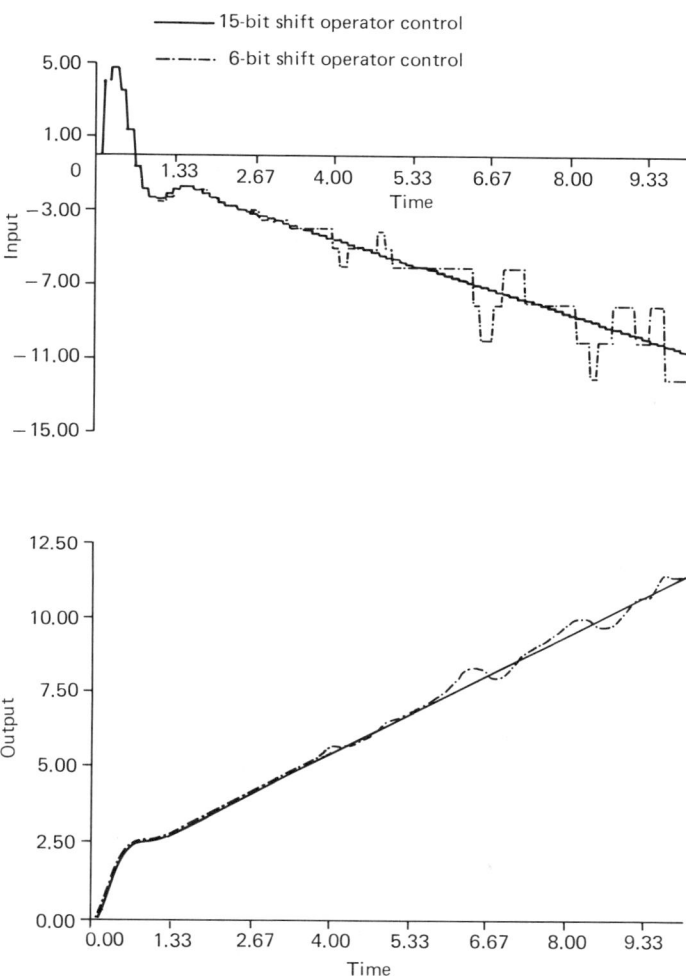

Figure 14.5.4 System response with controller implemented in shift form.

Simulations have been conducted for the closed loop systems with each of these controllers implemented in finite word length, floating-point arithmetic. The reference signal in each case was

$$y^*(t) = 1 + t \qquad (14.5.13)$$

Figure 14.5.4 shows the results when using the shift operator with 6-bit mantissa, floating-point arithmetic. Figure 14.5.5 repeats the results for the delta operator form. It can be seen from the figures that the delta form with 6-bit implementation is essentially indistinguishable from the 15-bit implementation. However, the 6-bit shift implementation performs markedly differently to the high-word-length implementation.

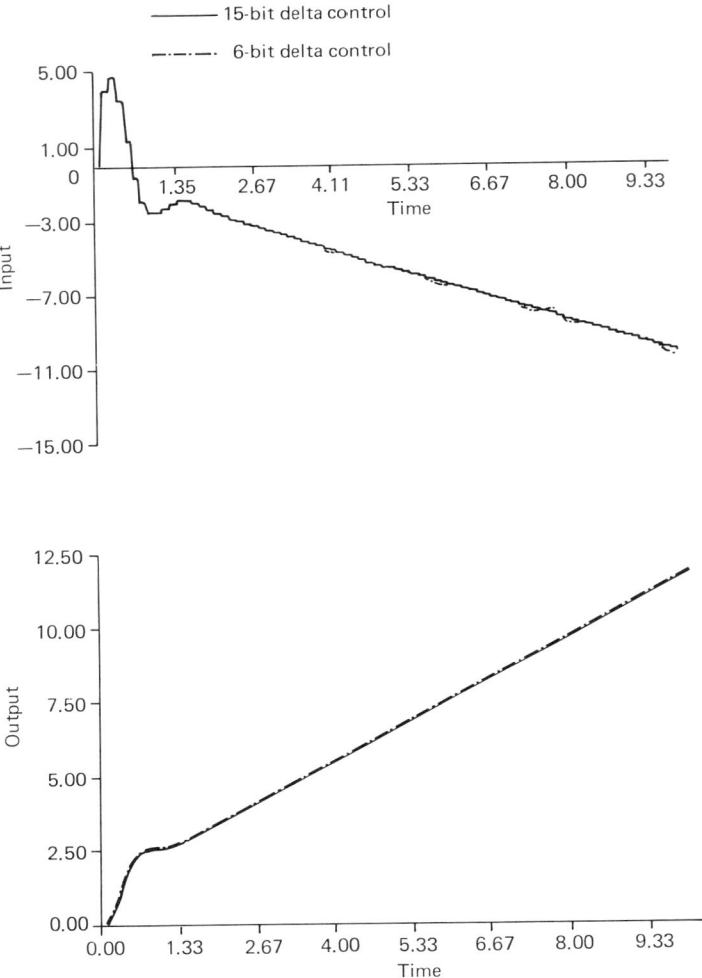

Figure 14.5.5 System response with controller implemented in delta form.

Similar conclusions apply to high-order systems, where the difference in performance between shift and delta becomes apparent at longer word lengths. ▽▽▽

Example 14.5.3. Example 14.5.1 Revisited

In Example 14.5.1 we compared three different implementations of a shift operator transfer function. Here we compare these results with delta operator implementations of the equivalent transfer function,

$$H_d(\gamma) = \frac{0.0597\gamma^2}{\gamma^4 + 1.2957\gamma^3 + 2.3111\gamma^2 + 1.1938\gamma + 0.8730} \quad (14.5.14)$$

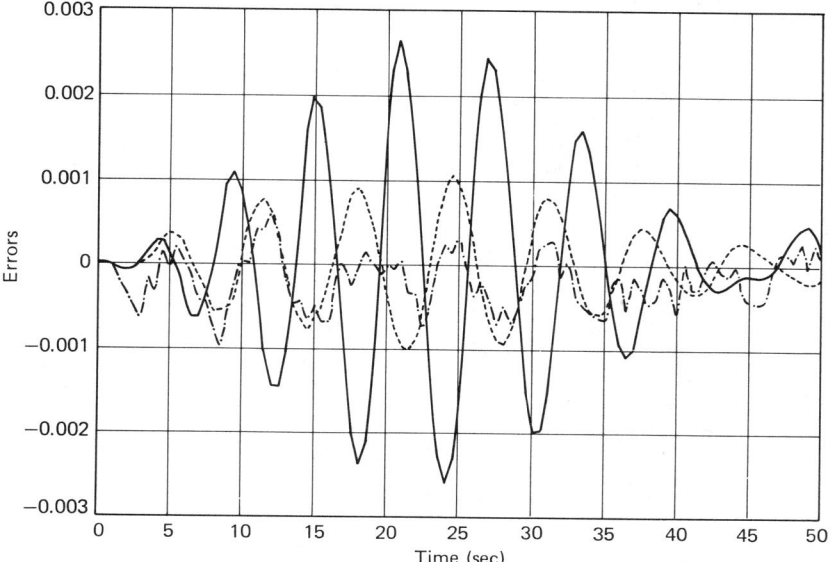

Figure 14.5.6 Numerical errors in step responses.

In this case a suitable cascade realization is

$$H_d(\gamma) = \left[\frac{0.2443\gamma}{\gamma^2 + 0.7562\gamma + 1.1318}\right]\left[\frac{0.2443\gamma}{\gamma^2 + 0.5394\gamma + 0.7713}\right] \quad (14.5.15)$$

and the parallel realization is

$$H_d(\gamma) = \frac{0.0131\gamma + 0.1964}{\gamma^2 + 0.7562\gamma + 1.1318} + \frac{-0.0131\gamma - 0.1338}{\gamma^2 + 0.574\gamma + 0.7713} \quad (14.5.16)$$

Figure 14.5.6 shows the numerical errors for the three different realizations, (14.5.14) to (14.5.16), when using 10-bit floating-point arithmetic. As in Figure 14.5.3, the solid line represents the errors for the direct realization, (14.5.14), the dashed line corresponds to the cascade, (14.5.15), and the dash–dot line corresponds to the parallel realization, (14.5.16).

We draw the following conclusions from Figures 14.5.3 and 14.5.6:

1. We see that the vertical axis scale in the two figures is not the same. Indeed, for the same realization, the delta operator is numerically superior to the shift operator (for example, a factor of 10 improvement for the direct realization!). Moreover, the numerical advantage of the delta operator increases as the sampling rate increases.
2. For high-order transfer functions, whether in delta or shift form, a cascade or parallel realization is generally superior to a direct realization in controller or observer canonical form. The superiority of the cascade or parallel realization is particularly apparent for the numerically poor shift operator case.

Sec. 14.5   Numerical Issues   479

14.6 SUMMARY

The key points covered in this chapter were:

- Discussion of the effect of sampling rate leading to the guideline that the sampling rate should be between 10 and 50 times the bandwidth.
- Discussion of the choice of antialiasing filter.
- Word length considerations in A/D and D/A converters.
- Methods for achieving bumpless transfer, and for avoiding integral windup.
- For high-order systems, the use of parallel or cascade realizations was advocated (Section 14.5.2).
- Numerical advantages of delta implementations over shift implementations were discussed in Section 14.5.3.

14.7 REFERENCES

Some of the issues addressed in this chapter are covered in other books on digital control, including:

ÅSTRÖM, K. J., and B. WITTENMARK (1984) *Computer Controlled Systems.* Prentice-Hall, Englewood Cliffs, N.J.

BIBBERO, R. J. (1977) *Microprocessors in Instruments and Control.* Wiley-Interscience, New York.

BOLLINGER, J. G., and N. A. DUFFIER (1988) *Computer Control of Machines and Process.* Addison-Wesley, Reading, Mass.

KATZ, P. (1981) *Digital Control Using Microprocessors.* Prentice-Hall, Englewood Cliffs, N.J.

Not many textbooks in the digital control area deal with numerical issues. This is unfortunate since these issues are frequently very important to the success or otherwise of digital control schemes. Very helpful background information is contained in most books on digital filtering; see for example:

RABINER, L. R., and B. GOLD (1975) *Theory and Application of Digital Signal Processing.* Prentice-Hall, Englewood Cliffs, N.J.

STANLEY, W. D., R. R. DOUGHERTY, and R. DOUGHERTY (1984) *Digital Signal Processing.* Prentice-Hall, Englewood Cliffs, N.J.

Also, books on numerical questions in linear algebra are very relevant to this topic; see for example:

HILDEBRAND, F. B. (1956) *Introduction to Numerical Analysis.* McGraw-Hill, New York.

WILKINSON, J. H. (1963) *Rounding Errors in Algebraic Processes.* Prentice-Hall, Englewood Cliffs, N.J.

The following papers give further information on the numerical aspects of delta operators in digital control:

AGARWAL, R. C., and C. S. BURRUS (1975) "New recursive digital filter structures having very low sensitivity and roundoff noise." *IEEE Trans. Circuits Systems*, Vol. CAS-22, No. 12, pp. 921–927.

AHMED, M. E., and P. R. BELANGER (1984) "Scaling and roundoff in fixed point implementation of control algorithms." *IEEE Trans. Ind. Electronics*, Vol. 31, No. 3, pp. 228–234.

BELANGER, P. R., and Y. GHOREIM (1980) "Fixed point arithmetic microprocessor implementation of self tuning regulators." Presented at Joint Automatic Control Conference, San Francisco.

KARWOSKI, R. J. (1979) "Introduction to the z-transform and its derivation." Tutorial paper, TRW psi Products, El Segundo, Calif.

MANTEY, P. E. (1969) "Eigenvalue sensitivity and state-variable selection." *IEEE Trans. Auto. Control*, Vol. AC-13, No. 3, pp. 263–269.

MIDDLETON, R. H., and G. C. GOODWIN (1987) "Improved finite word length characteristics in digital control using delta operators." *IEEE Trans. Auto. Control*, Vol. AC-31, No. 11, pp. 1015–1021.

MORONEY, P., A. S. WILLSKY, and P. K. HOUPT (1980) "The digital implementation of control compensators: The coefficient wordlength issue." *IEEE Trans. Auto. Control*, Vol. AC-25, pp. 621–630.

ORLANDI, G., and G. MARTINELLI (1984) "Low sensitivity recursive digital filters obtained via the delay replacement." *IEEE Trans. Circuits Systems*, Vol. CAS-31, No. 7, pp. 654–657.

TAN, C. I., and B. MCINNES (1982) "Adaptive digital control implemented using residue number systems." *IEEE Trans. Auto. Control*, Vol. AC-27, pp. 499–502.

TSCHAUNER, J. (1963) "Introduction à la théorie des systèmes échantillon." Dunod, Paris.

WILLIAMSON, D., and S. SRIDHARAN (1984) "Coefficient and state wordlength correction in digital Kalman filters." Technical Report, University of New South Wales, Australia.

14.8 PROBLEMS

1. Consider the following system:

$$H(s) = \frac{1}{s^2 + 0.1s + 1}$$

The objective is to design a control system having a closed loop bandwidth of approximately 1 Hz and zero steady-state error.

(a) Design a continuous time, unity feedback, control law to achieve this specification. Simulate the closed loop step response.

(b) Repeat part (a) for a digital control law with 2-Hz sampling rate.

(c) Repeat part (b) for a 10-Hz sampling rate. *Note:* In parts (b) and (c) the simulation of the step response should include the intersample behavior.

2. Repeat Problem 1(b) with the following sampling rates: 1 kHz and 1 MHz. (Note that in these cases, it is not necessary to compute the intersample response.) Explain why the performance is worse than for Problem 1(a).

3. For your solution to Problem 1(c), calculate (as in Example 14.3.1) the number of D/A converter bits required to give an output accuracy of 0.1. Simulate the closed loop system, with the quantizer, where the set point is (a) a unit step; and (b) a random input.
4. Repeat the design of Problem 1(c), except include a suitable antialiasing filter. Compare with the design of Problem 1(c).
5. Calculate (as in Table 14.3.1) the -3-dB frequency, gain at $0.9\omega_s$, effective deterioration in noise response, and gain and phase lag at $0.1\omega_s$ for the following antialiasing filter:

$$\frac{(\sqrt{2}\,\omega_s/5)s + (\omega_s^2/25)}{s^2 + (\sqrt{2}\,\omega_s/5)s + (\omega_s^2/25)}$$

6. Repeat the simulations presented in Figure 14.4.5 with the following choices of E_1:
   (a) $E_1(\gamma) = (\gamma + 1.59)(\gamma + 6.44)$
   (b) $E_1(\gamma) = (\gamma + 4)^2$
   (c) $E_1(\gamma) = (\gamma + 0.5)^2$
7. Why is it usual to disconnect integral action during start-up procedures?
8. Repeat Example 14.4.1 for the case where, in addition to the input amplitude limit, there is also an input slew rate limit of $\pm \frac{1}{2}$ unit per second.
*9. Prove Lemma B.1.1. *Hint:* First establish Equation (B.1.4); then show $J \cdot J^{-1} = I$ on noting that

$$P_i(r_j) = \begin{cases} 0, & i \neq j \\ \prod_{k \neq j}(r_k - r_j), & i = j \end{cases}$$

*10. Suppose we wish to add two floating-point numbers, a and b, that have a relative accuracy of $\pm \varepsilon$ [$\varepsilon = 2^{-(b+1)}$; $b =$ number of mantissa bits]. Show that the relative accuracy of the result is

$$\pm \frac{\varepsilon(|a| + |b|)}{|a + b|}$$

11. Consider a slow process where the desired closed loop time constant is about 30 min (that is, closed loop bandwidth $= 1/(2\pi 30 \times 60) = 8.8 \times 10^{-5}$ Hz). If the process is to be sampled once every 15 min, design a suitable antialiasing filter using analog hardware. Repeat the design using the arrangement in Figure 14.3.2, where $\Delta_2 = 15$ min and $\Delta_1 = 10$ sec.

15

An Industrial Case Study

15.1 INTRODUCTION

In this concluding chapter we present a practical case study that illustrates some of the material presented in this book. Our purpose in presenting this case study is twofold. First, we want to give some reassurance to the reader that the techniques that we have described really do relate to practical problems. Second, we want to illustrate the range of issues that arise in a realistic design example. We hope that the case study shows that success in real design problems depends both on having a deep understanding of the particular problem and access to a good tool kit of design aids. The problem we have chosen concerns roll eccentricity control for strip rolling mills. This example is of no particular significance other than it uses a range of control techniques and it illustrates the kind of considerations involved in a realistic problem.

15.2 ROLL ECCENTRICITY CONTROL FOR STRIP ROLLING MILLS

15.2.1 Introduction

The basic physical process of interest here is the rolling of material between preloaded cylindrical rolls to produce a flat strip with desired metallurgical properties. In metal strip production, the thickness tolerances now frequently required in subsequent manufacturing operations have been reduced over the last ten years to

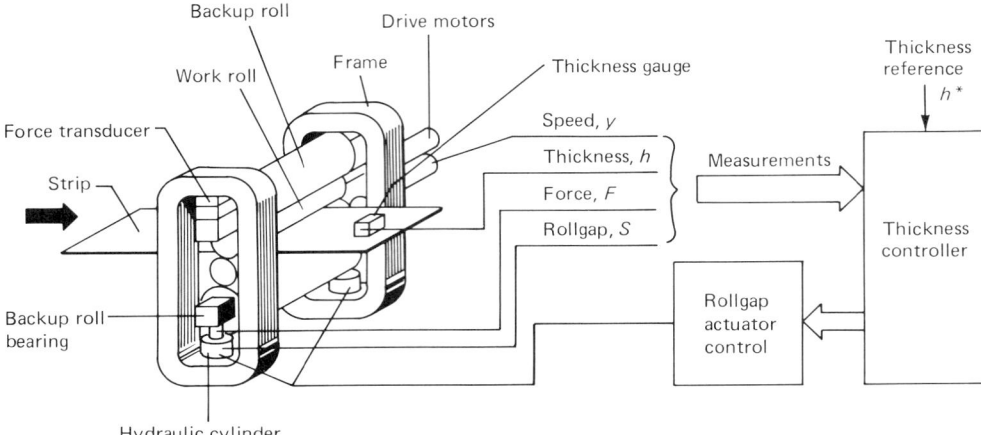

Figure 15.2.1 Rolling mill system components.

plus or minus 2% to 3%. Consequently, sources of thickness variation, such as roll eccentricity, that were previously of little practical significance have now become central issues. Coiled metal sheet products are typically produced in widths ranging from 0.5 to 2.0 m and thicknesses between 0.05 and 10.0 mm. This covers products from thin aluminum foil to coiled steel plate. Significant research effort has been directed toward the design of thickness controllers for single and multistand mills. The essence of the problem is to control the measured center-line thickness of strip emerging from a rolling mill stand, as illustrated in Figure 15.2.1, when auxiliary measurements of roll force, rollgap position, and motor speed are available.

Control is commonly affected by changing the relative gap between the work rolls with a motor-driven screw or hydraulic cylinder acting on the top or bottom backup roll bearing. Usually, the bearing position is measured with respect to the support frame (the rollgap position). The separation of the work rolls is not equal to this measurement because of elastic deformations in the stand components, which are an order of magnitude greater than the thickness control tolerance.

15.2.2 Rolling Mill Model

Analysis of the physics of the process shows that the strip exit thickness h can be modeled as

$$h = D(F, W) + (S - S_0) + e \qquad (15.2.1)$$

where $D(F, W)$ is the elastic deformation of the stand components, W is the strip width, S is the rollgap position with respect to an arbitrary datum, S_0 is a constant, and e is the effective total eccentricity signal for the complete set of rolls in the mill. During rolling, the variations in roll force are typically less than 15% of the average value, and a linear model $\Delta F/M$ [for the nonlinear function $D(F, W)$] may be assumed. Equation (15.2.1), in linearized form, then becomes

$$\Delta F = M(\Delta h - e - \Delta S) \qquad (15.2.2)$$

where the mill modulus M is $-\{[\partial D(F, W)]/\partial F\}^{-1}$. If inertial effects are negligible, the roll force F must also satisfy a nonlinear plastic deformation equation for the material being rolled. The linear form of this equation is

$$\Delta F = \frac{\partial F}{\partial h} \Delta h + F_d \qquad (15.2.3)$$

where F_d is a force change due to external disturbances other than roll eccentricity. These two equations can be solved simultaneously for ΔF and Δh to give

$$\Delta h = \frac{M}{M+k} e + \frac{M}{M+k} \Delta S + \frac{1}{M+k} F_d \qquad (15.2.4)$$

$$\Delta F = -\frac{kM}{M+k} e - \frac{kM}{M+k} \Delta S + \frac{MF_d}{M+k} \qquad (15.2.5)$$

where

$$k = -\frac{\partial F}{\partial h}$$

A typical value for $M/(M+k)$ is $\frac{1}{3}$ for a cold rolling mill. Thus with no control action, that is, S is held constant ($\Delta S = 0$), e is reduced by a ratio of about one-third in Δh.

A simulation showing these effects is given in Figures 15.2.2 and 15.2.3. Figure 15.2.2 shows the eccentricity signal e and the disturbance term $F_d/(M+k)$. Figure 15.2.3 shows the total output response for $\Delta S = 0$. Note that the eccentricity signal has been diminished by roughly one-third, although the disturbance appears undiminished on the output.

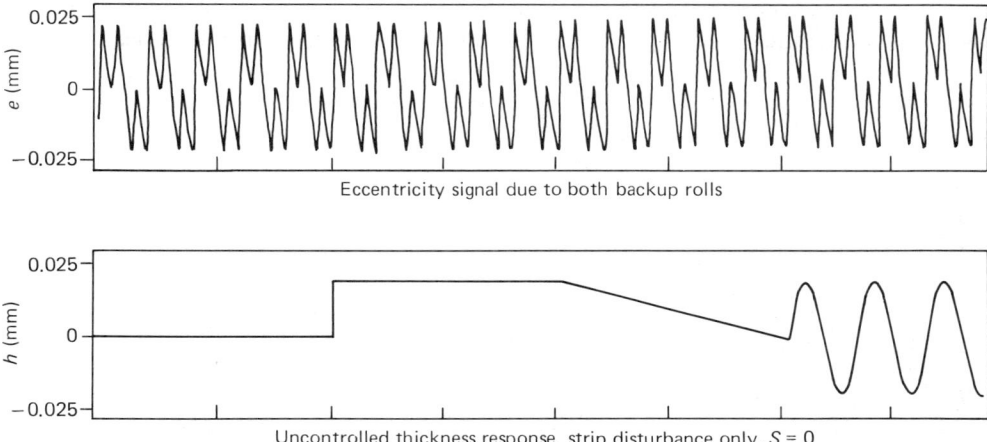

Figure 15.2.2 Eccentricity and disturbance signals.

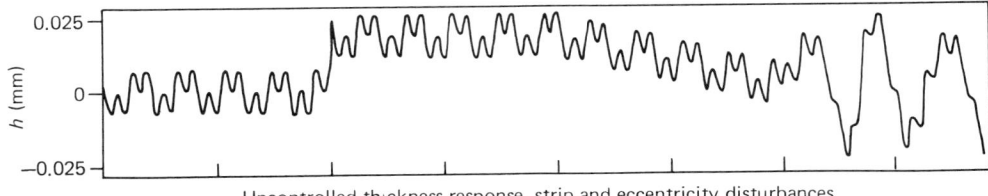

Uncontrolled thickness response, strip and eccentricity disturbances

Figure 15.2.3 Output response for $\Delta S = 0$.

15.2.3 Gaugemeter Control

A major conceptual advance in thickness control was the *gaugemeter* thickness error prediction technique. This approach estimated the instantaneous strip thickness from force and rollgap measurements. Specifically, from Equation (15.2.2) and ignoring e, we obtain the following estimate of the thickness

$$\Delta \hat{h}_g = \frac{1}{M} \Delta F + \Delta S \qquad (15.2.6)$$

Clearly, from (15.2.2), we have $\Delta \hat{h}_g$ is $\Delta h - e$ and thus the relationship between ΔS, e, F_d, and Δh_g is as shown in Figure 15.2.4.

Note that the gaugemeter estimate given in (15.2.6) depends only on ΔF and ΔS, thus avoiding the process time delay inherent in downstream thickness measurements. This gaugemeter technique is particularly important when the ratio of actuator response time to the transport delay time is less than 1. For the case of a mill with hydraulic actuators, this ratio typically varies between 0.05 and 0.75 as the mill is accelerated from the input coil threading speed to maximum speed.

One way of designing a feedback control system is to use high-gain feedback to force $\Delta \hat{h}_g$ to zero. This will reduce the effect of the disturbance F_d as desired. However, this gaugemeter technique, though beneficial in many ways, does have difficulties resulting from errors in the estimated strip thickness caused by roll

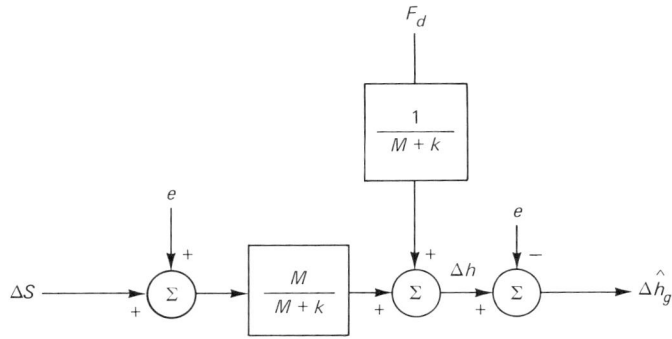

Figure 15.2.4 Gaugemeter thickness estimator.

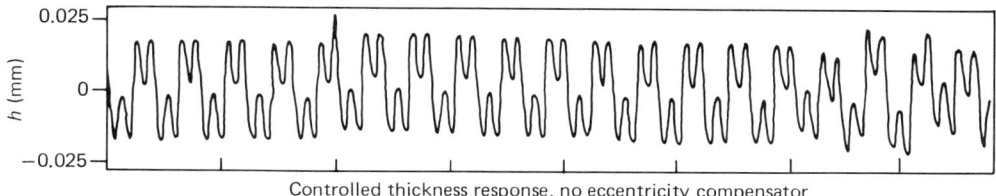

Figure 15.2.5 Gaugemeter control without eccentricity compensation.

eccentricity. This can be seen from Figure 15.2.4, where it is obvious that forcing $\Delta \hat{h}_g$ to zero implies that Δh is e. Actually, it can be seen from the figure that e is measurement noise on Δh; hence high-gain feedback from Δh_g will impress this noise on the output (see Chapter 13).

A simulation verifying this is shown in Figure 15.2.5. Comparing Figure 15.2.5 with Figure 15.2.3, we see that high-gain feedback from $\Delta \hat{h}_g$ has greatly reduced the effect of F_d but has magnified the effect of e by a factor of 3, as predicted.

15.2.4 Eccentricity Control Scheme

The only way out of the difficulties associated with the gaugemeter discussed is to build some form of observer or *soft sensor* for the unmeasurable eccentricity signal, using other measurements, including roll force, roll gap, and downstream thickness. The key to the development of this soft sensor is the fact that the eccentricity signal is nearly periodic, since it arises from the rotation of the work and backup rolls on the mill.

A characteristic of roll eccentricity disturbances is that they are not stationary. That is, although the initial waveform for each roll may be the result of grinding practices and bearing characteristics, the resultant waveform due to all rolls may change due to nonuniform thermal expansion, wear, and other factors.

For these reasons, it is preferable to try to identify the resultant eccentricity signal continuously, rather than before the start of rolling. Techniques for on-line eccentricity estimation have been published in numerous patents and papers. Although seemingly dissimilar, they all turn out to be special cases of the general scheme described next.

The component functions in the general thickness controller with eccentricity compensation are shown in Figure 15.2.6. The functions may be described briefly as follows:

Input signal synchronization: This is a sampling system for receiving mill signals at specified intervals of time or roll position and for aligning the digitized measurements in time (for example, a "time" or length shift is required for the downstream thickness measurement).

Eccentricity estimator: By suitably combining the set of input measurements for a particular point in time, an estimate of the roll eccentricity may be generated, as we show in detail later.

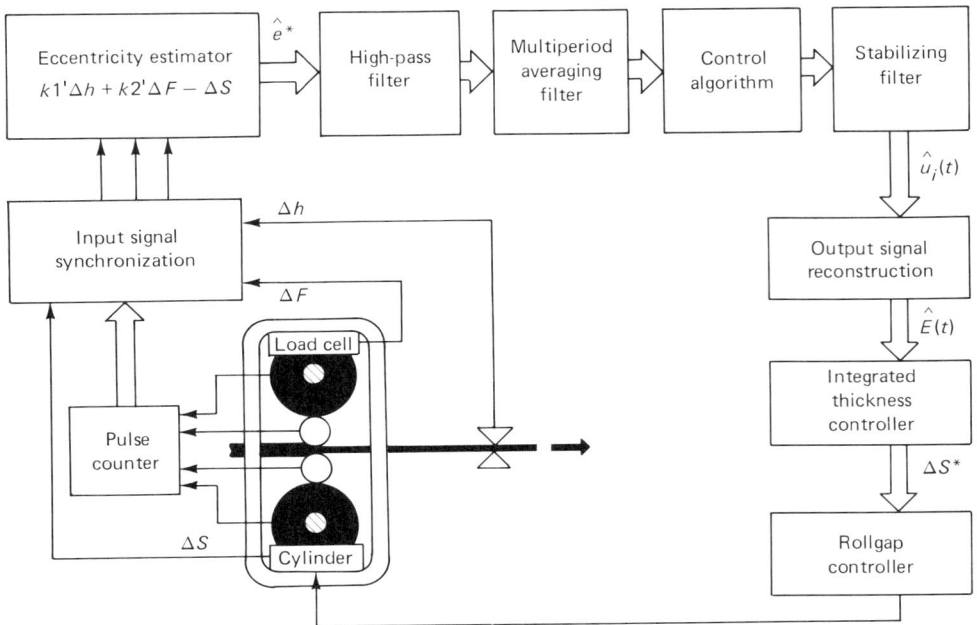

Figure 15.2.6 Structure of a roll eccentricity controller.

High-pass filter: A slow temporal drift in mill variables may occur during each roll period, and some form of high-pass filter is therefore necessary.

Multiperiod averaging filter: Processing the input signals separately for each roll period of interest enables an estimate to be generated for the eccentricity component originating from each roll.

Control algorithm: Once each eccentricity component is estimated, a controller may be designed to predict an output correction separately for each identified eccentricity component.

Stabilizing filter: A stabilizing filter may be required to reduce the sensitivity of the closed loop performance to small timing errors introduced in the input sampling or control output operations.

Output signal reconstruction: The filtered control output signals for each roll, estimated for the current roll angular positions, are combined and directed to the rollgap position actuator.

15.2.5 Design Considerations

In this section we give details of the design of the individual blocks just outlined.

Input-signal synchronization

When multiple rolls are involved, it is not possible to sample all the required input signals at the appropriate roll angular positions. Therefore, interpolation is required to provide values of the input signals synchronized to an appropriate length or roll circumference increment. Signals for the synchronization may be derived from analog tachogenerators or, preferably, pulse generators given by one or more of the rolls. Nominal roll diameters may be used to transform one set of roll angular positions to another in the absence of separate pulse generators; however, some allowance for interroll slip is desirable. An estimate of the strip exit slip ratio is also required to define the relative velocity of the strip and roll.

Eccentricity estimator

A number of measurements, including roll force, measured exit thickness, and rollgap position, contain information about the roll eccentricity effects. By combining two or more input signals, a soft sensor or observer for estimating the total roll eccentricity signal can be obtained. One such estimate can be obtained from (15.2.2) as

$$\hat{e} = \Delta h - \frac{\Delta F}{M} - \Delta S \qquad (15.2.7)$$

or, alternatively

$$\hat{e} = \Delta h - \Delta \hat{h}_g \qquad (15.2.8)$$

where $\Delta \hat{h}_g$ is the gaugemeter estimate of Δh. This expression is independent of the material being rolled.

When the thickness is not measured, as is the case for intermediate stands in a hot strip finishing mill, an alternative expression may be obtained from Equation (15.2.5), which can be rewritten as

$$e = -\frac{(1 + 1/a)\, \Delta F}{M} - \Delta S + F_d/Ma \qquad (15.2.9)$$

where $a = k/M$ (typically a will be near 2). This suggests the following alternate estimate of e:

$$\hat{e}' = -\frac{(1 + 1/a)\, \Delta F}{M} - \Delta S \qquad (15.2.10)$$

which is a function of ΔF and ΔS but not Δh.

Each of the two eccentricity estimates quoted, $\hat{e}$ and $\hat{e}'$, are dependent on the roll force measurement ΔF, which may in certain circumstances be corrupted by nonlinear frictional hysteresis effects. In such cases, an alternative eccentricity estimate is suggested by Equation (15.2.4):

$$\hat{e}'' = (1 + a)\, \Delta h - \Delta S \qquad (15.2.11)$$

A general form of instantaneous eccentricity estimated may therefore be obtained by combining $\hat{e}$, $\hat{e}'$, and $\hat{e}''$:

$$\hat{e}^* = k_1 \hat{e} + k_2 \hat{e}' + [1 - (k_1 + k_2)]\hat{e}'' \qquad (15.2.12)$$

or

$$\hat{e}^* = k_1' \Delta h + k_2' \Delta F - \Delta S \qquad (15.2.13)$$

where

$$k_1' = (1 - k_2)(1 + a) - k_1 a$$

and
$$k_2' = \frac{-[k_1 + k_2(1 + 1/a)]}{M}$$

The optimum values of the constants k_1 and k_2 are determined by the available measurements, accuracy of transducers, noise levels, and the precision with which the strip dependent parameter a can be calculated. Note that (15.2.13) gives a soft sensor for e, given the available measurements, that is, a subset of ΔF, ΔS, Δh.

High-pass filter

Since the measured thickness is the only input signal that is inherently measured in a deviational form, there is a bias present in the eccentricity estimate $\hat{e}^*$ that can be removed by subtracting the moving average of $\hat{e}^*$ as determined from a symmetrical window of data points spaced uniformly over a roll period, that is, a high-pass filter.

Multiperiod averaging filter

Having obtained a soft sensor signal for the instantaneous roll eccentricity, with low-frequency drift removed, the remaining estimation problem is to identify the individual components due to each roll and to enhance the signal-to-noise ratio. For the ith roll having a discrete period τ_i, the eccentricity e_i may be represented as a function of time t by the equation

$$e_i(t) = e_i(t - \tau_i), \qquad i = 1, \ldots, n \tag{15.2.14}$$

The composite eccentricity signal $\hat{e}^*(t)$ can be thought of as being the sum of the components $e_i(t)$ plus noise, that is,

$$\hat{e}^*(t) = \sum_{i=1}^{n} e_i(t) + N(t) \tag{15.2.15}$$

Our objective is therefore to estimate the components $e_i(T)$ for $0 \leq T < \tau_i$ from the composite signal $\hat{e}^*(t)$ for each of the n periods.

The required filter for a particular period needs to have a low gain at frequencies other than the corresponding roll rotation frequency and each of its harmonics. In principle, the design of such a filter could be achieved by any one of several methods, including Kalman filtering, notch filtering, and Fourier transformation. It will be shown that each of these methods leads to a similar end result.

(i) Kalman filter

The problem of estimating a single sinusoidal component buried in noise was studied in Chapter 10. There it was argued that a suitable near-optimal Kalman filter for this problem has the form

$$\rho \hat{x} = \begin{bmatrix} \dfrac{\cos \omega_0 \Delta - 1}{\Delta} & \dfrac{1}{\omega_0 \Delta} \sin \omega_0 \Delta \\ \dfrac{-\omega_0}{\Delta} \sin \omega_0 \Delta & \dfrac{\cos \omega_0 \Delta - 1}{\Delta} \end{bmatrix} \hat{x} + 2\xi\omega_0 \begin{bmatrix} \cos \omega_0 \Delta \\ \sin \omega_0 \Delta \end{bmatrix} (y - \hat{x}_1)$$

$$\tag{15.2.16}$$

This filter is time invariant, and hence we can evaluate the transfer function from y to the first component of the estimated state vector. This gives

$$\hat{x}_1(\delta) = F_\Delta(\delta) Y(\delta) \tag{15.2.17}$$

where

$$F_\Delta(\delta) = \frac{2\xi\omega_0 \delta(\cos \omega_0 \Delta) + 2\xi\omega_0[(\cos \omega_0 \Delta - 1)/\Delta]}{\delta^2 + \beta\delta + \alpha} \tag{15.2.18}$$

where

$$\beta = 2\left(\frac{1 - \cos \omega_0 \Delta}{\Delta}\right) + 2\xi\omega_0 \cos \omega_0 \Delta$$

$$\alpha = \frac{2(1 - \cos \omega_0 \Delta)(1 + 2\xi\omega_0 \Delta)}{\Delta^2} + \frac{2\xi(\sin \omega_0 \Delta)^2}{\Delta}$$

which in the limit, as $\Delta \to 0$, reduces to the following continuous time result:

$$F(s) = \frac{2\xi\omega_0 s}{s^2 + 2\xi\omega_0 s + \omega_0^2} \tag{15.2.19}$$

(ii) Notch Filter

The frequency response of the near-optimal filter given in Equation (15.2.19) can be easily seen to have a low gain at frequencies away from $f_0 = \omega_0/2\pi$. However, the frequency response at f_0 is exactly unity; that is, the filter gives the in-phase component of a single-frequency input at f_0. Also, the 3-dB bandwidth about the frequency f_0 is of order $\xi\omega_0/4\pi$. Therefore, it is apparent that the near-optimal filter has the characteristics of a conventional notch filter.

(iii) Fourier Analysis

An alternative method of extracting the eccentricity signal components corresponding to the frequency f_0 is to perform a Fourier analysis. This can be viewed as a correlation of the input signal with $\cos \omega_0 t$ and $\sin \omega_0 t$. This is easily evaluated using the following discrete time equation:

$$x(k+1) = \begin{bmatrix} x_1(k+1) \\ x_2(k+1) \end{bmatrix} = \alpha \begin{bmatrix} c & s \\ -s & c \end{bmatrix} \begin{bmatrix} x_1(k) \\ x_2(k) \end{bmatrix} + 2(1-\alpha)\begin{bmatrix} c \\ -s \end{bmatrix} \hat{e}^*(k) \tag{15.2.20}$$

where $c = \cos \omega_0 \Delta$, $s = \sin \omega_0 \Delta$. Solving this equation explicitly from the initial condition $x(0) = 0$ gives

$$x_1(k) = 2(1-\alpha) \sum_{l=1}^{k} \alpha^{l-1} \cos(l\omega_0 \Delta) \hat{e}^*(k-l) \tag{15.2.21}$$

$$x_2(k) = 2(1-\alpha) \sum_{l=1}^{k} \alpha^{l-1} \sin(l\omega_0 \Delta) \hat{e}^*(k-l) \tag{15.2.22}$$

Equations (15.2.21), (15.2.22) clearly generate the Fourier expansion of $\hat{e}^*(k-t)$ from which the normal Fourier coefficients can be recovered as

$$a = 2[x_1(t)\cos\omega_0 t - x_2(t)\sin\omega_0 t]$$

and (15.2.23)

$$b = 2[x_1(t)\sin\omega_0 t + x_2(t)\cos\omega_0 t]$$

This result reveals the one-to-one relationship between $x(t)$ and the Fourier coefficients a and b. Practically, it is more convenient to work directly with $x(t)$ rather than to evaluate explicitly the Fourier coefficients. Another important feature of Equations (15.2.21), (15.2.22) is that they have an embedded exponential data window.

The transfer function $F'(z)$ of the first component of the time invariant filter represented by Equation (15.2.20) is

$$F'(z) = \frac{2(1-\alpha)}{z^2 - 2\alpha c z + \alpha^2}[cz - \alpha] \quad (15.2.24)$$

or, in delta form,

$$F'_\Delta(\delta) = \frac{\frac{2(1-\alpha)}{\Delta}[\cos\omega_0\Delta]\delta + \frac{2(1-\alpha)}{\Delta^2}[\cos\omega_0\Delta - \alpha]}{\delta^2 + 2\delta\left\{\frac{1 - \alpha\cos\omega_0\Delta}{\Delta}\right\} + \left\{\frac{(1+\alpha^2) - 2\alpha\cos\omega_0\Delta}{\Delta^2}\right\}}$$

This result for $F'(z)$ is equal (up to terms in ξ^2) to that for $F(\delta)$ [see Equation (15.2.18)] with the choice

$$\alpha = 1 - \xi\omega_0\Delta \qquad \text{\textasciitilde}$$

(iv) Periodic Averaging Filter

Heuristic arguments suggest that a single period waveform of period τ can be estimated using the following discrete time *periodic averaging filter*, which acts on points separated by τ samples:

$$\hat{e}(t) = \alpha\hat{e}(t-\tau) + (1-\alpha)\hat{e}^*(t), \qquad 0 \le \alpha < 1 \quad (15.2.25)$$

Inspection of equation (15.2.25) shows that past data are given an exponential weighting. The parameter α affects the memory of the filter such that, if α is near 1, then the filter will have a long memory, good noise discrimination, and a slow response to dynamic changes. Conversely, if α is near 0, the filter will have a short memory with poor noise discrimination, but rapid adaptability.

Equation (15.2.25) is equivalent to a system of integrators in the time domain that operates on a periodic basis at fixed, uniform time displacements from an arbitrary point on the eccentricity cycle. Such a system of integrators is sometimes referred to as an *integral, rotating store* in recognition of the cyclic nature of the operation. Although motivated intuitively, this filter is exactly the same as that obtainable by combining filters of the form (15.2.19) or (15.2.24) for each harmonic component in the signal of period τ. This can be seen as follows: the filter in equation (15.2.18) can be approximately (for α near 1) written as

$$F(z) = \frac{D_s - D}{D_s}$$

where

$$D_s = z^2 - 2(1 - \xi\omega_0\Delta)cz + (1 - \xi\omega_0\Delta)^2, \qquad D = z^2 - 2cz + 1 \quad (15.2.26)$$

If all h harmonics of a signal of period τ, where $\tau = h\Delta/2$, are combined, then the resulting filter is

$$F_\tau(z) = \sum_{i=1}^{h} \frac{D_s^i - D^i}{D_s^i} \quad (15.2.27)$$

Ignoring terms of order ξ^2 or smaller and using Equation (15.2.26), this expression simplifies to the following surprising result:

$$F_\tau(z) = \frac{1 - \alpha}{z^\tau - \alpha}, \qquad \text{where } \alpha = (1 - \xi\omega_0)^{\tau/\Delta} \quad (15.2.28)$$

Note that this is exactly the transfer function of the periodic averaging filter given in Equation (15.2.26). ▽▽▽

Therefore, it may be concluded that the Kalman filter, notch filter, Fourier transform, and periodic averaging filter approaches are equivalent and reduce to identical mathematical operations on the data. Perhaps this result should not be surprising, since each approach is based on the assumed, periodic nature of the disturbance.

The extension of the single-period case to the multiple-period case is basically achieved by operating on $\hat{e}^*(t)$ by filters designed for different values of the periods τ_i. However, it helps to reduce interaction between the filters for different periods if the following algorithm is used:

$$\hat{e}_i(t) = \alpha_i \hat{e}_i(-\tau_i) + (1 - \alpha_i)\hat{e}^*(t) - k \sum_{\substack{j=1 \\ j \neq i}}^{n} \hat{e}_j(t - \tau_j), \qquad 0 < k < 1$$

$$(15.2.29)$$

Note that this form can be motivated from observer theory.

The summation term in Equation (15.2.29) removes the estimated components corresponding to other periods from the unfiltered eccentricity signal, prior to the application of the filtering procedure for a particular period. The choice $k = 1$ may give a marginally stable filter, and thus it is desirable to set k slightly less than 1.

Figure 15.2.27 shows a simulation of the application of these ideas to a four-roll mill. The top trace shows the composite eccentricity signal. The bottom

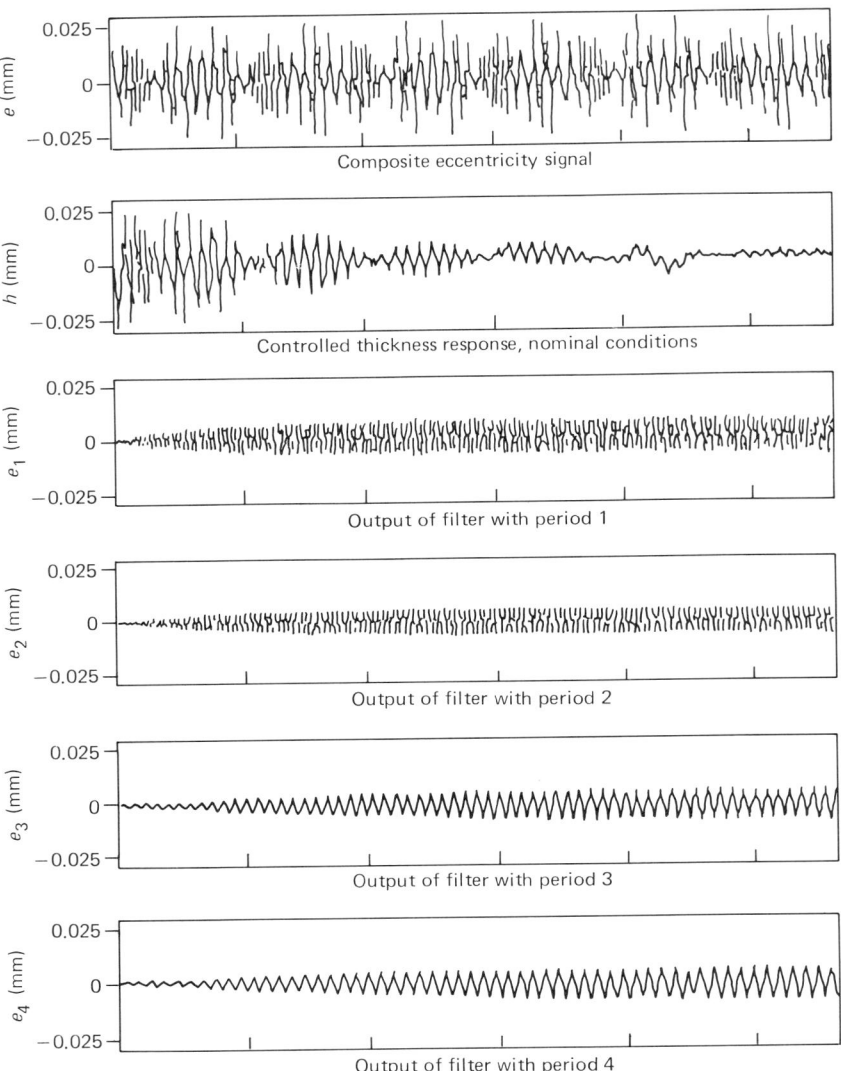

Figure 15.2.7 Simulation results for the case of an eccentricity disturbance from four rolls each having different periods.

four traces show the estimated individual eccentricity components corresponding to (15.2.13) to estimate $\hat{e}^*$, followed by (15.2.29) to extract the individual components.

Output signal reconstruction

If the measured value of Δh is used in determining $\hat{e}_1 \ldots \hat{e}_n$, then there will be a delay in the estimates corresponding to the delay inherent in the measurement of Δh. However, the periodicity of $e_1 \ldots e_n$ can be exploited to obtain an estimate of the composite eccentricity signal at the present time as follows:

$$\hat{e}(t) = \sum_{i=1}^{n} \hat{e}_i(t - \tau_i) \tag{15.2.30}$$

This equation was, in fact, a key motivation for isolating the individual eccentricity components.

We can then use Equation (15.2.31) to obtain an estimate of Δh based on the instantaneously available measurements of ΔF and ΔS; that is,

$$\Delta \hat{h} = \frac{1}{M} \Delta F + \Delta S + \hat{e} \tag{15.2.31}$$

The first two terms on the right side of (15.2.31) will be recognized as the gaugemeter estimate of Δh. However, use of these terms alone generally gives an unsatisfactory solution due to the presence of the eccentricity components, as we have seen earlier. The third term in (15.2.31) compensates for these components by use of the soft sensor for the unmeasured component e.

Control algorithm

Note that the thickness estimate given in (15.2.31) effectively eliminates the time delay inherent in the direct measurement of Δh. Thus the availability of $\Delta \hat{h}$ allows a wide bandwidth feedback control system to be designed that capitalizes on the fast actuator roll-positioning actuators. A slow integral loop from Δh gives zero steady-state tracking errors.

Figure 15.2.8 shows the simulated response under the same conditions as in Figures 15.2.3 and 15.2.5. The hydraulic system simulated was capable of responding to a 0.1-mm rollgap change in 0.06 sec. Key simulation parameters were:

- Mill modulus: 3.5 MN/mm
- Gaugemeter attenuation factor: 5.0
- Strip width: 1000 mm
- Strip plasticity constant a: 2.0
- Time delay to gauge: 0.4 sec
- Nominal backup roll period: 1.0 sec

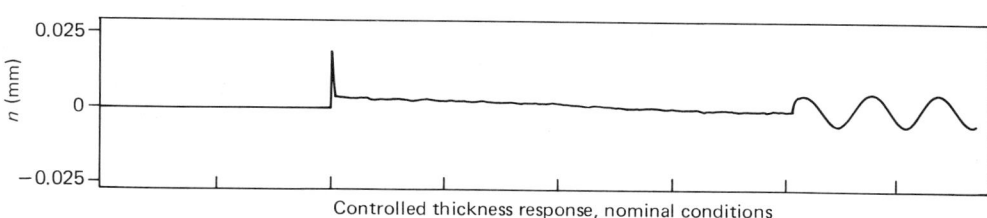

Figure 15.2.8 Simulation results using the eccentricity compensated gaugemeter estimate for feedback.

It can be seen that the eccentricity component has been completely eliminated, while the sinusoidal input disturbance has been significantly attenuated. Also, it can be seen that the system responds rapidly to step input disturbances.

15.2.6 Experimental Results

The scheme described has been patented and is available commercially. The trade name AUSREC is used to describe this approach to gauge control. It has been tested in several locations. In a typical application, the stand 1 eccentricity control system is implemented in a pair of DEC PDP 11/23 computers, one for gaugemeter and one for eccentricity control functions. Common measurements of force, hydraulic pressure, and cylinder position from each side of the mill are used, together with a thickness measurement from a fast-response thickness gauge located 2 m after the stand. A 1200-pulse/rev generator is connected to the work roll drive for tracking. Proximity switches are mounted near the ends of the backup roll necks to provide one synchronization pulse per revolution.

The applications software was written in FORTRAN 77 with a small amount of coding in MACRO Assembler to interface with the I/O hardware. One real-time task with several interrupt-triggered routines handles all the signal processing and calculations necessary to produce the individual eccentricity estimates and generate the correctly phased eccentricity compensation output signal from three individual eccentricity stores, one for each backup roll and one for the combined work roll eccentricity.

The number of elements in the work roll store is a predefined constant, nominally 12. The number of elements in each backup store is determined by the backup roll diameter so that each element corresponds approximately to the same elemental length of strip. A typical backup store contains 28 to 30 elements. An interpolation procedure is used for updating the store elements and for reconstructing the composite eccentricity signal for the compensation output.

The analog sampling is synchronized with the roll speed by causing an input conversion to be triggered after a fixed number of work roll pulses have been counted. With 12 elements per work roll store, a hardware counter is configured to trigger a DMA analog input once every 100 pulses. The completion of the analog input in turn triggers the software routine to process the data, update the stores, and generate the next output.

When new rolls are installed, the periodic filter constants for those rolls are initialized to a lower value to allow faster initial updating of the eccentricity stores for those rolls. Several fault-tolerant features have been built into the system. Each time a backup roll synchronizing pulse is received, a check is made on the number of work roll pulses received. If this is outside the expected range, then a synchronization fault is recognized, and the data collected during the last backup roll rotation is rejected. Temporary synchronization faults, possibly due to roll slippage or spurious pulses, are therefore prevented from corrupting the stores.

It is also possible to operate the system without one or both backup roll pulses. Successful tests were performed without the backup roll marker pulses, giving only a minor change in performance.

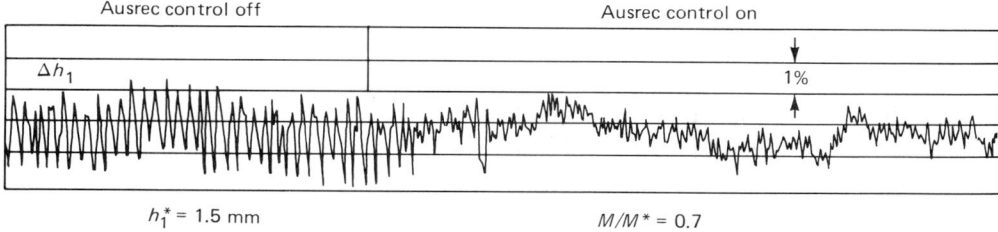

Figure 15.2.9 Measured eccentricity control result from a five-stand rolling mill.

Store updating is inhibited below a minimum mill speed and during periods of rapid acceleration or deceleration. A switch is provided to allow for selection of either load cells or hydraulic pressure for the force measurement. A real-time graphical display of any of the three eccentricity stores may be selected, and any common system variable may be displayed in chart-recorder format on the VDU screen.

Typical results from the rolling of 0.25-mm tinplate are shown in Figure 15.2.9 in the form of a stand 1 exit thickness variation chart. The trace has the $\pm 1\%$ tolerance limits marked to indicate the performance achieved. Limited tests conducted using the estimate $\hat{e}'$ [Equation (15.2.10)] gave results nearly equal to those for the thickness measurement based estimates.

15.3 SUMMARY

This chapter described an industrial case study that illustrates how digital control and estimation may be used in a realistic situation. Some of the key points made were:

- It is important to understand the physics of the problem and the objective of control.
- In gauge control in rolling mills, the performance of the control system is limited by the measurement time delay between the rolls and the exit thickness meter.
- The time delay can be eliminated by using an observer for the exit thickness that employs instantaneous measurements of roll force and screw down position.
- However, roll eccentricity prevents the design objectives from being achieved unless extra steps are taken.
- The main characteristic of the eccentricity signal is that it arises from the rotation of the rolls and hence is nearly periodic. This periodicity can be exploited to design an observer that removes the eccentricity components from the estimated exit thickness.
- The final design brings together the estimation of exit thickness with the estimation of eccentricity, leading to a wide-bandwidth controller that achieves high-tolerance gauge control.

15.4 REFERENCES

Background papers on roll eccentricity control are:

BLAIN, P. (1948) "Contribution à l'étude des laminoirs à bandes." *Rev. Metallurgie*, p. 45.

CREUSOT LOIRE (1970) British Patent 1319327.

DAVY-LOEWY (1974) German Patent 2416867.

EDWARDS, W. J. (1978) "Design of entry strip thickness controls for tandem cold mills." *Automatica*, Vol. 14, pp. 429–441.

FAPIANO, D. J., and D. E. STEEPER (1984) "Control of strip thickness in hot rolling." *Iron Steel Engineer*, Nov., pp. 34–43.

IMAI, I., and others (1973) "FARE detector and control system for elimination of roll eccentricity." *Isikawagima-Harima Engineering Rev.*, Vol. 13, No. 2, pp. 189–198.

MIKE, T., S. TAKAI, and T. YANAGUCHI (1984) "A new roll eccentricity compensation system." *Nisshin Steel Technical Report No. 51*, p. 77.

NAKAZATO, Y., and others (1983) "Improvements in cold rolling of extra-thin gauge strip." *Kawasaki Steel Technical Report 7*, pp. 44–54.

SIEMENS (1987) "New method compensator backup roll eccentricity affording closer tolerances." Siemens Technical Publication.

TEOH, E. K., and others (1984) "An improved thickness controller for a rolling mill." *Proc. 9th IFAC World Congress*, Budapest, pp. 1741–1746.

The AUSREC scheme for roll eccentricity is described in:

EDWARDS, J., P. THOMAS, and G. C. GOODWIN (1987) "Roll eccentricity control for strip rolling mill." Special Case Study Paper, 10th IFAC World Congress, Munich, Vol. 2, pp. 200–211.

An alternative application of multiperiod filtering is described in:

GOODWIN, G. C., R. J. EVANS, R. L. LEAL and R. A. FEIK (1986) "Sinusoidal disturbance rejection with application to helicopter flight data estimation." *IEEE ASSP*, Vol. 34, No. 3, pp. 479–485.

APPENDIX A

Hardware Aspects of Digital-To-Analog Interfacing

A.1 D/A CONVERSION

D/A conversion can be readily achieved by use of a simple resistor ladder network. A typical circuit is shown in Figure A.1.1. It is readily seen that

$$V_7' = \frac{1}{3}\left\{\sum_{i=0}^{7} 2^{7-i} V_i\right\}$$

$$= \frac{V_r}{3}\left\{\sum_{i=0}^{7} 2^{7-i} b_i\right\} \quad \text{(A.1.1)}$$

where b_i denotes the ith bit and is zero or one.

Usually the switches shown in Figure A.1.1 would be implemented using electronic devices, for example, field effect transistors on an integrated circuit. Note that the D/A conversion process is essentially instantaneous.

A.2 A/D CONVERSION

As discussed in Chapter 3, the basic idea of an A/D converter is a sampler. There are several classes of A/D converter. The simplest and fastest is sometimes referred to as a *flash converter*. The basic circuit for this is shown in Figure A.2.1. The disadvantage of a flash converter is obviously its complexity and cost.

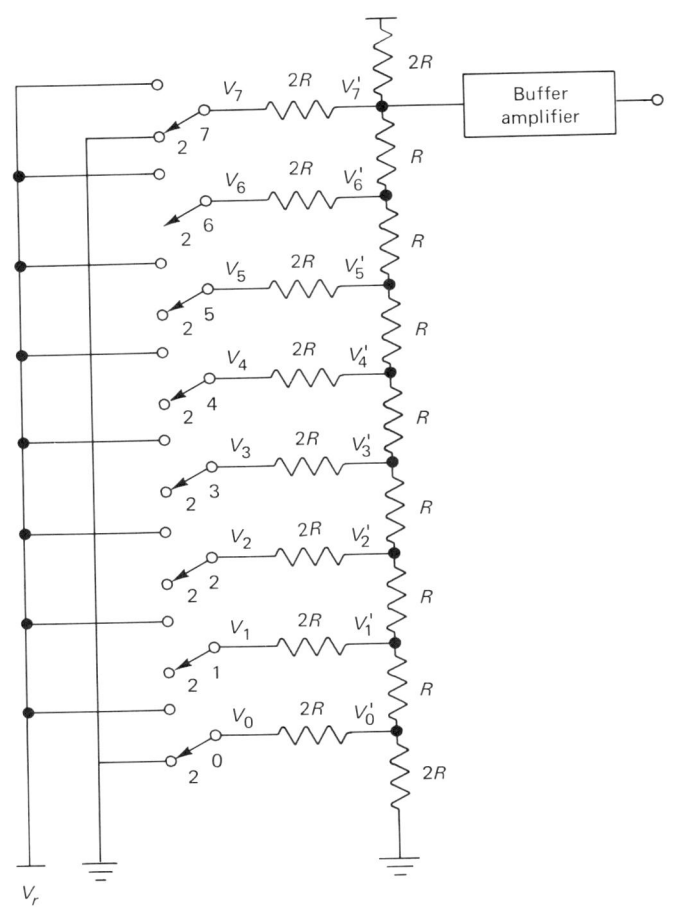

Figure A.1.1 Typical D/A converter circuit.

An alternative scheme is to use a D/A converter in a feedback arrangement as shown in Figure A.2.2. The principle of operation for this type of converter is that the search logic alters the output of the D/A converter until V_f is within half a bit of V_i. This search can be performed in a number of ways, for example, by exhaustive search, in which case the search logic is simply a counter plus comparator. Alternatively, successive approximation can be used, which corresponds to the use of a binary search.

It is clear from this brief description that in A/D conversion it will take some time before the converter output is a good representation of the signal. For this reason, typical A/D converters require a START pulse to initiate conversion. The A/D hardware usually includes a sample-and-hold circuit, which holds the analog signal as soon as the START pulse is sent. The A/D then issues a CONVERSION

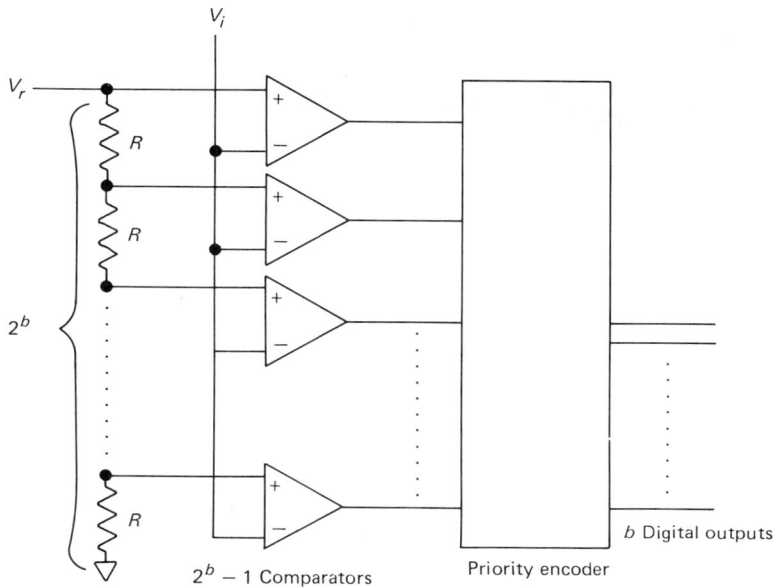

Figure A.2.1 Flash converter.

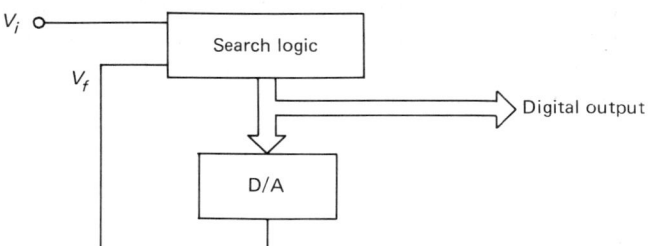

Figure A2.2 A/D converter using D/A feedback.

COMPLETE pulse when the A/D output can be read. The computer usually checks to see if the CONVERSION COMPLETE flag has been set before issuing a READ command to reach the A/D output. Alternatively, when conversion has been completed, the A/D may generate an interrupt to communicate with the computer.

A.3 REFERENCES

For more information on A/D and D/A converters, see:

HOROWITZ, P., and W. HILL (1980) *The Art of Electronics*. Cambridge University Press, New York.

SIMPSON, R. E. (1987) *Introductory Electronics for Scientists and Engineers*. Allyn and Bacon, Newton, Mass.

APPENDIX B

Numerical Issues in Control

In this appendix we give a more detailed treatment of finite word length characteristics of delta and shift operator models. Throughout we will use transfer function (or, equivalently, controller or observer state space) models. Note that when using cascade or parallel realizations (see Section 14.5) this analysis applies to each of the blocks.

B.1 COEFFICIENT REPRESENTATION

One issue that concerns the practical implementation of digital control algorithms is the effect of coefficient round-off on the overall control scheme. Let $p(x)$ denote a polynomial in either δ or q representing the numerator or denominator of the controller transfer function. Suppose we wish to represent the coefficients of $p(x)$ with a fixed number of bits. This will introduce an error in each coefficient, say of a maximum of ε. Given this maximum perturbation of the coefficients of a polynomial, approximate bounds on the corresponding variation in the roots of the polynomial can be evaluated.

Since, in general, the exact relationship between a change in coefficients and a change in roots is complicated, we will use the first derivative of the roots of the polynomial with respect to the vector of coefficients as an approximate measure of the sensitivity. Clearly, this will be accurate for small perturbations. The following result will be useful in deriving bounds on the root sensitivity.

Lemma B.1.1

Given any monic polynomial $p(x)$ of degree n having coefficients a_i = coefficient of x^i and zeros r_i ($i = 1, \ldots, n$), assumed distinct, then defining $p_i(x)$, $\underline{a}$ and $\underline{r}$ by

$$p_i(x) \cdot (x - r_i) = p(x) \tag{B.1.1}$$

$$\underline{a}^T = [\, a_0 \quad a_1 \quad \ldots \quad a_{n-1} \,] \tag{B.1.2}$$

$$\underline{r}^T = [\, r_1 \quad r_2 \quad \ldots \quad r_n \,] \tag{B.1.3}$$

we have

$$J = \frac{\partial \underline{a}}{\partial \underline{r}} = -[\, p_{ki} \,] = [\, J_{ik} \,] \tag{B.1.4}$$

(where p_{ki} is the coefficient of x^{k-1} in p_i) and

$$J^{-1} = -\left[r_i^{k-1} \prod_{\substack{l \neq i}}^{n} (r_i - r_l)^{-1} \right] = [\,(J^{-1})_{ki}\,] \tag{B.1.5}$$

Proof: See Problem 14.9. ▽▽▽

Now using J as in (B.1.4)

$$\tilde{\underline{a}} \approx J \cdot \tilde{\underline{r}} \tag{B.1.6}$$

where $\tilde{\underline{a}}$ and $\tilde{\underline{r}}$ denote small vector changes in the coefficient vector and the zero vector, respectively.

The appropriate interpretation of Equation (B.1.6) will depend on whether fixed-point or floating-point arithmetic is used. Also, it is important that scaling be incorporated to ensure that maximum use is made of the available word length. In the subsequent analysis we use the following notation: a', a'' denote coefficients for the delta and shift forms, respectively, r', r'' denote the corresponding zeros, and $\tilde{r}'$ will denote the perturbation (due to coefficient rounding) in the Z domain of the zero of a δ model; that is, $\tilde{r}' = \Delta \tilde{r}'$.

For the sake of illustration, we will analyze only the case of floating-point arithmetic. However, the same qualitative conclusions hold for fixed-point arithmetic; see the references given in Chapter 14.

Floating-point arithmetic is equivalent to a form of scaling in which the scaling is the exponent. Thus the scaling factors are $n_i = n(a_i) = 2^{[\log_2 |a_i| + 1]}$, where $[x]$ denotes the largest integer strictly less than x. To simplify the argument, we assume that Δ is an integer power of 2. A similar result holds for general Δ. Thus let a_i^* denote the mantissa of the floating-point coefficient a_i. Then (B.1.6) becomes

$$\tilde{r} \approx J^{-1} N \tilde{a}^* \tag{B.1.7}$$

where $N = \text{diag}\{n_i\}$. Thus the maximum possible change in zero location can be approximately bounded as

$$\|\tilde{r}\|_\infty \leq \|J^{-1} N\| \, \|\tilde{a}^*\|_\infty \tag{B.1.8}$$

(i) Shift operator form

From (B.1.8), we have

$$\|\tilde{r}''\|_\infty \leq \varepsilon \sup_i \left\{ \prod_{k \neq i}^n |r_i'' - r_k''|^{-1} \sum_{j=0}^{n-1} n(a_j'') |r_i''|^j \right\} \quad \text{(B.1.9)}$$

(ii) Delta form

$$\|\tilde{r}'\|_\infty \leq \varepsilon \sup_i \left\{ \prod_{k \neq i}^n |r_i' - r_k'|^{-1} \sum_{j=0}^{n-1} n(a_j') \Delta^{n-j} |r_i'|^j \right\} \quad \text{(B.1.10)}$$

(iii) Comparison

To compare (B.1.9) and (B.1.10), we need to express the errors in the same domain. Thus we let $\tilde{r}'$ be the perturbation of the delta form roots in the Z-domain; that is, $\tilde{r}' = \Delta \tilde{r}'$. Also, since we are considering the same controller in two different forms, it follows that the roots are related by

$$r_i' = \frac{r_i'' - 1}{\Delta} \quad \text{(B.1.11)}$$

(where r_i' is the δ-domain root and r_i'' is the Z-domain root). Using (B.1.11) in (B.1.10) then gives

$$\|\tilde{\tilde{r}}'\|_\infty \leq \varepsilon \sup_i \left\{ \prod_{k \neq i}^n |r_i'' - r_k''|^{-1} \sum_{j=0}^{n-1} n(a_j') \Delta^{n-j} |r_i'' - 1|^j \right\} \quad \text{(B.1.12)}$$

Inspection of (B.1.9) and (B.1.12) shows that comparison of delta and shift models depends on the relative magnitudes of $n(a_j') \Delta^{n-j}$ (delta) and $n(a_j'')$ (shift). Now a_j'' is related to the sum of products of the zeros; that is,

$$|a_j''| = \sum_{i_1 \neq i_2 \neq \ldots i_{n-j}} |r_{i_1}'' \ldots r_{n-j}''| \quad \text{(B.1.13)}$$

where we have assumed that all zeros have the same sign. Similarly,

$$\Delta^{n-j} |a_j'| = \Delta^{n-j} \sum_{i_1 \neq i_2 \neq \ldots i_{n-j}} |r_{i_1}'| \ldots |r_{i_{n-j}}'| \quad \text{(B.1.14)}$$

$$= \sum_{i_1 \neq i_2 \neq \ldots i_{n-j}} \Delta^{n-j} \left|\frac{r_{i_1}'' - 1}{\Delta}\right| \cdots \left|\frac{r_{i_{n-j}}'' - 1}{\Delta}\right| \quad \text{(B.1.15)}$$

$$= \sum_{i_1 \neq i_2 \neq \ldots i_{n-j}} |r_{i_1}'' - 1| \cdots |r_{i_{n-j}}'' - 1| \quad \text{(B.1.16)}$$

So that we can compare these expressions, we will make the assumption that the roots are closer to $1 + j0$ (in the Z plane) than they are to the origin; that is,

$$|r_k'' - 1| < |r_k''| \quad \text{(B.1.17)}$$

Note that this assumption will almost always be satisfied provided the sampling rate is chosen appropriately (see Section 14.2). Under these conditions we have from (B.1.13) and (B.1.16) on using (B.1.17) that

$$|a'_j| \Delta^{n-j} < |a''_j| \tag{B.1.18}$$

Since $n(\cdot)$ is a monotonic function, (14.5.26) implies

$$n(a'_j) \leq n(a''_j \Delta^{j-n}) = n(a''_j) \Delta^{j-n} \tag{B.1.19}$$

when Δ is a power of 2. Using (B.1.19) and (B.1.17), we see that the bound in (B.1.12) for the delta operator case is always less than the corresponding bound in (B.1.9) for the shift operator case. This shows that the delta operator form has lower root sensitivity to numerical errors than does the shift operator form. This is illustrated in the following example.

Example B.1.1

Consider the case where we wish to control a continuous time system with transfer function $H(s) = [s(1 + s)]^{-1}$, with a discrete time controller with a sampling rate of 10 Hz. It can be shown that this gives rise to a discrete time transfer function of

$$H''(q) = \frac{0.0048374q + 0.0046788}{q^2 - 1.90484q + 0.90484} \tag{B.1.20}$$

or

$$H'(\delta) = \frac{0.048374\delta + 0.95162}{\delta^2 + 0.95162\delta} \tag{B.1.21}$$

Now suppose we wish to assign the closed loop poles to $s = -0.5, -1,$ and -2; that is, $z = 0.95162, 0.90484,$ and 0.81269. The controller calculations have been performed as follows: The "exact" $(2n - 1) \times (2n - 1)$ eliminant matrix (where n = order) is calculated and all entries in this matrix rounded to a finite-word-length floating-point number. Similarly, the vector of coefficients of the desired closed loop polynomials is calculated and all entries are rounded to a finite-word-length floating-point number. Using this vector and matrix, an accurate solution to the pole

TABLE B.1.1 FINITE-WORD-LENGTH POLE ASSIGNMENT USING SHIFT OPERATORS FOR EXAMPLE B.1.1

Number of Bits	Controller Numerator	Controller Denominator	Closed Loop Poles
2	$0.0z + 0.0$	$z - 0.75$	$1.0000, 0.9048, 0.7500$
5	$2.5000z + 0.859$	$z - 0.7813$	$1.0161 \pm j0.2616, 0.6417$
8	$0.8438z - 0.4336$	$z - 0.7734$	$0.9682 \pm j0.1178, 0.7378$
12	$0.9240z - 0.8215$	$z - 0.7744$	$0.9329 \pm j0.0251, 0.8089$
16	$0.9293z - 0.4811$	$z - 0.7745$	$0.9516, 0.9043, 0.8189$
20	$0.9290z - 0.8405$	$z - 0.7745$	$0.9511, 0.9051, 0.8187$
25	$0.9290z - 0.8406$	$z - 0.7745$	$0.9512, 0.9049, 0.8187$
30	$0.9290z - 0.8406$	$z - 0.7745$	$0.9512, 0.9049, 0.8187$

TABLE B.1.2 FINITE-WORD-LENGTH POLE ASSIGNMENT USING DELTA OPERATORS FOR EXAMPLE B.1.1

Number of Bits	Controller Numerator	Controller Denominator	Closed Loop Poles
2	$1.0000\delta + 0.7500$	$\delta + 2.0000$	$-0.3738, -1.3131 \pm j0.4304$
5	$1.0625\delta + 0.9063$	$\delta + 2.2500$	$-1.5400, -1.2623, -0.4408$
8	$0.9336\delta + 0.8828$	$\delta + 2.2500$	$-1.7978, -0.9645, -0.4845$
12	$0.9285\delta + 0.8840$	$\delta + 2.2559$	$-1.8140, -0.9505, -0.4879$
16	$0.9290\delta + 0.8841$	$\delta + 2.2555$	$-1.8127, -0.9517, -0.4877$
20	$0.9290\delta + 0.8841$	$\delta + 2.2555$	$-1.8127, -0.9516, -0.4877$
25	$0.9290\delta + 0.8841$	$\delta + 2.2555$	$-1.8127, -0.9516, -0.4877$
30	$0.9290\delta + 0.8841$	$\delta + 2.2555$	$-1.8127, -0.9516, -0.4877$

assignment equation is calculated using double-precision arithmetic on a VAX 11/750 (55-bit mantissa, 7-bit exponent). This solution is then rounded to a vector of finite-word-length floating-point numbers, which are considered to be the control law as calculated using finite word length. With this finite-word-length control law, the resultant closed loop poles are calculated. Table B.1.1 shows the result for the shift operator version of the problem for various numbers of mantissa bits, and Table B.1.2 shows the same results for the delta version of the problem. Note that ill conditioning of the pole assignment equation was removed from the problem by double precision for this part of the calculation.

As a measure of performance, consider the number of bits required to specify the continuous time version of the closed loop poles to an accuracy of $x\%$; that is, for 1% accuracy the shift operator pole near $z = 0.90484$ ($s = -1$) should be within the range $[e^{-\Delta(1+1\%)}, e^{-\Delta(1-1\%)}]$ and so on for the other poles and for the delta version. We see from the tables that the shift formulation needs 16 mantissa bits for 1% accuracy and 13 bits for 10% accuracy, whereas the delta formulation requires only 9 bits for 1% accuracy and 6 bits for 10% accuracy. Thus the delta form has a 7-bit advantage over shift in this example.

B.2 FREQUENCY RESPONSE SENSITIVITY

In this section we examine the sensitivity of the evaluation of the frequency response of a system for delta and shift models. Consider the following delta domain model:

$$H'(\gamma) = \frac{B'(\gamma)}{A'(\gamma)} = \frac{b'_{n-1}\gamma^{n-1} + \cdots + b'_0}{\gamma^n + a'_{n-1}\gamma^{n-1} + \cdots + a'_0} \tag{B.2.1}$$

which is equivalent to the following shift domain model

$$H''(z) = \frac{B''(z)}{A''(z)} = \frac{b''_{n-1}z^{n-1} + \cdots + b''_0}{z^n + a''_{n-1}z^{n-1} + \cdots + a''_0} \tag{B.2.2}$$

If we consider the evaluation of the transfer function to be performed using floating-point arithmetic, then multiplication and division can be performed with very small relative error [typically, $\leq 2^{-(b+1)}$, where b is the number of mantissa bits]. The main errors thus occur in the evaluation of the polynomials, which involves various additions. In particular, when adding two numbers of similar magnitude, but opposite sign, large numerical errors may result.

Thus we suggest the following measure of the numerical conditioning associated with the evaluation of a polynomial, $p(\cdot)$, at some complex number x when using floating-point arithmetic (see Problem 14.10 for further motivation for this measure):

$$\nu(p, x) = \frac{\sum_{i=0}^{n}|a_i' x^i|}{|p(x)|} \tag{B.2.3}$$

Hence for the denominator in (B.2.1) we have

$$\nu'(\omega) \triangleq \nu\left(A', \left(\frac{e^{j\omega\Delta} - 1}{\Delta}\right)\right) = \frac{\sum_{k=0}^{n}|a_k'((e^{j\omega\Delta} - 1)/\Delta)^k|}{|A'((e^{j\omega\Delta} - 1)/\Delta)|} \tag{B.2.4}$$

Similarly, for the denominator in (B.2.2) we have

$$\nu''(\omega) \triangleq \nu(A'', e^{j\omega\Delta}) = \frac{\sum_{k=0}^{n}|a_k'' e^{j\omega\Delta k}|}{|A''(e^{j\omega\Delta})|} \tag{B.2.5}$$

Since

$$A''(z) = \Delta^n\left(A'\left(\frac{z-1}{\Delta}\right)\right)$$

we can show that

$$\frac{\nu'(\omega)}{\nu''(\omega)} = \frac{\Delta^n \sum_{k=0}^{n}|a_k'| |e^{j\omega\Delta} - 1/\Delta|^k}{\sum_{k=0}^{n}|a_k''|} \tag{B.2.6}$$

$$= \frac{\sum_{k=0}^{n}|\Delta^{n-k} a_k'|(2 - 2\cos\omega\Delta)^{k/2}}{\sum_{k=0}^{n}|a_k''|} \tag{B.2.7}$$

Using the argument in Section B.1, Equation (B.1.18), we then have that for $\omega\Delta < \pi/3$ [that is, $\omega < \frac{1}{6}\omega_s$, which implies $(2 - 2\cos\omega\Delta) < 1$], and again assuming that the roots are closer to $1 + j0$ in the Z plane than to the origin,

$$\nu'(\omega) < \nu''(\omega) \tag{B.2.8}$$

In other words, for frequencies less than one-sixth of the sampling frequency, the delta operator form gives a less sensitive frequency response evaluation. Note that the converse is not true; $\omega\Delta > \pi/3$ does *not* imply $\nu'(\omega) > \nu''(\omega)$. We also remark that the frequency response well beyond the closed loop bandwidth is in general unimportant in controller design.

B.3 ROUND-OFF NOISE

In most finite-word-length implementations of a digital control algorithm, errors will be introduced into the system due to the finite number of bits used to store and calculate intermediate quantities, for example, controller states. Under suitable conditions on the input to a discrete time system (that there be sufficient noise and/or input variation), the errors introduced due to finite word lengths may be considered as an almost stochastic process (see references given in Chapter 14, for example, Wilkinson, 1963, or Rabiner and Gold, 1975, pp. 30). Thus the term *round-off noise* has arisen. In this section a theoretical comparison of the round-off noise performance of first-order shift and delta sections will be examined.

Consider a finite-word-length implementation of the transfer function $H(\rho) = (n_1\rho + n_0)/(\rho + d_0)$ as shown in Figure B.3.1, where ρ is an operator (either q or δ). Let the desired delta transfer function be

$$H(\delta) = \frac{b_1\delta + b_0}{\delta + a_0} \tag{B.3.1}$$

Then the equivalent shift transfer function is

$$H(z) = \frac{b_1 z + [b_0\Delta - b_1]}{z + [a_0\Delta - 1]} \tag{B.3.2}$$

If each operation is considered to produce an answer that is accurate to a fractional error of at most 2^{-b} (where b is the number of bits in the mantissa and it is assumed that the exponent contains sufficient bits to preclude underflow or overflow), then a model including round-off noise is shown in Figure B.3.2, where ε_i are rectangularly distributed random variables having zero mean and range $[-2^{-b}, 2^{-b}]$ and are mutually independent. Thus the effect of each noise source on the output may be considered seperately.

The noise sources, ε_1 and ε_7, are unavoidable noise sources due to the input and output quantization and will have the same effect on the output in either the

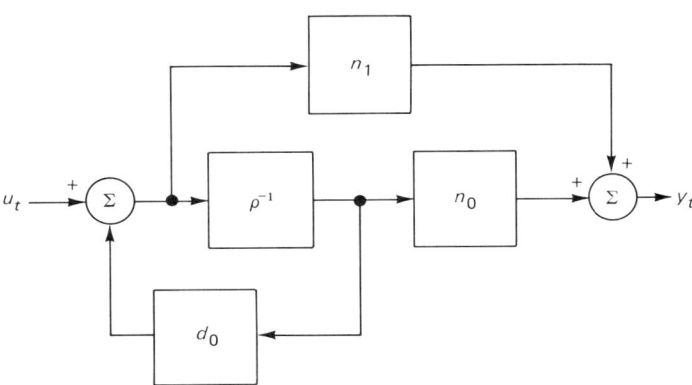

Figure B.3.1 Implementation of a first-order transfer function.

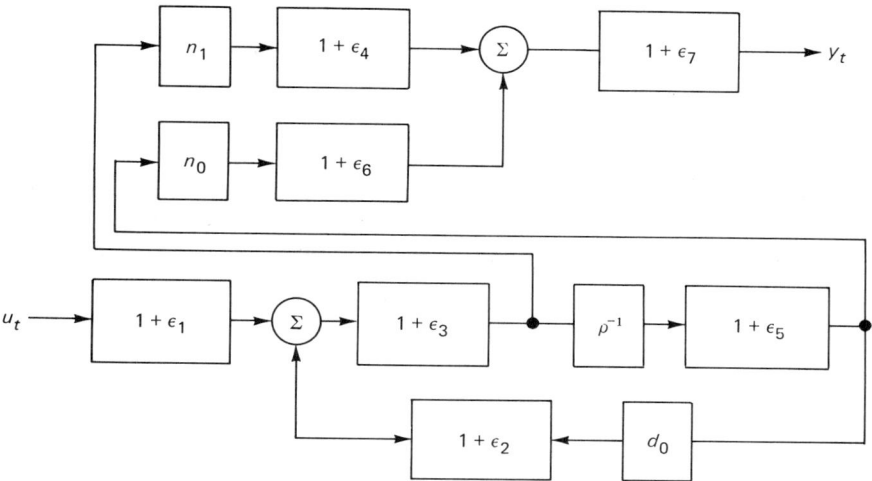

Figure B.3.2 Floating-point finite-word-length implementation of a first-order transfer function showing round-off noise sources (ε_i).

shift or delta formulation. The noise output due to ε_2 will be

$$y_{2_t} = -\left[\frac{d_0 u_t}{(p + d_0)}\right]\{H(p)\varepsilon_{2_t}\} \qquad (B.3.3)$$

where the term in parentheses will be called the *filtered error* and the term in brackets will be called the *gain term* (which depends on u_t). From (B.3.3) it can be shown that the filtered error is the same for both shift and delta formulations. However, provided the roots are closer to $1 + j0$ in the Z plane than to the origin, then the gain term for the delta form will always be less than that for the equivalent shift form. Thus, as far as ε_2 is concerned, the delta form of controller gives less output noise.

The noise output due to ε_3 will be

$$y_{3_t} = -\left[\frac{p u_t}{p + d_0}\right]\{H(p)\varepsilon_{3_t}\} \qquad (B.3.4)$$

Once again the filtered error is the same for shift and delta forms of the controller. However, the gain term will be different. It can be shown that, provided u_t has a signal bandwidth less than one-tenth of the sampling frequency, the delta form gives less output noise to ε_3 than the shift form. The noise outputs due to ε_4 to ε_6 are, respectively,

$$y_{4_t} = \left[\frac{n_1 p u_t}{p + d_0}\right]\varepsilon_{4_t} \qquad (B.3.5)$$

$$y_{5_t} = \left[\frac{(n_0 - d_0 n_1) u_t}{p + d_0}\right]\left\{\frac{p\varepsilon_{5_t}}{(p + d_0)}\right\} \qquad (B.3.6)$$

and
$$y_{6_t} = \left[\frac{n_0 u_t}{p + d_0}\right]\varepsilon_{6_t} \tag{B.3.7}$$

Note that under the standing assumption (B.1.11) the output noise due to all but ε_5 will be no worse for delta than for shift. The fifth noise source is always worse for delta since for shift operators it does not arise, the backward shift operation being performed without error. This analysis shows that round-off noise, due to almost all effects, is worse in shift form than in delta form. The examples in Section 14.5 illustrate this claim.

APPENDIX C

Numerical Issues in Optimal Filtering

In Chapter 10, we saw that the delta form of the optimal filter had the advantage over the shift form that it converged to the continuous case as the sampling period went to zero. Here we show that the delta form has the secondary advantage that its numerical properties are superior to the shift form. As previously, this is basically because the delta form centers the calculations by shifting the origin to the point $1 + j0$ in the Z plane.

C.1 THE DISCRETE RICCATI DIFFERENCE EQUATION (DRDE)

The key difference between the shift and delta form of the DRDE is the way in which the transition matrix is expressed. There are changes in the other matrices, but these amount to scaling by Δ, which does not significantly affect the floating-point representation. We will thus concentrate on the merits, or otherwise, of replacing A_q (shift) by $\Delta A_\delta + I$ (delta) in the solution of the DRDE. We will describe the floating-point implementation a matrix $X = \{x_{ij}\}$ (which can denote either A_q or A_δ) as $[X]_{\text{FP}}$, where

$$[X]_{\text{FP}} = \{x_{ij}(1 + \varepsilon_{ij})\} \qquad \text{(C.1.1)}$$

In the following lemma, we consider the effect of the preceding floating-point errors on the evaluation of A_q and $\Delta A_\delta + I$. Note that this is a simple extension of the results in Appendix B regarding coefficient representation.

Lemma C.1.1

$$\left\|[A_q]_{FP} - A_q\right\|_F \leq \varepsilon \|A_q\|_F \triangleq L_q \tag{C.1.2}$$

$$\left\|I + \Delta[A_\delta]_{FP} - A_q\right\|_F \leq \varepsilon \Delta \|A_\delta\|_F \triangleq L_\delta \tag{C.1.3}$$

where

$$\varepsilon = \underset{i,j}{\text{Max}} |\varepsilon_{ij}| \leq 2^{-b} \tag{C.1.4}$$

and where $\|\cdot\|_F$ denotes the Fröbenius norm.

Proof: The result is immediate on noting that from (C.1.1) we have

$$\left\|[X]_{FP} - X\right\|_F \leq \underset{i,j}{\text{Max}} |\varepsilon_{ij}| \|X\|_F \tag{C.1.5}$$

▽▽▽

To compare the results in Lemma C.1.1, we will use the spectral radius as a lower bound for the norm of A_q and A_δ since the spectral radius is less than or equal to any consistent norm. Then

$$L_q \triangleq \varepsilon \|A_q\|_F \geq \varepsilon \sigma[A_q] = \varepsilon \underset{i}{\text{Max}} |e^{\lambda_i \Delta}| \tag{C.1.6}$$

$$L_\delta \triangleq \varepsilon \Delta \|A_\delta\|_F \geq \varepsilon \Delta \sigma[A_\delta] = \varepsilon \underset{i}{\text{Max}} |e^{\lambda_i \Delta} - 1| \tag{C.1.7}$$

where $\sigma[\cdot]$ denotes spectral radius and $\{\lambda_i\}$ denotes the eigenvalues of the continuous time A matrix.

For fast sampling (relative to the time constants of the system), we have $|e^{\lambda_i \Delta}| \simeq 1$ for $i = 1, \ldots, n$, in which case the lower bound on L_q given in (C.1.6) is much greater than the lower bound on L_δ given in (C.1.7). Inspection of (C.1.2), (C.1.3) then suggests that the delta form leads to a smaller error than the shift form. To be more precise about this comparison, we need to consider specific structures for the matrices A_q and A_δ. For example, if we use the Jordan canonical form, then the comparison relies on the errors in representing the diagonal terms of A_q and $A_q - I$, which correspond to the eigenvalues in this case. Clearly, whenever the eigenvalues of A_q are closer to unity than zero, then the eigenvalues are better represented in the delta form. Thus, whenever rapid sampling is used, the delta form is superior.

C.2 THE ALGEBRAIC RICCATI EQUATION

Most methods for solving the Algebraic Riccati Equation (ARE) start from the Hamiltonian matrix. We will use the well-known solution method based on eigenvectors (see Chapter 10). For simplicity, we will assume that the eigenvalues of the Hamiltonian matrices are distinct.

Let us assume that $M \in \mathbb{R}^{2n}$ is a Hamiltonian matrix with distinct eigenvalues; then there exists a nonsingular matrix X such that

$$X^{-1}MX = \left[\begin{array}{c|c} \Lambda_1 & 0 \\ \hline 0 & \Lambda_2 \end{array}\right] \quad (C.2.1)$$

where $\Lambda_1, \Lambda_2 \in \mathbb{R}^n$ are diagonal matrices and where Λ_1 contains only the stable eigenvalues. If

$$X = \left[\begin{array}{c|c} X_{11} & X_{12} \\ \hline X_{21} & X_{22} \end{array}\right],$$

then $\begin{bmatrix} X_{11} \\ X_{21} \end{bmatrix}$ will span the stable subspace and the stabilizing solution for the ARE can be evaluated as $P = X_{21}X_{11}^{-1}$. Clearly, the sensitivity of the solution method depends on the sensitivity of the eigenvectors. To carry out the analysis we will use a perturbation method.

We consider a model for the perturbations as in (C.1.1). This leads to

$$[M_q]_P = M_q + E_q \quad (C.2.2)$$
$$[M_\delta]_P = M_\delta + E_\delta \quad (C.2.3)$$

where $[M_q]_P, [M_\delta]_P$ denote the perturbed matrices and

$$\|E_q\|_F \leq \varepsilon \|M_q\|_F \quad (C.2.4)$$
$$\|E_\delta\|_F \leq \varepsilon \|M_\delta\|_F \quad (C.2.5)$$

To state the main result of this section, we need the following preliminary result.

Lemma C.2.1

Hamiltonian matrices M_δ and M_q have the same eigenvectors.

Proof: Immediate from the definition of the delta form of the matrices given in Lemma 10.7.1 and the corresponding result for the shift form. ▽▽▽

We then have the following result.

Lemma C.2.2

$$\|x_i - [x_{iq}]_P\|_2 \leq \frac{\varepsilon \|M_q\|_F}{\Omega_q} \triangleq U_q \quad (C.2.6)$$

$$\|x_i - [x_{i\delta}]_P\|_2 \leq \frac{\varepsilon \|M_\delta\|_F}{\Omega_\delta} \triangleq U_\delta \quad (C.2.7)$$

where $\{x_i\}$ denotes the eigenvectors of M_δ and M_q, and $\{[x_{iq}]_P\}, \{[x_{i\delta}]_P\}$ denote the perturbed eigenvectors of M_q and M_δ, respectively. Ω_q, Ω_δ denote a lower bound on the minimum distance between the ith eigenvalue and any other eigenvalue for M_q and M_δ, respectively.

Proof: From Stewart (1973), we have

$$\|x_i - [x_i]_P\|_2 \leq \frac{\alpha}{\Omega} \tag{C.2.8}$$

where α is the two norm of the perturbation matrix and

$$\Omega \leq \min_{\substack{j \\ j \neq i}} \{|\lambda_i - \lambda_j|\}, \qquad j = 1, 2, \ldots, 2n \tag{C.2.9}$$

The result then follows on using (C.2.4), (C.2.5) and on noting that

$$\|\cdot\|_2 \leq \|\cdot\|_F \tag{C.2.10}$$

▽▽▽

To compare the results in (C.2.6), (C.2.7), we will use the spectral radius σ as in Section C.1. We note that

$$U_q \geq \frac{\varepsilon \sigma(M_q)}{\Omega_q} \tag{C.2.11}$$

$$U_\delta \geq \frac{\varepsilon \Delta \sigma(M_\delta)}{\Omega_q} \tag{C.2.12}$$

where we have used the fact that $\Omega_\delta = \Omega_q/\Delta$ since $\lambda_q = \Delta \lambda_\delta + 1$.

For fast sampling, we have

$$\sigma(M_q) \simeq \max_i |1 + \beta_i \Delta| \tag{C.2.13}$$

$$\sigma(M_\delta) \simeq \max_i |\beta_i \Delta| \tag{C.2.14}$$

where $\{\beta_i\}$ denotes the eigenvalues of the continuous time Hamiltonian matrix M_c.

Thus, for fast sampling (relative to the time constants of M_c, which are the regulator or filter time constants), $|\beta_i \Delta| \ll 1$, and hence the lower bound on U_q given in (C.2.11) is much greater than the lower bound in U_δ given in (C.2.12). This suggests that the delta form leads to smaller error than does the shift form. As for the DRDE, the result can be made more explicit if specific canonical forms are used for the matrices.

An alternative expression for the perturbed eigenvector is given in Golub and Van Loan (1983) as

$$[x_i]_P = x_i + \sum_{j=1}^{2n} \frac{y_j^T E x_i}{(\lambda_i - \lambda_j) y_j^T x_j} x_j + O(\varepsilon^2) \tag{C.2.15}$$

where

$E \triangleq \varepsilon F$ = perturbation matrix (E_q or E_δ) with $\|F\|_2 = 1$

y_j = left eigenvector associated with λ_j

$O(\varepsilon^2)$ = residual of a Taylor expansion around $\varepsilon = 0$

By neglecting $O(\varepsilon^2)$, we get

$$\|x_i - [x_i]_P\|_2 \leq \frac{K\|E\|_F}{\Omega} \tag{C.2.16}$$

where K is a positive constant that is the same for delta and shift forms, and Ω is as defined in (C.2.9).

Comparing (C.2.16) with (C.2.6), (C.2.7), we see that the same conclusion applies.

C.3 REFERENCES

General results on numerical analysis and perturbation techniques are contained in:

GOLUB, G. H., and C. F. VAN LOAN (1983) *Matrix Computations*. Johns Hopkins University Press, Baltimore, Md.

STEWART, G. W. (1973) *Introduction to Matrix Computations*. Academic Press, New York.

WILKINSON, J. H. (1963) *Rounding Errors in Algebraic Processes*. Prentice-Hall, Englewood Cliffs, N.J.

Results on finite-word-length effects in Kalman filtering may be found in:

ARNOLD, W., and A. J. LAUB (1984) "Generalized eigenproblem algorithms and software for algebraic Riccati equations." *Proc. IEEE*, Vol. 72, No. 12, pp. 1746–1754.

GELB, A. (1974) *Applied Optimal Estimation*. MIT Press, Cambridge, MA.

PAPPAS, T., A. J. LAUB, and N. R. SANDELL (1980) "On the numerical solution of the discrete-time algebraic Riccati equation." *IEEE Trans. Auto. Control*, Vol. AC-25, No. 4, pp. 631–641.

SALGADO, M. E., R. H. MIDDLETON, and G. C. GOODWIN (1988) "Connection between continuous and discrete Riccati equations with applications to Kalman filtering." *Proc. IEE*, Part D, Vol. 135, No. 1, pp. 28–34.

SRIDHARAN, S. (1985) Ph.D. Thesis, University of New South Wales, Australia.

VAN DOOREN, P. (1981) "A generalized eigenvalue approach for solving Riccati equations." *SIAM J. Sci. Stat. Compt.*, Vol. 2, No. 2, pp. 121–135.

WILLIAMSON, D. (1985) "Finite word length design of digital Kalman filters for state estimation." *IEEE Trans. Auto. Control*, Vol. AC-30, No. 10, pp. 930–939.

APPENDIX D

CAD Software for Control System Design

Throughout this book, various complicated mathematical operations have been used. In practice, computer software packages are an invaluable aid to the control system designer. Listed next are some of the features of a control CAD package that we feel are important:

- Matrix algebra (for example, matrix exponentials, eigenvalue analysis, controller/observer forms)
- System transformations from one description to another (for example, continuous to discrete, transfer function to/from state space)
- Classical control design aids (for example, Nyquist diagrams, root locus, bode plots)
- Modern control design aids (for example, pole-placement algorithms, LQ regulator design, H^∞ design)
- Flexible simulation capability, including nonlinear simulation, hybrid (continuous and discrete element) simulation
- System identification and parameter estimation packages
- Real-time code generation (some packages are available that produce real-time code segments from a block diagram of a controller)

Many good CAD packages are available; however, few give good coverage of all the points listed. At the time of writing this text, several people are extending these packages to handle the δ operator approach for discrete time systems. This will be beneficial in view of the superior numerical properties and greater insight afforded by this operator compared with the shift operator.

In summary, we suggest that the selection of a suitable CAD package can turn control system design from a laborious mathematical chore into a stimulating, interactive exercise.

APPENDIX E

Real-Time Software

It has been said that research in any area is 1% inspiration and 99% perspiration. A similar remark applies to digital control. In particular, a real control problem usually involves 49% understanding the process and objectives, 49% creating a suitable software environment for control, and 2% solving for the control law. In view of this, it would be inappropriate in a book on digital control not to alert the reader to some of the issues involved in real-time software. This is a large topic in its own right, and therefore we will confine our remarks to some commonly used terms, some simple aspects of real-time programming, and some observations on proper control laws.

E.1 TERMINOLOGY

Some of the key concepts used in real-time operating systems are:

Busy Wait. A busy wait is a loop that performs no other function than to wait until some event has occurred. An example code fragment is

```
           ⋮
    { busy wait follows }
    repeat { nothing }
    until event_occurs;
           ⋮
```

Note that the busy wait wastes CPU time. Unless we have no other processing or are positive that the event will occur shortly, the busy wait ought to be avoided where possible. The use of polling, interrupts, or semaphore waits is recommended.

Polling. Polling is a form of busy wait where we are waiting for one or more of (say) *n* events to occur. A typical code fragment follows:

```
{ polling code section }
repeat { forever }
    FOR  i := 1 to n DO
              If event_occurred [i] THEN
                    perform_action [i]
until false; { poll forever }
```

Interrupts. Interrupts are in essence a hardware generated procedure (that is, subroutine) call. Because interrupts are often generated by external hardware, they are rarely synchronized with the program currently running, and it is normally assumed that they can occur at any time. This means that the hardware must store the processor status, as well as the program counter. Most computers allow interrupts to be disabled at critical sections of the main code (see later).

Tasks. A task (sometimes referred to as a process) denotes the execution of a program. It is analogous to the operation of cooking in a culinary context or to the state trajectories in a control theory context. Note that more than one task may use the same subroutine (procedure) at one time, just as we may cook more than one cake at a time using the same recipe.

An example of several tasks using one subroutine is where we have multiple PID controllers implemented on a single computer. The action of computing the control for each loop is a task, yet clearly the PID subroutine may be shared.

Scheduler. Real-time operating systems contain a subroutine (procedure) that allows switching between tasks. This is termed the scheduler. The scheduler is often called as an interrupt routine when a line frequency clock tick occurs. It may also be called from semaphore operations described next.

Semaphore. A semaphore is a record, associated with a resource, containing two variables:

1. An integer value describing the number of tasks (if any) which are currently waiting for access to the resource
2. A list (for example, linked list) of tasks waiting for use of the resource

Examples of resources are input/output buffers and variables shared between tasks. Semaphores allow protection of critical sections (see later) and synchronization between tasks. The two basic semaphore operations are described next.

Wait (semaphore). When a task requires access to a resource, a semaphore wait should be executed. The semaphore wait checks to see if the resource is available (that is, semaphore value < 0), and, if so, the semaphore value is incremented and the task is allowed to continue. If the resource is not available, the semaphore value is incremented, the task is placed on the semaphore wait list, and the scheduler is called. The scheduler will restart another task since the current task cannot continue. The wait operation *must* be indivisible (that is, uninterruptable) and should *not* be called from an interrupt routine.

Signal (semaphore). This is the opposite operation to wait. When a task has finished using a resource or when an interrupt denoting availability of a resource occurs, a semaphore signal should be executed. This decreases the semaphore value and may call the scheduler to restart a task blocked on the semaphore. This instruction must also be indivisible.

Critical Sections. If more than one task alters a common variable, and the alteration is divisible in any of the tasks, then a critical section exists. (An alteration is said to be divisible if an interrupt can be serviced after alteration has begun, but before alteration is complete. In most higher-level languages, virtually all instructions are divisible.) One solution to a critical section problem is to make all critical sections indivisible by disabling interrupts prior to the section and enabling interrupts after leaving the section. A better solution is to use semaphore operations. However, the semaphore operations are themselves critical sections, and so interrupt disable/enable must be used when implementing the semaphore operations.

E.2 REAL-TIME PROGRAMMING

The main difficulties that arise with real-time programming in a control or estimation context occur when we wish to perform several tasks simultaneously. This will generally be the case, for example, if we are controlling one or more loops as well as performing other tasks, such as keyboard monitoring or data display, with a single processor. One way of coping with multiple tasks is illustrated next for the case of multiple loops.

For simplicity, we assume that the control laws are strictly proper. (The case of proper control laws will be treated in Section E.3.)

```
{ Multiloop controller, real time clock interrupt occurs every delta seconds, interrupt
routine executes "signal (real_time_clock)" }
Task multiloop_controller;
        VAR   i:Integer;
BEGIN
Repeat { forever }
        wait (real_time_clock);
        FOR i := 1 to n DO send_control_signal(i);
        FOR i := 1 to n DO read_from_AD(i,y,y_star)
        { read_from_AD(i) does a busy wait till conversions are complete };
        FOR i := 1 to n DO calculate_next_control(i)
Until false
END_task;
```

Note that in this scheme the multiple tasks have effectively been combined into one task, controlling every loop. Note also that the delay between the interrupt occurrence and the time when the A/D converter is read is $n \times$ (D/A time) + $(i - 1) \times$ (A/D time). Thus, unless the bandwidths (and thus sampling rate) are low, this structure may place difficult constraints on the analog interface timing.

An alternative structure that uses a multitasking arrangement is shown next.

```
{ Multiloop, Multitasking controller. Interrupt occurs every delta / n seconds where
there are n control loops.}
Task multiloop_multitask;
VAR i:  INTEGER
BEGIN
        Repeat { forever }
        wait (real_time_clock)
        i := (i mod n) + 1;
        send_control_signal (i);
        read_from_AD (i,y,y-star);
        calculate_next_control (i)
        Until false
END_task;
```

Note that in the multitasking arrangement the A/D converter is always read after the D/A conversion. One compromise in this arrangement is that interrupts now occur n times more often than in the previous arrangement, and so any overheads associated with servicing the interrupt will be multiplied.

Multirate controllers, those where different sampling rates are required within one computer, also raise some issues. Clearly, if the computations for all loops can be performed within the shortest sampling period, there is little difficulty. In other cases, however, the controller may need to be implemented as a multitasking system, with faster loops being given a higher priority than slower loops.

E.3 PROPER CONTROL LAWS

Another issue that is of some relevance to timing considerations is the distinction between proper and strictly proper control laws. We will say that a (possibly nonlinear) control law that maps sequences of sampled plant outputs y_k and reference signals y_k^* to a sequence of plant inputs u_k is proper if we can write

$$u_k = f(k, u_{k-1}, u_{k-2}, \ldots, y_k, y_{k-1}, \ldots, y_k^*, y_{k-1}^*, \ldots) \tag{E.3.1}$$

Note that in general, a control law will be impossible to implement unless it can be written in this form, since the current value of u will depend on future values of y and/or y^*.

We will also call a control law *strictly* proper if f in (E.3.1) does *not* depend on y_k or y_k^*.

Note that for finite-dimensional linear systems, the notions of proper and strictly proper have the following simple interpretations. Any controller of the form

$$\delta x_c = A_c x_c + B_c y + B_c' y^* \tag{E.3.2}$$

$$u = C_c x_c + D_c y + D_c' y^* \tag{E.3.3}$$

is proper. Moreover, it is strictly proper if $D_c = D'_c = 0$. Alternatively, if we use operator notation, any controller of the form

$$L(\delta)u = -P(\delta)y + G(\delta)y^* \tag{E.3.4}$$

is proper if and only if $\partial(P)$, $\partial(G) \leq \partial(L)$ and is strictly proper if and only if $\partial(P)$, $\partial(D) < \partial(L)$.

From an implementation viewpoint, a strictly proper control law is usually simpler. The following Pascal-like routine shows how a strictly proper controller may be programmed.

```
task strictly_proper_controller;
BEGIN
        repeat { forever }
        wait (real_time_clock);   {wait for clock interrupt to occur }
        send_to_DA(u);  { send current control signal to D / A }
        read_from_AD (y,ystar); { get current y and y* }
        update_control (y,ystar,u,state)
        {update states, and calculate future control signal }
        until false {control is complete }
END_task;
```

The timing relevant to the preceding strictly proper control law is given in Figure E.3.1. Note that accurate timing of the control signal and the sampling of the outputs can be assured provided the interrupt latency (time required to service an interrupt) and the scheduler latency are small compared with the sampling period. In particular, the control signal is sent t_{lat} after each sampling period, while the outputs are sampled $t_{\text{lat}} + t_{\text{D/A}}$ after each sampling period. In cases where a memory mapped D/A scheme is used, $t_{\text{D/A}}$ is generally the time required for one memory move operation and is thus on the order of microseconds. Note also that

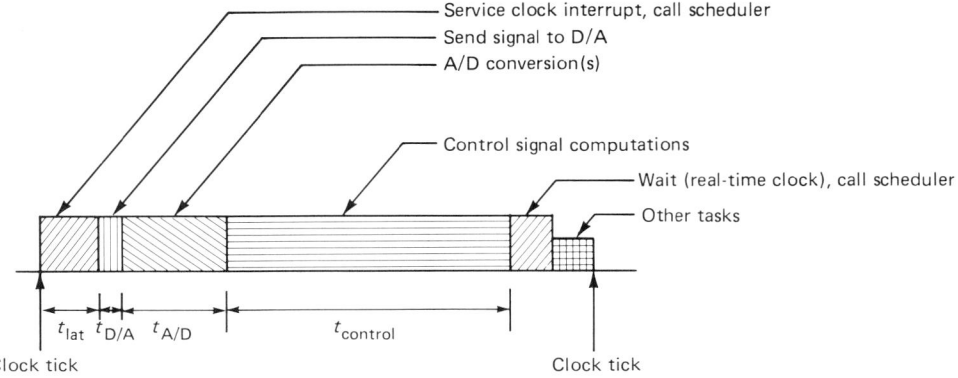

Figure E.3.1 Timing for strictly proper control law.

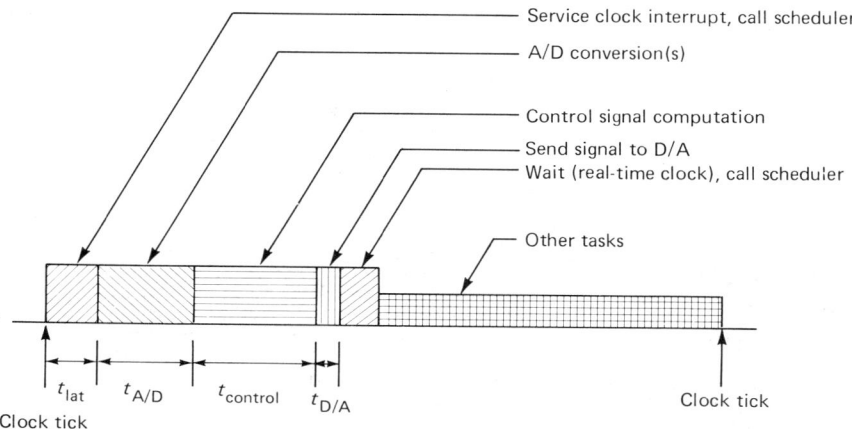

Figure E.3.2 Timing for proper control laws.

we have almost one complete sampling period in which to compute the control signal.

By contrast, Figure E.3.2 shows the timing for a typical proper control law, corresponding to the use of a routine with the following structure.

```
task proper_controller;
BEGIN
    repeat { forever }
    wait (real_time_clock); { semaphore wait for real time clock tick }
    read_from_AD (y,ystar); { get current y and y* }
    update_control (y,ystar,u,state);
    { update controller states and calculate current control }
    send_to_DA (u) { send control signal }
    until false
END_task;
```

Note that in this case the control signal is only sent $t_{lat} + t_{A/D} + t_{control}$ after each clock tick. Thus, in this case, we should ensure that all these times are much smaller than the sampling interval. Sometimes this may create a difficult timing constraint, in which case we suggest the following modification.

First, we note that any proper, linear control function can be split as follows:

$$u = -K_y y + K_* y^* + H(\delta) \begin{bmatrix} y \\ y^* \end{bmatrix} \quad (E.3.5)$$

where $H(\delta)$ is a strictly proper operator. The control algorithm for this new structure is given next.

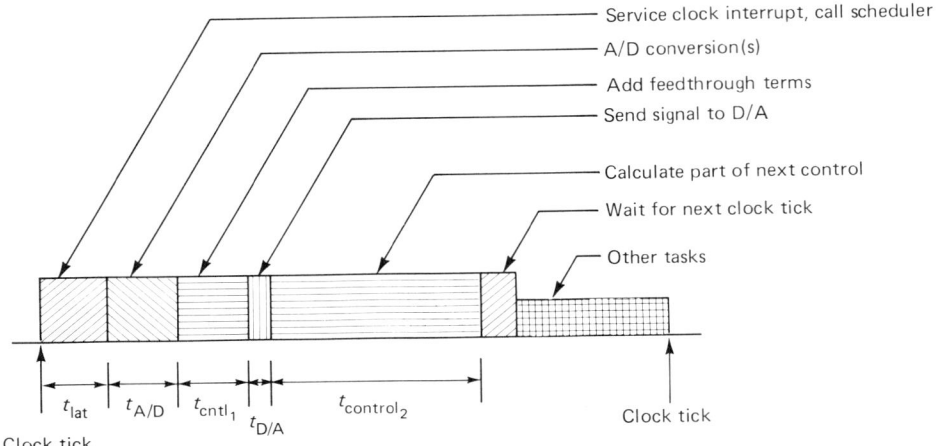

Figure E.3.3 Timing for better proper control laws.

```
task better_proper_controller;
BEGIN
  repeat { forever }
    wait (real_time_clock); { semaphore wait for real time clock tick }
    read_from_AD (y,ystar); { get current y and y* }
    u := u − ky * y + kstar * ystar;
    { add feedthrough terms to current u }
    send_to_DA(u); { send control signal }
    partial_control_update (u,ystar,u,state)
        { find strictly proper part of next control signal }
  until false
END_task;
```

The timing for this algorithm structure is given in Figure E.3.3. Note that in this case the timing constraints placed on the control signal computations are relieved by splitting the computations into strictly proper (less critical) and proper (more critical) parts. Note that this type of splitting is particularly important if a complicated control algorithm (for example, adaptive control), which takes a significant proportion of the sampling period, is used.

APPENDIX F

Convergence Analysis of a Typical Adaptive Control Algorithm

F.1 THE ALGORITHM

Consider the illustrative adaptive control algorithm described in Section 12.10.2. We will assume that the system is modeled as in (12.8.3); that is,

$$y = \left[H_0(1 + \overline{H}) + \overline{\overline{H}}\right]u = \frac{B_a}{A_a}u \qquad (F.1.1)$$

where

$$H_0 = \frac{B}{A} \qquad (F.1.2)$$

is the nominal model. The model (F.1.1) can be rewritten as

$$Ay = Bu + \eta \qquad (F.1.3)$$

where

$$\eta = \left[B\overline{H} + A\overline{\overline{H}}\right]u \qquad (F.1.4)$$

Applying the filter GS/EJQ gives the model (12.10.3); that is,

$$\bar{y} = \phi^T\theta + \eta_f \qquad (F.1.5)$$

where

$$\eta_f = \left[B\overline{H} + A\overline{\overline{H}}\right]u_f \qquad (F.1.6)$$

525

and

$$u_f = \frac{1}{E}\bar{u}, \qquad \bar{u} = \frac{GS}{JQ} \qquad (F.1.7)$$

We now base the dead zone function in the parameter estimator on the overbound $\varepsilon\omega(t)$ described in the following result

Lemma F.1.1

Consider the form of the unmodelled response given in Equation (F.1.4). Provided $\overline{H}$ and $\overline{\overline{H}}$ are stable and strictly proper, there exists constants $\sigma_0 \in (0, 1/\Delta)$, $\varepsilon \geq 0$, and a constant vector v such that

$$|\eta_f(t)| \leq \varepsilon\omega(t), \qquad \text{for all } t \qquad (F.1.8)$$

where

$$\omega(t) = \sup_{0 \leq \tau \leq t} \left\{ |v^T x(\tau)| E(-\sigma_0, t-\tau) \right\} \qquad (F.1.9)$$

x is the following state vector:

$$x \triangleq \left[\rho^{n+s+g-1} z_f', \ldots, z_f', \rho^{n-1} u_f, \ldots, u_f \right]^T \qquad (F.1.10)$$

and

$$z_f' \triangleq \frac{1}{FQ} z, \qquad F \triangleq EJ, \qquad z \triangleq y - y^* \qquad (F.1.11)$$

Proof (outline): We introduce an arbitrary Hurwitz polynomial $V = \rho^{n-1} + v_{n-2}\rho^{n-2} + \cdots$, v_0 of degree $n-1$. In view of Equation (F.1.6), η_f can then be expressed as

$$\eta_f = \left[\frac{B\overline{H} + A\overline{\overline{H}}}{V} \right] V u_f \qquad (F.1.12)$$

Then, if we define

$$v = [0, \ldots, 0, 1, v_{n-2}, \ldots, v_0]^T$$

it follows that $V u_f = v^T x$. The remainder of the proof is straightforward as in Lemma 12.8.1 (see Problem 12.24). ▽▽▽

To implement the dead zone algorithm, we shall assume that constants ε, v, σ_0 are known such that (F.1.8) is satisfied. The convergence properties of the dead zone parameter estimator were given in Lemma 12.8.2. From this result, we recall that the key properties are:

(i) $V = \tilde{\theta}^T P^{-1} \tilde{\theta}$ is a nonincreasing function.
(ii) $\tilde{f} = \dfrac{f(k\varepsilon\omega(t), e(t))}{\left(1 + \phi(t)^T \phi(t)\right)^{1/2}} \in \mathscr{L}_2^+$
(iii) $\rho\hat{\theta} \in \mathscr{L}_2^+$

The specification of the adaptive controller is then completed by using the estimated parameters $\hat{\theta}$ in the certainty equivalence adaptive control law (12.10.7) to (12.10.10).

F.2 ANALYSIS OF THE ALGORITHM

The key equations that we will use in the subsequent convergence analysis are summarized next.

- The parameter estimator (12.8.11) through (12.8.14), modified as in Section 12.6, to ensure that the design identity (12.10.7) can always be solved
- The prediction error:

$$e(t) = \bar{y}(t) - \phi(t)^T \hat{\theta}(t) \tag{F.2.1}$$

- The design identity (12.10.7):

$$\hat{A}\hat{L}GS + \hat{B}\hat{P} = \hat{A}^* \tag{F.2.2}$$

- The feedback control law (12.10.9), (12.10.10):

$$\bar{u} = (E - \hat{L})u_f - \hat{P}z_f', \qquad u = \frac{JQ}{GS}\bar{u} \tag{F.2.3}$$

The control law equation and the prediction error equation may be combined into the following closed loop equation:

$$\rho x(t) = \overline{A}(t)x(t) + B_1(e + r) \tag{F.2.4}$$

where

$$\overline{A}(t) \triangleq \begin{bmatrix} -\hat{c}_{n+g+s-1} & & -\hat{c}_0 & \hat{b}_{n-1} & & \hat{b}_0 \\ 1 & 0 & 0 & 0 & & 0 \\ 0 & & & & & \\ 0 & 0 & 1 & 0 & 0 & & 0 \\ -\hat{p}_{n+g+s-1} & & -\hat{p}_0 & -\hat{l}_{n-1} & & -\hat{l}_0 \\ 0 & & 0 & 1 & 0 & 0 \\ & & & 0 & & \\ 0 & & 0 & 0 & 1 & 0 \end{bmatrix} \tag{F.2.5}$$

$$\hat{C}(\rho) \triangleq \rho^{n+g+s} + \hat{c}_{n+g+s-1}\rho^{n+g+s-1} + \cdots + \hat{c}_0$$

$$= G(\rho)\hat{A}(\rho)S(\rho) \tag{F.2.6}$$

$$B_1^T \triangleq [1, \ 0 \ \ldots \ 0] \tag{F.2.7}$$

$$r \triangleq \hat{A}\left(\frac{GS}{FQ}y^*\right) \tag{F.2.8}$$

The following lemma will be used in the analysis of the overall adaptive scheme, as described in the preceding equations.

Lemma F.2.1. Gronwall's Lemma

Suppose $m(t)$ and $k(t)$ are nonnegative functions of time, c is a nonnegative constant, and

$$m(t) \leq c + \int_0^t k(\tau) m(\tau) \, d\tau, \quad \text{for all } t$$

Then

$$m(t) \leq c \exp\left\{ \int_0^t k(\tau) \, d\tau \right\}$$

Proof: Let

$$\overline{m}(t) = c + \int_0^t k(\tau) \overline{m}(\tau) \, d\tau$$

Then, since $\rho \overline{m}(t) = k(t) \overline{m}(t)$ and $\overline{m}(0) = c$, we have

$$\overline{m}(t) = c \exp\left\{ \int_0^t \frac{1}{\Delta} \ln(1 + \Delta k(\tau)) \, d\tau \right\}$$

$$\leq c \exp\left\{ \int_0^t k(\tau) \, d\tau \right\}$$

Clearly, also,

$$m(t) \leq \overline{m}(t) \qquad \triangledown\triangledown\triangledown$$

The analysis of the closed loop system described will depend on the properties of the parameter estimator given in Lemma 12.8.2 and on the following properties of the homogeneous part of (F.2.4).

Lemma F.2.2

Consider the following homogeneous linear time varying system

$$\rho \bar{x}(t) = \overline{A}(t) \bar{x}(t) \tag{F.2.9}$$

Provided

(i) $\overline{A}(t)$ is bounded
(ii) $S_T^{T+t} \| \rho \overline{A}(\tau) \|^2 \leq k_0 t + k_1$, $\forall t, T$, where k_0 is sufficiently small.
(iii) The eigenvalues of $\overline{A}$ are strictly inside the stability boundary for all t.

Then (F.2.9) is exponentially stable.

Proof (outline): For simplicity, we treat only the continuous time case ($\rho = d/dt = D$). The discrete case is left to an exercise (see Problem 12.26).

Choose $\Gamma = \Gamma^T > 0$ and let $\Omega(t)$ denote the positive definite symmetric solution to

$$\bar{A}^T(t)\Omega(t) + \Omega(t)\bar{A}(t) = -\Gamma \tag{F.2.10}$$

Then using (i) to (iii) we can show (see Problem 12.26) that $\Omega(t)$ is uniformly positive definite, bounded, and $\exists k_2$ s.t.

$$\|\dot{\Omega}(t)\|_2 \leq k_2 \|\dot{\bar{A}}(t)\|_2 \tag{F.2.11}$$

Now consider the system

$$D(w(t)) = (\bar{A}(t) - \tfrac{1}{2}\Omega^{-1}(t)\dot{\Omega}(t))w(t) \tag{F.2.12}$$

and define

$$V(t) = w^T(t)\Omega(t)w(t) \tag{F.2.13}$$

we then have

$$\dot{V}(t) = -w^T(t)\Gamma w(t) \tag{F.2.14}$$

which establishes the exponential stability of (F.2.12). Rewriting (F.2.9) gives

$$D\bar{x}(t) = (\bar{A}(t) - \tfrac{1}{2}\Omega^{-1}(t)\dot{\Omega}(t))\bar{x}(t) + \tfrac{1}{2}\Omega^{-1}(t)\dot{\Omega}(t)\bar{x}(t) \tag{F.2.15}$$

The solution to (F.2.15) can be written

$$\bar{x}(t) = \bar{\Phi}(t,T)\bar{x}(T) + \int_T^t \bar{\Phi}(t,\tau)\tfrac{1}{2}\Omega^{-1}(\tau)\dot{\Omega}(\tau)\bar{x}(\tau)\,d\tau \tag{F.2.16}$$

where $\bar{\Phi}(t,\tau)$ is the state transition matrix of the system in (F.2.12). In view of the exponential stability of $\bar{\Phi}$ established in (F.2.13), (F.2.14), and using Schwarz's inequality, we have

$$\|\bar{x}(t)\|^2 \leq c_0 e^{-2\sigma(t-T)}\|\bar{x}(T)\|_2^2 + \int_T^t c_1 e^{-\sigma(t-\tau)}\|\dot{\Omega}(\tau)\|_2^2 \|\bar{x}(\tau)\|_2^2\,d\tau \tag{F.2.17}$$

for some $c_0, c_1, \sigma > 0$, and thus

$$\|\bar{x}(t)\|_2^2 e^{\sigma t} \leq c_0 \|\bar{x}(T)\|_2^2 + \int_T^t c_1 \|\dot{\Omega}(\tau)\|_2^2 e^{\sigma\tau}\|\bar{x}(\tau)\|_2^2\,d\tau \tag{F.2.18}$$

Using Gronwall's lemma, we then have

$$\|\bar{x}(t)\|_2^2 e^{\sigma t} \leq c_0 \|\bar{x}(T)\|_2^2 \exp\left\{\int_T^t c_1 k_2^2 \|\dot{\bar{A}}(\tau)\|_2^2\,d\tau\right\}$$

$$\leq c_0 \|\bar{x}(T)\|_2^2 \exp(c_1 k_2^2 k_1) \exp(k_0 c_1 k_2^2 t) \tag{F.2.19}$$

From (F.2.19), it is clear that, provided $k_0 < \sigma/(c_1 k_2^2)$, (12.2.9) is exponentially stable. ▽▽▽

We will now complete the stability analysis of the adaptive algorithm. (An alternative normalized algorithm is examined in an exercise; see Problem 12.27.)

Theorem F.2.1

(i) Provided the plant undermodeling is sufficiently small, that is, provided ε [in (F.1.8)] is sufficiently small, then the adaptive control law, applied to the plant, (F.1.1), is globally stable in the sense that y and u (and hence all states) are bounded for all finite initial states and any bounded y^*.

(ii) If, in addition, y^* is purely deterministic, satisfying $Sy^* = 0$, then it follows that

$$\lim |y(t) - y^*(t)| = 0 \qquad (F.2.20)$$

Proof: The proof is based on Equation (F.2.4). The various terms in this equation are dealt with as follows: $\overline{A}(t)$ is exponentially stable; r is bounded and thus a BIBS argument can be used; e can be decomposed into a sum of terms in f and $e - f$; the term in f is dealt with by Gronwall's lemma since $\tilde{f} \in \mathcal{L}_2^+$; and the term in $e - f$ is handled by a small gain type of argument since $|e - f| < k\varepsilon\omega$.
The algorithm ensures there are no finite escapes.

1. Noting that the eigenvalues of $\overline{A}(t)$ are the $2n + g + s$ zeros of $\hat{A}^*$ it follows that the conditions of Lemma F.2.2 are satisfied provided $\hat{A}^*$ is chosen appropriately.

From Equation (F.2.4),

$$x(t) = \Phi(t,0)x(0) + \int_0^t \Phi(t,\tau)B_1(e(\tau) + r(\tau)) \qquad (F.2.21)$$

where $\Phi(t, \tau)$ is the state transition matrix corresponding to $\overline{A}(t)$.

In view of Lemma F.2.2 and since r and x_0 are bounded, we have

$$\|x(t)\| \leq k_0 + \int_0^t k_1 E(-\sigma, t-\tau)|e(\tau)| \qquad (F.2.22)$$

for some $k_0, k_1, \sigma > 0$. From (12.8.12) and using Schwarz's inequality, it follows that

$$\|x(t)\| \leq k_0 + \int_0^t k_1 E(-\sigma, t-\tau)(k\varepsilon\omega(\tau)) \, d\tau + \frac{k_1}{\sqrt{\sigma}} \left\{ \int_0^t f^2(\tau) \right\}^{1/2} \qquad (F.2.23)$$

$$\leq k_0' + \frac{k_1 k\varepsilon \|v\|}{\sigma} \sup_{0 \leq \tau \leq t} \|x(t)\| + \frac{k_1}{\sqrt{\sigma}} \left\{ \int_0^t f^2(\tau) \right\}^{1/2} \qquad (F.2.24)$$

where we have used (F.1.9). Since the right side of (F.2.24) is monotonic nondecreasing in t, it follows that

$$\left\{ \sup_{0 \leq \tau \leq t} \|x(\tau)\| \right\} \leq k_0' + \frac{k_1 k\varepsilon}{\sigma} \|v\| \left\{ \sup_{0 \leq \tau \leq t} \|x(\tau)\| \right\}$$

$$+ \frac{k_1}{\sqrt{\sigma}} \left\{ \int_0^t f^2(\tau) \right\}^{1/2} \qquad (F.2.25)$$

Then, provided $\varepsilon < \sigma/(k_1\beta\|v\|)$, we can show that (see Problem 12.28)

$$\|x(t)\|^2 \leq k_2 + k_3 \int_0^t f^2(\tau) \tag{F.2.26}$$

$$= k_2 + k_3 \left(\int_0^t \tilde{f}^2(\tau) + \int_0^t \tilde{f}^2(\tau)\|\phi(\tau)\|^2 \right) \tag{F.2.27}$$

Since $\tilde{f} \in \mathcal{L}_2^+$ (Lemma 12.8.2) and $\|\phi(\tau)\| \leq \|x(\tau)\|$, it follows using Gronwall's lemma (Lemma F.2.1) that $x(t)$ is bounded. x bounded implies ϕ, ω, e bounded, and thus ρx is bounded in view of (F.2.4). Thus $\bar{u}$ and y are bounded.

To show u is bounded, we need to write the composite model (F.1.11) in terms of an input–output transfer function, as in

$$y = \frac{B_a}{A_a} u \tag{F.2.28}$$

where

$$\frac{B_a}{A_a} = \frac{B}{A}(1 + \overline{H}) + \overline{\overline{H}} \tag{F.2.29}$$

For the internal model controller to make sense, it is necessary that there be no cancellation between S (in the controller) and B_a (the numerator of overall plant). We therefore make this assumption in the sequel.

We next introduce a Hurwitz A_a^+ of the same degree as A_a; then GS/JQ and B_a/A_a^+ will be coprime in the ring of rational, stable, proper transfer functions. That is, there exist stable, proper, transfer functions χ, Λ such that

$$\chi \frac{GS}{JQ} + \Lambda \frac{B_a}{A_a^+} = 1 \tag{F.2.30}$$

$$\chi \frac{GS}{JQ} u + \Lambda \frac{B_a}{A_a^+} u = u \tag{F.2.31}$$

or

$$\chi \bar{u} + \Lambda \left(\frac{A_a}{A_a^+} \right) y = u \tag{F.2.32}$$

and thus u is bounded, since $\bar{u}$ and y are bounded.

2. Equation (F.2.4) holds in this case, also; however, since $Sy^* = 0$, we have $r = 0$ in view of (F.2.8); and since ϕ is bounded (as established in part 1), $\tilde{f} \in \mathcal{L}_2^+$ implies $f \in \mathcal{L}_2^+$. We then have

$$\|x(t)\| \leq k_1 E(-\sigma, t)\|x_0\| + \int_0^t k_1 E(-\sigma, t - \tau)|e(\tau)| \tag{F.2.33}$$

$$\leq k_1 E(-\sigma, t)\|x_0\| + \int_0^t k_1 E(-\sigma, t - \tau)|f(\tau)|$$

$$+ k_1 \varepsilon k \int_0^t k_1 E(-\sigma, t - \tau) \sup_{0 \leq T \leq \tau} \left\{ E(-\sigma_0, t - \tau)|v^T x(T)| \right\}$$

$$\tag{F.2.34}$$

The result in part 1 relied on $\varepsilon < \sigma/(k_1 k \|v\|)$; and so we select λ such that $0 < \varepsilon k_1 k \|v\| < \lambda < \sigma$. We now rearrange the third term on the right side of (F.2.34) as

$$k_1 \varepsilon k \int_0^t E(-\sigma, t-\tau) \sup_{0 \le T \le \tau} \{ E(-\sigma_0, t-\tau) |v^T x(T)| \}$$

$$\le k_1 \varepsilon k \|v\| \int_0^t E(-\lambda, t-\tau) E(-(\sigma-\lambda), t-\tau)$$

$$\times \sup_{0 \le T \le \tau} \{ E(-\sigma_0, t-\tau) \|x(T)\| \} \qquad \text{(F.2.35)}$$

$$\le \left(\frac{k_1 \varepsilon k \|v\|}{\lambda} \right) E(-\lambda_0, t) \sup_{0 \le \tau \le t} \{ E(\lambda_0, \tau) \|x(\tau)\| \} \qquad \text{(F.2.36)}$$

(where $\lambda_0 = \min\{(\sigma - \lambda), \sigma_0\}$). Using (F.2.36) in (F.2.34), we have

$$E(\lambda_0, t)\|x(t)\| \le k_1 \|x_0\| + \int_0^t k_1 E(\lambda_0, \tau) |f(\tau)|$$

$$+ \frac{k_1 \varepsilon k \|v\|}{\lambda} \sup_{0 \le \tau \le t} \{ E(\lambda_0, \tau) \|x(\tau)\| \} \qquad \text{(F.2.37)}$$

Since the right side of (F.2.37) is monotonic nondecreasing in t, and $k_1 \varepsilon k \|v\| < \lambda$, we then have

$$\sup_{0 \le \tau \le t} \{ E(\lambda_0, t)\|x(t)\| \} \le k_4 \|x_0\| + k_4 \int_0^t E(\lambda_0, \tau) |f(\tau)| \qquad \text{(F.2.38)}$$

and thus

$$\|x(t)\| \le k_4 E(\lambda_0, t)\|x_0\| + k_4 \int_0^t E(-\lambda_0, t-\tau)|f(\tau)| \qquad \text{(F.2.39)}$$

Since $\lambda_0 > 0$ and $f \in \mathscr{L}_2^+$, it follows from (F.2.39) that (see Problem 12.29)

$$\lim_{t \to \infty} x(t) = 0 \qquad \text{(F.2.40)}$$

Thus, from (12.9.8), it follows that $\omega(t) \to 0$. Since $f \in \mathscr{L}_2^+$ and ρf is bounded, we have $f(t) \to 0$ (see Problem 12.30). So $e(t) \to 0$ and from (12.9.29) we have

$$\lim_{t \to \infty} (\rho x(t)) = 0 \qquad \text{(F.2.41)}$$

From (F.2.41) and (F.2.40), it follows that $z(t) = y(t) - y^*(t) \to 0$. ▽▽▽

Index

A

A/D conversion, 27, 28, 464, 499
Adaptive control, 394, 525
Additional measurements, 438
Adjoint variable, 315
Algebraic Riccati equation, 284, 314, 342, 512
Aliasing, 28
All stabilizing controllers, 236, 249
All state estimators, 293
Amplitude limit, 421
Analog interface hardware, 461
Analog-to-digital conversion, 27, 28, 464, 499
Antialiasing filter, 27, 279, 461
Asymptotic stability, 129, 205
AUSREC, 496

B

Balanced realizations, 202
Bandwidth, 433
Bezout identity, 245
Bilinear systems, 355
Bilinear transformation, 115
Bode diagram, 111
Bounded input, bounded output stability, 205
Bumpless transfer, 471
Busy wait, 518
Butterworth filter, 44
Butterworth response, 116

C

CAD software, 516
Canonical form
    for observable systems, 198
    for reachable systems, 186
    Jordan, 37
Canonical structure, 201
Cascade realization, 474, 479
Cayley-Hamilton theorem, 38
Certainty equivalence, 394

Circle criterion, 149
Classical control, 128
Coloured noise, 369
Compensation, 159
Conditional expectation, 265
Constant trace, 392
Continuous time models, 7
Control, optimal, 309, 311
Controllability, 179, 180
   form, 187
   gramian, 191
   matrix, 181
Controller form, 188
Convergence analysis, 525
Co-state, 315
Coupled tanks, 24
Covariance matrix, 360
Critical section, 520

D

D/A conversion, 27, 32, 464, 499
Dead zone, 379, 526
Decalcification plant, 18
Delta
   function, 66
   operator, 43
   optimal filter, 275
   transform, 63, 81
Design, 416
Design guidelines, 433
Detectability, 196
Deterministic disturbance, 20, 34, 368
Diagonal dominance, 167
Digital-to-analog conversion, 27, 32, 464, 499
Diophantine equation, 277
Discrete Kalman filter, 269
Discrete models, 27, 32, 41, 46
Discrete transfer function, 87
Disturbance, 7, 20
   deterministic, 20
   sinusoidal, 35, 42, 43
Duality, 312

E

Eliminant matrix, 227, 364, 505
Euclid's algorithm, 229
Exponential
   data weighting, 388
   forgetting and resetting algorithm, 390
   stability, 205

F

Feedback, 128
Feedforward, 441
Filtering, 264
Finite-dimensional systems, 8
Finite word length, 502
Flash converter, 499
Fourier analysis, 491
Fractional representation, 234
Frequency
   folding, 29
   response, 109
   response sensitivity, 506
Frequency domain
   approximation, 114
   stability criteria, 143

G

Gain margin, 146, 323
Game theory, 335
Gaugemeter, 486
Generalized
   derivative, 65
   Hamiltonian matrix, 285, 316
   transform, 66, 68
Gershgorin bands, 167
Gradient algorithm, 390
Gronwall's lemma, 528

H

Hamiltonian matrix, 286, 316, 330, 342, 512
   generalized, 285, 316
Hankel singular values, 202
Hardware, 499
High-gain feedback, 417
H_∞
   design, 335
   optimization, 299, 330

I

Implementation issues, 456
Impulse response, 87
Industrial case study, 483
Infinite horizon LQ, 313
Innovations sequence, 268
Input, 7
Input innovations, 247
Input-output equivalence, 201
Input-output nonlinearities, 12
Integral windup, 467
Interactor matrix, 251
Internal model principle, 153, 325
Internal variable, 7
Interrupt latency, 522
Interrupts, 519
Inverse transformation, 76
Inverted pendulum, 10, 15, 435

J

Joint Markov model, 269
Jordan canonical form, 37
Jury's criterion, 136

K

Kalman filter, 263, 490
   continuous, 281
   delta, 275
   discrete, 269

Kalman Filter (*cont.*)
   properties, 283
   unified, 281

L

Lagrange multiplier, 336
Laplace transform, 55, 80
   properties, 56
Lead-lag, 164
Least squares, 356, 358
Left matrix fraction, 199
Linear approximation, 17
Linear regression, 352
Linear state space models, 12
Linear time varying systems, 528
Linearization, 15
Loop transfer recovery, 301
LQ optimal control, 310
Lyapunov equation, 194
Lyapunov stability, 208

M

Matrix exponential, 36, 66
Model reference control, 232, 251
Models
   continuous time, 7
   discrete time, 27, 32, 41
   input-output, 8, 19
   linear state space, 12
   state space, 8
Multioutput, 440
Multiperiod averaging filter, 490
Multirate, 440
Multirate controllers, 521
Multitasking, 521
Multivariable frequency domain stability, 166

N

Narrow-band filter, 290
Noise, 367
Nonlinear model, 355

Nonlinearities, 12
　input-output, 12
　output feedback, 14
Notation, 28, 65, 81
Notch filter, 491
Numerical integration, 113
Numerical issues, 45, 231, 283, 472, 476, 502, 511
Nyquist
　contour, 144
　criterion, 145
　diagram, 111, 323, 424

O

Observability, 179, 196
　form, 198
　gramian, 199
　matrix, 196
Observer, 220, 263
Observer form, 198
On-line estimation, 450
Optimal control, 309, 311
Optimal filtering, 511
　properties, 283
Optimal state estimation, 263
Output, 7
Output feedback, 221
Output feedback nonlinearities, 14
Output innovations, 246

P

Padé approximation, 122
Parallel realization, 474, 479
Parameter
　convergence, 362
　estimation, 272, 351
　time variations, 387
Parseval's theorem, 83
Partial state, 190
PBH observability test, 197
PBH reachability test, 184
Periodic averaging filter, 492
Persistent excitation, 364

Persistent spanning, 362
Perturbation method, 513
Perturbed eigenvector, 514
Per-unit values, 195
Phase advance compensation, 159
Phase lag compensation, 163
Phase margin, 146, 323
Pole assignment, 219
Poles and zeros, 90
　sampled data, 94
Polling, 519
Polynomial interpolation, 31
Pontryagin's minimum principle, 315
Prediction, 264
Prediction error, 361
Principle of optimality, 310
Prior knowledge, 366

R

Rapid sampling, 460
Rational approximations to delays, 121
Reachability, 180
Reachable decomposition, 183
Real-time programming, 518, 520
Real-time software, 518
Reconstructibility, 196
Rectangular rule approximation, 114
Recursive least squares, 358
Recursive optimal estimator, 268
Regression form, 236
Relative degree, 97, 252
Resetting algorithm, 390
Return difference, 151, 323
Riccati equation
　algebraic, 284, 314, 342, 512
　difference, 279
　differential, 281
Right matrix fraction, 189
Ring structure, 242
Rise time, 154
Robustness
　considerations in observer design, 299
　feedback systems, 147

Roll eccentricity control, 483
Rolling mill, 118, 483
Root locus, 138
Root sensitivity, 505
Round-off noise, 508
Routh criterion, 134
Runge-Kutta, 40

S

Saddlepoint, 336
Sampling, 28
    rapid, 460
    rate, 456
    slow, 457
Scheduler, 519
    latency, 522
Sector nonlinearity, 148
Semaphore, 519, 520
Sensitivity function, 151, 423
Separation theorem, 328
Sequential processing, 268
Settling time, 154
Shannon reconstruction theorem, 31, 458
Shift operator, 42
Signal (semaphore), 520
Signal reconstruction, 30
Sine wave, 289
Sinusoidal disturbance, 35, 42, 43
Slew rate, 421
Slow sampling, 457
Small gain argument, 530
Smith predictor, 241
Smoothing, 264
Soft sensor, 487
Spectral density, 275
Spectral factorization, 291, 321
Stability, 128, 205
    asymptotic, 205
    boundary, 110
    bounded input, bounded output, 205
    exponential, 205
    Lyapunov, 208
    nonlinear systems, 204
    regions, 130

Stability (*cont.*)
    tests, 132
    time varying systems, 528
Stabilizable, 180
Stabilizing solution, 285, 314
State estimation, 263
State space models, 8
State variable feedback, 219
Step invariance, 88, 117
Stochastic optimal control, 327
Strong solutions, 285, 314
Switching regulator, 47
Sylvester's theorem, 227
Symplectic matrix, 316
System type, 152

T

Tasks, 519
Test, PBH, 184
Test problem for advanced control, 445
Time delay, 118, 120, 430
Time domain analysis, 179
Tracking, 128, 150
Trade-offs, 416
Transfer functions, 85
    for delay, 119
Transforms
    delta, 63
    Laplace, 55
    properties, 71
    techniques, 54
    Z, 59
Transient, 128
    performance, 154
Trapezoidal rule approximation, 114
Two-point boundary value problem, 338

U

Uncertainty, 450
Undermodeling, 376
    estimation, 381
Undershoot, 422

Unified
  Kalman filter, 281
  notion, 67
  transform theory, 65
Unstable zeros, 422

W

Wait (semaphore), 519
White noise, 369

Wide sense conditional expectation, 265
Wiener processes, 276

Z

Z-transforms, 59, 60, 81
Zero-order hold discrete time equivalence, 90